Lecture Notes in Computer Scien

Commenced Publication in 1973
Founding and Former Series Editors:
Gerhard Goos, Juris Hartmanis, and Jan van Leeuwen

Wolfram Kahl   Timothy G. Griffin (Eds.)

Relational and Algebraic Methods in Computer Science

13th International Conference, RAMiCS 2012
Cambridge, UK, September 17-20, 2012
Proceedings

 Springer

Volume Editors

Wolfram Kahl
McMaster University
Department of Computing and Software
1280 Main Street West
Hamilton, ON, L8S 4K1, Canada
E-mail: kahl@mcmaster.ca

Timothy G. Griffin
University of Cambridge
Computer Laboratory
15 JJ Thomson Avenue
Cambridge, CB3 0FD, UK
E-mail: tgg22@cam.ac.uk

ISSN 0302-9743   e-ISSN 1611-3349
ISBN 978-3-642-33313-2   e-ISBN 978-3-642-33314-9
DOI 10.1007/978-3-642-33314-9

Springer Heidelberg Dordrecht London New York

Library of Congress Control Number: 2012946209

CR Subject Classification (1998):
F.4.1-3, I.1.1-4, I.2.3, F.3.1-2, D.2.4, G.2.2, D.1.1, D.3.2

LNCS Sublibrary: SL 1 – Theoretical Computer Science and General Issues

Typesetting: Camera-ready by author, data conversion by Scientific Publishing Services, Chennai, India

Printed on acid-free paper

Springer is part of Springer Science+Business Media (www.springer.com)

Preface

This volume contains the proceedings of the 13th International Conference on Relational and Algebraic Methods in Computer Science (RAMiCS 13). The conference took place in Cambridge, UK, September 17–20, 2012, and was the second conference using the RAMiCS title, but the 13th in a series that started out using the name "Relational Methods in Computer Science" with the acronym RelMiCS. From 2003 to 2008, RelMiCS conferences were held as joint events with "Applications of Kleene Algebras" (AKA) conferences, motivated by the substantial common interests and overlap of the two communities. The purpose of the RAMiCS conferences continues to be bringing together researchers from various subdisciplines of computer science, mathematics, and related fields who use the calculus of relations and/or Kleene algebra as methodological and conceptual tools in their work.

The call for papers invited submissions in the general area of relational and algebraic methods in computer science, adding special focus on formal methods for software engineering, logics of programs, and links with neighboring disciplines. This focus was also realized in the choice of the following three invited talks:

Formalized Regular Expression Equivalence and Relation Algebra in Isabelle, by Alexander Krauss (Munich, Germany) on joint work with Tobias Nipkow:

> We present the Isabelle formalization of an elegant equivalence checker for regular expressions. It works by constructing a bisimulation relation between (derivatives of) regular expressions. By mapping expressions to binary relations, an automatic and complete proof method for (in)equalities of binary relations over union, composition and (reflexive) transitive closure is obtained, which adds a practically useful decision procedure to the automation toolbox of Isabelle/HOL.
>
> Ongoing extensions of this work to partial derivatives and extended regular expressions will be covered in the end of the talk.

Algebraic Laws of Concurrency and Separation, by Peter O'Hearn (London, UK):

> This talk reports on ongoing work — with Tony Hoare, Akbar Hussain, Bernhard Möller, Rasmus Petersen, Georg Struth, Ian Wehrman, and others — drawing on ideas from Kleene algebra and concurrent separation logic. The approach we are taking abstracts from syntax or particular models. Message passing and shared memory process interaction, and strong (interleaving) and weak (partial order) approaches to sequencing, are accommodated as different models of the same core algebraic axioms. The central structure is that of an ordered bimonoid,

two monotone monoids over the same poset representing parallel and sequential composition, linked by an algebraic version of the exchange law
from 2-categories. Rules of program logic, related to Hoare and Separation logics, flow at once from this structure: one gets a generic program
logic from the algebra, which holds for a range of concrete models.

Using Relation Algebraic Methods in the Coq Proof Assistant, by Damien Pous
(Grenoble, France):

If reasoning in a point-free algebraic setting can help on the paper, it
also greatly helps in a proof assistant. We present a Coq library for working with binary relations at the algebraic, point-free, level. By combining
several automatic decision procedures (e.g., for Kleene algebra and residuated semirings) and the higher-order features of Coq, this library allows
us to formalize various theorems in a very simple way: we benefit from
the expressiveness of an interactive theorem prover and from the comfort
of automation for decidable fragments of relation algebra.

The body of this volume is made up of 23 contributions by researchers from all
over the world, selected by the Program Committee from 39 relevant submissions.
Each submission was reviewed by at least three Program Committee members;
the Program Committee did not meet in person, but had over one week of intense
electronic discussions.

The conference included, for the fifth time now, a PhD program; this gave
PhD students the opportunity to present their work in progress (not included
in this volume) interleaved with the general conference program. Motivated by
the presence of the PhD program, but also integrated with the remainder of the
conference, were two invited tutorials, each delivered in two installments:

Kleene Algebra with Tests, by Dexter Kozen (Cornell, USA):

Kleene algebra with tests (KAT) is a versatile algebraic system for reasoning about the equivalence of low-level imperative programs. It has
been shown to be useful in verifying compiler optimizations and performing static analysis. This three-hour tutorial introduces the basic
definitions, results, applications, and extensions of KAT. Possible topics
include: basic examples of equational reasoning; history and relation to
classical systems such as Hoare logic; common models, including language, relational, trace, and matrix models; axiomatizations and their
completeness and complexity; and applications.

The Isabelle/HOL Theorem Prover, by Lawrence C. Paulson (Cambridge, UK):

Isabelle/HOL is an interactive theorem prover for higher-order logic,
with powerful automation both to prove theorems and to find counterexamples. There are two sophisticated user interfaces, and a flexible language (Isar) in which to express proofs. In two separate one-hour sessions,
the tutorial introduces two aspects of Isabelle usage.

1. The verification of functional programs using induction and simplification. Isabelle's specification language includes recursive datatypes and functions. This executable fragment of HOL, although lacking in syntactic niceties, is sufficient for writing and verifying substantial functional programs.
2. Inductively defined sets and relations. In computer science, this fundamental concept is frequently used to define the operational semantics of programming languages. Isabelle has expressive mechanisms for defining sets inductively and for reasoning about them.

The tutorial is designed to give participants a glimpse at the main features of Isabelle, equipping them to learn more through self-study. In two hours, nothing can be covered in depth, but many things can be mentioned: the various kinds of automation; axiomatic type classes and locales, both of which allow abstract theory development followed by instantiation; the extensive built-in modeling mechanisms and libraries.

We are very grateful to the members of the Program Committee and the external referees for their care and diligence in reviewing the submitted papers. We would like to thank the members of the RAMiCS Steering Committee for their support and advice especially in the early phases of the conference organization, and Peter Höfner for doing an excellent job in publicity. We are grateful to the University of Cambridge for hosting RAMiCS 2012 and to the Computer Laboratory for providing administrative support. We gratefully appreciate the excellent facilities offered by the EasyChair conference administration system. Last but not least, we would like to thank the British Logic Colloquium and Winton Capital Management for their generous financial support.

July 2012   Wolfram Kahl
                                                         Timothy G. Griffin

Organization

Organizing Committee

Conference Chair

Timothy G. Griffin   University of Cambridge, UK

Programme Chair

Wolfram Kahl   McMaster University, Canada

Publicity

Peter Höfner   NICTA, Australia

Program Committee

Rudolf Berghammer	Christian-Albrechts-Universität zu Kiel, Germany
Harrie de Swart	Erasmus University Rotterdam, The Netherlands
Jules Desharnais	Université Laval, Canada
Marc Frappier	University of Sherbrooke, Canada
Hitoshi Furusawa	Kagoshima University, Japan
Timothy G. Griffin	University of Cambridge, UK
Peter Höfner	NICTA Ltd., Australia
Ali Jaoua	Qatar University, Qatar
Peter Jipsen	Chapman University, USA
Wolfram Kahl	McMaster University, Canada
Larissa Meinicke	The University of Queensland, Australia
Bernhard Möller	Universität Augsburg, Germany
Peter O'Hearn	Queen Mary, University of London, UK
José Nuno Oliveira	Universidade do Minho, Portugal
Ewa Orłowska	National Institute of Telecommunications, Warsaw, Poland
Matthew Parkinson	Microsoft Research Cambridge, UK
Damien Pous	CNRS, Lab. d'Informatique de Grenoble, France
Holger Schlingloff	Fraunhofer FIRST and Humboldt University, Germany
Gunther Schmidt	Universität der Bundeswehr München, Germany
Renate Schmidt	University of Manchester, UK

Georg Struth   University of Sheffield, UK
George Theodorakopoulos   University of Derby, UK
Michael Winter   Brock University, Canada

Steering Committee

Rudolf Berghammer   Christian-Albrechts-Universität zu Kiel,
                                    Germany
Harrie de Swart   Erasmus University Rotterdam,
                                    The Netherlands
Jules Desharnais   Université Laval, Canada
Ali Jaoua   Qatar University, Qatar
Bernhard Möller   Universität Augsburg, Germany
Ewa Orłowska   National Institute of Telecommunications,
                                    Warsaw, Poland
Gunther Schmidt   Universität der Bundeswehr München,
                                    Germany
Renate Schmidt   University of Manchester, UK
Michael Winter   Brock University, Canada

Additional Referees

Andrea Asperti   Simon Foster   Anna Radzikowska
Stefan Bolus   Roland Glück   Ingrid Rewitzky
Thomas Braibant   Sergey Goncharov   Patrick Roocks
Wojciech Buszkowski   Jonathan Hayman   Paulo F. Silva
Jacques Carette   Dexter Kozen   Kim Solin
James Cheney   Hartmut Lackner   Toshinori Takai
Han-Hing Dang   Mirko Navara   John Wickerson
Nikita Danilenko   Koki Nishizawa
Ernst-Erich Doberkat   Alessandra Palmigiano

Sponsoring Institutions

University of Cambridge, Computer Laboratory
British Logic Colloquium
Winton Capital Management, London, UK

Table of Contents

Theoretical Foundations

Relations and Algorithms

Preference Relations

Properties of Specialised Relations

Transitive Separation Logic

Han-Hing Dang and Bernhard Möller

Institut für Informatik, Universität Augsburg, D-86159 Augsburg, Germany
{h.dang,moeller}@informatik.uni-augsburg.de

Abstract. Separation logic (SL) is an extension of Hoare logic by operations and formulas that not only talk about program variables, but also about heap portions. Its general purpose is to enable more flexible reasoning about linked object/record structures. In the present paper we give an algebraic extension of SL at the data structure level. We define operations that additionally to heap separation make assumptions about the linking structure. Phenomena to be treated comprise reachability analysis, (absence of) sharing, cycle detection and preservation of substructures under destructive assignments. We demonstrate the practicality of this approach with the examples of in-place list-reversal and tree rotation.

Keywords: Separation logic, reachability, sharing, strong separation.

1 Introduction

Separation logic (SL) as an extension of Hoare logic includes spatial operations and formulas that do not only talk about single program variables, but also about heap portions (*heaplets*). The original purpose of SL was to enable more flexible reasoning about linked object/record structures in a sequential version. This has also been extended to concurrent contexts [18]. The central connective of this logic is the *separating conjunction* $P_1 * P_2$ of formulas P_1, P_2. It guarantees that the set of resources characterised by the P_i are disjoint. This implies under some circumstances that an assignment to a resource of P_1 does not change any value of resources in P_2. Consider, however, the following counterexample:

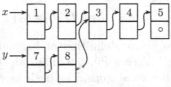

From the variables x and y two singly linked lists can be accessed. If we would run, e.g., an in-place list reversal algorithm on the list accessible from x, this would at the same time change the contents of the list accessible from y. This is because the lists show the phenomenon of *sharing*. Note that separating conjunction $*$ in the above situation alone would only guarantee that the cells with contents $1, \ldots, 5$ have different addresses than the ones with contents $7, 8$.

W. Kahl and T.G. Griffin (Eds.): RAMiCS 2012, LNCS 7560, pp. 1–16, 2012.

The purpose of the present paper is to define in an abstract fashion connectives stronger than $*$ that ensure the absence of sharing as depicted above. With this, we hope to facilitate reachability analysis within SL as, e.g., needed in garbage collection algorithms, or the detection and exclusion of cycles to guarantee termination in such algorithms. Moreover, we provide a collection of predicates that characterise structural properties of linked structures and prove inference rules for them that express preservation of substructures under selective assignments. Finally, we include abstraction functions into the program logic which allows very concise and readable reasoning. The approach is illustrated with two examples, namely in-situ list reversal and tree rotation.

2   Basics and Definitions

Following [4,10] we define abstract separation algebras that are used to represent the abstract structure of the underlying considered sets of resources.

Definition 2.1. A *separation algebra* is a partial commutative monoid $(H, \bullet, u)$. We call the elements of H *states*. The operation $\bullet$ denotes state combination and the *empty state* u is its unit. A partial commutative monoid is given by a partial binary operation satisfying the unity, commutativity and associativity laws w.r.t. the equality that holds for two terms iff both are defined and equal or both are undefined. The induced *combinability* or *disjointness* relation $\#$ is defined, for $h_1, h_2 \in H$ by $h_0 \# h_1 \Leftrightarrow_{df} h_0 \bullet h_1$ is defined.

As a concrete example we could instantiate states with heaplets, i.e., parts of the global heap in memory. Heaplets are modelled as partial functions $H =_{df} \mathbb{N} \rightsquigarrow \mathbb{N}$, i.e., from naturals to naturals for simplicity. The $\bullet$ operation then corresponds to union of partial functions and u denotes the empty set or the everywhere undefined function. The original disjointness relation for heaplets reads $h_0 \# h_1 \Leftrightarrow_{df} \ulcorner h_0 \urcorner \cap \ulcorner h_1 \urcorner = \emptyset$ for heaps h_0, h_1, where $\ulcorner h \urcorner$ is the domain of h. For more concrete examples we refer to [2].

Definition 2.2. Let $(H, \bullet, u)$ be a separation algebra. Predicates $P, Q, \ldots$ over H are elements of the powerset $\mathcal{P}(H)$. On predicates the separating conjunction is defined by pointwise lifting:

$$P * Q =_{df} \{h_1 \bullet h_2 : h_1 \in P, h_1 \in Q, h_1 \# h_2\}, \quad \mathsf{emp} =_{df} \{u\}.$$

We now abstract from the above definition of heaplets and replace them by elements of a *modal Kleene algebra* [6], our main algebraic structure. We will introduce its constituents in several steps. The basic layer is an *idempotent semiring* $(S, +, \cdot, 0, 1)$, where $(S, +, 0)$ forms an idempotent commutative monoid and $(S, \cdot, 1)$ a plain monoid.

A concrete example of an idempotent semiring is the set of relations. The natural order coincides with inclusion $\subseteq$, while $+$ abstracts $\cup$ and $\cdot$ abstracts relational composition $;$. The element 0 represents the empty relation while 1 denotes the identity relation.

To express reachability in this algebra we need to represent sets of nodes. Relationally, this can be done using subsets of the identity relation. In general semirings this can be mimicked by sub-identity elements $p \leq 1$, called *tests* [14,13]. Each of these elements is requested to have a complement relative to 1, i.e., an element $\neg p$ that satisfies $p + \neg p = 1$ and $p \cdot \neg p = 0 = \neg p \cdot p$. Thus, tests have to form a Boolean subalgebra. By this, $+$ coincides with the binary supremum $\sqcup$ and $\cdot$ with binary infimum $\sqcap$ on tests. Every semiring contains at least the greatest test 1 and the least test 0.

The product $p \cdot a$ means restriction of the domain of element a to starting nodes in p while multiplication $a \cdot p$ from the right restricts the range of a. By this, we can now axiomatise domain $\ulcorner \_$ and codomain $\_ \urcorner$ tests, following [6]. Note that, according to the general idea of tests, in the relation semiring these operations will yield sub-identity relations in one-to-one correspondence with the usual domain and range. For arbitrary element a and test p we have the axioms

$$a \leq \ulcorner a \cdot a \;,\quad \ulcorner(p \cdot a) \leq p \;,\quad \ulcorner(a \cdot b) = \ulcorner(a \cdot \ulcorner b) \;,$$
$$a \leq a \cdot a \urcorner \;,\quad (a \cdot p) \urcorner \leq p \;,\quad (a \cdot b) \urcorner = (a \urcorner \cdot b) \urcorner \;.$$

These imply additivity and isotony, among others, see [6]. Based on domain, we now define the *diamond* operation that plays a central role in our reachability analyses: $\langle a | \, p =_{df} (p \cdot a) \urcorner$. Since this is an abstract version of the diamond operator from modal logic, an idempotent semiring with it is called *modal*.

The diamond $\langle a | \, p$ calculates all immediate successor nodes under a, starting from the set of nodes p, i.e., all nodes that are reachable within one a-step, aka the *image* of p under a. This operation distributes through union and is strict and isotone in both arguments.

Finally, to be able to calculate reachability within arbitrarily many steps, we extend the algebraic structure to a modal *Kleene algebra* [12] by an iteration operator $*$. It can be axiomatised by the following unfold and induction laws:

$$1 + x \cdot x^* \leq x^* \;,\quad x \cdot y + z \leq y \Rightarrow x^* \cdot z \leq y \;,$$
$$1 + x^* \cdot x \leq x^* \;,\quad y \cdot x + z \leq y \Rightarrow z \cdot x^* \leq y \;.$$

This implies that a^* is the least fixed-point μ_f of $f(x) = 1 + a \cdot x$.

Next, we define the reachability function *reach* by

$$reach(p, a) =_{df} \langle a^* | \, p.$$

Among other properties, *reach* distributes through $+$ in its first argument and is isotone in both arguments. Moreover we have the *induction rule*

$$reach(p, a) \leq q \;\; \Leftarrow \;\; p \leq q \;\wedge\; \langle a | \, q \leq q.$$

3   A Stronger Notion of Separation

As we have described in Section 1, the standard separating conjunction $*$ alone often does not describe sufficient disjointness needed for program verification. Simple sharing patterns in data structures can not be excluded from the only use of $*$ as can be seen in the following examples: For addresses x_1, x_2, x_3 with

$$h_1 \quad \fbox{x_1} \longrightarrow x_3 \longleftarrow \fbox{x_2} \quad h_2 \qquad\qquad h_1 \quad \fbox{x_1} \qquad \fbox{x_2} \quad h_2$$

h_1 and h_2 satisfy the disjointness property since $\ulcorner h_1 \cap \ulcorner h_2 = \emptyset$. But still $h = h_1 \cup h_2$ does not appear very separated from the viewpoint of reachable cells, since in the left example both subheaps refer to the same address and in the right they form a simple cycle. This can be an undesired behaviour, since acyclicity of the data structure is a main correctness property needed for many algorithms working, e.g., on linked lists or tree structures. In many cases the separation expressed by $\ulcorner h_1 \cap \ulcorner h_2 = \emptyset$ is too weak. We want to find a stronger disjointness condition that takes such phenomena into account.

First, to simplify the description, for our new disjointness condition, we abstract from non-pointer attributes of objects, since they do not play a role for reachability questions. One can always view the non-pointer attributes of an object as combined with its address into a "super-address". Therefore we give all definitions in the following only on the relevant part of a state that affects the reachability observations.

Moreover, in reachability analysis we are only interested in the existence of paths and not in path labels etc. Hence, multiple, differently labelled links from one object to another are projected into a single non-labelled link. Such a projection function on labels can, e.g., be found in [7], which gives an algebraic approach for representing labelled graphs, based on fuzzy relations.

With this abstraction, a linked object structure can be represented by an *access relation* between object addresses. Again, we pass to the more abstract algebraic view by using elements from a modal Kleene algebra to stand for concrete access relations; hence we call them *access elements*. In the following we will denote access elements by $a, b, \ldots$.

Extending [8,17] we give a stronger separation relation $\#$ on access elements.

Definition 3.1. For access elements a_1, a_2, we define the *strong disjointness relation* $\#$ by setting $a = a_1 + a_2$ in

$$a_1 \, \# \, a_2 \Leftrightarrow_{df} reach(\ulcorner a_1, a) \cdot reach(\ulcorner a_2, a) = 0.$$

Intuitively, a is strongly separated into a_1 and a_2 if each address reachable from a_1 is unreachable from a_2 w.r.t. a, and vice versa. Note that since all results of the *reach* operation are tests, $\cdot$ coincides with their meet, i.e., intersection in the concrete algebra of relations.

This stronger condition rules out the above examples.

Clearly, $\#$ is commutative. Moreover, since we have for all p, b that $p \leq reach(p, b)$, the new separation condition indeed implies the analogue of the old one, i.e., both parts are disjoint: $a_1 \, \# \, a_2 \Rightarrow \ulcorner a_1 \cdot \ulcorner a_2 = 0$. Finally, $\#$ is downward closed by isotony of *reach*: $a_1 \, \# \, a_2 \wedge b_1 \leq a_1 \wedge b_2 \leq a_2 \Rightarrow b_1 \, \# \, b_2$.

It turns out that $\#$ can be characterised in a much simpler way. To formulate it, we define an auxiliary notion.

Definition 3.2. The *nodes* $\overline{\ulcorner a}$ of an access element a are given by $\overline{\ulcorner a} =_{df} \ulcorner a + a \urcorner$.

From the definitions it is clear that $\overline{\ulcorner a + b} = \overline{\ulcorner a} + \overline{\ulcorner b}$ and $\overline{\ulcorner 0} = 0$.

We show two further properties that link the nodes operator with reachability.

Lemma 3.3. *For an access element a we have*

1. $\ulcorner a \urcorner \leq reach(\ulcorner a, a)$,
2. $\ulcorner a \urcorner \cdot \ulcorner b \urcorner = 0 \Rightarrow reach(\ulcorner a, a + b) = \ulcorner a \urcorner$,
3. $\ulcorner a \urcorner = reach(\ulcorner a, a)$.

Proof

1. First, $\ulcorner a \urcorner \leq reach(\ulcorner a, a)$ by the reach induction rule from Section 2.
   Second, by a domain property, $\ulcorner a \urcorner = (\ulcorner a \cdot a)\urcorner = \langle a| \ulcorner a \urcorner \leq reach(\ulcorner a, a)$.
2. For ($\leq$) we know by diamond star induction that

   $$reach(\ulcorner a, a + b) \leq \ulcorner a \urcorner \Leftarrow \ulcorner a \urcorner \leq \ulcorner a \urcorner \wedge \langle(a + b)| \ulcorner a \urcorner \leq \ulcorner a \urcorner .$$

   $\ulcorner a \urcorner \leq \ulcorner a \urcorner$ holds by definition of $\ulcorner \urcorner$, while $\langle(a + b)| \ulcorner a \urcorner \leq \ulcorner a \urcorner$ resolves by diamond distributivity to $\langle a| \ulcorner a \urcorner \leq \ulcorner a \urcorner \wedge \langle b| \ulcorner a \urcorner \leq \ulcorner a \urcorner$. This is equivalent to $(\ulcorner a \cdot a)\urcorner \leq \ulcorner a \urcorner \wedge (\ulcorner a \cdot \ulcorner b \cdot b)\urcorner \leq \ulcorner a \urcorner$ by definition and a property of domain. Finally, the claim holds by $(\ulcorner a \cdot a)\urcorner \leq \ulcorner a \urcorner$ and $\ulcorner a \urcorner \cdot \ulcorner b \urcorner = 0$ by the assumption $\ulcorner a \urcorner \cdot \ulcorner b \urcorner = 0$. The direction ($\geq$) follows from Part 1, $a \leq a + b$ and isotony of *reach*.
3. This follows by setting $b = 0$ in Part 2. $\qquad\square$

Trivially, the first and last law state that all nodes in the domain and range of an access element a are reachable from $\ulcorner a$, while the second law denotes a locality condition. If the domain as well as the range of a second access element b are disjoint from both components of a then b does not affect reachability via a. Using these theorems we can give a simpler equivalent characterisation of $\circledast$.

Lemma 3.4. $a \circledast b \Leftrightarrow \ulcorner a \urcorner \cdot \ulcorner b \urcorner = 0.$

Proof. ($\Rightarrow$) From Lemma 3.3.1 and isotony of *reach* we infer $\ulcorner a \urcorner \leq reach(\ulcorner a, a) \leq reach(\ulcorner a, a + b)$. Likewise, $\ulcorner b \urcorner \leq reach(\ulcorner b, a + b)$. Now the claim is immediate.
($\Leftarrow$) Lemma 3.3.2 tells us $reach(\ulcorner a, a + b) \cdot reach(\ulcorner b, a + b) = \ulcorner a \urcorner \cdot \ulcorner b \urcorner$, from which the claim is again immediate. $\qquad\square$

Corollary 3.5. *For an access element a we always have* $0 \circledast a \Leftrightarrow a \circledast 0 \Leftrightarrow$ true.

By the use of Lemma 3.4, it is not difficult to derive the following result that will allow us to characterise the interplay of the new separation operation with standard separating conjunction.

Lemma 3.6. *The relation $\circledast$ is bilinear, i.e., it satisfies*

$$(a + b) \circledast c \Leftrightarrow a \circledast c \wedge b \circledast c \quad and \quad a \circledast (b + c) \Leftrightarrow a \circledast b \wedge a \circledast c.$$

As in standard SL, strong separation can be lifted to predicates.

Definition 3.7. For predicates P_1 and P_2, we define the *strongly separating conjunction* by $P_1 \circledast P_2 =_{df} \{a + b : a \in P_1, b \in P_2, a \circledast b\}$.

Recall that $\mathsf{emp} = \{0\}$. We have the following result.

Lemma 3.8. *Let S denote the set of predicates. Then $(S, \circledast, 0)$ is a separation algebra, i.e., $\circledast$ is commutative and associative and $P \circledast \mathsf{emp} = \mathsf{emp} \circledast P = P$.*

Proof. Commutativity is immediate from the definition. Neutrality of emp follows from Corollary 3.5 and by neutrality of 0 w.r.t. +.

For associativity, assume $a \in (P_1 \circledast P_2) \circledast P_3$, say $a = a_{12} + a_3$ with $a_{12} \oplus a_3$ and $a_{12} \in P_1 \circledast P_2$ and $a_3 \in P_3$. Then there are a_1, a_2 with $a_1 \oplus a_2$ and $a_{12} = a_1 + a_2$ and $a_i \in P_i$. From Lemma 3.4 we obtain therefore $\overline{a_{12}} \cdot \overline{a_3} = 0 \wedge \overline{a_1} \cdot \overline{a_2} = 0$. Moreover, by Lemma 3.6 the first conjunct is equivalent to $\overline{a_1} \cdot \overline{a_3} = 0 \wedge \overline{a_2} \cdot \overline{a_3} = 0$. Therefore we also have $a \in P_1 \circledast (P_2 \circledast P_3)$.

Hence we have shown that $(P_1 \circledast P_2) \circledast P_3 \subseteq P_1 \circledast (P_2 \circledast P_3)$. The reverse inequation follows analogously.   □

4   Relating Strong Separation with Standard SL

A central question that may arise while reading this paper is: why does classical SL get along with the weaker notion of separation rather than the stronger one?

We will see that some aspects of our stronger notion of separation are in SL implicitly welded into recursive data type predicates. To explain this, we first concentrate on singly linked lists. In [19] the predicate $list(x)$ states that the heaplet under consideration consists of the cells of a singly linked list with starting address x. Its validity in a heaplet h is defined by the following clauses:

$$h \models list(\mathsf{nil}) \Leftrightarrow_{df} h = \emptyset,$$
$$x \neq \mathsf{nil} \Rightarrow (h \models list(x) \Leftrightarrow_{df} \exists y : h \models [x \mapsto y] * list(y)).$$

Hence h has to be an empty list when $x = \mathsf{nil}$, and a list with least one cell at its beginning when $x \neq \mathsf{nil}$, namely $[x \mapsto y]$.

First, note that using $\circledast$ instead of $*$ would not work, because the heaplets used are obviously not strongly separate: their heaplets are connected by forward pointers to their successor heaplets.

To make our model more realistic, we now define the concept of *closed* relations and a special element that represents the improper reference nil.

Definition 4.1. A test p is called *atomic* iff $p \neq 0$ and $q \leq p \Rightarrow q = 0 \vee q = p$ for any other test q. We assume a special atomic test $\square$ that characterises the nil object. Then an access element a is called *proper* iff $\square \cdot \ulcorner a = 0$ and *closed* iff $a\urcorner \leq \ulcorner a + \square$.

Proper access elements do not link from the pseudo-reference $\square$ to another one. By closedness, there exist no dangling references in the access element a.

We summarise a few consequences of this.

Corollary 4.2. *If a_1 and a_2 are proper/closed then $a_1 + a_2$ is also proper/closed.*

Lemma 4.3. *For an access element a the following properties are equivalent:*

1. a is proper,   *2. $\square \cdot a = 0$,*   *3. $a = \neg\square \cdot a$.*

Proof. 1. implies 2. immediately by the definition of domain. To see that 2. implies 3. we calculate $a = \square \cdot a + \neg\square \cdot a = \neg\square \cdot a$. Finally, 3. implies 1. by $\square \cdot \ulcorner a = \square \cdot \ulcorner(\neg\square \cdot a) \leq \square \cdot \neg\square = 0$ since $\square$ is a test.   □

Lemma 4.4. *An access element a is closed iff $\overline{a} - \overline{a} \leq \square$.*

Proof. As tests form a Boolean subalgebra we immediately conclude $\overline{a} - \overline{a} \leq \square$
$\Leftrightarrow \overline{a} \cdot \neg \overline{a} \leq \square \Leftrightarrow \overline{a} \leq \overline{a} + \square$. $\square$

To prepare the further use of the special test $\square$, we redefine the strong disjointness relation into a weaker version. Since nil is frequently used as a terminator reference in data structures, it should still be allowed to be reachable.

Definition 4.5. For access elements a, a_1, a_2 and $a = a_1 + a_2$, we define the *stronger disjointness relation* $\circledast$ w.r.t $\square$ by

$$a_1 \circledast a_2 \Leftrightarrow_{df} \overline{\overline{a_1} \cdot \overline{a_2}} \leq \square.$$

Lemma 3.8 is not affected by this redefinition, i.e., $\circledast$ is still commutative and associative when based on this new version of $\circledast$.

Lemma 4.6. *For proper and closed a_1, a_2 with $\overline{a_1} \cdot \overline{a_2} = 0$ we have $a_1 \circledast a_2$.*
Proof. By distributivity and order theory we know

$$\overline{\overline{a_1} \cdot \overline{a_2}} \leq \square \Leftrightarrow \overline{a_1} \cdot \overline{a_2} \leq \square \wedge \overline{a_1} \cdot \overline{a_2} \leq \square \wedge \overline{a_1} \cdot \overline{a_2} \leq \square \wedge \overline{a_1} \cdot \overline{a_2} \leq \square.$$

The first conjunct holds by the assumption and isotony. For the second and analogously for the third we calculate $\overline{a_1 \cdot a_2} \leq \overline{a_1} \cdot (\overline{a_2} + \square) = \overline{a_1} \cdot \overline{a_2} + \overline{a_1} \cdot \square = 0 \leq \square$. The last conjunct again reduces by distributivity and the assumptions to $\square \cdot \square \leq \square$ which is trivial since $\square$ is a test. $\square$

Domain-disjointness of access elements is also ensured by the standard separating conjunction. Moreover, it can be shown, by induction on the structure of the *list* predicate, that all access elements characterised by its analogue are closed, so that the lemma applies. This is why for a large part of SL the standard disjointness property suffices.

5   An Algebra of Linked Structures

According to [20], generally recursive predicate definitions, such as the list predicate, are semantically not well defined in classical SL. Formally, their definitions require the inclusion of fixpoint operators and additional syntactic sugar. This often makes the used assertions more complicated; e.g., by expressing reachability via existentially quantified variables, formulas often become very complex. The direction we take rather tries to hide such additional information: it defines operations and predicates that implicitly include necessary correctness properties like the exclusion of sharing and reachability.

First, following precursor work in [16,17,7,8], we give some definitions to describe the shape of linked object structures, in particular of tree-like ones. We start by a characterisation of acyclicity.

Definition 5.1. Call an access element a *acyclic* iff for all atomic tests $p \neq \square$ we have $p \cdot \langle a^+ | p = 0$, where $a^+ = a \cdot a^*$.

Concretely, for an access relation a, each entry (x, y) in a^+ denotes the existence of a path from x to y within a. Atomicity is needed to represent a single node; the definition would not work for arbitrary sets of nodes.

A simpler characterisation can be given as follows.

Lemma 5.2. *a is acyclic iff for all atomic tests $p \neq \square$ we have $p \cdot a^+ \cdot p = 0$.*

Proof. $p \cdot \langle a^+ | p = 0 \Leftrightarrow (p \cdot a^+)^\daleth \cdot p = 0 \Leftrightarrow (p \cdot a^+ \cdot p)^\daleth = 0 \Leftrightarrow p \cdot a^+ \cdot p = 0.$ $\square$

Next, since certain access operations are deterministic, we need an algebraic characterisation of determinacy. We borrow it from [5]:

Definition 5.3. An access element a is *deterministic* iff $\forall p : \langle a | | a \rangle p \leq p$, where the dual diamond is defined by $|a\rangle p = {}^\ulcorner(a \cdot p)$.

A relational characterisation of determinacy of a is $a^\smile \cdot a \leq 1$, where $^\smile$ is the converse operator. Since in our basic structure, the semiring, no general converse operation is available, we have to express the respective properties in another way. We have chosen to use the well established notion of modal operators. This way our algebra works also for other structures than relations. The roles of the expressions $a^\smile$ and a are now played by $\langle a |$ and $|a\rangle$, respectively.

Now we define our model of linked object structures.

Definition 5.4. We assume a finite set L of *selector names*, e.g., left or right in binary trees, and a modal Kleene algebra S.

1. A *linked structure* is a family $a = (a_l)_{l \in L}$ of proper and deterministic access elements $a_l \in S$. This reflects that access along each particular selector is deterministic. The overall access element associated with a is then $\Sigma_{l \in L} a_l$, by slight abuse of notation again denoted by a; the context will disambiguate.
2. a is a *forest* if a is acyclic and has maximal in-degree 1. Algebraically this is expressed by the dual of the formula for determinacy, namely $\forall p : |a\rangle \langle a | p \leq p$. A forest is called a *tree* if ${}^\ulcorner\overline{a} = \langle a^* | r$ for some atomic test r; in this case r is called the *root* of the tree and denoted by $root(a)$.

Note that $\square$ is a tree, while 0 is not, since it has no root. But at least, 0 is a forest. It can be shown that the root of a tree is uniquely defined, namely

$$root(a) = \begin{cases} \square & \text{if } a = \square \\ {}^\ulcorner a - \overline{a}{}^\urcorner & \text{otherwise .} \end{cases}$$

Singly linked lists arise as the special case where we have only one selector next. In this case we call a tree *chain*.

We now want to define programming constructs and assertions that deal with linked structures. We start with expressions.

Definition 5.5. A *store* is a mapping from program identifiers to atomic tests. A *state* is a pair $\sigma = (s, a)$ consisting of a store s and an access element a. For an identifier i and a selector name l, the semantics of the expression i.l w.r.t. a state (s, a) with a being a linked structure is defined as

$$[\![i.l]\!]_{(s,a)} =_{df} \langle a_l | (s(i)).$$

Program *commands* are modelled as relations between states. The semantics of plain assignment is as usual: For identifier i and expression e we set

$$\text{i} := e \quad =_{df} \quad \{ ((s, a), (s[\text{i} \leftarrow p], a)) : p =_{df} [\![e]\!]_{(s,a)} \text{ is an atomic test} \}.$$

We will show below how to model assignments of the form $i.l := e$.

As already mentioned in Section 2, one can encode subsets or predicates as sub-identity relations. This way we can view state predicates S as commands of the form $\{(\sigma, \sigma) : \sigma \in S\}$. We will not distinguish predicates and their corresponding relations notationally. Following [13,4] we encode Hoare triples with state predicates S, T and command C as

$$\{S\}\, C\, \{T\} \Leftrightarrow_{df} S\,;C \subseteq C\,;T \Leftrightarrow S\,;C \subseteq U\,;T,$$

where U is the universal relation on states.

To treat assignments $i.l := e$, we add another ingredient to our algebra. Its purpose is to describe updates of access elements by adding or changing links.

Definition 5.6. Assuming atomic tests with $p \cdot q = 0 \wedge p \cdot \square = 0$, we define a *twig* by $p \mapsto q =_{df} p \cdot \top \cdot q$ where $\top$ denotes the greatest element of the algebra. The corresponding *update* is $(p \mapsto q)\,|\,a =_{df} (p \mapsto q) + \neg p \cdot a$.

Note, that by $p, q \neq 0$ also $p \mapsto q \neq 0$. Intuitively, in $(p \mapsto q)\,|\,a$, the single node of p is connected to the single node in q, while a is restricted to links that start from $\neg p$ only.

Now we can define the semantics of selective assignments.

Definition 5.7. For identifiers i, j and selector name l we set

$$i.l := j \ =_{df}\ \{\, ((s, a), ((s, s(i) \mapsto s(j)))\,|\,a) : s(i) \neq \square, s(i) \leq \lceil a \rceil \,\} \,.$$

In general such an assignment does not preserve treeness. We provide sufficient conditions for that in the form of Hoare triples in the next section.

6   Expressing Structural Properties of Linked Structures

Definition 6.1. The set $terms(a)$ of *terminal nodes* of a tree a is $terms(a) =_{df} \lceil a \rceil - \lceil a$, while the set of *linkable nodes* of the tree is $links(a) =_{df} terms(a) - \square$. A binary tree b is *linkable* iff $links(a) \neq 0$.

Assuming the *Tarski rule*, i.e., $\forall a : a \neq 0 \Rightarrow \top \cdot a \cdot \top = \top$, we can easily infer for a twig $(p \mapsto q)^\lceil = q$ and $\lceil(p \mapsto q) = p$.

Lemma 6.2. $\overline{p \mapsto q} = p + q$ and $terms(p \mapsto q) = q$ and $root(p \mapsto q) = p$.

Proof. The first result is trivial. $terms(p \mapsto q) = (p \mapsto q)^\lceil \cdot \neg \lceil(p \mapsto q) = q \cdot \neg p = q$ since $p \cdot q = 0 \Leftrightarrow q \leq \neg p$. The proof for *root* is analogous. $\square$

Definition 6.3. For an atomic test p and a predicate P we define the subpredicate $P(p)$ and its validity in a state (s, a) with a tree a by

$$P(p) =_{df} \{a : a \in P, root(a) = p\} \,, \qquad (s, a) \models P(i) \Leftrightarrow a \in P(s(i)) \,.$$

By this we can refer to the root of an access element a in predicates. A main tool for expressing separateness and decomposability is the following.

Definition 6.4. For linked structures a_1, a_2 we define *directed combinability* by

$$a_1 \triangleright a_2 \ \Leftrightarrow_{df}\ \lceil a_1 \cdot \overline{a_2} \rceil = 0 \wedge \lceil a_1 \rceil \cdot a_2^\lceil \leq \square \wedge root(a_2) \leq links(a_1) + \square \,.$$

This relation guarantees domain disjointness and excludes occurrences of cycles, since $\ulcorner a_1 \cdot \overline{\ulcorner a_2 \urcorner} = 0 \Leftrightarrow \ulcorner a_1 \cdot \ulcorner a_2 = 0 \wedge \ulcorner a_1 \cdot terms(a_2) = 0$. Moreover, it excludes links from non-terminal nodes of a_1 to non-root nodes of a_2. If a_1, a_2 are trees, it ensures that a_1 and a_2 can be combined by identifying some non-nil terminal node of a_1 with the root of a_2. Since a_1 is a tree, that node is unique, i.e., cannot occur more than once in a_1.

Note that by Lemma 4.6 the second conjunct above can be dropped when both arguments are singly-linked lists. We summarise some useful consequences of Definition 6.4.

Lemma 6.5. *If a is a tree then $a \triangleright 0$ and $a \triangleright \square$. Moreover, $\square \triangleright a \Rightarrow a = \square$.*

Lemma 6.6. *For trees a_1 and a_2, assume $a_1 \triangleright a_2$. Then $terms(a_1 + a_2) = (terms(a_1) - root(a_2)) + terms(a_2)$ and hence $links(a_1 + a_2) = (links(a_1) - root(a_2)) + links(a_2)$. Symmetrically, if $a_1 \neq 0$ then $root(a_1 + a_2) = root(a_1)$.*

Proof. By domain distributivity and De Morgan's laws we get

$$terms(a_1 + a_2) = \ulcorner a_1 \urcorner \cdot \neg \ulcorner a_1 \cdot \neg \ulcorner a_2 + \ulcorner a_2 \urcorner \cdot \neg \ulcorner a_1 \cdot \neg \ulcorner a_2 = terms(a_1) \cdot \neg \ulcorner a_2 + terms(a_2) \cdot \neg \ulcorner a_1 .$$

Since $terms(a_2) \leq \overline{\ulcorner a_2 \urcorner}$ by definition and $\overline{\ulcorner a_2 \urcorner} \leq \neg \ulcorner a_1$ by the assumption $a_1 \triangleright a_2$, the right summand reduces to $terms(a_2)$. To bring the left one into the claimed form, we first assume $a_2 \neq \square$ and calculate

$$terms(a_1) - root(a_2) = terms(a_1) \cdot (\neg \ulcorner a_2 + a_2 \urcorner)$$
$$= terms(a_1) \cdot \neg \ulcorner a_2 + terms(a_1) \cdot a_2 \urcorner = terms(a_1) \cdot \neg \ulcorner a_2 ,$$

since $terms(a_1) \cdot a_2 \urcorner = \ulcorner a_1 \urcorner \cdot \neg \ulcorner a_1 \cdot a_2 \urcorner = 0$ by $a_1 \triangleright a_2$. The case $a_2 = \square$ follows immediately.

For *root* we first assume $a_1 \notin \{\square, 0\}$, thus $a_1 + a_2 \neq \square$. Next, we calculate, symmetrically, $root(a_1 + a_2) = \ulcorner a_1 \cdot \neg \ulcorner a_1 \urcorner \cdot \neg \ulcorner a_2 \urcorner + \ulcorner a_2 \cdot \neg \ulcorner a_1 \urcorner \cdot \neg \ulcorner a_2 \urcorner$.

The first summand reduces to $\ulcorner a_1 \cdot \neg \ulcorner a_1 \urcorner = root(a_1)$, since $a_1 \oplus a_2$ implies $\ulcorner a_1 \cdot a_2 \urcorner = 0$, i.e., $\ulcorner a_1 \leq \neg \ulcorner a_2 \urcorner$. The second summand is, by definition, equal to $root(a_2) \cdot \neg \ulcorner a_1 \urcorner$. Since $a_1 \triangleright a_2$ implies $root(a_2) \leq terms(a_1)$ and hence $root(a_2) \leq \ulcorner a_1 \urcorner$, this summand reduces to 0. If $a_1 = \square$ then also $a_2 = \square$ by Lemma 6.5 and the claim follows.   □

Definition 6.7. We define the predicate tree $=_{df} \{a : a \text{ is a tree}\}$. For $P_1, P_2 \subseteq$ tree we define *directed combinability* $\oslash$ by

$$P_1 \oslash P_2 =_{df} \{a_1 + a_2 : a_i \in P_i, a_1 \triangleright a_2\} .$$

This allows, conversely, talking about decomposability: If $a \in P_1 \oslash P_2$ then a can be split into two disjoint parts a_1, a_2 such that $a_1 \triangleright a_2$ holds.

7   Examples

Using the just defined operations and predicates we give two concrete examples, namely in-situ list reversal and rotation of binary trees.

7.1   List Reversal

This example is mainly intended to show the basic ideas of our approach. The algorithm is well known, for variables i, j, k:

$$j := \square \; ; \; \text{while } (i \neq \square) \text{ do } (k := i.\text{next} \; ; \; i.\text{next} := j \; ; \; j := i \; ; \; i := k) \; .$$

Definition 7.1. We call a chain a a *cell* if $\ulcorner a$ is an atomic test.

Note that by $a = 0 \Leftrightarrow \ulcorner a = 0$, cells are always non-empty. This will be important for some of the predicates defined below.

Lemma 7.2. *For a cell a we have* $root(a) = \ulcorner a$*, hence* $\neg root(a) \cdot a = 0$*.*

Proof. By definition $root(a) \leq \ulcorner a$ and $root(a) \neq 0$. Thus $root(a) = \ulcorner a$. $\qquad\square$

Lemma 7.3. *Twigs* $p \mapsto q$ *are cells.*

Proof. By assumption, $\ulcorner(p \mapsto q) = p$ is atomic. Moreover, by properness, $reach(p, p \mapsto q) = \overline{p \mapsto q} = p + q$, acyclicity holds by $p \cdot q = 0$. It remains to show determinacy: for arbitrary tests s we have $q \cdot s \leq q \Rightarrow q \cdot s = 0 \lor q \cdot s = q \Leftrightarrow q \cdot s = 0 \lor q \leq s$. Hence, $\langle p \mapsto q | \, | p \mapsto q \rangle s \leq \langle p \mapsto q | \, p \leq q \leq s$. $\qquad\square$

Now, we define predicates $\mathsf{LIST} =_{df} \{\mathsf{list}, \mathsf{l\_cell}\}$ for singly linked lists by

$$\mathsf{list} =_{df} \{a : a \text{ is a chain with } links(a) = 0\},$$
$$\mathsf{l\_cell} =_{df} \{a : a \text{ is a cell with } links(a) \neq 0\}.$$

Lemma 7.4. *For predicates in* LIST*, the operator* $\oslash$ *is associative.*

Proof. Assume $a \in (P_1 \oslash P_2) \oslash P_3$ with $a = a_{12} + a_3 \land a_{12} \rhd a_3$ and $a_{12} \in P_1 \oslash P_2 \land a_3 \in P_3$. There exist a_1, a_2 with $a_1 \rhd a_2$ and $a_{12} = a_1 + a_2 \land a_i \in P_i$.

From the definitions we first know $\ulcorner a_1 \cdot \ulcorner a_2 = 0$ and $\ulcorner(a_1 + a_2) \cdot \ulcorner a_3 = 0 \Leftrightarrow \ulcorner a_1 \cdot \ulcorner a_3 = 0 \land \ulcorner a_2 \cdot \ulcorner a_3 = 0$. Hence, we can conclude $\ulcorner a_1 \cdot \ulcorner(a_2 + a_3) = 0$. Analogously, $\ulcorner a_1 \cdot (a_2 + a_3)\urcorner = 0$ and $a_1\urcorner \cdot (a_2 + a_3)\urcorner = 0$.

Finally, we show $root(a_3) \leq links(a_2)$ and hence $a_2 + a_3 \in P_2 \oslash P_3$: By the assumption $a_1 \rhd a_2$ and Lemma 6.6 we have $root(a_3) \leq links(a_1 + a_2) = (links(a_1) - root(a_2)) + links(a_2)$. Since a_1 is a chain, $links(a_1)$ is at most an atom, hence $links(a_1) - root(a_2) = 0$ by $a_1 \rhd a_2$, and we are done.

Moreover, $root(a_2 + a_3) \leq links(a_1)$ also follows from Lemma 6.6 and therefore $a \in P_1 \oslash (P_2 \oslash P_3)$. The reverse inclusion is proved analogously. $\qquad\square$

We give some further results used in the list reversal example.

Lemma 7.5. *If* $a \in \mathsf{list}$ *then* $\langle a^* | \, root(a) = \ulcorner a + \sqcap$*.*

Lemma 7.6. $\mathsf{l\_cell} \oslash \mathsf{list} \subseteq \mathsf{list}$*.*

Proof. Let $a_1 \in \mathsf{l\_cell}$ and $a_2 \in \mathsf{list}$ and assume $a_1 \rhd a_2$. We need to show $a_1 + a_2 \in \mathsf{list}$. Properness of $a_1 + a_2$ follows from Corollary 4.2. Using distributivity of domain, we get $\langle a_1 + a_2 | \, | a_1 + a_2 \rangle p \leq p$ since $\langle a_i | \, | a_i \rangle p \leq p$ as a_1, a_2 are deterministic and $\langle a_2 | \, | a_1 \rangle p \leq 0 \land \langle a_1 | \, | a_2 \rangle p \leq 0$ by $\ulcorner a_1 \cdot \ulcorner a_2 = 0$ which follows from the definition.

It remains to show $\langle (a_1 + a_2)^* | \ root(a_1 + a_2) = \ulcorner a_1 \urcorner + \ulcorner a_2 \urcorner + \square$. By definition we know $\ulcorner a_1 \cdot a_2 \urcorner = 0$, hence $a_2 \cdot a_1^* = a_2$. Finally, using $(a_1 + a_2)^* = a_1^* \cdot (a_2 \cdot a_1^*)^* = a_1^* \cdot a_2^*$ and Lemma 6.6 we have $\langle (a_1 + a_2)^* | \ root(a_1 + a_2) = \langle a_1^* \cdot a_2^* | \ root(a_1) = \langle a_1^* | \ root(a_1) + \langle a_2 \cdot a_2^* | \ root(a_1) = \ulcorner a_1 \urcorner + links(a_1) + \langle a_2 \cdot a_2^* | (\ulcorner a_1 \urcorner + links(a_1))$. By atomicity, definitions and Lemma 7.5, $\ulcorner a_1 \urcorner + root(a_2) + \langle a_2 \cdot a_2^* | \ root(a_2) = \ulcorner a_1 \urcorner + \langle a_2^* | \ root(a_2) = \ulcorner a_1 \urcorner + \ulcorner a_2 \urcorner + \square$. $\qquad\square$

Lemma 7.7. $p \neq \square \Rightarrow \mathsf{list}\,(p) \subseteq \mathsf{l\_cell}\,(p) \circledR \mathsf{list}$ *and* $\mathsf{list}\,(\square) = \{\square\}$.

Definition 7.8. Assume a set L of selector names. For $l \in L$, an l-*context* is a linked structure a over L with $a_l \in \mathsf{l\_cell}$, i.e., a structure with a "hole" as its l-branch. The corresponding predicate is $l\_context =_{df} \{a : a \text{ is an } l\text{-context}\}$.

Lemma 7.9. *For predicates Q, R and $l \in L$ we have*

$$\{ (l\_context(i) \circledD Q) \circledast R(j) \} \quad i.l := j \quad \{ (l\_context(i) \circledD R(j)) \circledast Q \},$$
$$\{ (Q \circledD l\_context(i)) \circledast R(j) \} \quad i.l := j \quad \{ Q \circledD (l\_context(i) \circledD R(j)) \},$$
$$\{ l\_context(i)) \circledD Q \} \quad k := i.l \quad \{ l\_context(i) \circledD Q(j) \}.$$

Proof. Assume a store s and $a \in (l\_context(p) \circledD Q) \circledast R(q)$. Thus, $a = a_1 + a_2 + a_3$ with $a_1 \in l\_context \wedge a_2 \in Q \wedge a_3 \in R$ and $a_1 \circledplus a_3 \wedge a_2 \circledplus a_3 \wedge a_1 \triangleright a_2 \wedge s(i) = root(a_1) \wedge s(j) = root(a_3)$. By $a_1 \triangleright a_2$ we get $\ulcorner a_1 \cdot a_2 \urcorner = 0 \wedge \ulcorner a_1 \urcorner \cdot a_2 \urcorner \leq \square \wedge root(a_2) = links(a_1)$.

We show $(root(a_1) \mapsto root(a_3)) + \neg root(a_1) \cdot a \in (l\_context(p) \circledD R(q)) \circledast Q$. First, $a_1 \circledplus a_3 \wedge a_1 \triangleright a_2$ implies $\neg root(a_1) \cdot a_3 = a_3 \wedge \neg root(a_1) \cdot a_2 = a_2$ and $\neg root(a_1) \cdot a_1 = 0$ by Lemma 7.2. Hence, $\neg root(a_1) \cdot a = a_2 + a_3$.

Further, we know $root(a_1), root(a_3) \neq 0$ by definition. Assume $a_3 \neq \square$, thus $root(a_3) \leq \ulcorner a \urcorner$ and $root(a_3) \neq \square$. Then $a_1 \circledplus a_3 \Rightarrow root(a_1) \cdot root(a_3) = 0$, Lemma 6.2 and assumptions imply $links(root(a_1) \mapsto root(a_3)) = root(a_3)$. Moreover, $\ulcorner (root(a_1) \mapsto root(a_3)) \cdot a_3 \urcorner = root(a_1) \cdot \ulcorner a_3 \urcorner \leq \ulcorner a_1 \cdot a_3 \urcorner = 0$ and $(root(a_1) \mapsto root(a_3))\urcorner \cdot a_3 \urcorner = root(a_3) \cdot a_3 \urcorner \leq 0$. Finally by Lemma 7.3 we have $(root(a_1) \mapsto root(a_3)) + a_3 \in l\_context(p) \circledD R(q)$.

It remains to show $(root(a_1) \mapsto root(a_3)) \circledplus a_2$ and $a_3 \circledplus a_2$. The latter follows from commutativity of $\circledplus$ while the former resolves to $root(a_1) \cdot \ulcorner a_2 \urcorner = 0$ and $root(a_3) \cdot \ulcorner a_2 \urcorner = 0$ by Lemma 6.2. We calculate $\ulcorner a_2 \urcorner = \langle a_2^* | \ root(a_2) = \ulcorner a_2 \urcorner + \square$ by Lemma 7.5. Hence, $root(a_1) \cdot \ulcorner a_2 \urcorner = root(a_1) \cdot (\ulcorner a_2 \urcorner + \square) = 0$ by $a_1 \triangleright a_2$ and a_1 linkable. Similarly, $root(a_3) \cdot \ulcorner a_2 \urcorner \leq 0$ by assumptions.

If $a_3 = \square$ then $root(a_3) = \square$ and $R = \{\square\}$; the proof for this case is analogous to the above one, except that $root(a_3) \cdot \ulcorner a_2 \urcorner \leq \square$ and $root(a_3) \cdot a_3 \urcorner \leq \square$.

Proofs for the remaining triples can be given similarly. $\qquad\square$

To prove functional correctness of in-situ reversal we introduce the concept of *abstraction functions* [9]. They are used, e.g., to state invariant properties.

Definition 7.10. Assume $a \in \mathsf{list}$ and an atom $p \in \ulcorner a \urcorner$. We define the abstraction function li_a w.r.t. a as well as the semantics of the expression $i^{\rightarrow}$ for a program identifier i as follows:

$$li_a(p) =_{df} \begin{cases} \langle\rangle & \text{if } p \cdot \ulcorner a \urcorner = 0, \\ \langle p\rangle \bullet li_a(\langle a | p) & \text{otherwise}, \end{cases} \qquad [\![i^{\rightarrow}]\!]_{(s,a)} =_{df} li_a(s(i)).$$

Here • stands for concatenation and $\langle\rangle$ denotes the empty word.

Now using Lemma 7.9 and Hoare logic proof rules for variable assignment and while-loops, we can provide a full correctness proof of the in-situ list reversal algorithm. The invariant of the algorithm is $I \Leftrightarrow_{df} (j^{\rightarrow})^{\ddagger} \bullet i^{\rightarrow} = \alpha$, where $\_^{\dagger}$ denotes word reversal. Note that $(s,a) \models I \Leftrightarrow [\![(j^{\rightarrow})^{\dagger} \bullet i^{\rightarrow}]\!]_{(s,a)} = s(\alpha)$ where α represents a sequence.

$\{ \text{ list}(i) \wedge i^{\rightarrow} = \alpha \}$
$\quad$ j := □ ;
$\quad \{ (\text{list}(i) \circledast \text{list}(j)) \wedge I \}$
$\quad$ while $(i \neq □)$ do $\Big($
$\qquad \{ ((\text{l\_cell}(i) \oslash \text{list}(i.\text{next})) \circledast \text{list}(j)) \wedge I \}$
$\qquad$ k := i.next ;
$\qquad \{ ((\text{l\_cell}(i) \oslash \text{list}(k)) \circledast \text{list}(j)) \wedge (j^{\rightarrow})^{\dagger} \bullet i \bullet k^{\rightarrow} = \alpha \}$
$\qquad \{ ((\text{l\_cell}(i) \oslash \text{list}(k)) \circledast \text{list}(j)) \wedge (i \bullet j^{\rightarrow})^{\dagger} \bullet k^{\rightarrow} = \alpha \}$
$\qquad$ i.next := j ;
$\qquad \{ ((\text{list}(i) \circledast \text{list}(k)) \wedge (i^{\rightarrow})^{\dagger} \bullet k^{\rightarrow} = \alpha \}$
$\qquad$ j := i; i := k;
$\qquad \{ (\text{list}(j) \circledast \text{list}(i)) \wedge I \}$
$\Big)$
$\{ \text{ list}(j) \wedge (j^{\rightarrow})^{\dagger} = \alpha \}$
$\{ \text{ list}(j) \wedge j^{\rightarrow} = \alpha^{\dagger} \}$

Each assertion consists of a structural part and a part connecting the concrete and abstract levels of reasoning. The same pattern will occur in the tree rotation algorithm in the next subsection.

Compared to [19] we hide the existential quantifiers that were necessary there to describe the sharing relationships. Moreover, we include all correctness properties of the occurring data structures and their interrelationship in the definitions of the new connectives. Quantifiers to state functional correctness are not needed due to use of the abstraction function. Hence the formulas become easier to read and more concise.

To further underpin practicability of our approach, one could e.g. change the first two commands in the while loop of the list reversal algorithm (inspired by [3]), so that the algorithm could possibly leave a memory leak. Then after the assignment i.next := j we would get the postcondition $(\text{l\_cell}(i) \oslash \text{list}(j)) \circledast \text{list}$. This shows that the memory part characterised by list cannot be reached from i nor from j due to strong separation. Moreover, there is no program variable containing a reference to the root of that part.

7.2   Tree Rotation

To model binary trees we use the selector names left and right. A *binary tree* is then a tree $(b_{\text{left}}, b_{\text{right}})$ over $\{\text{left}, \text{right}\}$. Setting $b =_{df} b_{\text{left}} + b_{\text{right}}$, we define an abstraction function tr similar to the $\rightarrow$ function above:

$$tr_b(p) =_{df} \begin{cases} \langle\rangle & \text{if } p \cdot \ulcorner b = 0 \text{ ,} \\ \langle tr_b(\langle b_{\text{left}}|\,p), p, tr_b(\langle b_{\text{right}}|\,p)\rangle & \text{otherwise ,} \end{cases}$$

$$[\![i^{\leftrightarrow}]\!]_{(s,b)} =_{df} tr_b(s(i)) \text{ .}$$

As an example, we now present the correctness proof of an algorithm for tree rotation as known from the data structure of AVL trees. We first give a "clean" version, in which all occurring subtrees are separated. After that we will show an optimised version, where, however, sharing occurs in an intermediate state. The verification of that algorithm would take a few more steps, and hence we will not include it because of space restrictions.

For abbreviation we define two more predicates, for $l \in \{\text{left}, \text{right}\}$, by

$$l\_\text{tree\_context} =_{df} \text{tree} \cap l\_\text{context} \text{ .}$$

Let T_l, T_k, T_l stand for trees and p, q denote atomic tests:

$\{ \text{tree}(i) \wedge i^{\leftrightarrow} = \langle T_l, p, \langle T_k, q, T_r\rangle\rangle \}$
$j := i.\text{right};$
$\quad\{ (\text{right\_tree\_context}(i)) \, \oslash \, \text{tree}(j) \wedge$
$\quad\quad i^{\leftrightarrow} = \langle T_l, p, \langle T_k, q, T_r\rangle\rangle \wedge j^{\leftrightarrow} = \langle T_k, q, T_r\rangle \}$
$i.\text{right} := \Box;$
$\quad\{ \text{tree}(i) \, \circledast \, \text{tree}(j) \wedge$
$\quad\quad i^{\leftrightarrow} = \langle T_l, p, \langle\rangle\rangle \wedge j^{\leftrightarrow} = \langle T_k, q, T_r\rangle \}$
$k := j.\text{left};$
$\quad\{ \text{tree}(i) \, \circledast \, ((\text{left\_tree\_context}(j)) \, \oslash \, \text{tree}(k)) \wedge$
$\quad\quad i^{\leftrightarrow} = \langle T_l, p, \langle\rangle\rangle \wedge j^{\leftrightarrow} = \langle T_k, q, T_r\rangle \wedge k^{\leftrightarrow} = T_k \}$
$j.\text{left} := \Box;$
$\quad\{ \text{tree}(i) \, \circledast \, \text{tree}(j) \, \circledast \, \text{tree}(k) \wedge$
$\quad\quad i^{\leftrightarrow} = \langle T_l, p, \langle\rangle\rangle \wedge j^{\leftrightarrow} = \langle\langle\rangle, q, T_r\rangle \wedge k^{\leftrightarrow} = T_k \}$
$j.\text{left} := i;$
$\quad\{ (\text{left\_tree\_context}(j) \, \oslash \, \text{tree}(i)) \, \circledast \, \text{tree}(k) \wedge$
$\quad\quad i^{\leftrightarrow} = \langle T_l, p, \langle\rangle\rangle \wedge j^{\leftrightarrow} = \langle\langle T_l, p, \langle\rangle\rangle, q, T_r\rangle \wedge k^{\leftrightarrow} = T_k \}$
$i.\text{right} := k;$
$\quad\{ \text{left\_tree\_context}(j) \, \oslash \, (\text{right\_tree\_context}(i) \, \oslash \, \text{tree}(k)) \wedge$
$\quad\quad j^{\leftrightarrow} = \langle\langle T_l, p, T_k\rangle, q, T_r\rangle \wedge i^{\leftrightarrow} = \langle T_l, p, T_k\rangle \wedge k^{\leftrightarrow} = T_k \}$

In particular, j now points to the rotated tree. The optimised version reads, without intermediate assertions, as follows:

$\{ i^{\leftrightarrow} = \langle T_l, p, \langle T_m, q, T_r\rangle\rangle \}$
$\quad j := i.\text{right};$
$\quad i.\text{right} := j.\text{left};$
$\quad j.\text{left} := i;$
$\{ j^{\leftrightarrow} = \langle\langle T_l, p, \langle\rangle\rangle, q, T_r\rangle \}$

8   Related Work

There exist several approaches to extend SL by additional constructs to exclude sharing or restrict outgoing pointers of disjoint heaps to a single direction. Wang et al. [22] defined an extension called *Confined Separation Logic* and provided a relational model for it. They defined various central operators to assert, e.g., that all outgoing references of a heap h_1 point to another disjoint one h_2 or all outgoing references of h_1 either point to themselves or to h_2.

Our approach is more general due to its algebraicity and hence also able to express the mentioned operations. It is intended as a general foundation for defining further operations and predicates for reasoning about linked object structures.

Another calculus that follows a similar intention as our approach is given in [3]. Generally, there heaps are viewed as labelled object graphs. Starting from an abstract foundation the authors define a logic e.g. for lists with domain-specific predicates and operations which is feasible for automated reasoning.

By contrast, our approach enables abstract derivations in a fully first-order algebraic approach, called pointer Kleene algebra [7]. The given simple (in-)-equational laws allow a direct usage of automated theorem proving systems as PROVER9 [15] or any other systems through the TPTP LIBRARY [21] at the level of the underlying separation algebra [11]. This supports and helpfully guides the development of domain specific predicates and operations. The assertions we have presented are simple and still suitable for expressing shapes of linked list structures without the need of any arithmetic as in [3]. Part of such assertions can be automatically verified using SMALLFOOT [1].

9   Conclusion and Outlook

A general intention of the present work was relating the approach of pointer Kleene algebra with SL. The algebra has proved to be applicable for stating abstract reachability conditions and the derivation of such. Therefore, it can be used as an underlying separation algebra in SL. We defined extended operations similar to separating conjunction that additionally assert certain conditions on the references of linked object structures. As a concrete example we defined predicates and operations on linked lists and trees that enabled correctness proofs of an in-situ list-reversal algorithm and tree rotation.

As future work, it will be interesting to explore more complex or other linked object structures such as doubly-linked lists or threaded trees. In particular, more complex algorithms like the *Schorr-Waite Graph Marking* or concurrent garbage collection algorithms should be treated.

Acknowledgements. We thank all reviewers for their fruitful comments that helped to significantly improve the paper. This research was partially funded by the DFG project *MO 690/9-1 AlgSep   Algebraic Calculi for Separation Logic*.

References

1. Berdine, J., Calcagno, C., O'Hearn, P.W.: A Decidable Fragment of Separation Logic. In: Lodaya, K., Mahajan, M. (eds.) FSTTCS 2004. LNCS, vol. 3328, pp. 97–109. Springer, Heidelberg (2004)

2. Calcagno, C., O'Hearn, P.W., Yang, H.: Local Action and Abstract Separation Logic. In: Proc. of the 22nd Symposium on Logic in Computer Science, pp. 366–378. IEEE Press (2007)
3. Chen, Y., Sanders, J.W.: Abstraction of Object Graphs in Program Verification. In: Bolduc, C., Desharnais, J., Ktari, B. (eds.) MPC 2010. LNCS, vol. 6120, pp. 80–99. Springer, Heidelberg (2010)
4. Dang, H.H., Höfner, P., Möller, B.: Algebraic separation logic. Journal of Logic and Algebraic Programming 80(6), 221–247 (2011)
5. Desharnais, J., Möller, B.: Characterizing determinacy in Kleene algebra. Information Sciences 139, 253–273 (2001)
6. Desharnais, J., Möller, B., Struth, G.: Kleene algebra with domain. ACM Transactions on Computational Logic 7(4), 798–833 (2006)
7. Ehm, T.: The Kleene algebra of nested pointer structures: Theory and applications, PhD Thesis (2003), http://www.opus-bayern.de/uni-augsburg/frontdoor.php?source_opus=89
8. Ehm, T.: Pointer Kleene Algebra. In: Berghammer, R., Möller, B., Struth, G. (eds.) RelMiCS/Kleene-Algebra Ws 2003. LNCS, vol. 3051, pp. 99–111. Springer, Heidelberg (2004)
9. Hoare, C.A.R.: Proofs of correctness of data representations. Acta Informatica 1, 271–281 (1972)
10. Hoare, C.A.R., Hussain, A., Möller, B., O'Hearn, P.W., Petersen, R.L., Struth, G.: On Locality and the Exchange Law for Concurrent Processes. In: Katoen, J.-P., König, B. (eds.) CONCUR 2011. LNCS, vol. 6901, pp. 250–264. Springer, Heidelberg (2011)
11. Höfner, P., Struth, G.: Can refinement be automated? In: Boiten, E., Derrick, J., Smith, G. (eds.) Refine 2007. ENTCS, vol. 201, pp. 197–222. Elsevier (2008)
12. Kozen, D.: A completeness theorem for Kleene algebras and the algebra of regular events. Information and Computation 110(2), 366–390 (1994)
13. Kozen, D.: Kleene algebra with tests. ACM Transactions on Programming Languages and Systems 19(3), 427–443 (1997)
14. Manes, E., Benson, D.: The inverse semigroup of a sum-ordered semiring. Semigroup Forum 31, 129–152 (1985)
15. McCune, W.W.: Prover9 and Mace4, http://www.cs.unm.edu/~mccune/prover9
16. Möller, B.: Some applications of pointer algebra. In: Broy, M. (ed.) Programming and Mathematical Method. NATO ASI Series, Series F: Computer and Systems Sciences, vol. 88, pp. 123–155. Springer (1992)
17. Möller, B.: Calculating with acyclic and cyclic lists. Information Sciences 119(3-4), 135–154 (1999)
18. O'Hearn, P.W.: Resources, Concurrency, and Local Reasoning. Theoretical Computer Science 375(1-3), 271–307 (2007)
19. Reynolds, J.C.: An introduction to separation logic. In: Broy, M. (ed.) Engineering Methods and Tools for Software Safety and Security, pp. 285–310. IOS Press (2009)
20. Sims, E.J.: Extending Separation Logic with Fixpoints and Postponed Substitution. Theoretical Computer Science 351(2), 258–275 (2006)
21. Sutcliffe, G., Suttner, C.: The TPTP problem library: CNF release v1.2.1. Journal of Automated Reasoning 21(2), 177–203 (1998)
22. Wang, S., Barbosa, L.S., Oliveira, J.N.: A Relational Model for Confined Separation Logic. In: Proc. of the 2nd IFIP/IEEE Intl. Symposium on Theoretical Aspects of Software Engineering, TASE 2008, pp. 263–270. IEEE Press (2008)

Unifying Lazy and Strict Computations

Walter Guttmann

Institut für Programmiermethodik und Compilerbau, Universität Ulm
walter.guttmann@uni-ulm.de

Abstract. Non-strict sequential computations describe imperative programs that can be executed lazily and support infinite data structures. Based on a relational model of such computations we investigate their algebraic properties. We show that they share many laws with conventional, strict computations. We develop a common theory generalising previous algebraic descriptions of strict computation models including partial, total and general correctness and extensions thereof. Due to non-strictness, the iteration underlying loops cannot be described by a unary operation. We propose axioms that generalise the binary operation known from omega algebra, and derive properties of this new operation which hold for both strict and non-strict computations. All algebraic results are verified in Isabelle using its integrated automated theorem provers.

1 Introduction

Previous works show that various sequential computation models can be unified by devising algebraic structures whose operations satisfy a common set of axioms [9,10,13,12,14]. This unified treatment covers partial-, total- and general-correctness models as well as extensions of them. It provides a common approximation order, a unified semantics of recursion and iteration, and common preconditions, correctness calculi and pre-post specifications. Particular results proved in the unified setting include complex program transformations, refinement laws and separation theorems, such as Kozen's while-program normalisation and Back's atomicity refinement [20,2].

So far, only models of strict computations have been considered for this unifying algebraic approach. By 'strict' we mean that a computation cannot produce a defined output from an undefined input. This matches the conventional execution of imperative programs: for example, if $A =$ while $true$ do skip is the endless loop or $A = (x := 1/0)$ aborts, then $A ; P = A$ for every program P. However, there are also models of non-strict computations, which can recover from undefined input [8]. In such models, for example, $A ; (x := 2) = (x := 2)$ holds for either of the above definitions of A, assuming the state contains the single variable x. As elaborated in [8], this makes it possible to construct and compute with infinite data structures.

So far, the investigation of such non-strict computations has been concerned with a relational model [8] and an operational semantics [21]. This paper extends the unifying algebraic approach to cover non-strict computation models

W. Kahl and T.G. Griffin (Eds.): RAMiCS 2012, LNCS 7560, pp. 17–32, 2012.

in addition to strict ones. We therefore provide axioms and obtain results which are valid across this wide range of models.

Section 2 recalls the relational model of non-strict computations. In Section 3 we recall the basic algebraic structures that describe sequential, nondeterministic computations. The key observation is that many axioms are valid in both strict and non-strict settings. Section 4 contributes new axioms which uniformly describe the endless loop in these models. They generalise previous unifying algebraic descriptions. In Section 5 we derive the unified semantics of recursion and loops from these new axioms. Due to the weaker setting, the underlying iteration can no longer be described by a unary operation. Section 6 therefore generalises the binary operation known from omega algebra [3]. We contribute axioms for this operation, which hold for both strict and non-strict computations, and establish a collection of its properties. They are applied to derive Back's atomicity refinement theorem, which is proved for the first time in a non-strict setting.

All algebraic results are verified in Isabelle [25] heavily using its integrated automated theorem provers. The proofs are omitted and can be found in the theory files available at http://www.uni-ulm.de/en/in/pm/staff/guttmann/algebra/.

2   A Relational Model of Non-strict Computations

Our previous work [8] describes in full detail a relational model of non-strict, sequential computations. We recall this model as far as it is necessary for our subsequent algebraic description. For simplicity we assume a homogeneous setting in which variables cannot be added to or removed from the state. The addition of a typing discipline which gives heterogeneous algebras with partial operations is orthogonal to our present aims.

2.1   States

The state of a sequential computation that models an imperative program is given by the values of its variables. Let the variables be $x_1, x_2, \ldots$ which we abbreviate as $\vec{x}$. Associate with each variable x_i its type or range D_i, which is the set of values the variable can take. Each set D_i contains two special elements ∞ and $\frac{1}{2}$ with the following intuitive meaning:

- If x_i has the value ∞ and this value is needed, the execution of the program does not terminate.
- If x_i has the value $\frac{1}{2}$ and this value is needed, the execution aborts.

Thus ∞ and $\frac{1}{2}$ represent the results of non-terminating and undefined computations, respectively. Each set D_i is partially ordered by $\preccurlyeq$ such that ∞ is the least element. For elementary types $\preccurlyeq$ is flat, treating $\frac{1}{2}$ like any other value different from ∞. The full theory [8] imposes additional structure on D_i to facilitate the construction of sum, product, function and recursive types.

Let $D_I = \prod_{i \in I} D_i$ denote the Cartesian product of the ranges of the variables x_i with $i \in I$ for an index set I. A state is an element $\vec{x}_I \in D_I$ where I indicates the variables comprising the state. The index set I is constant in our homogeneous setting. The partial order $\preccurlyeq$ is lifted componentwise to states.

2.2   Statements

Statements transform states into new states. We use x_i and x_i' to denote the values of the variable x_i before and after execution of a statement. A computation is modelled as a homogeneous relation $R \subseteq D_I \times D_I$ on the state space D_I. An element $(\vec{x}, \vec{x}') \in R$ intuitively means that the execution of R with input $\vec{x}$ may yield the output $\vec{x}'$. Several output values for the same input indicate non-determinism. Note that the componentwise lifted partial order $\preccurlyeq$ is a relation on states.

Computations may be specified by set comprehensions like $\{(\vec{x}, \vec{x}') \mid x_1'=x_2\}$ or $\{(\vec{x}, \vec{x}') \mid x_2 = 7\}$. We abbreviate such a comprehension by its constituent predicate, that is, $x_1'=x_2$ or $x_2=7$ for the preceding examples.

Programming constructs include the following ones:

- The program skip is modelled by the relation $\preccurlyeq$ in order to enforce an upper closure on the image of each state. This is typical for total-correctness approaches [7,17].
- The assignment $\vec{x} \leftarrow \vec{e}$ is the relation $\preccurlyeq \, ; \, (\vec{x}' = \vec{e}) \, ; \preccurlyeq$ which is the usual assignment composed with $\preccurlyeq$ to obtain upper closure. It follows that $\vec{x} \leftarrow \vec{\infty}$ is the universal relation.
- Sequential composition of computations P and Q is their relational composition $P \, ; Q$.
- Non-deterministic choice between computations P and Q is their union $P \cup Q$.
- The conditional if b then P else Q is the relation

$$(b=\infty \cap \vec{x} \leftarrow \vec{\infty}) \cup (b=\frac{1}{2} \cap \vec{x} \leftarrow \vec{\frac{1}{2}}) \cup (b=true \cap P) \cup (b=false \cap Q) \, .$$

  It distinguishes the four possible values of the condition b in the current state. An outcome of ∞ or $\frac{1}{2}$ is propagated to the whole state.
- The recursive specification $P = f(P)$ is solved as the greatest fixpoint νf of the function f with respect to the refinement order $\subseteq$. This, too, is typical for total correctness: for example, it follows that the endless loop is the universal relation, which absorbs any computation in a non-deterministic choice and suitably equals $\vec{x} \leftarrow \vec{\infty}$.

Expressions in assignments and conditionals are assumed to be $\preccurlyeq$-continuous. Several examples illustrate that computations in this model are non-strict:

- $(x_1, x_2 \leftarrow \frac{1}{2}, 2) \, ; \, (x_1 \leftarrow x_2) = (x_1, x_2 \leftarrow 2, 2)$,
- $(\vec{x} \leftarrow \vec{\infty}) \, ; \, (x_1, x_2 \leftarrow 2, 2) = (x_1, x_2, x_3.. \leftarrow 2, 2, \vec{\infty})$,
- $(\vec{x} \leftarrow \vec{\infty}) \, ; \, (\vec{x} \leftarrow \vec{e}) = (\vec{x} \leftarrow \vec{e})$ if all expressions e are constant.

Such computations are similar to Haskell's state transformers [22]. Our relational semantics suits sequential computations better than the λ-calculus, allows non-determinism and distinguishes non-termination from undefinedness.

The following consequences are shown in [8].

Theorem 1. *Consider the above programming constructs.*

1. *Every relation P composed of those constructs satisfies $\preccurlyeq\,;P = P = P\,;\preccurlyeq$.*
2. *Relations composed of those constructs are total.*
3. *Functions composed of those constructs are $\subseteq$-isotone.*
4. *Functions composed of those constructs without the non-deterministic choice are $\bigcap$-continuous.*

With Theorem 1.1 there are two ways to obtain a monoid structure with respect to sequential composition:

1. For the set of all relations, the identity relation is a neutral element.
2. For the set of relations $\{P \mid\, \preccurlyeq\,;P = P = P\,;\preccurlyeq\}$, which includes all those composed of the above programming constructs, the skip program $\preccurlyeq$ is a neutral element.

We treat the first structure in the present paper as this choice simplifies the representation of the conditional. An investigation of the second monoid structure is postponed.

In the remainder of this paper we provide an algebraic description of this computation model as well as other, strict ones. Basic and compound programs are represented by elements of algebraic structures. Operations of these structures represent some of the above programming constructs, the laws of which are specified as axioms:

- Section 3 axiomatises the operations $\cdot$ and $+$ for sequential composition and non-deterministic choice, which yields the refinement order $\leq$.
- Section 5 introduces a general approximation order $\sqsubseteq$, which instantiates to $\geq$ in this model, and axiomatises the $\sqsubseteq$-least fixpoint κf, which instantiates to the $\leq$-greatest fixpoint νf for recursion.

Conditional statements are not represented as above by intersection with condition relations, but by using test elements which act as filters in sequential composition. As usual, while-loops appear as special cases of recursion.

3   Basic Algebraic Structures for Sequential Computations

The model of non-strict computations presented in Section 2 is based on relations and relational operations. We therefore inherit many properties known from relation algebras [27,23] and similar structures. The same is true for a number of strict computation models including partial-, total- and general-correctness models and various extensions which differ in their treatment of finite, infinite and aborting executions as described in [12,14].

3.1   Lattice, Semiring and Domain

In particular, both strict and non-strict computations form an algebraic structure $(S, +, \curlywedge, \cdot, 0, 1, \top)$ such that $(S, +, \curlywedge, 0, \top)$ is a bounded distributive lattice and $(S, +, \cdot, 0, 1)$ is a semiring without the right zero law. Their common basis is a bounded join-semilattice, which is a structure $(S, +, 0)$ satisfying the axioms

$$x + (y + z) = (x + y) + z \qquad\qquad x + x = x$$
$$x + y = y + x \qquad\qquad 0 + x = x$$

The semilattice order $x \leq y \Leftrightarrow x + y = y$ has least element 0 and least upper bound $+$. A bounded distributive lattice $(S, +, \curlywedge, 0, \top)$ adds a dual bounded meet-semilattice $(S, \curlywedge, \top)$ along with distribution and absorption axioms:

$$x \curlywedge (y \curlywedge z) = (x \curlywedge y) \curlywedge z \qquad\qquad x \curlywedge x = x$$
$$x \curlywedge y = y \curlywedge x \qquad\qquad \top \curlywedge x = x$$

$$x + (y \curlywedge z) = (x + y) \curlywedge (x + z) \qquad\qquad x + (x \curlywedge y) = x$$
$$x \curlywedge (y + z) = (x \curlywedge y) + (x \curlywedge z) \qquad\qquad x \curlywedge (x + y) = x$$

The semilattice order has another characterisation $x \leq y \Leftrightarrow x \curlywedge y = x$, the greatest element $\top$ and the greatest lower bound $\curlywedge$. An idempotent semiring $(S, +, \cdot, 0, 1)$ without the right zero law – simply called *semiring* in the remainder of this paper – adds to a bounded join-semilattice $(S, +, 0)$ a monoid $(S, \cdot, 1)$, distribution axioms and a left annihilation axiom:

$$1 \cdot x = x \qquad\qquad x \cdot (y + z) = (x \cdot y) + (x \cdot z)$$
$$x \cdot 1 = x \qquad\qquad (x + y) \cdot z = (x \cdot z) + (y \cdot z)$$
$$x \cdot (y \cdot z) = (x \cdot y) \cdot z \qquad\qquad 0 \cdot x = 0$$

We conventionally abbreviate $x \cdot y$ as xy. Note that $x0 = 0$ is not an axiom, because in many computation models this law does not hold (although it does for our non-strict computations). The operations $+$, $\curlywedge$ and $\cdot$ are $\leq$-isotone.

In computation models, the operations $+$, $\curlywedge$ and $\cdot$ are instantiated by the non-deterministic choice, intersection and sequential composition. The constants 0, 1 and $\top$ correspond to the empty, identity and universal relations (in some models adapted to satisfy certain healthiness conditions). The order $\leq$ is the subset order and $x \leq y$ expresses that x refines y.

Strict and non-strict models furthermore support the operation d which describes the domain $d(x)$ of a computation x, that is, the initial states from which execution of x is enabled. Semirings with domain have been investigated in [5]; the following axioms for d are taken from [6]:

$$x = d(x)x \qquad d(x) \leq 1 \qquad d(xy) = d(xd(y))$$
$$d(0) = 0 \qquad d(x + y) = d(x) + d(y)$$

It follows that $(d(S), +, \cdot, 0, 1)$ is a bounded distributive lattice in which the operations $\cdot$ and $\curlywedge$ coincide. We also obtain the Galois connection $d(x) \leq d(y) \Leftrightarrow x \leq d(y)\top$ given by [1]. The explicit characterisation $d(x) = x\top \curlywedge 1$ is not a consequence of the axioms and, actually, it does not hold in some of the strict models. Elements of $d(S)$ represent tests in computation models. Boolean complements of tests can be added but are not required in the present paper.

3.2 Kleene Algebra, Omega Algebra and Itering

A Kleene algebra $(S, +, \cdot, {}^*, 0, 1)$ extends a semiring $(S, +, \cdot, 0, 1)$ with an operation * satisfying the following unfold and induction axioms [19]:

$$1 + yy^* \leq y^* \qquad z + yx \leq x \Rightarrow y^*z \leq x$$
$$1 + y^*y \leq y^* \qquad z + xy \leq x \Rightarrow zy^* \leq x$$

It follows that y^*z is the least fixpoint of $\lambda x.yx + z$ and that zy^* is the least fixpoint of $\lambda x.xy + z$. An omega algebra $(S, +, \cdot, {}^*, {}^\omega, 0, 1)$ extends a Kleene algebra with an operation $^\omega$ satisfying the following axioms [3,24]:

$$yy^\omega = y^\omega \qquad x \leq yx + z \Rightarrow x \leq y^\omega + y^*z$$

It follows that $y^\omega + y^*z$ is the greatest fixpoint of $\lambda x.yx + z$ and that $\top = 1^\omega$ is the greatest element. For models which require other fixpoints we use the following generalisation of Kleene algebras. An *itering* $(S, +, \cdot, {}^\circ, 0, 1)$ extends a semiring with an operation $^\circ$ satisfying the sumstar and productstar equations of [4] and two simulation axioms introduced in [13]:

$$(x + y)^\circ = (x^\circ y)^\circ x^\circ \qquad zx \leq yy^\circ z + w \Rightarrow zx^\circ \leq y^\circ(z + wx^\circ)$$
$$(xy)^\circ = 1 + x(yx)^\circ y \qquad xz \leq zy^\circ + w \Rightarrow x^\circ z \leq (z + x^\circ w)y^\circ$$

The following result gives five instances of iterings which cover several strict computation models [13]. The element L represents the endless loop as detailed in the Section 4.

Theorem 2. *Iterings have the following instances:*

1. *Every Kleene algebra is an itering using $x^\circ = x^*$.*
2. *Every omega algebra is an itering using $x^\circ = x^\omega 0 + x^*$.*
3. *Every omega algebra with $\top x = \top$ is an itering using $x^\circ = x^\omega + x^*$.*
4. *Every demonic refinement algebra [28] is an itering using $x^\circ = x^\omega$.*
5. *Extended designs [15,14] form an itering using $x^\circ = d(x^\omega)\mathsf{L} + x^*$.*

However, in Section 6 we shall see that the non-strict model of Section 2 is not a useful itering. Also the properties of L known from the strict models need to be generalised to hold in the non-strict model. Thus the topic of the remainder of this paper is to develop a theory which captures L, recursion and iteration in all models of Theorem 2 and in the non-strict model.

4   Infinite Executions

Many models of sequential computations, in particular our non-strict model of Section 2, can represent infinite executions in addition to finite ones. In some of these models, for example, general-correctness models, this requires the use of an approximation order $\sqsubseteq$ for recursion which is different from the refinement order $\leq$ and typically based on the Egli-Milner order. To express the approximation order, which we do in Section 5, we have to represent the computation which contains precisely all infinite executions, that is, the endless loop L.

In this section and in the following one we work in a structure S that is a bounded distributive lattice and a semiring and has a domain operation.

The element $L \in S$ satisfies rather different properties in different computation models; for example, $L = 0$ in partial-correctness models, $L = \top$ in total-correctness models and in our non-strict model, and neither holds in general correctness and for extended designs. We therefore carefully choose the following axioms for L so that they hold in all of these models:

(L1) $xL = x0 + d(x)L$

(L2) $d(L)x \leq xd(L)$

(L3) $d(L)\top \leq L + d(L0)\top$

(L4) $L\top \leq L$

(L5) $x0 \wedge L \leq (x \wedge L)0$

We reuse axioms (L1) and (L2) from our previous unifying treatments of strict computation models [10,14]. The axioms (L3) and (L4) generalise the property $Lx = L$ which holds in strict models, but not in our non-strict model. Finally, axiom (L5) is a property of algebras for hybrid systems [18] which satisfy $L = \top 0$, but the latter holds neither for extended designs nor in our non-strict model.

The following remarks describe the intuition underlying the axioms and their use in the subsequent development.

- Axiom (L1) expresses that intermediate states cannot be observed in infinite executions, which is typical for relational models. A computation x followed by an infinite execution essentially amounts to infinite executions starting from the initial states of x – contained in $d(x)L$ – up to aborting executions in strict models – contained in $x0$.
- Axiom (L2) is used to show that sequential composition, finite and infinite iteration are $\sqsubseteq$-isotone. All of our models satisfy either $d(L) = 0$ or $d(L) = 1$.
- Axiom (L3) is equivalent to the requirement that L is the least element of the approximation order $\sqsubseteq$ which we use for defining the semantics of recursion in Section 5. It is furthermore used to derive the semantics of iteration as a special case of recursion. It can equivalently be stated as $d(L)x \leq L + d(L0)x$.
- Axiom (L4) can equivalently be stated as $Lx \leq L$. This is a key weakening of the previously used axiom $Lx = L$, which is now recognised to characterise strict computations. The intuitive meaning of $Lx = L$ is that anything which is supposed to happen 'after' an infinite execution is ignored. Hoare [16] formulates this characteristic strictness property as $\top x = \top$, which is the same in models where $L = \top$, such as the designs of the Unifying Theories of Programming [17], but different in other models. We use axiom (L4) in particular to show that sequential composition, finite and infinite iteration are $\sqsubseteq$-isotone.
- Axiom (L5) is used in Section 5 to simplify the characterisation of solutions to recursive specifications. Intuitively, in strict models the operation $\cdot 0$ cuts away the finite executions of a computation (unless constraints of the model

force their presence), and λL keeps only the infinite executions. The axiom's impact is that these two operations commute.

Each of the axioms (L1)–(L3) is independent from the remaining axioms as counterexamples generated by Nitpick or Mace4 witness, but the dependence is unknown for (L4) and (L5). Axioms (L3)–(L5) can be strengthened to equalities as shown by the following lemma, along with further consequences of the above axioms.

Lemma 3

1. $x0 \curlywedge L = (x \curlywedge L)0 = d(x0)L$
2. $Lx \le L = LL = L\top = L\top L = Ld(L)$
3. $d(L)x = (x \curlywedge L) + d(L0)x$
4. $d(x\top 0)L \le d(xL)L = d(x)L = x\top \curlywedge L \le xL \le x0 + L$
5. $xd(y)L = x0 + d(xy)L$
6. $x \le y \Leftrightarrow x \le y + L \wedge x \le y + d(y0)\top$
7. $x = x\top \Rightarrow d(L0)x \le d((x \curlywedge L)0)\top$

The following Venn diagram shows for each of several properties of L in which computation models it holds (P/T/G/E/N = partial correctness/total correctness/general correctness/extended designs/non-strict):

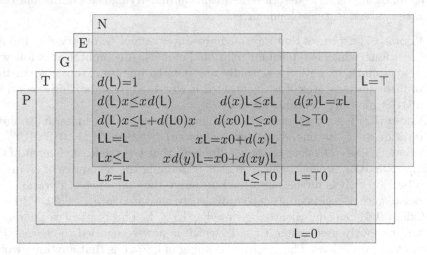

Only the properties that hold in all models are suitable for our unifying approach, while the others are useful for specialised theories.

5 Recursion

In many models of sequential computations the semantics of the recursive specification $x = f(x)$ is defined as the least fixpoint of the function f with respect to a given approximation order. However, quite different approximation orders are used in the individual models, for example, $\leq$ in partial correctness, $\geq$ in total correctness and in our non-strict model, and variants of the Egli-Milner order in general correctness and for extended designs. We reuse the following approximation relation $\sqsubseteq$ from our previous unifying treatment of strict computation models [10]:

$$x \sqsubseteq y \Leftrightarrow x \leq y + \mathsf{L} \wedge d(\mathsf{L})y \leq x + d(x0)\top .$$

Intuitively, the part $x \leq y + \mathsf{L}$ states that executions may be added and infinite executions may be removed, and the part $d(\mathsf{L})y \leq x + d(x0)\top$ states that this may happen only if x has infinite executions. However, this consideration is somewhat simplified and we must verify that $\sqsubseteq$ instantiates to the correct approximation order for our strict and non-strict models. This is shown in [10] for partial-, total- and general-correctness models; for extended designs this follows by a calculation similar to the one in [14]; finally, our non-strict model has $\mathsf{L} = \top$ and $d(\mathsf{L}) = 1$ and $x0 = 0$, whence $x \sqsubseteq y$ reduces to $x \geq y$ as required.

Because of the weaker axioms (L1)–(L5), new proofs are needed for the following results. The first shows that $\sqsubseteq$ is indeed a partial order.

Theorem 4. *The relation $\sqsubseteq$ is a partial order with least element L. The operations $+$ and $\cdot$ and $\lambda\mathsf{L}$ are $\sqsubseteq$-isotone.*

In each of our models, the semantics of the recursion $x = f(x)$ is the $\sqsubseteq$-least fixpoint κf of the function $f : S \to S$. Additionally, let μf denote the $\leq$-least fixpoint of f and νf the $\leq$-greatest one, as specified by the following properties:

$$\begin{array}{ll} f(\kappa f) = \kappa f & f(x) = x \Rightarrow \kappa f \sqsubseteq x \\ f(\mu f) = \mu f & f(x) = x \Rightarrow \mu f \leq x \\ f(\nu f) = \nu f & f(x) = x \Rightarrow \nu f \geq x \end{array}$$

These laws hold if the respective fixpoint exists, which typically requires additional properties of the function f or the structure S, from which we abstract.

Let $x \sqcap y$ denote the $\sqsubseteq$-greatest lower bound of x and y, provided it exists. We can then generalise our previous characterisations of κf [10,14] to the present setting which covers extended designs and our non-strict model in addition to partial-, total- and general-correctness models.

Theorem 5. *Let $f : S \to S$ be $\leq$- and $\sqsubseteq$-isotone, and assume that μf and νf exist. Then the following are equivalent:*

1. *κf exists.*
2. *κf and $\mu f \sqcap \nu f$ exist and $\kappa f = \mu f \sqcap \nu f$.*
3. *κf exists and $\kappa f = (\nu f \curlywedge \mathsf{L}) + \mu f$.*
4. *$d(\mathsf{L})\nu f \leq (\nu f \curlywedge \mathsf{L}) + \mu f + d(\nu f 0)\top$.*
5. *$d(\mathsf{L})\nu f \leq (\nu f \curlywedge \mathsf{L}) + \mu f + d(((\nu f \curlywedge \mathsf{L}) + \mu f)0)\top$.*
6. *$(\nu f \curlywedge \mathsf{L}) + \mu f \sqsubseteq \nu f$.*
7. *$\mu f \sqcap \nu f$ exists and $\mu f \sqcap \nu f = (\nu f \curlywedge \mathsf{L}) + \mu f$.*
8. *$\mu f \sqcap \nu f$ exists and $\mu f \sqcap \nu f \leq \nu f$.*

To obtain the semantics of a recursion more easily, this theorem reduces the calculation of κf to that of μf and νf, which are extremal fixpoints with respect to $\leq$ instead of the more complex $\sqsubseteq$. The characterisation (4) can be included in the above theorem due to axiom (L5), without which only the more involved characterisation (5) is available. They are helpful because they do not use $\sqsubseteq$ or κf. By requiring the stronger property (2) of the following result, we can generalise the additional characterisations given in [10,14] to the present setting, too.

Corollary 6. *Let $f : S \to S$ be $\leq$- and $\sqsubseteq$-isotone, and assume that μf and νf exist. Then the following are equivalent and imply the statements of Theorem 5:*

1. *κf exists and $\kappa f = d(\nu f 0)\mathsf{L} + \mu f$.*
2. *$d(\mathsf{L})\nu f \leq \mu f + d(\nu f 0)\top$.*
3. *$d(\nu f 0)\mathsf{L} + \mu f \sqsubseteq \nu f$.*
4. *$\mu f \sqcap \nu f$ exists and $\mu f \sqcap \nu f = d(\nu f 0)\mathsf{L} + \mu f$.*

We instantiate these results about general recursion to the special case of iteration using the Kleene star and omega operations of Section 3.2. The loop while p do w is characterised by the unfolding equation

$$\text{while } p \text{ do } w = \text{if } p \text{ then } (w \text{ ; while } p \text{ do } w) \text{ else skip} .$$

Representing the conditional in terms of the non-deterministic choice, this recursion has the form $x = yx + z$, where y and z depend on the condition p and the body w of the loop. The following result shows how to obtain the $\sqsubseteq$-least fixpoint of such a linear characteristic function. It assumes that S is also an omega algebra.

Corollary 7. *Let $y, z \in S$ and $f : S \to S$ with $f(x) = yx + z$. Then $\kappa f = (y^\omega \curlywedge \mathsf{L}) + y^* z = d(y^\omega)\mathsf{L} + y^* z$.*

Finally, we obtain that the various iteration operations are $\sqsubseteq$-isotone, whence by Theorem 4 also the while-loop construct is $\sqsubseteq$-isotone. For this, S has to be also a Kleene algebra, an omega algebra or an itering, respectively.

Theorem 8. *Let $x, y \in S$ such that $x \sqsubseteq y$. Then $x^* \sqsubseteq y^*$ and $x^\omega \sqsubseteq y^\omega$ and $x^\circ \sqsubseteq y^\circ$.*

6   A Binary Operation for Iteration

In strict computation models that form an itering, the iteration underlying loops can be described in terms of a unary operation. Namely, $\mathsf{L}z = \mathsf{L}$ holds in such models, whence for $f(x) = yx + z$ we obtain

$$\kappa f = d(y^\omega)\mathsf{L} + y^* z = d(y^\omega)\mathsf{L}z + y^* z = (d(y^\omega)\mathsf{L} + y^*)z = y^\circ z$$

using the unary operation $y^\circ = d(y^\omega)\mathsf{L} + y^*$ which specialises to the itering operations of Theorem 2 in partial-, total- and general-correctness models [14].

However, this is not the case in our non-strict computation model. In particular, in this model

$$\kappa f = \nu f = y^\omega + y^* z \neq (y^\omega + y^*)z$$

in general, as can be seen by setting $z = 0$ and observing that $x0 = 0$ for every x. The instance $z = 0$ arises for infinite loops such as while $true$ do skip. The semantics of this loop is $\top$ and not 0 in the non-strict model. Thus loops cannot be represented in the form $y^\circ z$ in this model, no matter how the unary operation $^\circ$ is defined.

In the non-strict model, this problem is solved by using the binary iteration operation of omega algebra, namely $y \star z = y^\omega + y^* z$ [3]. In fact, the non-strict model of Section 2 is an omega algebra in which $x0 = 0$ holds for every x – in contrast to the strict models of total and general correctness as well as extended designs, which do not satisfy this right annihilation property.

On the other hand, omega algebra's binary operation $y \star z = y^\omega + y^* z$ does not describe iteration in several strict computation models, because in general it differs from $y^\circ z = d(y^\omega)\mathsf{L} + y^* z$. A unified description of iteration for strict and non-strict models therefore requires a more general binary operation $\star$, like the one we introduce next.

A *binary itering* $(S, +, \cdot, \star, 0, 1)$ is a semiring $(S, +, \cdot, 0, 1)$ extended with a binary operation $\star$ satisfying the following axioms:

$$(x + y) \star z = (x \star y) \star (x \star z) \qquad\qquad x \star (y + z) = (x \star y) + (x \star z)$$
$$(xy) \star z = z + x((yx) \star (yz)) \qquad\qquad (x \star y)z \leq x \star (yz)$$
$$zx \leq y(y \star z) + w \Rightarrow z(x \star v) \leq y \star (zv + w(x \star v))$$
$$xz \leq z(y \star 1) + w \Rightarrow x \star (zv) \leq z(y \star v) + (x \star (w(y \star v)))$$

These axioms generalise the itering axioms by appropriately composing to an iteration of the form y° a continuation z. The distributivity axiom $x \star (y + z) = (x \star y) + (x \star z)$ and the semi-associativity axiom $(x \star y)z \leq x \star (yz)$ have to be added here, while for the unary operation they follow from the corresponding properties of $\cdot$. The sumstar equation and the first simulation axiom generalise theorems of [3].

To understand the computational meaning of the two simulation axioms they can be seen as generalising the basic simulation laws $zx \leq yz \Rightarrow z(x \star v) \leq y \star (zv)$ and $xz \leq zy \Rightarrow x \star (zv) \leq z(y \star v) + (x \star 0)$, where $x \star 0$ is needed since $\star$ may capture infinite iterations of x. These are similar to simulation laws known in Kleene and omega algebras, where they follow from even simpler induction axioms that characterise the $\leq$-least and $\leq$-greatest fixpoints of linear functions. Because $\star$ is intended for iteration in several computation models that require different fixpoints, we cannot use those induction axioms. However, we might expect a characterisation as the $\sqsubseteq$-least fixpoint of a linear function using a unified approximation order $\sqsubseteq$ such as the one in Section 5. For general-correctness models such properties are shown in [11].

Note that full associativity $(x \star y)z = x \star (yz)$ does not hold in our non-strict model as witnessed by setting $x = 1$ and $z = 0$:

$$(1 \star y)0 = 0 \neq \top = \top + 0 = 1^\omega + 1^* 0 = 1 \star 0 = 1 \star (y0)$$

since binary iteration is $y \star z = y^\omega + y^* z$ in this model. It is therefore not obvious how to generalise the axioms and other formulas from iterings to binary iterings. For example, $y^\circ z$ could be translated to $y \star z$ or to $(y \star 1)z$, and similar options are available for each occurrence of $^\circ$ in a formula. In particular for the axioms, these choices have a critical impact: certain combinations might yield a formula that fails in some target computation models, while another choice might yield a formula too weak to derive a useful theory.

An *extended binary itering* is a binary itering which satisfies the additional axiom

$$w(x \star (yz)) \le (w(x \star y)) \star (w(x \star y)z) .$$

In the special case $w = 1$, it is a substitute for associativity by replacing $x \star (yz)$ with $(x \star y)z$ at the expense of iterating $x \star y$.

The following result shows that binary iterings indeed capture both the non-strict and the strict models.

Theorem 9. *Binary iterings have the following models:*

1. *Every itering is an extended binary itering using $x \star y = x^\circ y$.*
2. *Every omega algebra is an itering using $x \star y = x^\omega + x^* y$.*
3. *Every omega algebra with the additional axiom $x \le x\top x\top$ is an extended binary itering using $x \star y = x^\omega + x^* y$.*

Part 1 covers the five strict computation models of Theorem 2, including partial, total and general correctness as well as extended designs. Parts 2 and 3 cover our non-strict model of Section 2. In particular, this model satisfies the property $x \le x\top x\top$. It is a weakening of the 'Tarski rule' of relation algebra, according to which $x = 0$ or $\top x\top = \top$ holds for every x [26], and can equivalently be stated in each of the following forms in omega algebras:

$$
\begin{array}{lll}
x\top = x\top x\top & x\top = (x\top)^\omega & xy^\omega = (xy^\omega)^\omega \\
x\top \le x\top x\top & x\top \le (x\top)^\omega & xy^\omega \le (xy^\omega)^\omega \\
x \le x\top x\top & x \le (x\top)^\omega &
\end{array}
$$

It implies the law $x^{\omega\omega} = x^\omega$, but not vice versa.

The following result shows a selection of properties which hold in binary iterings and therefore in all of the above computation models. This collection and subsequent ones form a reference to guide the axiomatisation and facilitate program reasoning as in Corollary 13.

Theorem 10. *Let S be a binary itering and $p, w, x, y, z \in S$. Then the following properties 1–56 hold.*

1. $0 \star x = x$
2. $x \le x \star 1$
3. $y \le x \star y$
4. $xy \le x \star y$
5. $x(x \star y) = x \star (xy)$
6. $x(x \star y) \le x \star y$
7. $(x \star 1)y \le x \star y$
8. $(xx) \star y \le x \star y$
9. $x \star x \le x \star 1$
10. $x \star (x \star y) = x \star y$

11. $(x \star x) \star y = x \star y$

12. $(x(x \star 1)) \star y = x \star y$

13. $(x \star 1)(y \star 1) = x \star (y \star 1)$

14. $1 \star (x \star y) = (x \star 1) \star y$

15. $x \star (1 \star y) = (x \star 1) \star y$

16. $((x \star 1) \star 1) \star y = (x \star 1) \star y$

17. $x \star y = y + x(x \star y)$

18. $x \star y = y + (x \star (xy))$

19. $y + xy + (x \star (x \star y)) = x \star y$

20. $(1 + x) \star y = (x \star 1) \star y$

21. $(x0) \star y = x0 + y$

22. $x + y \le x \star (y \star 1)$

23. $x \star (x + y) \le x \star (1 + y)$

24. $(xx) \star ((x+1)y) \le x \star y$

For example, property 5 exchanges $\cdot$ with $\star$ and property 10 shows that iteration is transitive. Properties 17 and 18 are unfold laws for the operation $\star$.

25. $(xy) \star (xz) = x((yx) \star z)$

26. $(x \star (y \star 1)) \star z = (y \star (x \star 1)) \star z$

27. $(x \star (y \star 1)) \star z = x \star ((y \star (x \star 1)) \star z)$

28. $(y(x \star 1)) \star z = (y(y \star (x \star 1))) \star z$

29. $x \star (y(z \star 1)) = (x \star y)(z \star 1)$

30. $(x + y) \star z = x \star (y \star ((x + y) \star z))$

31. $(x + y) \star z = (x + y) \star (x \star (y \star z))$

32. $(x + y) \star z \le (x \star (y \star 1)) \star z$

33. $(x + y0) \star z = x \star (y0 + z)$

34. $x \star z \le (x + y) \star z$

35. $(xy) \star z \le (x + y) \star z$

36. $x \star (y \star z) \le (x + y) \star z$

37. $x \star (y \star z) \le ((x \star y) \star z) + (x \star z)$

38. $x \star ((y(x \star 1)) \star z) \le (x + y) \star z$

39. $x \star ((y(x \star 1)) \star z) \le ((x \star 1)y) \star (x \star z)$

40. $(w(x \star 1)) \star (yz) \le (x \star w) \star ((x \star y)z)$

Property 25 corresponds to the sliding law of Kleene algebra [19].

41. $x \le y \Rightarrow x \star z \le y \star z$

42. $y \le z \Rightarrow x \star y \le x \star z$

43. $x \le y \Rightarrow x \star (y \star z) = y \star z$

44. $x \le y \Rightarrow y \star (x \star z) = y \star z$

45. $1 \le x \Rightarrow x(x \star y) = x \star y$

46. $1 \le z \Rightarrow x \star (yz) = (x \star y)z$

47. $x \le z \star y \wedge y \le z \star w \Rightarrow x \le z \star w$

48. $x \le z \star 1 \wedge y \le z \star w \Rightarrow xy \le z \star w$

49. $yx \le x \Rightarrow y \star x \le x + (y \star 0)$

50. $yx \le xy \Rightarrow (xy) \star z \le x \star (y \star z)$

51. $yx \le xy \Rightarrow y \star (x \star z) \le x \star (y \star z)$

52. $yx \le xy \Rightarrow (x + y) \star z = x \star (y \star z)$

Properties 41 and 42 state that $\star$ is $\le$-isotone. Property 46 shows that $\star$ and $\cdot$ associate if the continuation z is above 1. Properties 51 and 52 correspond to basic simulation and separation laws of omega algebra.

53. $yx \le x(y \star 1) \Rightarrow y \star (x \star z) \le x \star (y \star z) = (x + y) \star z$

54. $yx \le x(x \star (1 + y)) \Rightarrow y \star (x \star z) \le x \star (y \star z) = (x + y) \star z$

55. $y(x \star 1) \le x \star (y \star 1) \Leftrightarrow y \star (x \star 1) \le x \star (y \star 1)$

56. $p \le pp \wedge p \le 1 \wedge px \le xp \Rightarrow p(x \star y) = p((px) \star y) = p(x \star (py))$

Properties 53 and 54 sharpen the simulation and separation laws. Property 56 is useful to import and preserve tests in iterations (which can be introduced, for example, using the domain operation of Section 3.1).

It follows that $y \star z$ is a fixpoint of $\lambda x.yx + z$ and that $z(y \star 1)$ is a prefixpoint of $\lambda x.xy + z$. Moreover, if a binary itering has a greatest element $\top$, it satisfies $x \star \top = \top = \top(x \star 1)$.

Properties 10, 17 and 36 of the preceding theorem appear in [3], and properties 53 and 54 generalise theorems therein.

In extended binary iterings, and therefore in all of our computation models, we can add the following properties.

Theorem 11. *Let S be an extended binary itering and $w, x, y, z \in S$. Then the following properties 1–15 hold.*

1. $y((x + y) \star z) \leq (y(x \star 1)) \star z$
2. $w(x \star (yz)) \leq (w(x \star y)) \star z$
3. $w((x \star (yw)) \star z) = w(((x \star y)w) \star z)$
4. $(x \star w) \star (x \star (yz)) = (x \star w) \star ((x \star y)z)$
5. $(w(x \star y)) \star z = z + w((x + yw) \star (yz))$
6. $x \star ((y(x \star 1)) \star z) = y \star ((x(y \star 1)) \star z)$
7. $x \star 0 = 0 \Rightarrow (x \star y)z = x \star (yz)$
8. $(x + y) \star z = x \star ((y(x \star 1)) \star z)$
9. $(x + y) \star z = ((x \star 1)y) \star (x \star z)$
10. $(x + y) \star z = (x \star y) \star ((x \star 1)z)$
11. $(x(y \star 0)) \star 0 = x(y \star 0)$
12. $(x \star w) \star (x \star 0) = (x \star w) \star 0$

Property 7 gives another condition under which $\star$ and $\cdot$ associate. Property 8 is the slided version of the sumstar law of Kleene algebra.

13. $w((x \star (yw)) \star (x \star (yz))) = w(((x \star y)w) \star ((x \star y)z))$
14. $(y(x \star 1)) \star z = (y \star z) + (y \star (yx(x \star ((y(x \star 1)) \star z))))$
15. $x \star ((x \star w) \star ((x \star y)z)) = (x \star w) \star ((x \star y)z)$

It is unknown whether properties 6–9 of the preceding theorem hold in binary iterings. All the other properties do not follow in binary iterings as counterexamples generated by Nitpick or Mace4 witness.

On the other hand, there are properties which are characteristic for the strict or non-strict settings and therefore not suitable for a unifying theory.

Theorem 12. *The following properties 1–6 hold in the model of Theorem 9.1 – extended by $\top$ for the last two – but not in the models of Theorems 9.2 and 9.3.*

1. $(x \star y)z = x \star (yz)$
2. $(x \star 1)y = x \star y$
3. $(x \star 1)x = x \star x$
4. $(x + y) \star z = ((x \star 1)y) \star ((x \star 1)z)$
5. $(x\top) \star y = y + x\top y$
6. $\top \star y = \top y$

The following properties 7–12 hold in the models of Theorems 9.2 and 9.3, but not in the model of Theorem 9.1 extended by $\top$.

7. $1 \star x = \top$
8. $\top \star x = \top$
9. $x(1 \star y) \leq 1 \star x$
10. $x = yx \Rightarrow x \leq y \star 1$
11. $x = z + yx \Rightarrow x \leq y \star z$
12. $x \leq z + yx \Rightarrow x \leq y \star z$

The following properties 13–14 hold in the model of Theorem 9.3, but neither in the model of Theorem 9.2 nor in the model of Theorem 9.1 extended by $\top$.

13. $(x\top) \star z = z + x\top$
14. $x\top = x\top x\top$

We thus have the following four variants of the sumstar property, which coincide in our strict models:

- $(x + y) \star z = (x \star y) \star (x \star z)$ (binary itering axiom)
- $(x + y) \star z = ((x \star 1)y) \star (x \star z)$ (Theorem 11.9)
- $(x + y) \star z = (x \star y) \star ((x \star 1)z)$ (Theorem 11.10)
- $(x + y) \star z = ((x \star 1)y) \star ((x \star 1)z)$ (Theorem 12.4)

In contrast to the first three, however, Theorem 12.4 does not hold in the non-strict model. This exemplifies the difficulty in generalising from iterings to binary iterings.

Our final result applies Theorems 10 and 11 to derive Back's atomicity refinement theorem [2,28]. Because we generalise it to extended binary iterings, it is valid in our non-strict and in several strict computation models. Whether it holds in binary iterings is unknown.

Corollary 13. *Let S be an extended binary itering and $b, l, q, r, s, x, z \in S$ such that*

$$s = sq \qquad rb \le br \qquad rl \le lr \qquad bl \le lb \qquad r \star q \le q(r \star 1)$$
$$x = qx \qquad qb = 0 \qquad xl \le lx \qquad ql \le lq \qquad q \le 1$$

Then

$$s((x + b + r + l) \star (qz)) \le s((x(b \star q) + r + l) \star z) .$$

7 Conclusion

Strict and non-strict computation models can be unified algebraically. This includes a common approximation order, a common semantics of recursion and an operation describing iteration. Based on this unified treatment common refinement results can be derived.

An issue for future work is how to choose the set of computations that are represented by the algebraic description. For the strict models, only computations satisfying certain healthiness conditions are included. As a consequence, for example, the unit of sequential composition is not the identity relation but modified so as to satisfy the healthiness conditions. For the non-strict model, all computations are included in this paper. It remains to be investigated whether they can be restricted according to Theorem 1.

Acknowledgement. I thank the anonymous referees for valuable comments.

References

1. Aarts, C.J.: Galois connections presented calculationally. Master's thesis, Department of Mathematics and Computing Science, Eindhoven University of Technology (1992)
2. Back, R.J.R., von Wright, J.: Reasoning algebraically about loops. Acta Inf. 36(4), 295–334 (1999)
3. Cohen, E.: Separation and Reduction. In: Backhouse, R., Oliveira, J.N. (eds.) MPC 2000. LNCS, vol. 1837, pp. 45–59. Springer, Heidelberg (2000)

4. Conway, J.H.: Regular Algebra and Finite Machines. Chapman and Hall (1971)
5. Desharnais, J., Möller, B., Struth, G.: Kleene algebra with domain. ACM Transactions on Computational Logic 7(4), 798–833 (2006)
6. Desharnais, J., Struth, G.: Internal axioms for domain semirings. Sci. Comput. Program. 76(3), 181–203 (2011)
7. Gritzner, T.F., Berghammer, R.: A relation algebraic model of robust correctness. Theor. Comput. Sci. 159(2), 245–270 (1996)
8. Guttmann, W.: Imperative abstractions for functional actions. Journal of Logic and Algebraic Programming 79(8), 768–793 (2010)
9. Guttmann, W.: Partial, Total and General Correctness. In: Bolduc, C., Desharnais, J., Ktari, B. (eds.) MPC 2010. LNCS, vol. 6120, pp. 157–177. Springer, Heidelberg (2010)
10. Guttmann, W.: Unifying Recursion in Partial, Total and General Correctness. In: Qin, S. (ed.) UTP 2010. LNCS, vol. 6445, pp. 207–225. Springer, Heidelberg (2010)
11. Guttmann, W.: Fixpoints for general correctness. Journal of Logic and Algebraic Programming 80(6), 248–265 (2011)
12. Guttmann, W.: Unifying Correctness Statements. In: Gibbons, J., Nogueira, P. (eds.) MPC 2012. LNCS, vol. 7342, pp. 198–219. Springer, Heidelberg (2012)
13. Guttmann, W.: Algebras for iteration and infinite computations. Acta Inf. (to appear, 2012)
14. Guttmann, W.: Extended designs algebraically. Sci. Comput. Program. (to appear, 2012)
15. Hayes, I.J., Dunne, S.E., Meinicke, L.: Unifying Theories of Programming That Distinguish Nontermination and Abort. In: Bolduc, C., Desharnais, J., Ktari, B. (eds.) MPC 2010. LNCS, vol. 6120, pp. 178–194. Springer, Heidelberg (2010)
16. Hoare, C.A.R.: Theories of Programming: Top-Down and Bottom-Up and Meeting in the Middle. In: Wing, J.M., Woodcock, J., Davies, J. (eds.) FM 1999. LNCS, vol. 1708, pp. 1–27. Springer, Heidelberg (1999)
17. Hoare, C.A.R., He, J.: Unifying theories of programming. Prentice Hall Europe (1998)
18. Höfner, P., Möller, B.: An algebra of hybrid systems. Journal of Logic and Algebraic Programming 78(2), 74–97 (2009)
19. Kozen, D.: A completeness theorem for Kleene algebras and the algebra of regular events. Information and Computation 110(2), 366–390 (1994)
20. Kozen, D.: Kleene algebra with tests. ACM Trans. Progr. Lang. Syst. 19(3), 427–443 (1997)
21. Lai, A.Y.C.: Operational Semantics and Lazy Execution. Forthcoming PhD thesis, University of Toronto (expected 2012)
22. Launchbury, J., Peyton Jones, S.: State in Haskell. Lisp and Symbolic Computation 8(4), 293–341 (1995)
23. Maddux, R.D.: Relation-algebraic semantics. Theor. Comput. Sci. 160(1-2), 1–85 (1996)
24. Möller, B.: Kleene getting lazy. Sci. Comput. Program. 65(2), 195–214 (2007)
25. Nipkow, T., Paulson, L.C., Wenzel, M.: Isabelle/HOL. LNCS, vol. 2283. Springer, Heidelberg (2002)
26. Schmidt, G., Ströhlein, T.: Relationen und Graphen. Springer (1989)
27. Tarski, A.: On the calculus of relations. The Journal of Symbolic Logic 6(3), 73–89 (1941)
28. von Wright, J.: Towards a refinement algebra. Sci. Comput. Program. 51(1-2), 23–45 (2004)

Foundations of Coloring Algebra with Consequences for Feature-Oriented Programming

Peter Höfner[1,3], Bernhard Möller[2], and Andreas Zelend[2]

[1] NICTA, Australia
[2] Universität Augsburg, Germany
[3] University of New South Wales, Australia

Abstract. In 2011, simple and concise axioms for feature compositions, interactions and products have been proposed by Batory et al. They were mainly inspired by Kästner's Colored IDE (CIDE) as well as by experience in feature oriented programming over the last decades. However, so far only axioms were proposed; consequences of these axioms such as variability in models have not been studied. In this paper we discuss the proposed axioms from a theoretical point of view, which yields a much better understanding of the proposed algebra and therefore of feature oriented programming. For example, we show that the axioms characterising feature composition are isomorphic to set-theoretic models.

1 Introduction

Over the last years *Feature Oriented Programming* and *Feature Oriented Software Development* (e.g. [7,8]) have been established in computer science as a general programming paradigm that provides formalisms, methods, languages, and tools for building maintainable, customisable, and extensible software. In particular, *Feature Orientation* (FO) has widespread applications from network protocols [7] and data structures [9] to software product lines [21]. It arose from the idea of level-based designs, i.e., the idea that each program (design) can be successively built up by adding more and more levels (features). Later, this idea was generalised to the abstract concept of features. A *feature* reflects an increment in functionality or in the software development.

Over the years, FO was more and more supported by software tools. Examples are FeatureHouse [2], the AHEAD Tool Suite [5], GenVoca [11] and Colored IDE (CIDE) [20]. As shown in several case studies, these tools can be used for large-scale program synthesis (e.g. [2,21,19,20]).

Although the progress over the recent past in the area of FO was quite impressive, the mathematical structure and the mathematical foundations were studied less intensively. Steps towards a structural description and analysis are done with the help of feature models. A *feature model* is a (compact and) structural representation for use in FO. With respect to (software) product lines it describes all possible products in terms of features. Feature models were first introduced in

W. Kahl and T.G. Griffin (Eds.): RAMiCS 2012, LNCS 7560, pp. 33–49, 2012.
© Springer-Verlag Berlin Heidelberg 2012

the Feature-Oriented Domain Analysis method (FODA) [17]. Since then, feature modelling has been widely adopted by the software product line community and a number of extensions have been proposed. A further step towards an abstract description of FO was AHEAD [8]. It expresses hierarchical structures as nested sets of equations. Recently, several purely algebraic approaches were developed:

(a) *Feature algebra* [3] captures many of the common ideas of FO, such as introductions, refinements, or quantification, in an abstract way. It serves as a formal foundation of architectural metaprogramming [6] and automatic feature-based program synthesis [14]. The central notion is that of a *feature structure forest* that captures the hierarchical dependence in large products or product lines. Features may be added using the operation of forest superimposition, which allows a stepwise structured buildup.

(b) *Coloring algebra* [10] (CA) captures common ideas such as feature composition, interaction and products. It does not use an explicit tree or forest structure. Rather, the connection between product parts is made through *variation points* at which features or their parts may be inserted or deleted. Next to composition, the algebra also takes feature interaction into account by defining operators for determining conflicts and their repairs.

(c) *Delta modeling* [13] is not centred around a program and its structure. Rather it describes the building history of a product as a sequence of modifications, called *deltas*, that are incrementally applied to an initial product (e.g., the empty one). Conflict resolution is performed using special deltas.

The present paper builds on the second algebraic structure and combines CA with ring theory. Although most of the presented mathematics is well known, the relationship to FO is new and leads to new insights into the mathematical and structural understanding of FO. Starting with a brief recapitulation of the axioms of CA and their motivation in Section 2, we derive some basic properties for CA in Section 3, where we also discuss their relationship to FO. In Section 4, we present one of the main contributions of the paper, namely that in finite models feature composition as axiomatised in the algebra is always isomorphic to symmetric difference on sets of so called base colors. These base colors are studied more closely in Section 5; we show that many properties are already determined by them. In the following Section 6, we analyse small models of CAs, give a generic set-theoretic model and discuss a possible representation theorem for CA. Before summarising the paper in Section 8, we present another model of CA, which is useful for feature oriented software development (Section 7).

2 The Coloring Algebra

CA was introduced by Batory et al. [10], inspired by Kästner's CIDE [20]. In CIDE, a source document is painted in different colors, one color per feature. Insertion (Composition) of feature f into feature g yields a piece of code with fragments in various colors. Therefore the terms "feature" and " color" become synonymous and we will switch freely between the two in the sequel.

CA offers operations for feature composition $(+)$, feature interaction $(\cdot)$ as well as full interaction (cross-product) $(\times)$[1]. To illustrate the main ideas behind these operations we give an example named fire-and-flood control [10]. Assume a library building that is equipped with a fire control (*fire*). When a sensor of the control detects a fire, the sprinkling system gets activated. Later the library owner wants to retrofit a flood control system (*flood*) to protect the books and documents from water damage. When the system detects water on the floor, it shuts off the water main valve. If installed separately, both features operate as intended. However, when both are installed, they interact in a harmful way. *fire* activates the sprinklers, after a few moments the flood control shuts off the water and the building burns down. Algebraically the described system can be expressed by *flood* + *fire*. When using both systems, the interaction *flood · fire* of both features has to be considered: for example *flood · fire* could prioritise *fire* over *flood*. The entire system is then established by the cross-product

$$flood \times fire = flood \cdot fire + flood + fire \ .$$

Feature Composition. Every program offers a set of features, which (hopefully) satisfy the specified requirements. A feature can be nearly everything: a piece of code, part of some documentation, or even an entire program itself. Such "basic" features may be composed to form a program. In CA the order of composition does not matter[2]. Moreover, CA assumes an "empty feature" 0. Let F be an abstract set of features. Then feature composition $+ : F \times F \to F$ is a binary operator which is associative and commutative.

A crucial point for algebras covering FO is how multiple instances of features are handled. There are three possible solutions:

(a) multiple occurrences are allowed, but do not add more information;
(b) duplicates are removed, so that each feature occurs at most once; or
(c) a feature is removed if it is already present, i.e., composition is *involutory*:

$$\forall f : f + f = 0 \ . \tag{1}$$

The latter is the design decision taken in CA.[3] In a monoid satisfying Equation (1) every element is its own inverse; therefore such a monoid is also known as *Boolean group* (e.g., [12]). In particular we have $f = g \Leftrightarrow f + g = 0$ and $+$ is *cancellative*, i.e., $f + g = f + h \Leftrightarrow g = h$. Every Boolean group is commutative:

$$0 = f + f = f + 0 + f = f + g + g + f \ ,$$

which implies $f + g = g + f$. Hence the axiom of commutativity can be skipped.

---

[1] The original notation in [10] for $+$ and $\cdot$ and the element 0 below was $\cdot$, # and 1, respectively; we have changed that for a more direct connection with ring theory.
[2] This is a design decision and in contradiction to some other approaches such as [4], but follows approaches such as CIDE.
[3] A discussion on the usefulness of this axiom is given in [10]; the aim of the present paper is to discuss consequences of the axioms and not the axioms themselves.

Feature Interaction is a commutative and associative operator $\cdot : F \times F \to F$. On the one hand it might introduce additional features to yield a "consistent" and "executable" program. On the other hand, it might also list features of f and g that have to be removed. This is, for example, the case if f and g contradict each other. We follow the usual notational convention that $\cdot$ binds tighter than $+$.

CA assumes that, next to commutativity and associativity, feature interaction satisfies the following two axioms:

$$f \cdot 0 = 0 , \qquad (2) \qquad\qquad f \cdot f = 0 . \qquad (3)$$

Equation (2) expresses that no feature is in contradiction with the empty one; Equation (3) states that every feature is consistent with itself.

Moreover, CA assumes that feature interaction distributes over composition:

$$f \cdot (g + h) = f \cdot g + f \cdot h . \qquad (4)$$

Full Interaction is the "real" composition of two features, i.e., two features are composed under simultaneous repair of conflicts. In sum, full interaction is defined as $f \times g =_{df} f \cdot g + f + g$.

As mentioned, in CA features are abstractly viewed as colors. Therefore we now combine the above requirments into an abstract algebraic definition of CAs.

Coloring Algebra. A *CA* is a structure $(F, +, \cdot, 0)$ such that $(F, +, 0)$ is a (commutative) involutive group and $(F, \cdot)$ is a commutative semigroup satisfying Equations (3) and (4). Equation (2) follows from the other axioms, in particular distributivity. Elements of such an algebra are called *colors*. Following the above motivational discussion, an element h is called a *repair* iff $\exists f, g : h = f \cdot g$.

Mathematically, the definition means that a CA is an involutive and commutative ring without multiplicative unit. We list a few straightforward properties of full interaction.

Lemma 2.1. *Assume a CA* $(F, +, \cdot, 0)$ *and* $f, g, h \in F$, *then the following equations hold:* $(g + h) \times f = (g \times f) + (h \times f)$, $f \times 0 = f$, $f \times g = g \times f$, $(f \times g) \times h = f \times (g \times h)$ *and* $(g + h) \times f = (g \times f) + (h \times f)$.

3   First Consequences

We list a couple of interesting further properties and explain their interpretation in FO. All proofs can be found automatically by Prover9 [22]. Hence, we only present those that help understanding the structure of CA.

3.1   Basic Properties of Interaction

For the following lemmas, we assume a CA $(F, +, \cdot, 0)$ and $f, g, h \in F$.

Lemma 3.1. *A repair cannot introduce new conflicts, i.e.,* $f \cdot g = h \Rightarrow f \cdot h = 0$.

Lemma 3.2. *The repair of three elements* f, g, h *satisfies an exchange law:*

$$(f + g) \cdot (f + h) = (f + g) \cdot (g + h) = (f + h) \cdot (g + h) = f \cdot g + f \cdot h + g \cdot h .$$

It is easy to see that 0 is the unique fixpoint of $f \cdot x = x$. From this we get

Lemma 3.3. *A repair does not delete one of its components entirely, i.e., if* $f \neq 0$ *then* $f \cdot g \neq f$ *and if* $f + g \neq 0$ *then* $f \cdot g \neq f + g$.

Note that the precondition of the second statement is equivalent to $f \neq g$.

Lemma 3.4. *Colors cannot repair each other in "cycles", i.e.,*

(a) No non-trivial color is its own repair: $f \cdot g = f \Rightarrow f = 0$.
(b) Repairs are mutually exclusive: $f \cdot h_1 = g \wedge g \cdot h_2 = f \Rightarrow f = 0$.
(c) Part (b) can be extended to finite chains:

$$f \cdot h_1 = h_2 \wedge \left(\bigwedge_{i=1}^{n} h_{3i-1} \cdot h_{3i} = h_{3i+1} \right) \wedge h_{3n+1} \cdot h_{3n+2} = f \Rightarrow f = 0 .$$

Proof

(a) From $f \cdot g = f$ we infer $f \cdot g \cdot g = f \cdot g$. By absorption and strictness the left hand side reduces to 0, so that we have $0 = f \cdot g = f$. The claim also follows from Lemma 3.3.
(b) The assumptions yield $f = g \cdot h_2 = f \cdot h_1 \cdot h_2$ and Part (a) shows the claim.
(c) Straightforward induction on n.   □

Moreover, inserting the consequence $f = 0$ into the antecedents of Parts (b) and (c), strictness implies that all colors occurring at the right hand side of an equation of (b) and (c), namely f, g and h_{3i+1}, are equal to 0.

The absence of cycles makes the divisibility relation w.r.t. $\cdot$ into a strict partial order on non-empty colors: we define, with $F^+ =_{df} F - \{0\}$,

$$f < g \Leftrightarrow_{df} f, g \in F^+ \wedge \exists h \in F : f \cdot h = g .$$

Lemma 3.5. *Composition* $+$ *and interaction* $\cdot$ *are not isotone w.r.t.* $<$.

Proof. The smallest counterexample has 8 elements.

+	0	1	2	3	4	5	6	7
0	0	1	2	3	4	5	6	7
1	1	0	3	2	5	4	7	6
2	2	3	0	1	6	7	4	5
3	3	2	1	0	7	6	5	4
4	4	5	6	7	0	1	2	3
5	5	4	7	6	1	0	3	2
6	6	7	4	5	2	3	0	1
7	7	6	5	4	3	2	1	0

·	0	1	2	3	4	5	6	7
0	0	0	0	0	0	0	0	0
1	0	0	0	0	2	2	2	2
2	0	0	0	0	0	0	0	0
3	0	0	0	0	2	2	2	2
4	0	2	0	2	0	2	0	2
5	0	2	0	2	2	0	2	0
6	0	2	0	2	0	2	0	2
7	0	2	0	2	2	0	2	0

Assume a CA with $F = \{0, \ldots, 7\}$ and operations defined in the tables given on the left. Then, $1 < 2$, but $1 + 1 = 0 \not< 3 = 1 + 2$ and $1 \cdot 2 = 0 \not< 0 = 2 \cdot 2$.   □

3.2 Interaction Equivalence and Ideals

It is useful to group colors according to their behaviour under interaction. To achieve this we define an equivalence relation $\sim$ by

$$f \sim g \Leftrightarrow_{df} \forall h : f \cdot h = g \cdot h .$$

The equivalence class of f under $\sim$ is denoted by $[f] =_{df} \{g \mid f \sim g\}$. Elements $f \in [0]$ are (as usual) called *annihilators*, since $\forall g \in F : f \cdot g = 0$.

We will show that annihilators play a central rôle for the construction of models for the $\cdot$ operator. First, we set up a connection with the strict order $<$.

Lemma 3.6. *An element of F^+ is an annihilator iff it is maximal w.r.t. $<$.*

Proof. By contraposition of Equation (2), we get $f \cdot g \neq 0 \Rightarrow f \neq 0 \wedge g \neq 0$.
($\Rightarrow$) Assume $f \neq 0$ to be maximal but not an annihilator. Then there is an element g with $f \cdot g \neq 0$. The above remark yields $g \neq 0$ and therefore $f < f \cdot g$, which is a contradiction to the maximality of f.
($\Leftarrow$) Assume $f \neq 0$ to be non-maximal but an annihilator. Then there is an element $g \neq 0$ with $f < g$, and by definition $\exists h : f \cdot h = g$. Since f is an annihilator, we get $g = 0$, which yields a contradiction.   □

For finite $F \neq \{0\}$ there exists at least one maximal element in F^+ and hence a non-zero annihilator. It is well known that the set $[0]$ of annihilators is a ring ideal, i.e., is closed under $+$ and under $\cdot$ with arbitrary elements of F. It even forms a subtractive ideal (e.g. [1]), i.e., $f \in [0] \wedge f + g \in [0] \Rightarrow g \in [0]$.

Lemma 3.7. *If $f + g$ is an annihilator then $f \cdot g = 0$ and $f \cdot h = g \cdot h$ for $h \in F$.*

Proof. The first claim can be shown by $0 = (f + g) \cdot g = (f \cdot g) + (g \cdot g) = f \cdot g$. The second one by $0 = (f + g) \cdot h = (f \cdot h) + (g \cdot h) \Leftrightarrow (f \cdot h) = (g \cdot h)$.   □

Next we link annihilators with the equivalence relation $\sim$.

Lemma 3.8

(a) *Composition is cancellative w.r.t. $\sim$, i.e., $f + g \sim f + h \Leftrightarrow g \sim h$.*
   In particular, $f \sim f + g \Leftrightarrow g \sim 0$.
(b) *$f \sim g + h \Leftrightarrow f + g \sim h$. In particular, $f \sim g \Leftrightarrow f + g \sim 0$.*
(c) *$\sim$ is a congruence w.r.t. $+$ and $\cdot$.*
(d) *$[f] = \{f \cdot g \mid g \in [0]\}$.*

4   Models—Feature Composition

So far we have only looked at some basic foundations and properties for FO, most of them well known in mathematics, but unknown for FO. Let us now turn to some concrete models. Looking at the literature, we note that a concrete model has only been sketched [10].

Let us first look at feature composition $(F, +, 0)$. Involution (Equation (3)) expresses that every element has an inverse, namely itself. Therefore every element has order 2. By the classification of finitely generated Abelian groups, any finite 2-group is a power of $\mathbb{Z}_2$ (the two element group); hence there is exactly one finite model satisfying these axioms for each of the cardinalities 2, 4, 8, ... This immediately follows from the Kronecker Basis Theorem (e.g. [18]).

Theorem 4.1. *Every finite algebra satisfying the axioms for feature composition is isomorphic to a model that can be obtained by using symmetric difference on a power set of a finite set.*

Due to the nature of software engineering, the set F of colors, i.e., the set of all possible combinations of features is always finite, so that the assumption of Theorem 4.1 is satisfied in that context. Moreover, by this theorem there is no need to distinguish between the abstract CA and the set model any longer. Hence we can freely use more operators and relations, such as set union $\cup$, intersection $\cap$ or subset $\subseteq$ in every finite model of CA. Both, the neutral element of CA (0) and the empty set ($\emptyset$) will be denoted by 0 in the remainder. In particular, we have $0 \subseteq f$ for all $f \in F$. The theorem states that there are generic models: assume a set B of *base colors*. Then $(2^B, \triangle, 0)$ satisfies the axioms for feature composition, where $\triangle$ is the symmetric difference of sets, defined, for $M, N \in 2^B$, as

$$M \triangle N =_{df} (M \cup N) - (M \cap N) .$$

The greatest element B of 2^B is denoted by $\top$.

Lemma 4.2. *In general, neither $+$ nor $\cdot$ is isotone w.r.t. $\subseteq$. Therefore, none of these operations distributes over $\cup$ or $\cap$.*

Proof. To show the first claim we give a counterexample in $(2^B, \triangle, 0)$. Let $a = \{1\}$ and $b = \{1, 2\}$. Then $a \subseteq b$ holds, but $a \triangle b = \{2\}$ is not a subset of $b \triangle b = 0$. To prove that $\cdot$ is not isotone, we use contraposition and calculate again in the set model $(2^B, \triangle, 0)$. Consider a finite model of CA. Assume that the above implication holds and that $\cdot$ is not trivial, i.e., there are base colors a, b with $a \cdot b \neq 0$. Then, since $a, b \subseteq a + b$ (as a and b are base colors), by isotony we would have $a \cdot b \subseteq (a + b) \cdot (a + b) = 0$. Since 0 is the least element w.r.t. $\subseteq$, we get $a \cdot b = 0$ i.e., a contradiction. $\square$

We can enrich the set algebra to a first (albeit not very interesting) model for CA by assuming that there are no interactions at all between sets, i.e., for sets $M, N \in 2^B$, we define $M \cdot N =_{df} 0$. Then $(2^B, \triangle, \cdot, 0)$ forms a CA. In a later section we will discuss more sophisticated models and will show how these could be constructed systematically.

5 Base Colors

By Theorem 4.1, we can use set-theoretical knowledge for FO. In particular we can assume that every element of a finite CA is finitely generated, i.e., there is a set B of base colors (base features) from which all other colors are built. In general we call a color f of a CA F *base* iff it is isomorphic to a singleton set[4]; the set of all base colors is again denoted by B. If F is finitely generated, every element is a sum of base colors, i.e., for all $f \in F$

$$f = \sum_{i \in I} a_i$$

for an index set I and base colors $a_i \in B$.

---

[4] In set theory singleton sets of base colors are also called *atoms*.

In the remainder of the paper we use $a, b, c, \ldots$ to denote base colors and $f, g, h, \ldots$ for arbitrary colors. Moreover, we assume that sums (of base colors) are *reduced*, i.e., if $f = \sum_{i \in I} b_i$ then $b_i \neq b_j$ for all $i, j \in I$ with $i \neq j$.

Due to distributivity of $\cdot$ over $+$ it is possible to reduce general interaction to the one between base colors only. More precisely, assume two finitely-generated colors $f = \sum_{i \in I} a_i$ and $g = \sum_{j \in J} b_j$, then

$$f \cdot g = \left(\sum_{i \in I} a_i \right) \cdot \left(\sum_{j \in J} b_j \right) = \sum_{i \in I} \sum_{j \in J} (a_i \cdot b_j) . \tag{5}$$

Hence only the interaction (conflicts) of base colors has to be considered. This reduces the number of possible models.

6 Models for Coloring Algebra

6.1 Small Models

Let us now look at some possible models; we will construct them systematically. The most trivial one was already given at the end of Section 4; it had *no* interaction at all. The next example has *exactly one non-trivial* repair h. Formally, we calculate in a CA $(F, +, \cdot, 0)$ with distinguished base colors $a, b \in B$ and $h \in F$ satisfying $a \cdot b = h$ and $c \cdot d = 0$ for all other base colors $c, d \in B - \{a, b\}$.

By Lemma 3.4, a, b and h must be different. By Lemma 3.3, h differs from $a + b$, hence this example has at least three base colors. Vice versa this means that all models with at most two base colors can only have the trivial interaction.

By Equation (5), we can determine all interactions:

$$f \cdot g = \begin{cases} h & \text{if } C \\ 0 & \text{otherwise} , \end{cases} \quad \text{where} \quad C \Leftrightarrow_{df} \begin{cases} (a \subseteq f \wedge b \not\subseteq f \wedge b \subseteq g) \vee \\ (a \not\subseteq f \wedge b \subseteq f \wedge a \subseteq g) \vee \\ (a \subseteq g \wedge b \not\subseteq g \wedge b \subseteq f) \vee \\ (a \not\subseteq g \wedge b \subseteq g \wedge a \subseteq f) . \end{cases}$$

The first case in the definition of $f \cdot g$ describes all situations where either a occurs in f and b in g, or vice versa. Therefore the repair has to be introduced. However, it forbids the case, where a and b occur in both f and g: in this setting the repair is "introduced twice" and therefore does not show up.

Let us now assume that we have two repairs for base colors in our model, i.e., there are base colors $a, b, c, d \in B$ with $a \cdot b = h_1 + h_2$, $c \cdot d = h_1 + h_3$, and $a_1 \cdot a_2 = 0$ otherwise ($a_1 \in B - \{a, b\}$, $a_2 \in B - \{c, d\}$). Note that we do not require c, d to be different from a, b. Moreover, we assume that the repairs contain a common part; the case of disjoint repairs is just a special case ($h_1 = 0$). Using again Equation (5), we can determine all interactions:

$$f \cdot g = \begin{cases} h_1 + h_2 & \text{if } (c \not\subseteq f \wedge d \not\subseteq f) \vee (c \not\subseteq g \wedge d \not\subseteq g) \wedge C \\ h_1 + h_3 & \text{if } (a \not\subseteq f \wedge b \not\subseteq f) \vee (a \not\subseteq g \wedge b \not\subseteq g) \wedge C[a/c, b/d] \\ h_2 + h_3 & \text{if } C \wedge C[a/c, b/d] \\ 0 & \text{otherwise} . \end{cases}$$

Here the formula $C[a/c, b/d]$ is C with a replaced by c and b by d. These two examples show how the interaction operation can be derived from interaction on base colors. Of course one has to keep in mind when defining the interaction on base colors that some equations such as $f \cdot g = f$ are not possible (see Section 3).

Table 1. Number of Models for CA

#base colors/ #colors	#interact. (up to iso.)	# CA (up to iso.)
1/2	1	1
2/4	2	1
3/8	557	2
4/16		2

However, the examples give rise to the conjecture that there cannot be many different models for CA, since everything can be reduced to base colors and the variety there is limited. To underpin this conjecture, we generated all models of a particular size using Mace4, a counterexample generator [22]. The results are presented in Table 1. Of course generation of algebras with Mace4 requires isomorphism checking; although this is offered by the tool suite, it is resource intensive. Hence we could also determine numbers up to algebras of size 16. The table shows (a) the number of possible algebras (up to isomorphism) when only the axioms for interaction ($\cdot$) are used, and (b) the number of CAs.

6.2   A General Model for Coloring Algebra

In this chapter we show how to construct a general model for CA. The presented model is based on sets only and can already be applied straight away to FO when assuming that a color (feature) is a set of base colors. Later we will present a model that is even more practicable for FO, based on variation points.

As before we assume a set B of base colors and set F as 2^B. As we have seen, the only possibility for composition is $f + g = f \bigtriangleup g$. (cf. Theorem 4.1).

Furthermore, Equation 5 shows that interaction needs only be considered at the level of base colors. Next to that, we assume that the interaction of base colors is again a base color. In sum, we assume an associative interaction operator $\circ$ on B, i.e., $(B, \circ)$ is a semigroup, and a special element $e \in B$ that satisfies the annihilation properties $e \circ a = e = a \circ a$ for all $a \in B$. A structure $(B, \circ, e)$ with these properties is called a *base color semigroup*.

Based on that, feature interaction ($\cdot$) can be defined as

$$f \cdot g =_{df} \bigtriangleup_{a \in f} \bigtriangleup_{b \in g} \iota(a \circ b),$$

where the injection $\iota : B \to F$ is given by $\iota(e) = 0$ and $\iota(a) = \{a\}$ for $a \in B - \{e\}$. By associativity and commutativity of $\bigtriangleup$ this is well defined. Within $\cdot$, multiple occurrences of the same $\circ$-product cancel out so that at most one of them is left.

By this remark we immediately obtain $f \cdot f = 0$, since every product $a \circ a = e$ and hence $\iota(a \circ a) = 0$, while for any two different colors $a, b \in f$ we have both products $\iota(a \circ b)$ and $\iota(b \circ a)$ in the $\bigtriangleup$-aggregation, which by commutativity of $\circ$ are equal and hence cancel out. Another straightforward consequence of the definition is $f \cdot 0 = 0$.

Lengthy but straightforward calculations show that interaction, as defined earlier in this section, is associative commutative and distributes over $+$ (Δ) [15].

Thus we have defined a concrete model of a CA, solely based on a set of base colors and an abstract interaction operation $\circ$ on them. In the remainder of this section, we will give a possible concrete definition of this abstract operation.

For that, we assume a finite set P of *pigments*. The idea is to define base colors as certain sets of pigments. As before, sets of base colors will be used to define colors. Formally, a *set of base colors* is a non-empty subset $B \subseteq 2^P$ that is downward closed: $a \in B \wedge b \subseteq a \Rightarrow b \in B$. Hence every set of base colors contains $\emptyset$. The set B is called *full* iff $B = 2^P$.

For two base colors (sets of pigments) $a, b \in B$ we define a non-conflict predicate *noconf* by

$$noconf(a,b) \Leftrightarrow_{df} a \neq \emptyset \wedge b \neq \emptyset \wedge a \cap b = \emptyset .$$

Intuitively, two base colors do not show any conflict, if they do not share a common resource (pigment). The condition $a \neq \emptyset \wedge b \neq \emptyset$ is needed to exclude the empty base color, which by definition has no conflict with any other base color, but should also not interact with any. As we will see in the next definition, we have to distinguish these two behaviours.

The interaction $\circ : B \times B \rightarrow B$ of base colors is defined by

$$a \circ b =_{df} \begin{cases} a \cup b & \text{if } noconf(a,b) \wedge a \cup b \in B , \\ \emptyset & \text{otherwise.} \end{cases}$$

This definition entails $a \circ a = \emptyset = \emptyset \circ a$ for all $a \in B$. Hence $\emptyset$ (as element of B) plays the rôle of the annihilating element e above. Moreover, $\circ$ is commutative and associative. A proof can be found in [15]. Hence $(B, \circ, \emptyset)$ is a base color semigroup that can be used to create a CA, according to the definitions for composition and interaction given earlier in this section.

The definitions given above immediately entail the following property.

Lemma 6.1. *If $a \in B$ is $\subseteq$-maximal in B then a is an annihilator.*

Let us illustrate the construction with an example.

Example 6.2. Assume three pigments r, g, b and the full set of base colors, i.e., $B = 2^{\{r,g,b\}}$. Using the interaction operation $\circ$ on base colors, we get for example

$$\{r\} \circ \{g, b\} = \{r, g, b\} \qquad \text{and} \qquad \{r\} \circ \{r, g\} = \emptyset = \{r, g\} \circ \{g, b\} .$$

We also illustrate the CA over $(B, \circ, \emptyset)$. For convenience and readability we leave out set braces for colors (elements of $F = 2^B$), that consist of only one base color. For example rgb is used instead of $\{\{r, b, g\}\}$. We can now define a color that consists of all base colors containing the pigment r as

$$red =_{df} \{\{r\}, \{r, g\}, \{r, b\}, \{r, g, b\}\} = r + rg + rb + rgb .$$

As an example how interaction $\cdot$ on colors works, consider

$$(r + rg + rb + rgb) \cdot (b + rb + g)$$
$$= rb + \emptyset + rg + rgb + \emptyset + \emptyset + \emptyset + \emptyset + rbg + \emptyset + \emptyset + \emptyset$$
$$= rb + rg .$$

To conclude the example we briefly resume the discussion on the relationship to CIDE [20]. As mentioned, in that tool code can be colored. Singly colored code comprises the features that a customer can choose. But code can also be endowed with more than one color. If a code fragment is, for example, marked with the colors red and blue, that fragment is the repair of the two singly colored fragments with colors red and blue. Similarly, code fragments marked with three colors are repairs of three features, etc. By the full interaction operator, a repair is then composed with the two conflicting features into a new one, where by cancellativity of composition the conflicting parts are removed and supplemented by new ones if necessary.

Let us assume that a customer chooses the pigments r and g from the above set P. The one can automatically compute their repair: $r \cdot g = rg$. To create and deliver the final product full interaction ($\times$) can be used: $r \times g = rg + r + g$. It can be seen that the repair rg has indeed been added. If now a customer wants have the additional feature b, the model of CA can determine all the bits needed for the final product;

$$r \times g \times b = (rg + r + g) \times b$$
$$= (rg + r + g) \cdot b + rg + r + g + b$$
$$= rgb + rb + gb + rg + r + g + b .$$

We see that now not only the singly colored fragments are used, but also all repairs.   □

6.3   Towards a General Representation Theorem

We have presented a representation theorem for composition, and defined and discussed a number of possible interaction operations. So far, we have not found a complete representation theorem for CA; in this section we present a couple of useful properties that are hopefully steps towards one.

We investigate *generating systems* w.r.t. interaction $\cdot$, i.e., subsets $G \subseteq F$ such that every element in the image set of $\cdot$ equals a combination $g_1 \cdot \ldots \cdot g_n$ for some $n \in \mathbb{N}$ and $g_i \in G$. A generating system is *minimal* if no proper subset of it is a generating system. In a finite algebra such systems always exist.

Lemma 6.3. *Let G be a minimal generating system for $\cdot$. Then no two distinct elements of G can be related by $\sim$.*

Proof. A relation $g_1 \sim g_2$ would mean that in every $\cdot$-product of the above form g_1 could be replaced by g_2 without changing its value. Hence one of g_1, g_2 could be omitted from G while still yielding a generating system, in contradiction to the choice of G.   □

Theorem 6.4. *Let G be a minimal generating system for $\cdot$. Then the elements of G form a system of representatives for the equivalence classes of $\sim$. Since $\sim$ is a congruence, the set of these classes can be made into a quotient semiring by defining $[f] + [g] =_{df} [f + g]$ and $[f] \cdot [g] =_{df} [f \cdot g]$.*

Proof. Every element lies in its own equivalence class, while, by the previous lemma, different generators lie in different classes. This shows the first claim. The second one is standard semiring theory. □

When analysing the constructed and the generated models, we also looked more closely at the structure of the equivalence classes generated by the relation $\sim$. We made the following observations that are underlying the conjectures below on the representation theorem:

(a) The smallest element (w.r.t. $\subseteq$) of each class is a base color.
(b) All other elements are formed by composition with every possible combination of the annihilating base colors.

This and some observations presented earlier motivates us to state the following

Conjectures. The first conjecture is that the assumption we made at the beginning of Section 6.2 is always true, namely that the repair of two base colors is *always* a base color itself. If this conjecture is correct, the presented model might be *the* generic model, i.e., all interaction operations of all models of CA are always isomorphic to such a model. In particular, the only freedom to define interaction is given by the underlying semigroup of pigments. That would also imply that in FO only repairs of really simple fragments have to be considered and the number of possible CAs could be determined by the number of semigroups satisfying the additional annihilation requirements. This latter claim is underpinned by our third conjecture stating that in case of a non-trivial interaction operation, the elements $f \in 2^N$ form a system of representatives for the equivalence classes, where $[f] = f + [0] =_{df} \{f + g \mid g \in [0]\}$ and N is the set of all non-annihilating base colors.

The last conjecture is partially substantiated by the following property, which is immediate from Lemma 3.8(b):

Lemma 6.5. *The equivalence classes $[f]$ under $\sim$ are closed under composition with annihilators:*

$$[f] + [0] \subseteq [f].$$

7   A Model Based on Variation Points

Although the set-theoretic model given in Sections 4 and 6 is interesting and already covers a lot of the aspects of CIDE, it is, of course, not fully adequate for FO, since it does not take details of the program structure, such as classes and objects, into account. To get a handle on such aspects, in FO *variation points* are used. "A *variation point* identifies a location at which a variable part may occur.

It locates the insertion point for the variants and determines the characteristics (attributes) of the variability" [23]. Variation points are also called *extension points* (e.g., [20]) or *hot spots* (e.g., [16]).

The model we present here can be used directly for FO, as it is based on variation points and code fragments. We assume disjoint sets VP of variation points and C of code fragments. Variation points might, e.g., be given as line numbers before or after which further elements can be inserted.

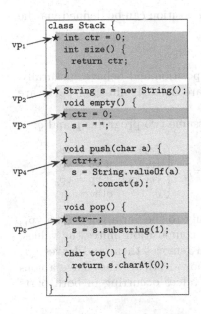

```
class Stack {
  ★ int ctr = 0;
    int size() {
      return ctr;
    }

  ★ String s = new String();
    void empty() {
  ★ ctr = 0;
      s = "";
    }
    void push(char a) {
  ★ ctr++;
      s = String.valueOf(a)
        .concat(s);
    }
    void pop() {
  ★ ctr--;
      s = s.substring(1);
    }
    char top() {
      return s.charAt(0);
    }
}
```

vp₁, vp₂, vp₃, vp₄, vp₅

Fig. 1. The Counted Stack With Variation Points [10]

An illustrative example with a "Counted Stack" is presented in Figure 1; it was taken from [10]. Due to lack of space we skip an explanation of the details; the figure should just give an impression how things look like.

Our model will be based on the set model presented before. Therefore we only consider code fragments that commute with each other. For example, we could look at entire methods, i.e., code fragments that start with something like "void empty(){" and end with "}" and may contain variation points from VP as well as code fragments from C. Note that a code fragment can contain variation points again.

A *program* is now a (total) function $p: VP \to 2^{VP \cup C}$.[5] Its semantics is as follows: if, for a variation point $\mathbf{vp}$ the value $p(\mathbf{vp})$ is not the empty set then $p(\mathbf{vp})$ is installed at $\mathbf{vp}$; otherwise $\mathbf{vp}$ remains empty. This construction is well defined, since by the standard isomorphism $(A^B)^C \simeq A^{C \times B}$ for function spaces we have $(2^{VP \cup C})^{VP} \simeq 2^{VP \times (VP \cup C)}$, where a program p is represented by the relation $\{(\mathbf{vp}, \mathbf{vqc}) \mid \mathbf{vqc} \in p(\mathbf{vp})\}$.

The empty program e is the empty function, i.e., $e(\mathbf{vp}) = \emptyset$ for all $\mathbf{vp} \in VP$. A program *white* with a method body at variation point $\mathtt{start}$ is given as

$$white(\mathbf{vp}) = \begin{cases} \{\,\mathtt{class\ Stack}\{\ \mathbf{vp}_1\ \mathbf{vp}_2\ \}\,\} & \text{if } \mathbf{vp} = \mathtt{start} \\ \emptyset & \text{otherwise}\,. \end{cases}$$

This coincides with the "white" part of Fig. 1, while the "darkgray" fragment is

$$darkgray(\mathbf{vp}) = \begin{cases} \{\,\mathtt{int\ ctr\ =\ 0;\ \ldots\ ctr;}\}\,\} & \text{if } \mathbf{vp} = \mathbf{vp}_1 \\ \{\,\mathtt{ctr\ =\ 0;}\,\} & \text{if } \mathbf{vp} = \mathbf{vp}_3 \\ \{\,\mathtt{ctr++;}\,\} & \text{if } \mathbf{vp} = \mathbf{vp}_4 \\ \{\,\mathtt{ctr--;}\,\} & \text{if } \mathbf{vp} = \mathbf{vp}_5 \\ \emptyset & \text{otherwise}\,. \end{cases}$$

---

[5] An isomorphic model uses partial functions which are undefined for empty variation points.

To build a program from a given function p, we just replace all occurrences of each vp by its value $p(\text{vp})$. This yields a function with no variation points in its values. Of course, this is only possible if the values of p do not depend on each other cyclically. If we now choose one variation point as the **start** point, a program (with filled variation points) has been derived.

By this simple algorithm for program derivation, the presented model is particularly interesting for FO. The remaining question is how to define feature composition and feature interaction.

Similar to the set-theoretic model, feature composition can be defined via the symmetric difference:

$$(p + q)(\text{vp}) =_{df} p(\text{vp}) \Delta q(\text{vp}) .$$

This definition satisfies the laws for feature composition and behaves naturally. Identical parts of the values $p(\text{vp})$ and $q(\text{vp})$ are deleted, differing parts are collected in the result set.

Let us explain this with a simple example. Assume two programs p and q:

$$p(\text{vp}) =_{df} \begin{cases} \{v_2\} & \text{if } \text{vp} = \text{vp}_1 \\ \{v_3\} & \text{if } \text{vp} = \text{vp}_2 \\ \{v_2, v_4\} & \text{if } \text{vp} = \text{vp}_3 \\ \emptyset & \text{otherwise ,} \end{cases} \quad q(\text{vp}) =_{df} \begin{cases} \{v_3\} & \text{if } \text{vp} = \text{vp}_1 \\ \{v_3\} & \text{if } \text{vp} = \text{vp}_2 \\ \{v_1, v_4\} & \text{if } \text{vp} = \text{vp}_3 \\ \emptyset & \text{otherwise .} \end{cases}$$

Both programs assign non-trivial information only to the variation points vp_1, vp_2 and vp_3. The values at vp_1 are disjoint; the composition unites the values. The values at vp_2 are identical; the composition removes them and leaves vp_2 "unset". The values at vp_3 are neither disjoint nor identical—they have a non-empty intersection; the composition deletes all values occurring in both parts and retains the rest. Formally this means

$$(p + q)(\text{vp}) =_{df} \begin{cases} \{v_2, v_3\} & \text{if } \text{vp} = \text{vp}_1 \\ \emptyset & \text{if } \text{vp} = \text{vp}_2 \\ \{v_1, v_2\} & \text{if } \text{vp} = \text{vp}_3 \\ \emptyset & \text{otherwise .} \end{cases}$$

The same construction can be applied to implement feature interaction:

$$(p \cdot q)(\text{vp}) =_{df} p(\text{vp}) \cdot q(\text{vp}) .$$

In the concrete model this turns into $(p \cdot q)(\text{vp}) =_{df} \Delta_{a \in p(\text{vp})} \Delta_{b \in q(\text{vp})} \iota(a \circ b)$. To complete this definition we need to specify the underlying base color semigroup. We choose the pigment set $P =_{df} VP \cup C$ and the full base color set $B =_{df} 2^P$.

In this paper we have shown that, for a given size, each definition of $+$ is isomorphic to that of symmetric difference in the set-model; by this we do not have much freedom to define composition and the presented definition seems to be canonical. In contrast to that, the interaction operation $\cdot$ offers much more flexibility. Of course, we cannot give a compact definition, since interaction really depends on implementation details and therefore on the source code. However, as we have shown in Section 5, only the interaction between base colors has to be defined, interaction for arbitrary elements then lifts by Equation (5). We can even do better and give the programmer some guidelines on how repairs should be defined by the lemmas given in Section 3. The most important of these is Lemma 3.4 which can easily be implemented and offers a quick consistency check on interactions.

8 Conclusion and Outlook

We have carried out a careful analysis of coloring algebra (CA) [10]. The study has yielded several interesting and sometimes surprising results.

First, we have presented a series of properties for FO. Most of them could be proven fully automatically using an automated theorem prover such as Prover9.

Second, we have used Mace4 not only to falsify conjectures (as we do regularly), but also for the generation of finite models. Doing this, and by creating and analysing models by hand, it turned out that there exist only very few models of CA, up to isomorphism. This was a surprise: when the algebra was designed, it was believed that the operation for feature interaction (·) offers a lot of freedom, and probably the composition operation does so as well. However, we have shown that composition is always isomorphic to symmetric difference in a set model. By this representation theorem more operations, such as set union and intersection, could be introduced in CA for free. This has allowed us the definition of base colors, which come into play naturally as the "smallest" features available. We also gave a derivation towards a representation theorem for CA in general. So far we could not entirely prove our conjecture. This is not surprising, since representation theorems are generally hard to prove. However, the lemmas presented indicate that our conjecture holds.

As the last contribution of the paper, we have given a concrete model for FO. It is based on functions over sets and is more useful for FO than the generic set-theoretical one. To show this, we have given a simple algorithm to describe how elements of this model can be transformed into executable programs.

In sum, the analysis, even without the missing representation theorem, has yielded deeper insights for CA and will hopefully lead to a much better understanding of the basics underlying feature oriented programming and feature oriented software development.

Acknowledgement. We are grateful to Peter Jipsen for pointing out Kronecker's base theorem. Part of the work was carried out during a sponsored visit of the second author at NICTA. The work of the third author was funded by the German Research Foundation (DFG), project number MO 690/7-2 FeatureFoundation.

References

1. Allen, P.J.: A fundamental theorem of homomorphisms for semirings. Proceedings of the American Mathematical Society 21(2), 412–416 (1969)
2. Apel, S., Kästner, C., Lengauer, C.: FeatureHouse: Language-independent, automated software composition. In: 31st International Conerence on Software Engineering (ICSE), pp. 221–231. IEEE Press (2009)
3. Apel, S., Lengauer, C., Möller, B., Kästner, C.: An Algebra for Features and Feature Composition. In: Meseguer, J., Roşu, G. (eds.) AMAST 2008. LNCS, vol. 5140, pp. 36–50. Springer, Heidelberg (2008)
4. Apel, S., Lengauer, C., Möller, B., Kästner, C.: An algebraic foundation for automatic feature-based program synthesis. Sc. Comp. Prog. 75(11), 1022–1047 (2010)
5. Batory, D.: Feature-oriented programming and the AHEAD tool suite. In: ICSE 2004: 26th International Conference on Software Engineering, pp. 702–703. IEEE Press (2004)
6. Batory, D.: From implementation to theory in product synthesis. ACM SIGPLAN Notices 42(1), 135–136 (2007)
7. Batory, D., O'Malley, S.: The design and implementation of hierarchical software systems with reusable components. ACM Transactions Software Engineering and Methodology 1(4), 355–398 (1992)
8. Batory, D., Sarvela, J.N., Rauschmayer, A.: Scaling step-wise refinement. In: ICSE 2003: 25th International Conference on Software Engineering, pp. 187–197. Proceedings of the IEEE (2003)
9. Batory, D., Singhal, V., Sirkin, M., Thomas, J.: Scalable software libraries. ACM SIGSOFT Software Engineering Notes 18(5), 191–199 (1993)
10. Batory, D., Höfner, P., Kim, J.: Feature interactions, products, and composition. In: 10th ACM International Conference on Generative Programming and Component Engineering (GPCE 2011), pp. 13–22. ACM Press (2011)
11. Batory, D., Singhal, V., Thomas, J., Dasari, S., Geraci, B., Sirkin, M.: The GenVoca model of software-system generators. IEEE Software 11(5), 89–94 (1994)
12. Bernstein, B.: Sets of postulates for boolean groups. Annals of Mathematics 40(2), 420–422 (1939)
13. Clarke, D., Helvensteijn, M., Schaefer, I.: Abstract delta modeling. In: Visser, E., Järvi, J. (eds.) GPCE, pp. 13–22. ACM (2010)
14. Czarnecki, K., Eisenecker, U.: Generative Programming: Methods, Tools, and Applications. Addison-Wesley (2000)
15. Höfner, P., Möller, B., Zelend, A.: Foundations of coloring algebra with consequences for feature-oriented programming. Tech. Rep. 2012-06, Institut für Informatik der Universität Augsburg (2012)
16. Johnson, R., Foote, B.: Designing reusable classes. Journal of Object Oriented Programming 1(2), 22–35 (1988)
17. Kang, K.C., Cohen, S.G., Hess, J.A., Novak, W.E., Peterson, A.S.: Feature-oriented domain analysis (FODA) feasibility study. Technical Report CMU/SEI-90-TR-21, Carnegie-Mellon University Software Engineering Institute (1990)
18. Kargapolov, M., Merzliakov, I.: Fundamentals of the Theory of Groups. Graduate texts in mathematics. Springer (1979)
19. Kästner, C., Apel, S., Batory, D.: A case study implementing features using AspectJ. In: Software Product Lines, 11th International Conference (SPLC), pp. 223–232. IEEE Computer Society (2007)

20. Kästner, C.: Virtual Separation of Concerns: Toward Preprocessors 2.0. Ph.D. thesis, University of Magdeburg (2010)
21. Lopez-Herrejon, R.E., Batory, D.: A Standard Problem for Evaluating Product-Line Methodologies. In: Bosch, J. (ed.) GCSE 2001. LNCS, vol. 2186, pp. 10–24. Springer, Heidelberg (2001)
22. McCune, W.W.: Prover9 and Mace4, http://www.cs.unm.edu/~mccune/prover9 (accessed July 12, 2012)
23. Reinhartz-Berger, I., Tsoury, A.: Experimenting with the Comprehension of Feature-Oriented and UML-Based Core Assets. In: Halpin, T., Nurcan, S., Krogstie, J., Soffer, P., Proper, E., Schmidt, R., Bider, I. (eds.) BPMDS 2011 and EMMSAD 2011. LNBIP, vol. 81, pp. 468–482. Springer, Heidelberg (2011)

Towards an Algebra for Real-Time Programs

Brijesh Dongol[1], Ian J. Hayes[2], Larissa Meinicke[2], and Kim Solin[2]

[1] Department of Computer Science, The University of Sheffield, UK
[2] School of Information Technology and Electrical Engineering,
The University of Queensland, Brisbane, Australia
b.dongol@sheffield.ac.uk, ian.hayes@itee.uq.edu.au,
{l.meinicke, k.solin}@uq.edu.au

Abstract. We develop an algebra for an interval-based model that has been shown to be useful for reasoning about real-time programs. In that model, a system's behaviour over all time is given by a stream (mapping each time to a state) and the behaviour over an interval is determined using an interval predicate, which maps an interval and a stream to a Boolean. Intervals are allowed to be open/closed at either end and adjoining (i.e., immediately adjacent) intervals do not share any common points but are contiguous over their boundary. Values of variables at the ends of open intervals are determined using limits, which allows the possible piecewise continuity of a variable at the boundaries of an interval to be handled in a natural manner. What sort of an algebra does this model give rise to? In this paper, we take a step towards answering that question by investigating an algebra of interval predicates.

1 Introduction

Over the years, algebraic abstractions of models in several different problem domains have been developed. This paper is focused on an interval-based model that has been shown to be useful for reasoning about real-time systems [4–8]. The model is inspired by the duration calculus [19], however, unlike the duration calculus, two adjoining intervals (i.e., intervals that immediately follow each other) do not share a point of overlap. Hence, intervals may be open/closed/infinite at either end and may also be empty [17]. Using our more general notion of an interval, we develop *interval predicates*, which may be used to describe the real-time behaviour of a system over time. Interval predicates map an interval and a stream (trace over all times) to a Boolean [10, 15].

The algebra for interval predicates that we obtain has many similarities to the algebraisation of the duration calculus presented by Höfner and Möller [14]; in particular, the basic structure is a weak quantale. However, because streams encode the complete (real-time) behaviour, interval predicates allow one to express properties such as sampling [2, 6], non-deterministic expression evaluation [11], and neighbourhoods [19] in a straightforward manner. Unlike Höfner and Möller [13], because the intervals we consider may be open or closed at either end, the domain operator in our model is trivial.

This paper is structured as follows. First, Section 2 presents interval predicates and clarifies some of the differences from the duration calculus. Section 3 shows that interval predicates form a weak quantale and iteration is discussed in Section 4. Section 5

W. Kahl and T.G. Griffin (Eds.): RAMiCS 2012, LNCS 7560, pp. 50–65, 2012.

presents compositional methods for reasoning about interval predicates and Section 6 presents temporal rules and predicates outside an interval directly in the model.

2 Interval Predicates

We let *Intv* denote the set of all continuous intervals, i.e., contiguous subsets of $\mathbb{R}$. Formally, we have

$$Intv \;\widehat{=}\; \{\Delta \subseteq \mathbb{R} \mid \forall t, t' : \Delta, t'' : \mathbb{R} \bullet t \le t'' \le t' \Rightarrow t'' \in \Delta\}$$

where $a : A$ means that a is a member of A (sometimes denoted $a \in A$). Using '.' for function application, we let $\mathsf{lub}.U$ and $\mathsf{glb}.U$ denote the *least upper* and *greatest lower* bounds of a set U, respectively. We use ∞ to denote the least upper bound of $\mathbb{R}$, $-\infty$ to denote the greatest lower bound of $\mathbb{R}$, and note that $\infty, -\infty \notin \mathbb{R}$. Thus, for any interval Δ, $\mathsf{lub}.\Delta, \mathsf{glb}.\Delta \in \mathbb{R} \cup \{-\infty, \infty\}$. The *length* of a non-empty interval Δ is given by $\ell.\Delta \;\widehat{=}\; \mathsf{lub}.\Delta - \mathsf{glb}.\Delta$. For the empty interval $\varnothing$, $\ell.\varnothing = 0$.

We must often reason about two *adjoining* intervals, i.e., intervals that immediately precede or follow another. Thus, for $\Delta_1, \Delta_2 \in Intv$, we define

$$\Delta_1 \propto \Delta_2 \;\widehat{=}\; \begin{aligned}(\Delta_1 \ne \varnothing \wedge \Delta_2 \ne \varnothing) \Rightarrow \\ ((\mathsf{lub}.\Delta_1 = \mathsf{glb}.\Delta_2) \wedge (\Delta_1 \cup \Delta_2 \in Intv) \wedge (\Delta_1 \cap \Delta_2 = \varnothing))\end{aligned}$$

Note that if $\Delta_1 \propto \Delta_2$, the intervals Δ_1 and Δ_2 must be disjoint. Intervals may be open, closed or half-open, however, if $\Delta_1 \propto \Delta_2$ and both Δ_1 and Δ_2 are non-empty, then either Δ_1 is right-closed and Δ_2 is left-open, or Δ_1 is right-open and Δ_2 is left-closed [5–7, 17]. This is unlike the duration calculus [19] (and consequently Höfner and Möller's hybrid algebra [14]) where adjoining intervals share a single point of overlap. From the definition it is clear that, for any Δ, both $\varnothing \propto \Delta$ and $\Delta \propto \varnothing$ hold. Given that seq.X defines the set of possibly infinite sequences of type X, we let part.Δ denote the set of all (possibly infinite) partitions of interval Δ, which is defined as follows:

$$\mathsf{part}.\Delta \;\widehat{=}\; \{\delta : \mathsf{seq}.Intv \mid \Delta = \bigcup \mathsf{ran}.\delta \wedge \forall i : \mathsf{dom}.\delta \setminus \{0\} \bullet \delta.(i-1) \propto \delta.i\}$$

We assume variable names are taken from the set *Var*, a *state space* over a set of variables $V \subseteq Var$ is given by $\Sigma_V \;\widehat{=}\; V \to Val$ and a *state* is a member of Σ_V, i.e., a total function mapping variables in V to values in *Val*. A *stream* of behaviours over V is given by $Stream_V \;\widehat{=}\; \mathbb{R} \to \Sigma_V$. To facilitate reasoning about specific parts of a stream, we use *interval predicates*, which have type $IntvPred_V \;\widehat{=}\; Intv \to (Stream_V \to \mathbb{B})$.

We assume pointwise lifting of operators on stream and interval predicates in the normal manner, e.g., if p_1 and p_2 are interval predicates, Δ is an interval and s is a stream, we have $(p_1 \wedge p_2).\Delta.s = (p_1.\Delta.s \wedge p_2.\Delta.s)$. When reasoning about properties of programs, we would like to state that whenever a property p_1 holds over any interval Δ and stream s, a property p_2 also holds over Δ and s. Hence, we define universal implication for $p_1, p_2 \in IntvPred_V$ and $\Delta \in Intv$ as

$$p_1.\Delta \Rrightarrow p_2.\Delta \;\widehat{=}\; \forall s : Stream_V \bullet p_1.\Delta.s \Rightarrow p_2.\Delta.s$$
$$p_1 \Rrightarrow p_2 \;\widehat{=}\; \forall \Delta : Intv \bullet p_1.\Delta \Rrightarrow p_2.\Delta$$

We say $p_1 \equiv p_2$ holds iff both $p_1 \Rightarrow p_2$ and $p_2 \Rightarrow p_1$ hold.

The interval predicate below holds iff the given interval is empty, where the underlying stream is implicit in both sides of the definition.

$$\text{Empty}.\Delta \mathrel{\widehat{=}} \Delta = \varnothing$$

We note that unlike the duration calculus, the least interval is the empty interval; the least allowable interval in the duration calculus is a point interval.

We define some additional interval predicates as follows, where the stream is once again implicit.

$$\text{Inf}.\Delta \mathrel{\widehat{=}} \text{lub}.\Delta = \infty \qquad\qquad \text{True}.\Delta \mathrel{\widehat{=}} true$$
$$\text{Fin}.\Delta \mathrel{\widehat{=}} \text{lub}.\Delta \in \mathbb{R} \cup \{-\infty\} \qquad \text{False}.\Delta \mathrel{\widehat{=}} false$$

Note that an interval Δ may have infinite length even if $\text{Fin}.\Delta$ holds if $\text{glb}.\Delta = -\infty$. However, we rarely consider such intervals because the programs we model are assumed to take place after some initialisation. We are however interested in reasoning about programs that execute forever and those that terminate (see Section 5.2).

The *chop* operator is a basic operator on interval predicates that allows a given interval to be split into two parts (cf. [6, 16, 19]). For $p_1, p_2 \in IntvPred$ and $\Delta \in Intv$, we define:

$$(p_1 \mathrel{;} p_2).\Delta \mathrel{\widehat{=}} (\exists \Delta_1, \Delta_2 \bullet \langle \Delta_1, \Delta_2 \rangle \in \text{part}.\Delta \wedge p_1.\Delta_1 \wedge p_2.\Delta_2) \vee (\text{Inf} \wedge p_1).\Delta$$

Thus, $(p_1 \mathrel{;} p_2).\Delta$ holds iff Δ may be split into adjoining intervals Δ_1 and Δ_2 such that both $p_1.\Delta_1$ and $p_2.\Delta_2$ hold, or the least upper bound of Δ is ∞ and p_1 holds for Δ. We allow the second disjunct $(\text{Inf} \wedge p_1).\Delta$ because, for example, p_1 may model a non-terminating command, in which case p_2 will never be able to execute.

When applying the chop operator, the subintervals into which the given interval is split are disjoint. Thus, unlike the duration calculus [19] we are able to state that there are two adjoining intervals, say Δ_1 and Δ_2, such that a state predicate c holds for all times in Δ_1 and $\neg c$ holds for all times in Δ_2. In the duration calculus, everywhere properties must be approximated using an "almost everywhere" operator, which allows a state predicate to be false in a set of measure zero.

This results in differences in the model that we obtain against that of Höfner and Möller [14]. In particular, assuming that '$\mathrel{\overset{\circ}{,}}$' denotes the duration calculus chop, Höfner and Möller have $p \mathrel{\overset{\circ}{,}} q \equiv p \wedge q$ for tests p and q (a test is an interval predicate that only holds in point intervals). This equivalence holds in the duration calculus because $(p \mathrel{\overset{\circ}{,}} q).\{t\}$ for $t \in \mathbb{R}$ holds iff $p.\{t\}$ and $q.\{t\}$ holds. Such an equivalence cannot hold in our model because if $p \mathrel{;} q$ holds, the intervals in which p and q hold must be disjoint. Hence, $(p \mathrel{;} q).\{t\}$ implies that either $p \Rightarrow \text{Empty}$ or $q \Rightarrow \text{Empty}$ holds. The intended use of tests in [14] is to enable modelling of if-then-else programming constructs, where a test models a guard evaluation (that takes place instantanously). Modelling guard evaluation using tests may lead to unimplementable specifications; to accurately model real-time programs, one must assume that guard evaluation takes time [4–7]. Moreover, as will be elaborated below, the only set of elements in our model that would satisfy the axioms usually used for tests in the abstract setting, is the set $\{\text{False}, \text{Empty}\}$.

3   Algebra

3.1   Basic Definitions

A *weak dioid* (or *weak idempotent semiring*) is a structure $(A, +, ; , 0, 1)$ where $(A, +, 0)$ is an idempotent commutative monoid, $(A, ; , 1)$ is a monoid, ';' is left strict and ';' distributes over '+'. To simplify the notation, we omit ';' when appropriate. Thus, a weak dioid satisfies the following axioms.

$$a + (b + c) = (a + b) + c \tag{1}$$
$$a + b = b + a \tag{2}$$
$$a + 0 = a \tag{3}$$
$$a + a = a \tag{4}$$
$$a(bc) = (ab)c \tag{5}$$
$$1a = a = a1 \tag{6}$$
$$0a = 0 \tag{7}$$
$$a(b + c) = ab + ac \tag{8}$$
$$(a + b)c = ac + bc \tag{9}$$

A *natural order* '$\leq$' is defined on A where $a \leq b$ iff $a + b = b$. For the order '$\leq$', 0 is the least element and $a + b$ is the *join* of a and b.

Definition 1 ([14]). *A weak dioid S is a* weak quantale *iff S is a complete lattice under the natural order '$\leq$' and ';' is universally disjunctive in its left argument.*

Note that a weak quantale is not required to be universally disjunctive in its right argument, although it must be (finitely) disjunctive in its right argument because every weak quantale is also a weak semiring.

Definition 2 ([14]). *A weak dioid $(A, +, ; , 0, 1)$ is* Boolean *iff $(A, +, \curlywedge, ^-, 0)$ is a Boolean algebra with* meet *operator '$\curlywedge$' and every element $a \in A$ has a complement $\bar{a}$.*

For any Boolean algebra, we have the *shunting* rule $a \curlywedge b \leq c \Leftrightarrow a \leq \bar{b} + c$. Furthermore, by monotonicity of ';', for any Boolean weak dioid, both of the following hold.

$$a(b \curlywedge c) \leq ab \curlywedge ac \tag{10}$$
$$(a \curlywedge b)c \leq ac \curlywedge bc \tag{11}$$

A Boolean weak dioid forms the basis of a *Boolean weak quantale* in the natural way.

3.2   The Algebra of Interval Predicates

We are now ready to prove the following proposition.

Proposition 3. *(IntvPred, $\vee$, ; , False, Empty) is a Boolean weak quantale.*

Proof. We begin by showing that $(IntvPred, \Rightarrow)$ forms a complete Boolean lattice. Recall that $IntvPred$ is the set $Intv \rightarrow (Stream \rightarrow \mathbb{B})$ and that for $p, q \in IntvPred, p \Rightarrow q$ if and only if for all intervals Δ and for all streams s, one has that $p.\Delta.s \Rightarrow q.\Delta.s$. Now, since $(\mathbb{B}, \Rightarrow)$ is a complete lattice and the ordering $\Rightarrow$ is a pointwise extension of $\Rightarrow$, the pointwise-extension lemma applied twice gives that $(IntvPred, \Rightarrow)$ is also a complete Boolean lattice.[1]

Next, we show that ';' is universally disjunctive in its left argument. This is done by the following derivation.

$$((\bigvee_{i \in I} q_i) \,;\, p).\Delta$$
$\equiv$ definitions
$$(\exists \Delta_1, \Delta_2 \bullet \langle \Delta_1, \Delta_2 \rangle \in \mathsf{part}.\Delta \wedge (\bigvee_{i \in I} q_i).\Delta_1 \wedge p.\Delta_2) \vee (\mathsf{Inf} \wedge (\bigvee_{i \in I} q_i)).\Delta$$
$\equiv$ lifting
$$(\exists \Delta_1, \Delta_2 \bullet \langle \Delta_1, \Delta_2 \rangle \in \mathsf{part}.\Delta \wedge (\bigvee_{i \in I} q_i.\Delta_1) \wedge p.\Delta_2) \vee (\mathsf{Inf}.\Delta \wedge (\bigvee_{i \in I} q_i.\Delta))$$
$\equiv$ pull out $\bigvee$ from both disjuncts, assume i not free in $p.\Delta_2$ (otherwise rename)
$$\bigvee_{i \in I}(\exists \Delta_1, \Delta_2 \bullet \langle \Delta_1, \Delta_2 \rangle \in \mathsf{part}.\Delta \wedge q_i.\Delta_1 \wedge p.\Delta_2) \vee \bigvee_{i \in I}(\mathsf{Inf}.\Delta \wedge q_i.\Delta)$$
$\equiv$ combine the disjuncts
$$\bigvee_{i \in I}((\exists \Delta_1, \Delta_2 \bullet \langle \Delta_1, \Delta_2 \rangle \in \mathsf{part}.\Delta \wedge q_i.\Delta_1 \wedge p.\Delta_2) \vee (\mathsf{Inf} \wedge q_i).\Delta)$$
$\equiv$ definitions
$$(\bigvee_{i \in I}(q_i \,;\, p)).\Delta$$

The proof that $(IntvPred, \vee, ;\, , \mathsf{False}, \mathsf{Empty})$ satisfies weak dioid axioms is given in Appendix A.

Remarks. Note that we get

$$(p \,;\, \mathsf{False}).\Delta$$
$\equiv$ definitions
$$(\exists \Delta_1, \Delta_2 \bullet \langle \Delta_1, \Delta_2 \rangle \in \mathsf{part}.\Delta \wedge p.\Delta_1 \wedge \mathsf{False}.\Delta_2) \vee (\mathsf{Inf}.\Delta \wedge p.\Delta)$$
$\equiv$ logic and definitions
$$\mathsf{Inf}.\Delta \wedge p.\Delta$$
$\equiv$ definitions
$$(\mathsf{Inf} \wedge p).\Delta$$

so we have the property

$$(p \,;\, \mathsf{False}) \equiv \mathsf{Inf} \wedge p \tag{12}$$

By (12), we cannot have universal disjunctivity in the right argument. In particular, for $I = \varnothing$, assuming disjunctivity, we get

$$(p \,;\, \mathsf{False}) \equiv (p \,;\, \bigvee_{i \in \varnothing} q_i) \equiv \bigvee_{i \in \varnothing}(p \,;\, q_i) \equiv \mathsf{False}$$

which we *know* does not hold. Similar reasoning shows that neither $(\mathsf{True} \,;\, p) \equiv \mathsf{True}$ nor $(p \,;\, \mathsf{True}) \equiv \mathsf{True}$ holds. Nevertheless, the following property does hold:

$$p \vee \mathsf{Inf} \Rightarrow (\mathsf{True} \,;\, p) \tag{13}$$

---

[1] The pointwise-extension lemma says the following [1, p. 42]: "Let B be a poset and A any set. Then the pointwise extension of B to $A \rightarrow B$ is also a poset. Furthermore, this poset is a lattice (complete, bounded, distributive, boolean lattice) if B is a lattice (complete, bounded, distributive, boolean lattice)."

4 Iteration

4.1 Basic Properties

By Knaster-Tarski, any monotone function from a complete lattice to itself has a complete lattice of fixpoints. This means that we can define

$$a^* \mathrel{\hat{=}} (\mu z \cdot az + 1) \qquad a^\omega \mathrel{\hat{=}} (\nu z \cdot az + 1) \qquad a^\infty \mathrel{\hat{=}} (\nu z \cdot az)$$

with the intuition that $*$ is a finite iteration, ∞ is an infinite iteration, and ω is an iteration that is either finite or infinite. From the definitions, one immediately has the following unfolding rules,

$$a^* = aa^* + 1 \qquad a^\omega = aa^\omega + 1 \qquad a^\infty = aa^\infty$$

as well as the following induction rules,

$$az + 1 \leq z \ \Rightarrow \ a^* \leq z \tag{14}$$
$$z \leq az + 1 \ \Rightarrow \ z \leq a^\omega \tag{15}$$
$$z \leq az \ \Rightarrow \ z \leq a^\infty \tag{16}$$

Consider the functions $(\lambda z \cdot pz \vee \mathsf{Empty})$ and $(\lambda z \cdot pz)$. It is clear that these are monotone with respect to $\Rightarrow$. Since *IntvPred* is a complete lattice, each of the induction rules above above hold for interval predicates. For any interval predicate p, both $p \Rightarrow \mathsf{Empty}^\omega$ and $p \Rightarrow \mathsf{Empty}^\infty$ hold (by the above induction rules). In particular, this means that $\mathsf{True} \Rightarrow \mathsf{Empty}^\infty$ and $\mathsf{True} \Rightarrow \mathsf{Empty}^\omega$. Since True is the greatest element with respect to $\Rightarrow$, one then gets that $\mathsf{Empty}^\infty \equiv \mathsf{Empty}^\omega \equiv \mathsf{True}$.

4.2 Derived Properties

Since our structure is a (Boolean) weak quantale, we can immediately apply the following slightly less general version of Höfner and Möller's Lemma 3.5(1) [14].

Proposition 4. *Let* $(A, +, ; , 0, 1)$ *be a weak quantale and let* $a, b, c \in A$. *Then,*
$$b + ac \leq c \ \Rightarrow \ a^*b \leq c$$

The fact that our structure is a completely distributive lattice follows from the pointwise-extension lemma, so we can also immediately apply the following proposition of Höfner and Möller [14] (Lemma 3.5(2)).

Proposition 5. *Let* $(A, +, ; , 0, 1)$ *be a weak quantale and let* $a, b, c \subset A$ *and suppose that the underlying lattice is completely distributive. Then,*
$$c \leq ac + b \ \Rightarrow \ c \leq a^\infty + a^*b$$

However, the strong iteration (ω) induction rule cannot be generalised in the usual way. Indeed, for interval predicates p, q and r, consider $r \Rightarrow pr \vee q \ \Rightarrow \ r \Rightarrow p^\omega q$ with $p \equiv \mathsf{Empty}$ and $q \equiv \mathsf{False}$. Then one gets $r \Rightarrow r \ \Rightarrow \ r \Rightarrow \mathsf{True}; \mathsf{False}$, that is,

$$r \Rightarrow \mathsf{True}; \mathsf{False},$$

which is not true in general (for example, take r to be the predicate denoting the set of finite intervals). This can also be used to show why $(\nu z \cdot pz \vee q) \equiv p^\omega q$ does not hold. Nevertheless, it can be proved that $p^\omega \equiv p^* \vee p^\infty$ holds. That is, p^ω denotes a finite or infinite iteration of p.

We use $a^+ \mathrel{\widehat=} aa^*$ to denote the *positive iteration* of interval predicate a, i.e., a^+ holds iff a is iterated at least once a finite number of times. We have the following $^+$-induction law and unfolding laws:

$$az + a \leq z \Rightarrow a^+ \leq z \tag{17}$$
$$a^\infty = a^+ a^\infty \tag{18}$$

5   Compositional Reasoning

5.1   Splits/Joins

Proofs of interval predicates may be composed/decomposed if the interval predicate under consideration joins/splits, respectively [4, 5, 7, 8, 10]. Informally speaking, an interval predicate p splits if given that it holds over an interval Δ, p holds over all subintervals of Δ. An interval predicate p joins if p holds in an interval Δ whenever p^ω holds in Δ. For example, recalling that $\ell.\Delta$ denotes the length of interval Δ, interval predicate $\ell \leq 42$ splits but does not join and $\ell > 42$ joins but does not split. We present an algebraic characterisation of splits and joins. Unlike Höfner and Möller who only distinguish between joining (which they call submodular) and modular elements (which both split and join) [14], we distinguish between splitting, joining and modular elements.

Definition 6. *Suppose* $(A, +, ; , 0, 1)$ *is a Boolean weak quantale. We say* $a \in A$ *splits iff* $\forall b, c \colon A \cdot a \curlywedge bc \leq (a \curlywedge b)(a \curlywedge c)$*, say* a *joins iff* $\forall b, c \colon A \cdot (a \curlywedge b)(a \curlywedge c) \leq a \curlywedge bc$ *and that* a *is* modular *iff* a *both splits and joins.*

Elements that split and join may be distributed in and out of finite, possibly infinite and positive iterations as described by the following lemma. The distribution result for positive iteration for modular elements given in [14] may be derived from our lemma below.

Lemma 7. *Suppose* $a \in A$ *where* $(A, +, ; , 0, 1)$ *is a Boolean weak quantale.*

(1) If a splits, then for any $b \in A$, $a \curlywedge b^* \leq (a \curlywedge b)^*$ holds.
(2) If a splits, then for any $b \in A$, $a \curlywedge b^\omega \leq (a \curlywedge b)^\omega$ holds.
(3) If a joins, then for any $b \in A$, $(a \curlywedge b)^+ \leq a \curlywedge b^+$ holds.

Proof (1)

$$\begin{aligned}
&a \curlywedge b^* \leq (a \curlywedge b)^* \\
= \quad & \text{shunting} \\
&b^* \leq \overline{a} + (a \curlywedge b)^* \\
\Leftarrow \quad & \text{(14) $*$-induction} \\
&b(\overline{a} + (a \curlywedge b)^*) + 1 \leq \overline{a} + (a \curlywedge b)^* \\
= \quad & \text{shunting} \\
&a \curlywedge (b(\overline{a} + (a \curlywedge b)^*) + 1) \leq (a \curlywedge b)^*
\end{aligned}$$

The left hand side of $\leq$ simplifies as follows:

$$
\begin{aligned}
&a \curlywedge (b(\bar{a} + (a \curlywedge b)^*) + 1) \\
=\quad &\text{distribute } \curlywedge \text{ over } + \\
&(a \curlywedge b(\bar{a} + (a \curlywedge b)^*) + (a \curlywedge 1)) \\
\leq\quad &a \text{ splits, } a \curlywedge 1 \leq 1 \\
&(a \curlywedge b)(a \curlywedge (\bar{a} + (a \curlywedge b)^*) + 1) \\
\leq\quad &\text{distribute } \curlywedge \text{ over } +, z \curlywedge \bar{z} = 0 \text{ and } 0 + z = z \\
&(a \curlywedge b)(a \curlywedge (a \curlywedge b)^*) + 1 \\
\leq\quad &z \leq y \curlywedge z \\
&(a \curlywedge b)(a \curlywedge b)^* + 1 \\
=\quad &* \text{ unfolding} \\
&(a \curlywedge b)^*
\end{aligned}
$$

Proof (2)

$$
\begin{aligned}
&a \curlywedge b^\omega \leq (a \curlywedge b)^\omega \\
=\quad &(15) \ \omega\text{-induction} \\
&a \curlywedge b^\omega \leq (a \curlywedge b)(a \curlywedge b^\omega) + 1
\end{aligned}
$$

The left hand side simplifies as follows.

$$
\begin{aligned}
&a \curlywedge b^\omega \\
=\quad &\omega \text{ unfolding} \\
&a \curlywedge (bb^\omega + 1) \\
\leq\quad &\text{distribute } \curlywedge \text{ over } + \\
&(a \curlywedge bb^\omega) + (a \curlywedge 1) \\
\leq\quad &a \text{ splits, } a \curlywedge 1 \leq 1 \\
&(a \curlywedge b)(a \curlywedge b^\omega) + 1
\end{aligned}
$$

Proof (3)

$$
\begin{aligned}
&(a \curlywedge b)^+ \leq a \curlywedge b^+ \\
\Leftarrow\quad &(17) + \text{induction} \\
&(a \curlywedge b)(a \curlywedge b)^+ + \\
&\qquad (a \curlywedge b) \leq a \curlywedge b^+ \\
\Leftarrow\quad &a \text{ joins} \\
&(a \curlywedge bb^+) + (a \curlywedge b) \leq a \curlywedge b^+ \\
=\quad &\text{distribute } \curlywedge \text{ over } + \\
&a \curlywedge (bb^+ + b) \leq a \curlywedge b^+ \\
=\quad &+ \text{ unfolding} \\
&true
\end{aligned}
$$

□

Note that the property $(a \curlywedge b)^* \leq a \curlywedge b^*$ does not hold in general even if a joins because the $*$ on the left hand side of $\leq$ may reduce to 1, whereas a must hold on the right hand side of $\leq$.

Definition 8. *Suppose $(A, +, ; , 0, 1)$ is a Boolean weak quantale. We say $a \in A$ ω-joins iff $\forall b, c : A \bullet (a \curlywedge b)^\omega \leq a \curlywedge b^\omega$ and $*$-joins iff $\forall b, c : A \bullet (a \curlywedge b)^* \leq a \curlywedge b^*$.*

Lemma 9. *Suppose $a \in A$ where $(A, +, ; , 0, 1)$ is a Boolean weak quantale.*

(1) If a $*$-joins then $(a \curlywedge b)^* \leq a \curlywedge b^*$.
(2) If a ω-joins then $(a \curlywedge b)^\omega \leq a \curlywedge b^\omega$.

Proof (1). By monotonicity of $*$, $(a \curlywedge b)^* \leq a^* \curlywedge b^*$ and by assumption $a^* \leq a$, $a^* \curlywedge b^* \leq a \curlywedge b^*$.

Proof (2). This proof has exactly the same structure as the proof of (a), but uses monotonicity of ω in place of $*$.

□

5.2   Distinguishing between Different Forms of Iteration

Abstractions of interval predicates Inf and Fin are defined like Höfner and Möller [14]. For a Boolean weak quantale $(A, +, ; , 0, 1)$, and $a \in A$, we say a is *purely infinite* iff $a0 \leq a$ and *purely finite* iff $a0 = 0$. One can define the largest purely infinite and purely finite elements INF and FIN, respectively as follows [14], where $a \in A$

$$a \leq \text{INF} \Leftrightarrow a0 \leq a \qquad a \leq \text{FIN} \Leftrightarrow a0 \leq 0$$

Thus, elements INF and FIN correspond to interval predicates Inf and Fin, respectively. Element FIN is modular [14]. We prove that FIN *-joins as follows.

$$
\begin{aligned}
& (\text{FIN} \curlywedge a)^* \\
= \quad & \text{unfolding} \\
& (\text{FIN} \curlywedge a)^+ + 1 \\
\leq \quad & \text{Lemma 7 (3) and FIN joins} \\
& (\text{FIN} \curlywedge a^+) + 1 \\
= \quad & 1 \leq \text{FIN} \\
& (\text{FIN} \curlywedge a^+) + (\text{FIN} \curlywedge 1) \\
= \quad & \text{distributivity and folding} \\
& \text{FIN} \curlywedge a^*
\end{aligned}
$$

Further properties of INF and FIN are given by the lemma below.

Lemma 10

$$a = (\text{INF} \curlywedge a) + (\text{FIN} \curlywedge a) \tag{19}$$
$$\text{INF} ; a = \text{INF} \tag{20}$$
$$\text{INF} \curlywedge a = (\text{INF} \curlywedge a)^\infty + (\text{INF} \curlywedge a)^+ \tag{21}$$

Unlike [18], which only distinguishes between finite and infinite iteration, we are able to distinguish between termination and Zeno iteration for finite intervals, and between divergence and infinite iteration for infinite intervals. Höfner and Möller present several other relationships between FIN and INF in [14, Lemma 3.6].

$$
\begin{aligned}
\text{Term}\, a &\cong \text{FIN} \curlywedge a^* & \text{Zeno}\, a &\cong \text{FIN} \curlywedge a^\infty \\
\text{Diverge}\, a &\cong \text{INF} \curlywedge a^+ & \text{NonTerm}\, a &\cong \text{INF} \curlywedge a^\infty
\end{aligned}
$$

The following lemma relates the properties above. An alternative proof of (22) below is given in [14].

Lemma 11. *Suppose $(A, +, ; , 0, 1)$ is a Boolean weak quantale and $a \in A$. Then each of the following holds.*

$$\text{Term}\, a = (\text{FIN} \curlywedge a)^* \tag{22}$$
$$\text{Zeno}\, a = \text{FIN} \curlywedge (\text{FIN} \curlywedge a)^\infty \tag{23}$$
$$\text{Diverge}\, a \leq \text{NonTerm}\, a \tag{24}$$

Proof (22). The proof of $\text{Term}\, a \leq (\text{FIN} \curlywedge a)^*$ follows by part 1 of Lemma 7 because FIN splits and the proof of $(\text{FIN} \curlywedge a)^* \leq \text{Term}\, a$ holds by part 1 of Lemma 9 because FIN *-joins.

Proof (23)

$$\text{Zeno}\, a = \text{FIN} \curlywedge (\text{FIN} \curlywedge a)^\infty$$
$\Leftrightarrow$ expanding definition, logic
$$\text{FIN} \leq \frac{(\overline{a^\infty} + (\text{FIN} \curlywedge a)^\infty) \curlywedge}{((\text{FIN} \curlywedge a)^\infty + a^\infty)}$$
$\Leftarrow$ monotonicity of $\curlywedge$
$$\text{FIN} \leq \frac{(\overline{a^\infty} + (\text{FIN} \curlywedge a)^\infty) \curlywedge}{(\overline{a^\infty} + a^\infty)}$$
$\Leftrightarrow$ boolean algebra rules
$$\text{FIN} \leq (\overline{a^\infty} + (\text{FIN} \curlywedge a)^\infty)$$
$\Leftrightarrow$ shunting
$$\text{FIN} \curlywedge a^\infty \leq (\text{FIN} \curlywedge a)^\infty$$
$\Leftarrow$ (16) infinite induction
$$\text{FIN} \curlywedge a^\infty \leq (\text{FIN} \curlywedge a)(\text{FIN} \curlywedge a^\infty)$$
$\Leftarrow$ ∞-unfolding
$$\text{FIN} \curlywedge aa^\infty \leq (\text{FIN} \curlywedge a)(\text{FIN} \curlywedge a^\infty)$$
$\Leftarrow$ FIN splits
true

Proof (24)

$$\text{NonTerm}\, a$$
$=$ definition and (18)
$$\text{INF} \curlywedge a^+ a^\infty$$
$=$ (19)
$$\text{INF} \curlywedge$$
$$((\text{INF} \curlywedge a^+) + (\text{FIN} \curlywedge a^+))a^\infty$$
$=$ (9) distribution
$$\text{INF} \curlywedge$$
$$((\text{INF} \curlywedge a^+)a^\infty + (\text{FIN} \curlywedge a^+)a^\infty)$$
$\geq$ monotonicity
$$\text{INF} \curlywedge (\text{INF} \curlywedge a^+)a^\infty$$
$=$ (20)
$$\text{INF} \curlywedge \text{INF} \curlywedge a^+$$
$\geq$ logic, definition
$$\text{Diverge}\, a$$

$\square$

6   Properties in the Model

6.1   Temporal Properties

Safety and progress properties are often expressed using operators $\square$ (always) and $\diamond$ (sometime), respectively. These properties are often defined in interval logics using the chop operator [14, 16, 19]. We define $\square$ and $\diamond$ directly as functions of type $IntvPred \rightarrow IntvPred$, then provide a link to an algebraic definition that uses chop via Lemma 13 below. In particular, for an interval predicate p, we define:

$$(\square p).\Delta \;\widehat{=}\; \forall \Delta': Intv \bullet \Delta' \subseteq \Delta \Rightarrow p.\Delta' \tag{25}$$

$$(\diamond p).\Delta \;\widehat{=}\; \exists \Delta': Intv \bullet \Delta' \subseteq \Delta \land p.\Delta' \tag{26}$$

Thus, $(\square p).\Delta$ and $(\diamond p).\Delta$ hold iff p holds in all and some subinterval of Δ, respectively. The lemma below allows $\diamond$ (and consequently $\sqcap$) to be treated algebraically.

Lemma 12. *Interval predicate p splits if $p \Rightarrow \square p$.*

Lemma 13. *For any interval predicate p, $\diamond p \equiv \text{Fin}\,;\, p\,;\, \text{True}$.*

Proof. The '$\Leftarrow$' direction is trivial. For the '$\Rightarrow$' direction, we perform case analysis. If $p = \text{Empty}$, then we have $\diamond\text{Empty} \equiv \text{True} \equiv (\text{Fin}\,;\, \text{True}) \equiv (\text{Fin}\,;\, \text{Empty}\,;\, \text{True})$. If $p \neq \text{Empty}$, then we have the following derivation for some arbitrary interval Δ.

$(\Diamond p).\Delta$
$\equiv \exists \Delta': Intv \bullet \Delta' \subseteq \Delta \wedge p.\Delta'$
$\equiv \exists \Delta', \Delta_1, \Delta_2: Intv \bullet \langle \Delta_1, \Delta', \Delta_2 \rangle \in \mathsf{part}.\Delta \wedge p.\Delta'$
$\Rightarrow \quad p \neq \mathsf{Empty}$
$\quad \exists \Delta', \Delta_1, \Delta_2: Intv \bullet \langle \Delta_1, \Delta', \Delta_2 \rangle \in \mathsf{part}.\Delta \wedge (\mathsf{glb} \neq \infty \wedge p).\Delta'$
$\Rightarrow \quad \Delta_1 \propto \Delta'$ by definition of part and $\mathsf{True}.\Delta_2$ trivially holds
$\quad \exists \Delta', \Delta_1, \Delta_2: Intv \bullet \langle \Delta_1, \Delta', \Delta_2 \rangle \in \mathsf{part}.\Delta \wedge$
$$\qquad\qquad\qquad (\mathsf{lub} \neq \infty).\Delta_1 \wedge (\mathsf{glb} \neq \infty \wedge p).\Delta' \wedge \mathsf{True}.\Delta_2$$
$\equiv \quad$ definitions
$\quad (\mathsf{Fin}\ ;\ p\ ;\ \mathsf{True}).\Delta$ □

We can show that for any interval predicate p, $\Box p$ splits but does not necessarily join, and $\Diamond p$ joins, but does not necessarily split using the following example.

Example 14. It is trivial to show that $\Box p \equiv \Box\Box p$ holds, which by Lemma 12 implies that $\Box p$ splits. It is also trivial to show that $(\Diamond p)^\omega \Rightarrow \Diamond p$ holds and hence $\Diamond p$ joins. To see that $\Box p$ does not necessarily join, consider $p \equiv (\ell \leq 42)$. In particular, it is possible for both $\Box(\ell \leq 42).\Delta_1$ and $\Box(\ell \leq 42).\Delta_2$ to hold, where $\Delta_1 \propto \Delta_2$, but $\Box(\ell \leq 42).(\Delta_1 \cup \Delta_2)$ may not hold. Similarly, to see that $\Diamond p$ does not necessarily split, consider $p \equiv (\ell \geq 42)$.   □

6.2 Previous and Next Properties

Höfner and Möller [13] follow Henzinger's model and formalise behaviours of hybrid systems using *trajectories* [12, 14] mapping the times within an interval to values. To state properties outside the interval algebraically, they use a *domain* operator [3] from the carrier set to the Boolean subalgebra of tests, consisting of elements which are always below 1. The domain operator can be defined in any Boolean weak quantale [3], so this operator is also present in our framework. However, in our model, the constant 1 corresponds to Empty and the the only element below Empty is False, which means that in our model, the set of tests is the rather uninteresting set $\{\mathsf{False}, \mathsf{Empty}\}$. This also means that the domain operator cannot be used like in [14]. Instead, in our model, we have access to the stream of all behaviours (even outside) the given interval and hence, properties outside the given interval may be stated directly using interval predicates [4, 6, 8].

It is particularly useful to reason about properties immediately before and after an interval. In this paper, we present results for *next* properties; properties that hold in some interval that immediately follows the given interval. Similar results hold for interval predicates on the immediately preceding intervals, but these are omitted due to space considerations.

Definition 15. *For an interval predicate p and interval Δ, we define:*

$$(\mathsf{next}.p).\Delta \hat{=} \exists \Delta': Intv \bullet \Delta \propto \Delta' \wedge p.\Delta'$$

Lemma 16. *Suppose p and q are interval predicates.*

$$\mathsf{next}.p \wedge (p \Rightarrow q) \Rightarrow \mathsf{next}.q \qquad\qquad\qquad (27)$$

$$\neg\text{next}.p \Rightarrow \text{next}.\neg p \tag{28}$$

$$\text{next}.(p \lor q) \equiv \text{next}.p \lor \text{next}.q \tag{29}$$

Proof. Properties (27) and (29) are trivial to prove by expanding the definitions. We prove (28) as follows.

$$(\neg\text{next}.p).\Delta$$
$$\equiv \neg(\exists\Delta': Intv \bullet \Delta \propto \Delta' \land p.\Delta')$$
$$\equiv (\forall\Delta': Intv \bullet \Delta \propto \Delta' \Rightarrow (\neg p).\Delta')$$
$$\Rightarrow \quad \Delta \propto \varnothing \text{ for any interval } \Delta$$
$$\quad (\exists\Delta': Intv \bullet \Delta \propto \Delta' \land (\neg p).\Delta')$$
$$\equiv (\text{next}.\neg p).\Delta$$

Note that by monotonicity, $\text{next}.(p \land q) \Rightarrow \text{next}.p \land \text{next}.q$ holds, however, the implication does not necessarily hold in the reverse direction. To see this, consider $p \equiv (\ell = 42)$ and $q \equiv (\ell = 24)$.

Lemma 17. *For any interval predicate p, $\text{Fin} \Rightarrow (\text{next}.p)^+ = \text{next}.p$*

Proof. The proof of $\text{next}.p \Rightarrow (\text{next}.p)^+$ holds by definition of $^+$. For the other direction, we prove $\text{Fin} \land (\text{next}.p)^+ \equiv \text{next}.p$ as follows:

$$\text{Fin} \land (\text{next}.p)^+$$
$$\equiv \quad \text{definition}$$
$$\text{Fin} \land (\text{next}.p)^*(\text{next}.p)$$
$$\Rightarrow \quad \text{Fin splits}$$
$$(\text{Fin} \land (\text{next}.p)^*)(\text{Fin} \land (\text{next}.p))$$
$$\Rightarrow \quad \text{monotonicity of ';'}$$
$$\text{Fin} ; \text{next}.p$$
$$\Rightarrow \quad \text{expanding definitions}$$
$$\text{next}.p$$

The next operator allows one to state invariant-style properties, where p is guaranteed to hold in the next interval given that p holds in the current interval. Such lemmas allows properties over a larger interval to be decomposed over smaller intervals.

Lemma 18. *For any interval predicate p, both of the following hold.*

(1) If p splits then $(p \Rightarrow \text{next}.p)^+ \land \text{Fin} \Rightarrow (p \Rightarrow \text{next}.p)$.
(2) If p joins then $(p \land \text{next}.p)^+ \land \text{Fin} \Rightarrow (p \land \text{next}.p)$.

Proof (1)

$$(p \Rightarrow \text{next}.p)^+ \land \text{Fin} \Rightarrow (p \Rightarrow \text{next}.p)$$
$$= \quad \text{modus ponens}$$
$$(p \Rightarrow \text{next}.p)^+ \land p \land \text{Fin} \Rightarrow \text{next}.p$$
$$= \quad p \text{ splits, Lemma 7}$$
$$(\text{next}.p)^+ \land \text{Fin} \Rightarrow \text{next}.p$$
$$= \quad \text{Lemma 17}$$
$$\textit{true}$$

Proof (2)

$$(p \land \text{next}.p)^+ \land \text{Fin}$$
$$\Rightarrow \quad p \text{ joins, Lemma 7}$$
$$p \land (\text{next}.p)^+ \land \text{Fin}$$
$$\Rightarrow \quad \text{Lemma 17}$$
$$p \land \text{next}.p \land \text{Fin}$$

□

7   Conclusions and Future Work

This paper presents an algebra for interval predicates over the continuous domain of real numbers, which have been shown to be useful for reasoning about real-time programs [4–8]. As discussed throughout this paper, there are several important differences between this model and the duration calculus [19] used for interval-based reasoning of real-time programs. Our model shares the same algebra as the duration calculus as given by Höfner and Möller, and hence the algebraic results from [14] may be reapplied to interval predicates. However, unlike Höfner and Möller, who are able to use a domain operator to define properties outside the interval, we reason within the model directly using the algebraic properties that we have developed in the earlier sections of the paper. In particular, we exploit the splits and joins properties developed in Section 5.1.

Interval predicates may be used to reason about sampling errors [5–7] and about properties over multiple time-scales [4, 5, 7]. We aim to explore the algebraic properties of such interval predicates via mechanised proofs [9] as part of future work.

Acknowledgements. This research is supported by ARC Discovery Grant DP0987452 and EPSRC Grant EP/J003727/1. We thank our anonymous reviewers for their detailed comments and suggestions, which have helped improve this paper significantly.

References

1. Back, R.J., Von Wright, J.: Refinement Calculus: A Systematic Introduction. Springer-Verlag New York, Inc., Secaucus (1998)
2. Burns, A., Hayes, I.J.: A timeband framework for modelling real-time systems. Real-Time Systems 45(1), 106–142 (2010)
3. Desharnais, J., Möller, B., Struth, G.: Kleene algebra with domain. ACM Trans. Comput. Log. 7(4), 798–833 (2006)
4. Dongol, B., Hayes, I.J.: Approximating idealised real-time specifications using time bands. In: AVoCS 2011. ECEASST, vol. 46, pp. 1–16. EASST (2012)
5. Dongol, B., Hayes, I.J.: Deriving Real-Time Action Systems Controllers from Multiscale System Specifications. In: Gibbons, J., Nogueira, P. (eds.) MPC 2012. LNCS, vol. 7342, pp. 102–131. Springer, Heidelberg (2012)
6. Dongol, B., Hayes, I.J.: Deriving real-time action systems in a sampling logic. Science of Computer Programming (2012) (accepted October 17, 2011)
7. Dongol, B., Hayes, I.J.: Rely/Guarantee Reasoning for Teleo-reactive Programs over Multiple Time Bands. In: Derrick, J., Gnesi, S., Latella, D., Treharne, H. (eds.) IFM 2012. LNCS, vol. 7321, pp. 39–53. Springer, Heidelberg (2012)
8. Dongol, B., Hayes, I.J., Robinson, P.J.: Reasoning about real-time teleo-reactive programs. Technical Report SSE-2010-01, The University of Queensland (2010)
9. Guttmann, W., Struth, G., Weber, T.: Automating Algebraic Methods in Isabelle. In: Qin, S., Qiu, Z. (eds.) ICFEM 2011. LNCS, vol. 6991, pp. 617–632. Springer, Heidelberg (2011)
10. Hayes, I.J.: Towards reasoning about teleo-reactive programs for robust real-time systems. In: Proceedings of the 2008 RISE/EFTS Joint International Workshop on Software Engineering for Resilient Systems, pp. 87–94. ACM, New York (2008)
11. Hayes, I.J., Burns, A., Dongol, B., Jones, C.B.: Comparing models of nondeterministic expression evaluation. Technical Report CS-TR-1273, Newcastle University (2011)

12. Henzinger, T.A.: The theory of hybrid automata. In: LICS 1996, pp. 278–292. IEEE Computer Society, Washington, DC (1996)
13. Höfner, P., Möller, B.: Algebraic neighbourhood logic. J. Log. Algebr. Program. 76(1), 35–59 (2008)
14. Höfner, P., Möller, B.: An algebra of hybrid systems. J. Log. Algebr. Program. 78(2), 74–97 (2009)
15. Jones, C.B., Hayes, I.J., Jackson, M.A.: Deriving Specifications for Systems That Are Connected to the Physical World. In: Jones, C.B., Liu, Z., Woodcock, J. (eds.) Bjørner/Zhou Festschrift. LNCS, vol. 4700, pp. 364–390. Springer, Heidelberg (2007)
16. Moszkowski, B.C.: A complete axiomatization of interval temporal logic with infinite time. In: LICS, pp. 241–252 (2000)
17. Schobbens, P.-Y., Raskin, J.-F., Henzinger, T.A.: Axioms for real-time logics. Theor. Comput. Sci. 274(1-2), 151–182 (2002)
18. von Wright, J.: From Kleene Algebra to Refinement Algebra. In: Boiten, E.A., Möller, B. (eds.) MPC 2002. LNCS, vol. 2386, pp. 233–262. Springer, Heidelberg (2002)
19. Zhou, C., Hansen, M.R.: Duration Calculus: A Formal Approach to Real-Time Systems. EATCS: Monographs in Theoretical Computer Science. Springer (2004)

A   Interval Predicates Form a Weak Dioid

Axioms (1)–(4) are trivial to verify. To prove Axiom (5), we expand the definitions and simplify. The left hand side gives us the following.

$p(qr).\Delta$

$\equiv$ definitions

$(\exists \Delta_1, \Delta_2 \bullet \langle \Delta_1, \Delta_2 \rangle \in \text{part}.\Delta \land p.\Delta_1 \land qr.\Delta_2) \lor (\text{Inf} \land p).\Delta$

$\equiv$ definitions, expand $qr.\Delta_2$

$\begin{pmatrix} \exists \Delta_1, \Delta_2 \bullet \langle \Delta_1, \Delta_2 \rangle \in \text{part}.\Delta \land p.\Delta_1 \land \\ \qquad (\exists \Delta_3, \Delta_4 \bullet \langle \Delta_3, \Delta_4 \rangle \in \text{part}.\Delta_2 \land q.\Delta_3 \land r.\Delta_4) \lor (\text{Inf} \land q).\Delta_2 \end{pmatrix} \lor$
$(\text{Inf} \land p).\Delta$

$\equiv$ pull out existential, logic

$\begin{pmatrix} \exists \Delta_1, \Delta_2, \Delta_3, \Delta_4 \bullet \\ \quad (\langle \Delta_1, \Delta_2 \rangle \in \text{part}.\Delta \land p.\Delta_1 \land \langle \Delta_3, \Delta_4 \rangle \in \text{part}.\Delta_2 \land q.\Delta_3 \land r.\Delta_4) \lor \\ \quad (\langle \Delta_1, \Delta_2 \rangle \in \text{part}.\Delta \land p.\Delta_1 \land (\text{Inf} \land q).\Delta_2) \end{pmatrix} \lor$
$(\text{Inf} \land p).\Delta$

$\equiv$ logic

$\begin{pmatrix} \exists \Delta_1, \Delta_2, \Delta_3, \Delta_4, \bullet (\langle \Delta_1, \Delta_3, \Delta_4 \rangle \in \text{part}.\Delta \land p.\Delta_1 \land q.\Delta_3 \land r.\Delta_4) \lor \\ \qquad (\langle \Delta_1, \Delta_2 \rangle \in \text{part}.\Delta \land p.\Delta_1 \land (\text{Inf} \land q).\Delta_2) \end{pmatrix} \lor$
$(\text{Inf} \land p).\Delta$

$\equiv$ distribute existential over $\lor$, rename, logic

$(\exists \Delta_p, \Delta_q, \Delta_r \bullet \langle \Delta_p, \Delta_q, \Delta_r \rangle \in \text{part}.\Delta \land p.\Delta_p \land q.\Delta_q \land r.\Delta_r) \lor$ (A1)

$(\exists \Delta_p, \Delta_q \bullet \langle \Delta_p, \Delta_q \rangle \in \text{part}.\Delta \land p.\Delta_p \land (\text{Inf} \land q).\Delta_q) \lor$ (A2)

$(\text{Inf} \land p).\Delta$ (A3)

Similarly, the right hand side simplifies as follows:

$(pq)r.\Delta$
$\equiv$ as above
$(\exists \Delta_p, \Delta_q, \Delta_r \bullet \langle \Delta_p, \Delta_q, \Delta_r \rangle \in \text{part}.\Delta \wedge p.\Delta_p \wedge q.\Delta_q \wedge r.\Delta_r) \vee$ (B1)
$(\exists \Delta_p, \Delta_r \bullet \langle \Delta_p, \Delta_r \rangle \in \text{part}.\Delta \wedge (\text{Inf} \wedge p).\Delta_p \wedge r.\Delta_r) \vee$ (B2)
$(\exists \Delta_p, \Delta_q \bullet \langle \Delta_p, \Delta_q \rangle \in \text{part}.\Delta \wedge \text{Inf}.\Delta \wedge p.\Delta_p \wedge q.\Delta_q) \vee$ (B3)
$(\text{Inf} \wedge p).\Delta$ (B4)

To prove $p(qr).\Delta \Rightarrow (pq)r.\Delta$, it is trivial that $(A1) \vee (A3) \Rightarrow (pq)r.\Delta$ holds. For $(A2)$ we have:

$\quad (A2)$
$\equiv$ lifting
$\quad (\exists \Delta_p, \Delta_q \bullet \langle \Delta_p, \Delta_q \rangle \in \text{part}.\Delta \wedge p.\Delta_p \wedge \text{Inf}.\Delta_q \wedge q.\Delta_q)$
$\Rightarrow$ $\langle \Delta_p, \Delta_q \rangle \in \text{part}.\Delta$ and $\text{Inf}.\Delta_q$ implies $\text{Inf}.\Delta$
$\quad (B3)$

Similarly, to prove $(pq)r.\Delta \Rightarrow p(qr).\Delta$, it is clear that $(B1) \vee (B4) \Rightarrow p(qr).\Delta$. We prove the remaining cases as follows:

$\quad (B2)$
$\Rightarrow$ $\text{Inf}.\Delta_p$ and $\langle \Delta_1, \Delta_2 \rangle \in \text{part}.\Delta$
$\quad (\exists \Delta_p, \Delta_r \bullet \langle \Delta_p, \varnothing \rangle \in \text{part}.\Delta \wedge (\text{Inf} \wedge p).\Delta_p)$
$\equiv$ logic
$\quad (\exists \Delta_p \bullet \Delta_p = \Delta \wedge (\text{Inf} \wedge p).\Delta_p)$
$\equiv$ one-point rule
$\quad (A3)$

$\quad (B3)$
$\equiv$ $\text{Inf}.\Delta$ and $\langle \Delta_1, \Delta_2 \rangle \in \text{part}.\Delta$
$\quad (\exists \Delta_p, \Delta_q \bullet \langle \Delta_p, \Delta_q \rangle \in \text{part}.\Delta \wedge (\text{Inf}.\Delta_p \vee \text{Inf}.\Delta_q) \wedge p.\Delta_p \wedge q.\Delta_q)$
$\equiv$ distribute
$\quad (\exists \Delta_p, \Delta_q \bullet \langle \Delta_p, \Delta_q \rangle \in \text{part}.\Delta \wedge \text{Inf}.\Delta_p \wedge p.\Delta_p \wedge q.\Delta_q) \vee (A2)$
$\Rightarrow$ as with proof of $(B2)$ above
$\quad (A3) \vee (A2)$

The validity of Axiom (6) is proved as follows. We first show:

$\quad (\exists \Delta_1, \Delta_2 \bullet \langle \Delta_1, \Delta_2 \rangle \in \text{part}.\Delta \wedge p.\Delta_1 \wedge \text{Empty}.\Delta_2)$
$\equiv$ logic
$\quad (\exists \Delta_1 \bullet \langle \Delta_1, \varnothing \rangle \in \text{part}.\Delta \wedge p.\Delta_1)$
$\equiv$ definition of part and $\propto$
$\quad (\exists \Delta_1 \bullet (\Delta = \Delta_1 \cup \varnothing) \wedge p.\Delta_1)$
$\equiv$ logic
$\quad p.\Delta$

Then we perform some case analysis.

Assuming $(\neg\mathsf{Inf}).\Delta$:

$\quad (p \; ; \; \mathsf{Empty}).\Delta$

$\equiv \quad$ definition, assumption $(\neg\mathsf{Inf}).\Delta$

$\quad\quad \exists \Delta_1, \Delta_2 \bullet \langle \Delta_1, \Delta_2 \rangle \in \mathsf{part}.\Delta \wedge$

$\quad\quad\quad\quad p.\Delta_1 \wedge \mathsf{Empty}.\Delta_2$

$\equiv \quad$ calculation above

$\quad\quad p.\Delta$

Assuming $\mathsf{Inf}.\Delta$:

$\quad (p \; ; \; \mathsf{Empty}).\Delta$

$\equiv \quad$ definitions

$\quad\quad (\exists \Delta_1, \Delta_2 \bullet \langle \Delta_1, \Delta_2 \rangle \in \mathsf{part}.\Delta \wedge$

$\quad\quad\quad\quad p.\Delta_1 \wedge \mathsf{Empty}.\Delta_2) \vee (\mathsf{Inf} \wedge p.\Delta)$

$\equiv \quad$ calculation above, assumption $\mathsf{Inf}.\Delta$

$\quad\quad p.\Delta$

Now consider the other direction:

$\quad (\mathsf{Empty} \; ; \; p).\Delta$

$\equiv \quad$ definitions, assumption

$\quad\quad (\exists \Delta_1, \Delta_2 \bullet \langle \Delta_1, \Delta_2 \rangle \in \mathsf{part}.\Delta \wedge \mathsf{Empty}.\Delta_1 \wedge p.\Delta_2) \vee (\mathsf{Inf}.\Delta \wedge \mathsf{Empty}.\Delta)$

$\equiv \quad$ logic, second disjunct is false

$\quad\quad (\exists \Delta_2 \bullet \langle \varnothing, \Delta_2 \rangle \in \mathsf{part}.\Delta \wedge p.\Delta_2)$

$\equiv \quad$ definition of part and $\propto$

$\quad\quad (\exists \Delta_2 \bullet (\Delta = \varnothing \cup \Delta_2) \wedge p.\Delta_2)$

$\equiv \quad$ logic

$\quad\quad p.\Delta$

Axiom (7) follows from the following calculation.

$\quad (\mathsf{False} \; ; \; p).\Delta$

$\equiv \quad$ definitions

$\quad\quad (\exists \Delta_1, \Delta_2 \bullet \langle \Delta_1, \Delta_2 \rangle \in \mathsf{part}.\Delta \wedge \mathsf{False}.\Delta_1 \wedge p.\Delta_2) \vee (\mathsf{Inf}.\Delta \wedge \mathsf{False}.\Delta)$

$\equiv \quad$ logic

$\quad\quad \textit{false}$

$\equiv \quad$ definition

$\quad\quad \mathsf{False}.\Delta$

The validity of Axiom (8) is proved as follows.

$\quad p(q \vee r).\Delta$

$\equiv \quad$ definitions

$\quad\quad (\exists \Delta_1, \Delta_2 \bullet \langle \Delta_1, \Delta_2 \rangle \in \mathsf{part}.\Delta \wedge p.\Delta_1 \wedge (q \vee r).\Delta_2) \vee (\mathsf{Inf}.\Delta \wedge p.\Delta)$

$\equiv \quad$ definitions

$\quad\quad (\exists \Delta_1, \Delta_2 \bullet \langle \Delta_1, \Delta_2 \rangle \in \mathsf{part}.\Delta \wedge p.\Delta_1 \wedge (q.\Delta_2 \vee r.\Delta_2)) \vee (\mathsf{Inf}.\Delta \wedge p.\Delta)$

$\equiv \quad$ distributivity, split existential quantifier, duplicate last conjunct

$\quad\quad (\exists \Delta_1, \Delta_2 \bullet \langle \Delta_1, \Delta_2 \rangle \in \mathsf{part}.\Delta \wedge p.\Delta_1 \wedge q.\Delta_2) \vee (\mathsf{Inf}.\Delta \wedge p.\Delta) \vee$

$\quad\quad (\exists \Delta_1, \Delta_2 \bullet \langle \Delta_1, \Delta_2 \rangle \in \mathsf{part}.\Delta \wedge p.\Delta_1 \wedge r.\Delta_2) \vee (\mathsf{Inf}.\Delta \wedge p.\Delta)$

$\equiv \quad$ definitions

$\quad\quad pq.\Delta \vee pr.\Delta$

$\equiv \quad$ definitions

$\quad\quad (pq \vee pr).\Delta$

Finally, Axiom (9) is done similarly (expand definitions, distribute and regroup). This concludes the proof. $\qquad\qquad\square$

Automated Reasoning
in Higher-Order Regular Algebra

Alasdair Armstrong and Georg Struth

Department of Computer Science, The University of Sheffield, UK
{a.armstrong,g.struth}@dcs.shef.ac.uk

Abstract. We extend a large Isabelle/HOL repository for regular algebras towards higher-order variants based on directed sets and quantales, including reasoning based on general fixpoint properties and Galois connections. In this context we demonstrate that Isabelle's recent integration of automated theorem proving technology effectively supports higher-order reasoning. We present four case studies that underpin this claim: the calculus of Galois connections and fixpoints, action algebras and Galois connections, solvability conditions for regular equations and fixpoint fusion, and the implementation of formal language quantales.

1 Introduction

Regular algebras were originally conceived to axiomatise the equational theory of regular expressions, but their variants and extensions are now widely applied in computing. There are both first-order and higher-order variants which have their own advantages and limitations. Higher-order regular algebras often yield more accurate models and semantics; some of them support elegant abstract reasoning based on adjunctions and fixpoints of general isotone or continuous functions. Reasoning with first order variants, by contrast, is well supported by automated theorem proving (ATP) systems. Until recently, one could either ask for expressivity or automation, but not for both.

Recent advances in the integration of ATP tools into the higher-order interactive theorem proving (ITP) environment Isabelle/HOL [20], however, have dramatically changed this situation. This introduces a style of formalising mathematics which seems much simpler and more natural than traditional tactic based ITP. With the ATP approach, a large repository for first-order regular algebras[1] has already been implemented in Isabelle [12]. In this setting, Isabelle is predominantly used as a proof manager for engineering theory hierarchies and propagating theorems across it. More recent work [15,11] has started to push the boundary into the higher-order domain. ATP support remains surprisingly powerful and robust, for instance, when reasoning about set-theoretic models of regular algebras or higher-order properties such as Conway's powerstar axiom $x^* = (x^{n+1})^* \sum_{i=0}^{n} x^i$. But a systematic formalisation of higher-order regular algebras has so far not been undertaken.

---

[1] www.dcs.shef.ac.uk/~georg/isa

W. Kahl and T.G. Griffin (Eds.): RAMiCS 2012, LNCS 7560, pp. 66–81, 2012.
© Springer-Verlag Berlin Heidelberg 2012

This paper closes the gap between first and higher-order regular algebras in Isabelle/HOL. An ATP-based implementation of higher-order regular algebras, fixpoint calculi and Galois connections is presented. Both kinds of regular algebra are formally linked in Isabelle. Theorems proved about the former are therefore available for the latter. Instantiations of Galois connections yields theorems for free in the first-order domain. Powerful tools such as fixpoint fusion laws capture algebraic relationships in the higher-order domain that would fail in the first-order one. Detailed results are as follows.

We (re)implement basic lattice theory using ATP, including complete lattices, basic fixpoint theory, and the Knaster-Tarski theorem. Based on this we formalise the notion of Galois connection and prove some combined results for fixpoints, for example fixpoint fusion laws. After these preparations we formalise higher-order *-continuous Kleene algebras [18], for which the Kleene star must be postulated, and (unital) quantales, where the star is explicitly defined. We formalise the subclass relationships between unital quantales, *-continuous Kleene algebras and Kleene algebra, for which a large number of laws have already been implemented in the repository. All theorems proved for Kleene algebras are automatically propagated down the subclass hierarchy.

We provide four case studies on the interplay between first and higher-order reasoning using ATP. (1) We formalise large parts of the calculus of Galois connection from Aart's survey [1]. (2) We formalise Pratt's action algebra [21], which expands Kleene algebras by residuations axiomatised as (upper) adjoints. We obtain concrete properties of residuations automatically by instantiating abstract properties of adjoints and formally link action algebras with Kleene algebras and unital quantales. We also formalise Pratt's result that action algebras can be equationally axiomatised. (3) We formalise two conditions that imply (unique) solvability of fixpoint equations in regular algebra [10,23]. We formalise a higher-order proof of their equivalence using greatest fixpoint fusion, whereas no first-order proof is currently known. (4) Finally, we formalise that languages form quantales.

These case studies show that ATP in higher-order regular algebra is surprisingly powerful and robust. Most proofs of basic facts could be fully automated without user interaction. For more difficult facts, paper and pencil proofs could usually be automated quickly and easily, sometimes just by calling the simplifier with specific rules to eliminate higher-order structure. Overall we proved 278 facts for this paper, 244 fully automatically, and 25 by reconstructing paper and pencil proofs. Only 9 required serious user interaction beyond ATP.

The complete theory file can be found in our repository and utilised for more advanced applications. The design of useful libraries for regular algebras is another main application of our paper. As is so often the case when formalising mathematics, the majority of our proved statements are neither novel nor mathematically deep.

Our results also allow us to draw some more critical conclusions on theory engineering in Isabelle, in particular on name space problems and the locale mechanism, which we present at the end of this paper.

2   Automated Theorem Proving in Isabelle

Isabelle/HOL [20] is one of the most popular and well established theorem proving environments. It is widely used for formalising mathematics and in computing applications. As an ITP system, it has traditionally been based on reasoning with built-in rewrite-based simplifiers, theorem provers, special solvers and tactics. This often requires considerable user expertise and domain-specific knowledge of library functions and lemmas. More recently, external ATP systems and satisfiability modulo theories (SMT) solvers have been integrated via the Sledgehammer tool. When Sledgehammer is called on a proof goal, Isabelle uses a relevance filter to automatically gather verified hypotheses that are potentially useful for discharging it. It then passes tentative hypotheses and the goal to the external provers. On success, the external proof output is internally reconstructed by Isabelle to increase trustworthiness. Sledgehammer is complemented by tools such as Nitpick and Quickcheck, which search for counterexamples. For a recent overview of this technology see [4].

Apart from enabling ATP in an interactive setting, Isabelle also supports engineering theory hierarchies and documenting proofs in readable form. The first feature is provided by Isabelle's type classes and locales [3,16]. Type classes in Isabelle are similar to those in functional programming languages such as Haskell. Locales provide a module mechanism. The two concepts are closely linked and can often be used interchangeably. We use classes for simple algebraic specifications and locales for more complex parametric ones. Classes and locales provide a mechanism for theorem propagation: a theorem proved in a certain class is automatically valid in all subclasses; a model that belongs to a class is obviously an element of all superclasses. In our context, we show that unital quantales are subclasses—sublocales, in Isabelle parlance—of Kleene algebras. Therefore all theorems of Kleene algebra are automatically available for unital quantales, and all models of the unital quantale axioms are automatically available for Kleene algebra.

Experience shows that a high percentage of first-order calculational proofs in regular algebras can be fully automated. While previous papers [12,15,11] have predominantly explored full proof automation in regular algebra, we take a more pragmatic approach in this paper and formalise the most interesting theorems at paper and pencil granularity. This can be acheived with little effort using Isabelle's proof scripting language Isar. Our point is that interesting proofs can be documented and displayed as textbook proofs in Isabelle, with individual proof steps discharged automatically, while routine proofs can often be hidden by automation. The design of domain-specific libraries with useful lemmas is crucial in supporting this mathematically natural approach.

3   Lattices

The following sections follow our Isabelle theory file from the repository more or less sequentially. We start by formalising semilattices and lattices, as well as their complete variants, in Isabelle. This work is based on Wenzel's implementation of lattice theory in Isabelle[2], but, unlike previous approaches, uses ATP to show the difference between the new ATP-based and the traditional ITP approach. While Wenzel's proofs are necessarily often quite lengthy and perhaps overly detailed, ours are usually fully automatic.

We assume basic knowledge of lattice theory and briefly comment on some implementation details and proofs. Our main interest is in complete lattices, which require the implementation of infinite least upper and greatest lower bounds. As usual in Isabelle, this can be based on the pre-defined definite description operator. First we implement a predicate that encodes upper and least upper bounds within Isabelle's theory of order,

$$is\text{-}ub\ x\ A \leftrightarrow (\forall y \in A.y \leq x),$$
$$is\text{-}lub\ x\ A \leftrightarrow (\forall y \in A.y \leq x) \wedge (\forall y.(\forall z \in A.z \leq y) \rightarrow x \leq y).$$

We then prove that $is\text{-}lub\ x\ A \leftrightarrow (\forall z.(x \leq z \leftrightarrow (\forall y \in A.y \leq z)))$ and that least upper bounds are uniquely defined. We then define

$$\Sigma A = (\iota x.is\text{-}lub\ x\ A).$$

The right-hand side of this equation reads *the x that satisfies is-lub x A*. In Isabelle, the definite description operator ι returns an arbitrary value unless the formula has a unique solution. Reasoning with definite descriptions in Isabelle can be quite tedious. It requires demonstrating existence and uniqueness of the element under consideration as well as auxiliary facts for seamless reasoning. It is therefore essential to derive appropriate interface lemmas, for instance,

$$\exists z.is\text{-}lub\ z\ X \wedge (\forall z.is\text{-}lub\ z\ X \rightarrow z \leq x) \rightarrow \Sigma X \leq x,$$

that eliminate the higher-order structure in the consequent of formulas. A greatest lower bound operation Π can be defined dually. We have developed a small library for reasoning with least upper and greatest lower bounds. All proofs in this setting are fully automatic by ATP or SMT. We have defined $\sqcup$ and $\sqcap$ for binary joins and meets in Isabelle using Σ and Π.

We have formalised (complete) semilattices and lattices as classes in Isabelle. As usual, a *join semilattice* is a poset for which binary joins (or suprema) exist, whereas a *complete join semilattice* is a poset for which arbitrary joins exist. Meet semilattices and complete meet semilattices are obtained by duality. Proofs that, in join and meet semilattices, joins and meets are commutative and idempotent were fully automatic, with the proofs of associativity requiring only a small amount of user interaction.

Based on (complete) join and meet semilattices we have defined (complete) lattices as usual and implemented them as a type class that extends both (complete) join and meet semilattices. It is well known that every complete join semilattice is also a complete meet semilattice, hence a complete lattice. In Isabelle,

---

[2] `isabelle.in.tum.de/library/HOL/Lattice`

we were able to prove this essentially higher-order fact fully automatically. The element ΠA is defined in a join semilattice as $\Sigma\{y \mid \forall x \in A.y \leq x\}$ and reasoning about this set is needed in order to show that it yields a greatest lower bound.

In Isabelle, all theorems about complete semilattices are automatically available for complete lattices. We have also shown via a sublocale statement that every complete lattice is a lattice. Finally, we have formalised duality between join and meet semilattices. This yields theorems for free: theorems for meet semilattices can by obtained automatically from their join semilattice duals.

The definition of (co)-Heyting algebras, distributive lattices and boolean algebras, as well as of their complete variants, is straightforward from this basis.

In sum we have proved 49 theorems about orders and lattices by ATP in Isabelle. Of these, 43 were fully automatic, the rest needed simple user interaction.

4   Fixpoints and Galois Connections

The material in this section is again not entirely new. Theorems about fixpoints can be found in Isabelle's libraries, including fixpoint fusion laws which have already been implemented by Gammie in the study of the worker-wrapper transformation for functional programs, though in the context of the HOLCF theory of computable functions [13]. Again, our main contribution lies in automation.

The implementation of pre-fixpoints, post-fixpoints and fixpoints of functions again requires Isabelle's definite description operator. We have defined the predicates

$$is\text{-}fp \; x \; f \leftrightarrow f \; x = x,$$
$$is\text{-}lfp \; x \; f \leftrightarrow is\text{-}fp \; x \; f \wedge (\forall y.is\text{-}fp \; x \; f \rightarrow x \leq y),$$
$$is\text{-}gfp \; x \; f \leftrightarrow is\text{-}fp \; x \; f \wedge (\forall y.is\text{-}fp \; x \; f \rightarrow y \leq x)$$

for fixpoints and least and greatest fixpoints. Similar definitions can be given for (least and greatest) pre- and post-fixpoints. The definite description operator allows us to define

$$\mu f = \iota x.is\text{-}lfp \; x \; f \qquad \text{and} \qquad \nu f = \iota x.is\text{-}gfp \; x \; f.$$

We have proved some basic interface lemmas to reason with fixpoints. These include fixpoint unfold and induction laws for isotone functions and the fact that least fixpoints and least pre-fixpoints as well as greatest fixpoints and greatest post-fixpoints coincide and that they all are uniquely defined. We have then directly translated a textbook proof of the Knaster-Tarski theorem for least and greatest fixpoints of isotone functions over a complete lattice into Isabelle. Each proof step could easily be discharged automatically.

We have then shown some elementary facts from the fixpoint calculus over a complete lattice, for instance, that if f is isotone and $g \circ f = f \circ h$, then $is\text{-}fp \; x \; h$ implies $is\text{-}fp \; (f \; x) \; g$. Finally we have shown that if f and g are isotone and $f \leq g$, then $\mu f \leq \mu g$ and $\nu f \leq \nu g$.

Next we have implemented Galois connections over orders and (complete) lattices, including most of the material in Aart's survey article [1]. Formally, a *Galois connection* between two partial orders $(A_1, \leq_1)$ and $(A_2, \leq_2)$ is a pair of functions $f : A_1 \to A_2$ and $g : A_2 \to A_1$ that satisfy $f\ x \leq_2 y \leftrightarrow x \leq_1 g\ y$ for all $x \in A_1$ and $y \in A_2$. In this section we only mention the most fundamental concepts. Formalisation details are discussed in Section 8. We have defined the predicates

$$galois\text{-}connection\ f\ g \leftrightarrow \forall x, y.(f\ x \leq y \leftrightarrow x \leq g\ y),$$
$$lower\text{-}adjoint\ f \leftrightarrow \exists g.galois\text{-}connection\ f\ g,$$
$$upper\text{-}adjoint\ g \leftrightarrow \exists f.galois\text{-}connection\ f\ g,$$

where f and g are endofunctions of some given type $A \to A$. We have also implemented Galois connections between two different partial orders in Isabelle, but they are not needed in the context of our paper.

Equipped with both fixpoints and Galois connections, we have formalised some laws of the fixpoint calculus, such as rolling rules and fixpoint fusion laws (cf. [10]).

The rolling rules state that if f and g are adjoints, then $f(\mu(g \circ f)) = \mu(f \circ g)$. A dual rule holds for greatest fixpoints. We could formalise textbook proofs in four steps using ATP.

The least and greatest fixpoint fusion law state that, if f is a lower adjoint, h and k are isotone and $f \circ h = k \circ f$, then $f(\mu h) = \mu k$; if g is an upper adjoint, h and k are isotone and $g \circ h = k \circ g$, then $g(\nu h) = \nu k$. These proofs are essentially translations of paper and pencil proofs; individual steps are again by ATP.

Overall we have proved 42 laws about fixpoints; 32 fully automatically. Those about Galois connections will be discussed in Section 8.

All rules for least and greatest fixpoints in this section are order duals. Nevertheless we had to prove them separately, since it seems that locales are not able to capture duality for concepts defined outside of class and locale specifications. Our experience shows that this problem can be solved by using explicit carrier sets.

5   First-Order Regular Algebras

In this section we only consider Kleene algebras; for a formalisation of some other first-order variants of regular algebras in Isabelle see [11]. These structures are based on dioids or idempotent semirings.

Formally, a *dioid* is a structure $(S, +, \cdot, 0, 1)$ such that $(S, +, 0)$ is a commutative idempotent monoid (or semilattice) with least element 0, $(S, \cdot, 1)$ is a monoid. The two reducts interact by the distributivity laws $x(y + z) = xy + xz$ and $(x+y)z = xz + yz$, and the annihilation laws $x0 = 0$ and $0x = 0$. Because of the semilattice reduct, S can be endowed with a natural order $x \leq y \leftrightarrow x+y = y$ with least element 0. The operations $+$ and $\cdot$ are order preserving. Dioids and their variants have been implemented as classes in Isabelle and formally linked with join semilattices (with zero).

A *Kleene algebra* is a dioid S expanded by the operation $* : S \to S$ that satisfies the star unfold and induction laws

$$1 + xx^* \leq x^*, \qquad z + xy \leq y \Rightarrow x^*z \leq y$$

and their opposites, which are obtained by swapping the order of multiplication. Kleene algebras have also been implemented previously as classes, and a large number of facts have been proved in Isabelle by ATP, most of them fully automatically. In addition, the most important models of Kleene algebras have been implemented [12]: languages, binary relations, sets of traces and sets of paths. This provides a seamless transition between the axiomatic and semantic levels.

Finite powers x^i of an element x of a dioid have been implemented in Isabelle as a primitive recursive function, as have finite sums $\sum_{i=m}^{n} x^i$. In fact, Isabelle's library function *setsum* for finite sets can be used for this purpose. To reason about countably infinite sets of powers, as needed in the context of quantales, we have defined the set

$$powers \; x = \{y \mid \exists i.y = x^i\}.$$

We have also built a basic library that supports automated reasoning with these concepts. Laws such as $x^m x^n = x^{m+n}$ or $x^n x = x x^n$ can easily be proved using Isabelle's induction tactic. The base cases and induction steps of such laws are usually fully automatic. We have also derived induction laws similar to those for Kleene algebra: $xy \leq y \to x^n y \leq y$ and its dual with multiplication swapped.

6   Star-Continuous Kleene Algebras

Our first higher-order regular algebra is Kozen's $*$-continuous Kleene algebra [18], also called N-algebra by Conway [8]. The approach is similar to complete partial orders and assumes only the existence of certain infinite joins and of certain infinite distributivity laws that suffice for defining the star.

Formally, a dioid is a $*$-*continuous Kleene algebra* if the law

$$xy^*z = \Sigma\{xy^i z \mid i \in \mathbb{N}\}$$

holds. It combines the definition of y^* as a countable least upper bound of powers with an infinite left and right distributivity law. As in the Kleene algebra case this star is postulated and need not exist in arbitrary dioids.

In Isabelle we define a set-valued function similar to *powers*,

$$powers\text{-}c \; x \; y \; z = \{xwz \mid \exists i. \; w = y^i\},$$

in the class of dioids. We then define the class of $*$-continuous Kleene algebras as the class of dioids plus the following two axioms.

$$\forall x, y, z. \exists w. is\text{-}lub \; w \; (powers\text{-}c \; x \; y \; z)$$
$$xy^*z = \Sigma(powers\text{-}c \; x \; y \; z)$$

The second axiom implicitly uses Isabelle's definite description operator in the definition of Σ.

Since definitions and axioms are based on set comprehension, ATP systems often cannot deal with them directly. We have therefore provided an elimination rule to reduce them to predicate logic.

$$v \in \textit{powers-c } x\ y\ z \;\leftrightarrow\; \exists w.(v = xwz \wedge \exists i.w = y^i).$$

This law can be proved by Isabelle's simplifier. After a few auxiliary facts, we have proved the law

$$\{1\} \cup \textit{powers-c } x\ x\ 1 = \textit{powers } x$$

in dioids. It prepares the derivation of the star unfold axiom of Kleene algebra in $*$-continuous Kleene algebra. The proof of this lemma required a few steps, which could be discharged automatically. We have also proved that this star operation, as axiomatised in $*$-continuous Kleene algebras, is uniquely defined, as well as some further auxiliary lemmas, all of them automatically. From this basis we have proved fully automatically that the star unfold and star induction axioms of Kleene algebra are derivable in $*$-continuous Kleene algebra. Using Isabelle's sublocale mechanism we have then shown that the $*$-continuous Kleene algebras form a subclass of Kleene algebras. All laws derived for Kleene algebras elsewhere in the repository are thus available for the $*$-continuous variant. All in all, the development of $*$-continuous Kleene algebras from the dioid basis required 20 lemmas. All but two of them were fully automatic apart perhaps from calling the simplifier or Isabelle's induction tactic. In the context of $*$-continuous Kleene algebras, there is now the choice between performing proofs by explicit induction or by calculation using the first-order version of regular algebra.

7   Quantales

Quantales are even more expressive higher-order regular algebras which, under the name of S-algebra or standard Kleene algebra, have already been considered by Conway [8]. A quantale is essentially a dioid that is based on a complete join-semilattice (which is automatically a complete lattice) in which multiplication distributes from the left and right over arbitrary joins. Due to this continuity property and monotonicity of join and meet, all kinds of fixpoints exist and adjoints can be explicitly defined. In our context we are mainly interested in unital quantales with multiplicative units (if this unit does not exist it can be adjoined).

Formally, a *quantale* is a structure $(S, \leq, \cdot)$ such that $(S, \leq)$ is a complete lattice, $(S, \cdot)$ a semigroup and multiplication distributes over arbitrary joins,

$$x(\Sigma A) = \Sigma\{xy \mid y \in A\}, \qquad (\Sigma A)x = \Sigma\{yx \mid y \in A\}.$$

A quantale is *unital* if it has a left and right multiplicative unit 1.

We have proved some basic properties of quantales, most of them automatically. Since the functions $\lambda y.xy$ and $\lambda y.yx$ preserve arbitrary joins, they must be lower adjoints by abstract properties of Galois connections. We can explicitly define residuation operations for quantales,

$$x \to y = \Sigma\{z.xz \le y\} \qquad \text{and} \qquad x \leftarrow y = \Sigma\{z.zy \le x\},$$

and show in Isabelle

$$galois\text{-}connection \ (\lambda y.xy) \ (\lambda y.x \to y),$$
$$galois\text{-}connection \ (\lambda y.yx) \ (\lambda y.y \leftarrow x),$$

that is, they are upper adjoints for the above functions.

We have then formalised the fact that unital quantales form a subclass of dioids and have explicitly defined the star as

$$x^* = \Sigma(powers \ x).$$

We have then shown that this star satisfies the left and right continuity laws

$$x^*y = \Sigma(powers\text{-}c \ 1 \ x \ y) \qquad \text{and} \qquad yx^* = \Sigma(powers\text{-}c \ y \ x \ 1),$$

the proofs of which heavily rely on simplification. From these two laws is it then straightforward to show that unital quantales form a subclass of $*$-continuous Kleene algebras. By transitivity of Isabelle's locale mechanism, all theorems about Kleene algebras are now available in the unital quantale setting.

Finally, we have proved the fixpoint laws

$$\mu(\lambda y.1 + xy) = x^* = \mu(\lambda y.1 + yx)$$

fully automatically. This relates the star in unital quantales with the Knaster-Tarski theorem.

The unital quantales or S-algebras studied in this section can be contrasted with *closed semirings* that require only the existence of countable infima and suprema. Such semirings have been used, for instance, in the design of algorithms and combinatorial optimisation. In the context of Kleene algebras they have been studied by Kozen [17]. Closed semirings are strongly related to $*$-continuous Kleene algebras, since the star is defined in this setting with respect to indices ranging over $\mathbb{N}$. The implementation of countable least upper and greatest lower bounds in Isabelle is given in terms of its library function *setsum*; hence the definition of the class of closed semirings is straightforward. It seems similarly easy to situate this class between $*$-continuous semirings and unital quantales by using sublocale statements. Since this does not offer any substantial insights in this context, we leave this for future work.

In sum we proved 21 facts about quantales, 10 of them fully automatically.

8   Case Study 1: Galois Connections

We now return to Galois connections for our first case study on the automation of higher-order regular algebras. We have formalised most of the theorems in Aart's survey paper [1]. To our knowledge, most of these theorems are not yet available in Isabelle. We consider properties of Galois connections over partial orders, lattices and complete lattices and put particular emphasis on equational definitions of adjoints, for instance in terms of cancellation properties and join/meet preservation.

Isabelle's higher-order logic allows a smooth transition between first-order and higher-order concepts, for instance between pointwise and pointfree reasoning about functions. We made heavy use of this facility and found that ATP still works well at that level. For instance, we can lift the pointwise order on functions as $f \leq g \leftrightarrow \forall x. f\ x \leq g\ x$ and reason at that level.

First, we have derived standard properties of Galois connections over partial orders, such as the cancellation laws $f(g\ y) \leq y$ and $x \leq g(f\ x)$ as well as their pointfree counterparts, the fact that both adjoints are isotone, that $f \circ g \circ f = f$ and $g \circ f \circ g = g$, that $f \circ g$ and $g \circ f$ are idempotent, that Galois connections are preserved under the composition of lower and upper adjoints. Moreover, $f \circ g$ and $g \circ f$ are isotone, $g\ x = g\ y \leftrightarrow f(g\ x) = f(g\ y)$ and $f\ x = f\ y \leftrightarrow g(f\ x) = g(f\ y)$. All these proofs are more or less automatic. This holds, for example for proofs of statements such as the composition law

$$galois\text{-}connection\ f_1\ g_1 \land galois\text{-}connection\ f_2\ g_2$$
$$\Rightarrow galois\text{-}connection\ (f_1 \circ f_2)\ (g_1 \circ g_2).$$

Ore's definition of a Galois connection, that two isotone functions are adjoints if they satisfy the cancellation laws, was again fully automatic from our definition of a Galois connection.

We have also shown automatically that if f is a lower adjoint and g an upper adjoint, then $g(y) = \Sigma\{x \mid f\ x \leq y\}$ and $f(x) = \Pi\{y \mid x \leq g\ y\}$. The proofs that lower adjoints preserve all (existing) joins and upper adjoints preserve all (existing) meets required reasoning at the level of paper and pencil proofs.

Next we have implemented equivalent ways of formalising Galois connections, for instance that f and g are adjoints if and only if f is isotone and g satisfies the above least upper bound condition with respect to f if and only if g is isotone and f satisfies the above greatest lower bound condition with respect to g.

More properties can be obtained if the underlying order is a (complete) lattice, as in the quantale case. In the complete case, for instance, we have formalised the fact that f and g are adjoints if and only if f preserves meets and $g(y) = \Sigma\{x \mid f\ x \leq y\}$ if and only if g preserves joins and f satisfies the obvious meet condition and, similarly, that each isotone function that preserves all joins is a lower adjoint and each isotone function that preserves all meets is an upper adjoint. These proofs directly implement paper and pencil proofs.

Overall, we have implemented 57 facts about Galois connections; 48 proofs were fully automatic.

9 Case Study 2: Action Algebras

Our second case study deals with the application of higher-order reasoning in a first-order setting and the propagation of first-order properties into the higher-order domain. We consider the action algebras introduced by Pratt [21], which are Kleene algebras expanded by two operations of residuation. Their first axiomatisation is given by Galois connections, a second one is purely equational (even for the star). In some sense, therefore, action algebras are first-order shadows of quantales where residuation and the star can be explicitly defined.

Formally, an *action algebra* is a structure $(S, +, \cdot, \rightarrow, \leftarrow, 0, 1, ^*)$ such that $(S, +, \cdot, 0, 1)$ is a dioid, the operation $\lambda y.x \rightarrow y$ is an upper adjoint to $\lambda y.xy$ and $\lambda y.y \leftarrow x$ is an upper adjoint to $\lambda y.yx$. The star is defined in terms of the reflexive-transitive closure axioms

$$1 + x^* x^* + x \leq x^* \qquad \text{and} \qquad 1 + yy + x \leq y \rightarrow x^* \leq y.$$

The dioid and star axioms are reminiscent of a regular algebra that Boffa proved to be complete with respect to the equational theory of regular expressions [5,6]. He, however, uses the stronger looking star axioms

$$1 + x \leq x^*, \qquad x^* x^* = x^*, \qquad 1 + x \leq y \wedge yy = y \rightarrow x^* \leq y.$$

We have shown that dioids with both star axiom systems are interderivable, which yields a perhaps slightly more appealing new variant of Boffa's axioms. According to this result, Pratt's action algebra arises directly as an expansion of Boffa's regular algebra and is a fortiori complete. We have formalised our new version of Boffa's regular algebra as well as action algebras as classes in Isabelle.

First we have shown that unital quantales form a subclass of action algebras. Hence all theorems proved for action algebras are inherited by unital quantales. We have then proved the explicit Galois connections of action algebras by instantiating the abstract ones.

$$xy \leq z \leftrightarrow x \leq z \leftarrow y \quad \text{and} \quad xy \leq z \leftrightarrow y \leq x \rightarrow z.$$

Many properties of residuation in action algebra can now be derived by instantiation and, in particular, a fully equational axiomatisation of residuation can be obtained from that basis. Examples are the cancellation laws $y \leq x \rightarrow xy$ and $x(x \rightarrow y) \leq y$ and their duals with respect to opposition. We have proved about 20 simple facts about action algebras in Isabelle, all of them automatically, most of them about the interaction between residuation, multiplication and the star.

A particular example of laws that are not immediately related to the Galois connection, but have very simple automated proofs, are Pratt's pure induction laws, which are duals with respect to opposition:

$$(x \rightarrow x)^* \leq x \rightarrow x \qquad \text{and} \qquad (x \leftarrow x)^* \leq x \leftarrow x.$$

Interestingly, Pratt has shown that the Kleene algebra axioms are derivable from those of action algebra (we could easily formalise this fact in Isabelle). Since the

Kleene algebra axioms imply Boffa's (cf. [11]), the axioms of residuated Kleene algebras and action algebras are interdefinable.

One of Pratt's main results is the proof that, in action algebra, the Kleene star can be equationally characterised. Formally, an *equational action algebra* is a dioid expanded by the residuations $\rightarrow$ and $\leftarrow$ and the star * such that

$$x \rightarrow y \leq x \rightarrow (y + z), \qquad x(x \rightarrow y) \leq y, \qquad y \leq x \rightarrow xy,$$

$$y \leftarrow x \leq (y + z) \leftarrow x, \qquad (y \leftarrow x)x \leq y, \qquad y \leq yx \leftarrow x,$$

$$1 + x^*x^* + x \leq x^*, \qquad x^* \leq (x + y)^*, \qquad (x \rightarrow x)^* \leq x \rightarrow x.$$

Notice the duality between the axioms that do not mention the star.

By sublocale statements we have formalised the fact that the equational and the Horn axiomatisation of action algebras are are interderivable. Both sublocale proofs were simple and automatic.

In sum, we proved 24 facts about action algebras, all of which were automatic.

10   Case Study 3: Recursive Regular Equations

Arden's rule states that a if language L does not have the empty word property, that is, $\epsilon \notin L$, then the recursive language equation $X = LX \cup L'$ has the unique solution $X = L^*L'$ in X. A corresponding equation at the level of regular expressions can be used for obtaining regular expressions from automata [22]. However, absence of the empty word property must be inductively defined over regular expressions. Even more generally, we consider algebraic conditions for which *Arden's rule*

$$z + xy = y \rightarrow x^*z = y$$

is derivable in regular algebra. In the context of higher-order relation algebras without complementation, such conditions have been studied, for instance, by Doornbos, Backhouse and van der Woude [10]. These relation algebras are sub-classes of unital quantales in which additional infinite distributivity laws between joins and meets hold. Solvability conditions are expressed in terms of higher-order fixpoint conditions.

In the setting of first-order regular algebras, alternative first-order conditions can be given [23]. Arden's rule is related to the induction axiom of Kleene algebra, which can be strengthened to the formula $z + xy = y \rightarrow x^*z \leq y$. It is therefore quite obvious that it can be derived from the *strong deflationarity* condition

$$\forall y, z.(y \leq xy + z \Rightarrow y \leq x^*z)$$

on x. It is, however, not local in x, since x^* appears in the consequent. An alternative condition is *deflationarity* of x:

$$\forall y.(y \leq xy \Rightarrow y = 0),$$

but the proof that this condition implies Arden's rule is indirect and requires the presence of an $^\omega$-operator which is axiomatised as a greatest fixpoint of the function $\lambda y.xy + z$.

It is obvious by setting $z = 0$ that strong deflationarity implies deflationarity, but the converse direction can neither be proved nor refuted in Kleene algebra by means of Isabelle [23]. In higher-order relation algebra, however, a corresponding implication can be shown via greatest fixpoint fusion [10].

To apply this law we need to strengthen our notion of quantale. We define a *completely distributive quantale* to be a unital quantale that also satisfies the following infinite distributivity law between joins and meets.

$$x + \Pi Y = \Pi \{x + y \mid y \in Y\}.$$

Following Doornbos, Backhouse and van der Woude, we have then proved the following fixpoint fusion statement for Arden's rule in Isabelle.

$$\nu(\lambda y.xy + z) = x^* z + \nu(\lambda y.xy).$$

We proceeded by a series of simple lemmas to instantiate the preconditions required for greatest fixpoint fusion. These are that $\lambda y.x^* z + y$ is an upper adjoint, that $\lambda y.xy + z$ and $\lambda y.xy$ are isotone, and that

$$(\lambda y.x^* z + y) \circ (\lambda y.xy) = (\lambda y.xy + z) \circ (\lambda y.x^* z + y).$$

While the proofs of the first three statements were fully automatic, the commutativity condition required some initial simplification steps to eliminate the higher-order structure. The above fixpoint fusion statement was then fully automatic, too.

The proof that deflationarity and strong deflationarity are equivalent was thus automatic after splitting into two implications. This result is obviously interesting since it yields a simple algebraic condition for obtaining unique solutions for recursive linear regular equations. The question remains whether this result can be established in weaker, ideally first-order, regular algebras where fixpoint fusion is no longer applicable.

11   Case Study 4: Language Quantales

It has already been shown that Isabelle supports a seamless transition between abstract algebra and concrete models. Some important models of Kleene algebras—languages, binary relations, sets of traces—have been formalised using interpretation statements in Isabelle [12]. This makes the abstract algebraic layer available for reasoning in these models. This is very beneficial in practice: most of the derivation of Arden's rule in language theory, for instance, can be carried out in first-order regular algebra using this approach [11].

This section shows that also the quantale laws can easily be proved at the language level by automated reasoning. Our approach is based on Isabelle's standard implementation of languages in which words are represented as lists and word concatenation in terms of list append. As usual, the product of two languages X and Y can then be defined as $XY = \{xy : x \in X, y \in Y\}$, where xy stands for list append. The star of a language is defined, as usual, as $X^* = \Sigma_{i \geq 0} X^i$, where powers are given by a primitively recursive function.

Since ATP is rather fragile in this setting, we use rules for eliminating the higher-order structure such as

$$z \in XY \leftrightarrow \exists x, y . z = xy \wedge x \in X \wedge y \in y \quad \text{and} \quad x \in X^* \leftrightarrow \exists i . x \in X^i.$$

Both rules have been derived automatically. All dioid axioms could easily be proved by ATP from these definition. The Kleene algebra axioms have already been derived by ATP in [11]. They depend on the left and right continuity laws

$$X \cdot Y^* = \Sigma\{XY^i \mid i \in \mathbb{N}\} \quad \text{and} \quad Y^*X = \Sigma\{Y^iX \mid i \in \mathbb{N}\}$$

To derive the $*$-star continuity axiom $XY^*Z = \Sigma\{XY^iZ \mid i \in \mathbb{N}\}$ is equally simple. Even simpler is the derivation of the quantale laws

$$X \cdot (\Sigma A) = \Sigma\{XY \mid Y \in A\} \quad \text{and} \quad (\Sigma A) \cdot X = \Sigma\{YX \mid Y \in A\},$$

where A is a set of languages. We have verified the infinite distributivity laws $X \sqcap (\Sigma A) = \Sigma\{X \sqcap Y \mid Y \in A\}$ and its lattice dual $X \sqcup (\Pi A) = \Pi\{X \sqcup Y \mid Y \in A\}$ by similar means.

In principle, it should be possible to cast these theorems into formal interpretation statements in Isabelle. By formally establishing languages as unital quantales, it would then follow that they form also $*$-continuous Kleene algebras and Kleene algebras. We could not realise these interpretation statements, however, though all proof obligations were met. The reason seems to be limitations of Isabelle's sublocale mechanism which we discuss in the conclusion section.

12   Conclusion

We have formalised higher-order regular algebras, in particular $*$-continuous Kleene algebras and unital quantales, in Isabelle and derived some of their essential properties. We have also provided additional proof support in terms of Galois connections and fixpoints. Our main aim was to complement Isabelle's existing libraries for first-order regular algebras by higher-order variants, and to investigate the applicability of Isabelle's ATP and SMT integration in this setting, for which we used four case studies, and proved a total of 278 facts.

Our results on automated theorem proving are surprisingly positive and encouraging: Isabelle's Sledgehammer tool allowed us to automate even higher-order proof tasks involving infinite suprema, inductive data types, set comprehension and their combinations. That would be impossible with standalone ATP systems or SMT solvers. Without having considered the internals of Sledgehammer we can only speculate on the reasons for this success. The presence of various rules in Isabelle's libraries that eliminate higher-order structure and that are picked up automatically by Isabelle's relevance filter is certainly one of them. The development of additional theory transformations for ATP support in a higher-order setting seems an interesting research topic.

Our implementations also point at some apparent limitations of Isabelle for formalising mathematics. The first one concerns name spaces in theory hierarchies. Isabelle supports both programming-style and mathematical notation for operations, for instance *star* x versus x^*. The first notation supports qualified names such as *quantale.star*, the second one does not and is therefore prone to name clashes. For the sake of readability we have used mathematical notation in this paper though it often differs from that used in our Isabelle theory file.

Second, and more importantly, we were not always successful in using Isabelle's abstract type classes and locales in our theory hierarchy. First, we had to restrict Galois connections to one single order to avoid type unification problems when instantiating them in applications. Second, we could so far not successfully use interpretation statements in specifications that were not entirely given by classes, for instance in the case of semilattices where greatest upper bounds were *defined* in the context of orders. Third, we could not instantiate Galois connections and fixpoint fusion laws in situations where the order was defined on a subalgebra, for instance in modal Kleene algebras [9]. Isabelle's type system does not seem to support a smooth transition between algebras specified with and without explicit carrier sets. Fourth, dualities could so far not effectively be exploited, again, since locales do not seem to support duality based on properties that have been defined outside of them. Arguably, these problems would not exist in the presence of dependent types [14,2]; solving them is a main focus of our current work. Current work shows that an implementation of the entire approach with carrier sets solves most of these problems. Unfortunately, Isabelle's locale documentation is rather sparse when it comes to advanced theory engineering features. Therefore, a deeper evaluation of the potential and limitations of Isabelle for formalising mathematics seems difficult.

Our formalisation of higher-order regular algebras in Isabelle can be seen as part of a larger line of research. Beyond our own repository, for instance, Krauss and Nipkow have implemented a decision procedure for regular expressions in Isabelle [19]; Wu, Zhang and Urban have proved the Myhill-Nerode theorem in an algebraic setting [24]. Kozen's automata-based decision procedure for regular equations and libraries for Kleene algebras have been implemented by Braibant and Pous in Coq [7], however without access to ATP technology. In their approach, the advantages of dependent types are clearly apparent.

References

1. Aarts, C.J.: Galois connections presented calculationally. Master's thesis, Department of Mathematics and Computing Science, Eindhoven University of Technology (1992)
2. Asperti, A., Ricciotti, W., Sacerdoti Coen, C., Tassi, E.: Hints in Unification. In: Berghofer, S., Nipkow, T., Urban, C., Wenzel, M. (eds.) TPHOLs 2009. LNCS, vol. 5674, pp. 84–98. Springer, Heidelberg (2009)
3. Ballarin, C.: Tutorial to locales and locale interpretation. In: Contribuciones Científicas en honor de Mirian Andrés. Servicio de Publicaciones de la Universidad de La Rioja, Spain (2010)
4. Blanchette, J.C., Bulwahn, L., Nipkow, T.: Automatic Proof and Disproof in Isabelle/HOL. In: Tinelli, C., Sofronie-Stokkermans, V. (eds.) FroCoS 2011. LNCS, vol. 6989, pp. 12–27. Springer, Heidelberg (2011)

5. Boffa, M.: Une remarque sur les systèmes complets d'identités rationnelles. Informatique Théorique et Applications 24(4), 419–423 (1990)
6. Boffa, M.: Une condition impliquant toutes les identités rationnelles. Informatique Théorique et Applications 29(6), 515–518 (1995)
7. Braibant, T., Pous, D.: An Efficient Coq Tactic for Deciding Kleene Algebras. In: Kaufmann, M., Paulson, L. (eds.) ITP 2010. LNCS, vol. 6172, pp. 163–178. Springer, Heidelberg (2010)
8. Conway, J.H.: Regular Algebra and Finite Machines. Chapman and Hall (1971)
9. Desharnais, J., Struth, G.: Internal axioms for domain semirings. Science of Computer Programming 76(3), 181–203 (2011)
10. Doornbos, H., Backhouse, R.C., van der Woude, J.: A calculational approach to mathematical induction. Theor. Comput. Sci. 179(1-2), 103–135 (1997)
11. Foster, S., Struth, G.: Automated Analysis of Regular Algebra. In: Gramlich, B., Miller, D., Sattler, U. (eds.) IJCAR 2012. LNCS, vol. 7364, pp. 271–285. Springer, Heidelberg (2012)
12. Foster, S., Struth, G., Weber, T.: Automated Engineering of Relational and Algebraic Methods in Isabelle/HOL – (Invited Tutorial). In: de Swart, H. (ed.) RAMICS 2011. LNCS, vol. 6663, pp. 52–67. Springer, Heidelberg (2011)
13. Gammie, P.: The worker/wrapper transformation. Archive of Formal Proofs, 2009 (2009)
14. Garillot, F., Gonthier, G., Mahboubi, A., Rideau, L.: Packaging Mathematical Structures. In: Berghofer, S., Nipkow, T., Urban, C., Wenzel, M. (eds.) TPHOLs 2009. LNCS, vol. 5674, pp. 327–342. Springer, Heidelberg (2009)
15. Guttmann, W., Struth, G., Weber, T.: Automating Algebraic Methods in Isabelle. In: Qin, S., Qiu, Z. (eds.) ICFEM 2011. LNCS, vol. 6991, pp. 617–632. Springer, Heidelberg (2011)
16. Haftmann, F., Wenzel, M.: Local Theory Specifications in Isabelle/Isar. In: Berardi, S., Damiani, F., de'Liguoro, U. (eds.) TYPES 2008. LNCS, vol. 5497, pp. 153–168. Springer, Heidelberg (2009)
17. Kozen, D.: On Kleene Algebras and Closed Semirings. In: Rovan, B. (ed.) MFCS 1990. LNCS, vol. 452, pp. 26–47. Springer, Heidelberg (1990)
18. Kozen, D.: A completeness theorem for Kleene algebras and the algebra of regular events. Information and Computation 110(2), 366–390 (1994)
19. Krauss, A., Nipkow, T.: Proof pearl: Regular expression equivalence and relation algebra. J. Automated Reasoning 49(1), 95–106 (2012)
20. Paulson, L., Nipkow, T., Wenzel, M.: Isabelle (2011), http://www.cl.cam.ac.uk/research/hvg/Isabelle/index.html
21. Pratt, V.R.: Action Logic and Pure Induction. In: van Eijck, J. (ed.) JELIA 1990. LNCS, vol. 478, pp. 97–120. Springer, Heidelberg (1991)
22. Salomaa, A.: Two complete axiom systems for the algebra of regular events. J. ACM 13(1), 158–169 (1966)
23. Struth, G.: Left omega algebras and regular equations. J. Logic and Algebraic Programming (in press, 2012)
24. Wu, C., Zhang, X., Urban, C.: A Formalisation of the Myhill-Nerode Theorem Based on Regular Expressions (Proof Pearl). In: van Eekelen, M., Geuvers, H., Schmaltz, J., Wiedijk, F. (eds.) ITP 2011. LNCS, vol. 6898, pp. 341–356. Springer, Heidelberg (2011)

Towards Certifiable Implementation of Graph Transformation via Relation Categories

Wolfram Kahl*

McMaster University, Hamilton, Ontario, Canada
kahl@cas.mcmaster.ca

Abstract. The algebraic approach to graph transformation is a general framework for the definition of transformation mechanisms for complex structures that achieves its generality by using category-theoretic abstractions.

We present a framework for modular implementations of categoric graph transformation mechanisms that uses abstractions of relation categories as internal interfaces. Doing this in a dependently-typed programming language enables us to manage implementations of functionality together with their correctness proofs in the same language, thus progressing towards fully verified graph transformation system implementations.

1 Introduction

Graph transformation has most fruitfully been defined using the abstractions of category theory, giving rise to a whole literature on abstract high-level rewriting systems. Mechanised formalisations of category theory are most natural in systems of dependent types.

Actual implementation of categorically specified graph transformation naturally employs relational operations, and relational categories (such as variants of allegories and Kleene algebras) provide even more flexible support for graph transformation than the traditional approach restricting graph homomorphisms to functions.

The current paper is part of a project that endeavours to use the accepted mathematical definitions of relational categories, formulated and axiomatised precisely using dependent types, as programming interfaces in particular for the implementation of graph transformation systems.

1.1 The Double-Pushout (DPO) Approach to Graph Transformation

The central variant of what is known as the "algebraic approach to graph transformation" works most typically in categories of graph structures that are unary algebras (i.e., where all function symbols are unary) with conventional algebra

---

* This research is supported by the National Science and Engineering Research Council of Canada, NSERC.

W. Kahl and T.G. Griffin (Eds.): RAMiCS 2012, LNCS 7560, pp. 82–97, 2012.

homomorphisms. For "gluing" of different graph pieces in the course of transformation, the categoric concept of *pushout* is used; we show here a direct translation of the standard definition of "R and S are a pushout for P and Q" into the dependently-typed language of Agda [Nor07], which we will employ as mathematical notation throughout this paper; the basics of Agda are explained in Sect. 2.

```
record IsPushout {A B C D : Obj}
        (P : Mor A B) (Q : Mor A C)
        (R : Mor B D) (S : Mor C D) : Set (i ⊔ j ⊔ k) where
    field
        commutes : P ⨟ R ≈ Q ⨟ S
        universal : {Z : Obj} {R' : Mor B Z} {S' : Mor C Z}
                → P ⨟ R' ≈ Q ⨟ S'
                → ∃! _≈_ (λ U → R ⨟ U ≈ R' × S ⨟ U ≈ S')
```

$$
\begin{array}{ccc}
A & \xrightarrow{\;Q\;} & C \\
{\scriptstyle P}\downarrow & & \downarrow{\scriptstyle S} \\
B & \xrightarrow{\;R\;} & D \\
\end{array}
\qquad
\begin{array}{c}
S' \\
U \\
R' \\
Z
\end{array}
$$

Transformation rules are spans $\mathcal{L} \xleftarrow{\Phi_L} \mathcal{G} \xrightarrow{\Phi_R} \mathcal{R}$ consisting of left-hand side $\mathcal{L}$, right-hand side $\mathcal{R}$, gluing object $\mathcal{G}$, and two morphisms Φ_L and Φ_R that nowadays are frequently required to be monomorphism, see for example [CMR$^+$97]. Such a rule is applied to a start graph $\mathcal{A}$ via a "matching" homomorphism M by completing the following double-pushout diagram, and using $\mathcal{B}$ as result graph. More precisely, given a rule and a matching $M : \mathcal{L} \to \mathcal{A}$, first the "host graph" $\mathcal{H}$ together with the morphisms E and Ψ_L is constructed in such a way that the left square forms a pushout (such a *pushout complement* does not always exist even in standard graph categories). As a second step, the result graph $\mathcal{B}$ is obtained by calculating the pushout of Φ_R and E.

$$
\begin{array}{ccccc}
\mathcal{L} & \xleftarrow{\;\Phi_L\;} & \mathcal{G} & \xrightarrow{\;\Phi_R\;} & \mathcal{R} \\
{\scriptstyle M}\downarrow & & {\scriptstyle E}\downarrow & & \downarrow{\scriptstyle N} \\
\mathcal{A} & \xleftarrow{\;\Psi_L\;} & \mathcal{H} & \xrightarrow{\;\Psi_R\;} & \mathcal{B} \\
\end{array}
$$

For the study of properties of the resulting rewriting systems, including confluence and rule amalgamation properties, additional assumptions on the underlying category and on the class of monomorphisms allowed to be used as rule sides are useful; the currently most wide-spread framework for this is that of "adhesive HLR categories", see [EPPH06, EEPT06].

1.2   Categorical Implementation Structuring

The obvious requirements for an implementation of DPO graph transformation are the ability to represent, create, and exchange at least graphs and graph homomorphisms, and apply the constructions of pushout complements and pushouts. Calculating the composition of two homomorphisms is actually not required for just performing DPO rewriting steps, but will be useful for more complex operations. We can summarise this as:

Requirement 1 — Graph Category: *Represent graphs and graph homomorphisms as data, and implement pushouts and other categoric operations.*

The representation of graphs requires representations of their node end edge sets, and of the incidence relations between the two. Graph structures that can be considered as unary algebras can be implemented as diagrams over some base category of sets; the RATH-Agda libraries (see Sect. 3) include an implementation (with correctness proof) that derives pushouts in the diagram category from a choice of pushouts in the base category.

Requirement 2 — Base Category: *Represent sets and total functions as data, and implement pushouts and other categoric operations.*

Pushouts of mappings have local relational characterisations [Kaw90, Kah11a], which, however, do not provide any direct recipe for implementation. One well-known way to obtain pushouts from simpler constructions is calculating them from coproducts and coequalisers, where those exist in the category under consideration. Coequalisers of mappings F and G between sets can be calculated as quotient projections for the equivalence closure of a relation induced by F and G; this is a relational construction that cannot be usefully expressed in the categoric context normally used for graph transformation.

Since we want to avoid ad-hoc implementations of such constructions, we need to be able to switch from our base category to a relation category of sufficient expressiveness to allow formalisation and calculation of equivalence closures, and construction of quotients and subobjects:

Requirement 3 — Relation Category: *Represent sets and relations as data, and implement equivalence closure and other relational operations.*

Since relations are typically represented using different data structures than mappings, it will not be natural to use the actual mapping category of the chosen relation category as implementation for all aspects of categoric graph transformations. If a simpler or more compact implementation of mappings is chosen for the base category, this will then be equivalent (in the categoric sense) to the category of mappings in the relation category, and we need:

Requirement 4 — Interoperability: *Constructively prove equivalence of the base category with the category of mappings in the relation category.*

A typical setup for simple DPO graph transformation structured along the lines of the above requirements will satisfy Requirement 1 by deriving the category of graphs as a diagram category over the base category, and may include some implementations of complex relation operations that take some arguments directly as base category morphisms, or produce results directly in the base category without explicitly invoking the equivalence of Requirement 4 as part of the implementation.

1.3 Overview

We start with an introduction to essential features of the dependently-typed programming language (and proof checker) Agda2 (in the following just referred to as Agda) and its current standard library, and an overview of our RATH-Agda formalisation of categories of functions and relations in Sect. 3. In Sect. 4 we outline central features of an implementation of a verified relation category via parameterised sorted unique lists that are used simultaneously at two levels of the implementation. We proceed in Sect. 5 to define an implementation of reflexive-transitive closure (as essential ingredient of equivalence closure) in an way that is independent of the implementation of relations, purely in terms of our axiomatisation of the relevant relation categories.

The Agda source code from this project, including the modules discussed in this paper, is made available on-line at `http://relmics.mcmaster.ca/RATH-Agda/`.

2   Agda Notation

The Agda home page[1] states:

> **Agda is a dependently typed functional programming language.** It has inductive families, i.e., data types which depend on values, such as the type of vectors of a given length. It also has parametrised modules, mixfix operators, Unicode characters, and an *interactive* Emacs interface which can assist the programmer in writing the program.
> **Agda is a proof assistant.** It is an interactive system for writing and checking proofs. Agda is based on intuitionistic type theory, a foundational system for constructive mathematics developed by the Swedish logician Per Martin-Löf. It has many similarities with other proof assistants based on dependent types, such as Coq, Epigram, Matita and NuPRL.

Many Agda features will be explained when they are first used; here we only summarise a few *essential* aspects to make our use of Agda as the mathematical notation in the remainder of this paper more widely accessible.

Syntactically and "culturally", Agda frequently seems quite close to Haskell. However, the syntax of Agda is much more flexible: Almost any sequence of non-space characters is a legal lexeme, permitting the habit of choosing variable names for proof values that reflect their type, e.g., k≈e : k ≈K key e below in Sect. 4. Infix operators, and indeed mixfix operators of arbitrary arity, have names that contain underscore characters "_" in the positions of the first explicit arguments; below we use a binary infix operator _≈_ for morphism equality in categories, and a quaternary mixfix operator _:≈_:<_<:_ in the definition of SUList in Sect. 4 that actually takes also two implicit arguments, flagged by braces "{ ... }" surrounding their types, interspersed between its four explicit arguments. Such implicit argument positions are usually declared for arguments that supply the types to other arguments and can therefore be inferred from the latter; implicit arguments can be explicitly supplied using braces, as for example in ⊥ {A} {B} in the last line of the definition of USLCCZ' in Sect. 3.

Also, since Agda is strongly normalising and has no undefined values, the underlying semantics is quite different from that of Haskell. In particular, since

---

Agda is dependently typed, it does not have Haskell's distinction between terms, types, and kinds (the "types of the types"). The Agda constant Set_0 corresponds to the Haskell kind *; it is the type of all "normal" datatypes and is at the bottom of the hierarchy of type-theoretic universes in Agda, where universes Set ℓ are distinguished by universe indices ℓ : Level, where Level is an opaque special-purpose variant of the natural numbers; we write its maximum operator as _⊔_. This *universe polymorphism* is essential for being able to talk about both "small" and "large" categories or relation algebras, or, for another example, also for being able to treat diagrams of graphs and graph homomorphisms as graphs again.

3   RATH-Agda Categories

In [Kah11b], we presented a relatively fine-grained modularisation of sub-theories of division allegories, following our work on using semigroupoids to provide the theory of finite relations between infinite types, as they frequently occur as data structures in programming [Kah08], and on collagories [Kah11a] ("distributive allegories without zero morphisms" that are sufficient to contain an adhesive category of mappings) as foundation for relation-algebraic graph transformation. In this section, we present two monolithic definitions that provide appropriate foundations for most of the discussion in this paper. Each of these two definitions bundles a large number of theories of the RATH-Agda libraries summarised in [Kah11b].

We show here a monolithic definition of categories, which can be used as an alternative to the one within the fine-grained theory hierarchy of [Kah11b]. We follow the category-theoretic habit of not considering equality of objects, so Obj is a Set. Morphisms do have equality _≈_; for any two objects A and B we have Hom A B : Setoid j k, with the standard library providing an implementation of the standard type-theoretic concept of *setoid* as a carrier set together with an equivalence relation that is considered as equality on the carrier, much like the equality test == provided by the class Eq in Haskell.[2]

```
record Category' {i j k : Level} {Obj : Set i}
               (Hom : Obj → Obj → Setoid j k) : Set (i ⊔ j ⊔ k) where
    Mor : Obj → Obj → Set j
    Mor = λ A B → Setoid.Carrier (Hom A B)
    infix 4 _≈_ ; infixr 9 _⨾_
```

---

[2] We use variable names i, j, k, k_1, ℓA, ℓa, etc. for universe levels.

   We can write f ≈ g for two morphisms from carrier set Mor A B of the hom-setoid Hom A B since the object arguments A and B of _≈_ are declared implicit and can be derived from the type of f and g.

   The type of congruence of composition, ⨾-cong, is declared using a *telescope* introducing the seven named arguments A, B, C, f_1, f_2, g_1, and g_2 (here all implicit), which can be referred to in later parts of the type. The resulting *dependent function type* corresponds to "dependent products" frequently written using Π in other presentations of type theory.

$\_ \approx\_ = \lambda \{A\} \{B\} \to$ Setoid. $\_ \approx\_$ (Hom A B)
field $\_ \, _\S\_$: $\{A\ B\ C : Obj\} \to$ Mor A B $\to$ Mor B C $\to$ Mor A C
    $_\S$-cong   : $\{A\ B\ C : Obj\} \{f_1\ f_2 : Mor\ A\ B\} \{g_1\ g_2 : Mor\ B\ C\}$
          $\to f_1 \approx f_2 \to g_1 \approx g_2 \to (f_1 \,_\S\, g_1) \approx (f_2 \,_\S\, g_2)$
    $_\S$-assoc   : $\{A\ B\ C\ D : Obj\} \{f : Mor\ A\ B\} \{g : Mor\ B\ C\} \{h : Mor\ C\ D\}$
          $\to ((f \,_\S\, g) \,_\S\, h) \approx (f \,_\S\, (g \,_\S\, h))$
    Id   : $\{A : Obj\} \to$ Mor A A
    leftId   : $\{A\ B : Obj\} \to \{f : Mor\ A\ B\} \to (Id \,_\S\, f) \approx f$
    rightId   : $\{A\ B : Obj\} \to \{f : Mor\ A\ B\} \to (f \,_\S\, Id) \approx f$

As context for later sections, we now show a monolithic definition of upper semilattice categories (USLC) with converse (-C) and zero morphisms (-Z). In this unified setting, several of the axioms necessary in the constituent theories in the fine-grained theory hierarchy of [Kah11b] become derivable and therefore have not been included as "axioms" (record **field**s) here.

We approach allegories and Kleene categories via common primitives providing a local ordering on homsets. In locally ordered categories, "homsets" are partial orders, and a USLCCZ first of all contains a category that uses for its "homsets" the underlying setoids. The local poset ordering relations are again collected into a global parameterised relation, $\sqsubseteq$. We also add converse $\_ ^{\smile}$, binary join $\_ \sqcup \_$ in homsets, and least morphisms $\bot$ satisfying the zero laws (the right zero law follows via the converse laws from the left, and the equational versions follow via $\bot$-$\sqsubseteq$). As a result of $_\S$-$\sqcup$-subdistribL in the presence of converse, composition distributes over binary joins from both sides.

record USLCCZ′ $\{i\ j\ k_1\ k_2 : Level\} \{Obj : Set\ i\}$
               (Hom : Obj $\to$ Obj $\to$ Poset $j\ k_1\ k_2$) : Set ($i \sqcup j \sqcup k_1 \sqcup k_2$) **where**
field category : Category′ (λ A B $\to$ posetSetoid (Hom A B))
open Category′ category
infix 4 $\_ \sqsubseteq \_$; **infix** 10 $\_ ^{\smile}$; **infixr** 5 $\_ \sqcup \_$
$\_ \sqsubseteq \_ = \lambda \{A\} \{B\} \to$ Poset.$\_ \le \_$ (Hom A B)
field
  $_\S$-monotone : $\{A\ B\ C : Obj\} \{f\ f' : Mor\ A\ B\} \{g\ g' : Mor\ B\ C\}$
            $\to f \sqsubseteq f' \to g \sqsubseteq g' \to (f \,_\S\, g) \sqsubseteq (f' \,_\S\, g')$
  $\_ ^{\smile}$: $\{A\ B : Obj\}$ $\to$ Mor A B $\to$ Mor B A
  $^{\smile\smile}$: $\{A\ B : Obj\}$ $\{R : Mor\ A\ B\}$ $\to (R ^{\smile}) ^{\smile} \approx R$
  $^{\smile}$-involution : $\{A\ B\ C : Obj\} \{R : Mor\ A\ B\} \{S : Mor\ B\ C\} \to (R \,_\S\, S) ^{\smile} \approx S ^{\smile} \,_\S\, R ^{\smile}$
  $^{\smile}$-monotone : $\{A\ B : Obj\} \{R\ S : Mor\ A\ B\} \to R \sqsubseteq S \to (R ^{\smile}) \sqsubseteq (S ^{\smile})$
  $\_ \sqcup \_$: $\{A\ B : Obj\} \to$ Mor A B $\to$ Mor A B $\to$ Mor A B
  $\sqcup$-upper$_1$: $\{A\ B : Obj\} \{R\ S : Mor\ A\ B\} \to R \sqsubseteq (R \sqcup S)$
  $\sqcup$-upper$_2$: $\{A\ B : Obj\} \{R\ S : Mor\ A\ B\} \to S \sqsubseteq (R \sqcup S)$
  $\sqcup$-universal : $\{A\ B : Obj\} \{R\ S\ X : Mor\ A\ B\} \to R \sqsubseteq X \to S \sqsubseteq X \to (R \sqcup S) \sqsubseteq X$
  $_\S$-$\sqcup$-subdistribL : $\{A\ B\ C : Obj\} \{R_1\ R_2 : Mor\ A\ B\} \{S : Mor\ B\ C\}$
            $\to ((R_1 \sqcup R_2) \,_\S\, S) \sqsubseteq ((R_1 \,_\S\, S) \sqcup (R_2 \,_\S\, S))$
  $\bot$: $\{A\ B : Obj\} \to$ Mor A B
  $\bot$-$\sqsubseteq$: $\{A\ B : Obj\} \{R : Mor\ A\ B\} \to \bot \sqsubseteq R$
  leftZero$\sqsubseteq$: $\{A\ B\ C : Obj\} \{R : Mor\ B\ C\} \to (\bot \{A\} \{B\} \,_\S\, R) \sqsubseteq \bot$

Already in ordered categories with converse (OCCs), we have the standard relation-algebraic way of defining properties like univalence $((R ^{\smile} \, ; R) \sqsubseteq Id)$, totality $(Id \sqsubseteq (R \, ; R ^{\smile}))$, injectivity, etc., and deriving laws for them. Total and univalent morphisms (in *Rel*, these are the total functions) are called *mappings*; for morphisms that are known to be mappings we define the dependent sum type Mapping containing the morphism and a proof of its mapping properties:

isMapping : $\{A\ B\ :\ Obj\} \to$ Mor A B $\to$ Set k_2
isMapping R = isUnivalent R $\times$ isTotal R
record Mapping (A B : Obj) : Set $(i \cup j \cup k_2)$ **where field** mor : Mor A B
                                                                                                prf : isMapping mor

The mappings of a OCC J form a category MapCat J where the morphisms from A to B are the Mappings of J, that is, Mor (MapCat J) A B = Mapping J A B.

4   Implementing a Base Category of Concrete Relations

An only apparently natural choice of base category would use Set as the type of objects, and for two sets A and B, the standard Agda function type A $\to$ B for morphisms. This can be embedded into the category of concrete relations described in [Kah11b, Sect. 3]. Because of Agda's lack of extensionality, that would require, for morphisms in A $\to$ B, an equivalence formalised as pointwise propositional equality of functions, f ~ g = $\forall$ x $\to$ f x $\equiv$ g x.

However, using propositional equality is frequently very cumbersome, so it would be natural to switch from Set to Setoid, where equivalence of Setoid-respecting functions is defined via the equivalences of the two Setoids involved, and defines a Setoid of functions. However, like Set, also Setoid is more a reasoning type than an implementation type; both accommodate infinite carriers, which may be useful for the theory, but are not desired as node or edge sets of graphs that are to be manipulated by a machine. Even though functions between Sets (and Setoids) are executable, the corresponding relations are of pure reasoning types. Such relation types are sufficient where relation operations are needed only as part of correctness proofs of function implementations, as in AoPA [MKJ09]. Here, however, we require relation operations as part of our implementation, and therefore need a "programming representation" of relations.

The Agda standard library currently[3] contains AVL trees as a useful container datatype; however, it only includes proofs that the operations preserve the invariants, but no proofs that they satisfy their container semantics.

We base our first relation implementation library on a datatype "SUList" of sorted lists with unique keys, and parametrise the SUList implementation over the following items:

− a strict total order Key of key values; the carrier set of this will be referred to as K, and the equivalence and strict-ordering relations as _ $\approx$ K _ and _ <K _,

---

[3] As of version 0.6; http://wiki.portal.chalmers.se/agda/
pmwiki.php?n=Libraries.StandardLibrary

– an arbitrary Elem datatype,
– a key extraction function key : Elem → K.

In Agda, we introduce this parametrisation at the module level:

```
module Data.SUList.Core
  {ℓK ℓk₁ ℓk₂ : Level} (Key : StrictTotalOrder ℓK ℓk₁ ℓk₂)
  {ℓE : Level} (Elem : Set ℓE)
  (key : Elem → StrictTotalOrder.Carrier Key) where
```

We will show different instantiations of these parameters below.

The type SUList is indexed by a value of type Maybe K, which is nothing for empty lists and just k for non-empty lists starting with an element e satisfying k ≈K key e. Having the start key as part of the type of these lists makes it straightforward to accommodate the invariant of unique sortedness in the constructor for non-empty lists as an additional argument documenting that the start key of the tail list, unless nothing, is greater than the head key:[4]

```
data SUList : Maybe K → Set (ℓE ⊔ ℓK ⊔ ℓk) where
  []                 :                                      SUList nothing
  _:≈_:<_<:_ : (e : Elem) → {k : K} → k ≈K key e
      → {m : Maybe K} → (k<es : k <M m) → (es : SUList m) → SUList (just k)
```

(For technical reasons, using just (key e) directly as the type index for non-empty lists complicates proofs, so we use a separate value k that is is only equivalent to key e.)

The following elementship relation is at the core of the collection semantics of SUList-based datatypes, and is therefore the only definition, besides SUList itself, for which a certifier would need to inspect also the body.

For a key k_0 : K and a list es_0 : SUList m we have $k_0 \in es_0$ iff
– either k_0 is equal to the start key of es_0 (and therefore equal to the key of the first element),
– or k_0 is an element of the tail of the list.

The proof constructors ∈head and ∈tail for these two cases contain sufficient information for reconstructing the list es_0 from a proof of $k_0 \in es_0$, which is useful for being able to make such lists implicit arguments in later functions.

```
data _∈_ (k₀ : K) : {m : Maybe K} → SUList m → Set (ℓE ⊔ ℓK ⊔ ℓk₁ ⊔ ℓk₂) where
  ∈head : (e : Elem) → {k : K} → (k≈e : k ≈K key e) → k₀ ≈K k
      → {n : Maybe K} (k<n : k <M n) (es : SUList n) → k₀ ∈ e :≈ k≈e :< k<n <: es
  ∈tail  : (e : Elem) → {k : K} → (k≈e : k ≈K key e)
      → {n : Maybe K} (k<n : k <M n) {es : SUList n}
      → k₀ ∈ es                                  → k₀ ∈ e :≈ k≈e :< k<n <: es
```

----

[4] The standard library includes the following definition for the universe-polymorphic parameterised Maybe type:

```
data Maybe {a : Level} (A : Set a) : Set a where just : (x : A) → Maybe A
                                                nothing : Maybe A
```

As one utility function, we define a generalised intersection function that will be used in different instantiations to implement set intersection, relation intersection, and relation composition. This is therefore set in the context of three element types all with the same key type, and a function that maps two elements from the first two types with equivalent keys[5] to either nothing or an element of the third type with again an equivalent key:

module IntersectionWith (f : (e_1 : $Elem_1$) (e_2 : $Elem_2$) → key_1 e_1 ≈K key_2 e_2
$\qquad$ → Maybe (Σ [e_3 : $Elem_3$] key_1 e_1 ≈K key_3 e_3)) **where**

The start key of the intersection result depends not just on the start keys (and therewith the types) of the two argument lists, but on their inner structure (and on f), so the return type is a dependent pair containing the start key as first argument.

intersection : {m_1 : Maybe K} → $SUList_1$ m_1 → {m_2 : Maybe K} → $SUList_2$ m_2
$\qquad$ → Σ [m : Maybe K] $SUList_3$ m

For (m_3, es_3) = intersection es_1 es_2, we proved that k ∈ es_3 is logically equivalent to k ∈ es_1 ∧ k ∈ es_2 ∧ f e_1 e_2 ke_1≈ke_2 ≡ just (e_3, ke_1≈ke_3) where e_i = lookup k es_i and ke_1≈ke_j is a proof for key e_1 ≈key e_j. Although the generality of f makes the interface to the proof of this quite complicated (so that we omit the details here), it only contributes relatively little to the considerable complexity of the implementation of intersection, so that the possibility of using this same implementation as basis for both intersection and relation composition justifies the generalisation.

For an implementation of non-empty sets of elements of a strict total order Key, we instantiate SUList with Elem = K and key = id. A non-empty set is then a dependent pair consisting of a start key and a list starting with that (or an equivalent) key.

module Data.SUList.ListSet1 {ℓK ℓk_1 ℓk_2 : Level} (Key : StrictTotalOrder ℓK ℓk_1 ℓk_2)
$\quad$ **where module** Core = Data.SUList.Core Key K id
$\qquad$ $ListSet_1$: Set ($\ell K \cup \ell k$)
$\qquad$ $ListSet_1$ = Σ [k : K] SUList (just k)

Intersection for $ListSet_1$ uses a right-biased instance of the IntersectionWith parameter: If two equal elements are encountered, the second is returned:

fS : (k_1 k_2 : K) → k_1 ≈K k_2 → Maybe (Σ [k_3 : K] k_1 ≈K k_3)
fS k_1 k_2 k_1≈k_2 = just (k_2, k_1≈k_2)

The resulting intersection function may still return nothing, since the intersection of two non-empty lists can be empty:

intersection : $ListSet_1$ → $ListSet_1$ → Maybe $ListSet_1$

(A datatype ListSet that also comprises empty sets is implemented similarly.)

---

[5] The proof argument of type key_1 e_1 ≈K key_2 e_2 enforces this, while in weaker languages this might be encoded as a comment along the lines of "f e_1 e_2 will be called only if e_1 and e_2 have equal keys".

Our first relation implementation implements relations between strict total orders A and B by lists with keys from A and elements that contain a key from A and a non-empty set of B elements:

```
open ListSet1 B using () renaming (ListSet₁ to ℙB₁, ...)
Elem₀ = A₀ × ℙB₁
module Map = Data.SUList.Core A Elem₀ proj₁
```

While the inner lists need to be non-empty, the outer list may be empty, and a ListSetMap therefore contains an arbitrary type index for SUList:

```
ListSetMap : Set (ℓA ⊔ ℓa ⊔ ℓB ⊔ ℓb)
ListSetMap = Σ [ m : Maybe A₀ ] SUList m
```

The semantics is again encapsulated in the elementship relation: A proof of $(a, b) \in (\_, R)$ is a proof of $a \in R$ in the SUList elementship, together with a proof of $b \in$ lookup a R in the ListSet₁ membership:

```
_∈_ : A₀ × B₀ → ListSetMap → Set (ℓA ⊔ ℓa ⊔ ℓB ⊔ ℓb₁ ⊔ ℓb₂)
(a, b) ∈ (_, R) = Σ [ a∈R : Map._∈_ a R ] SetB._∈_ b (proj₂ (Map.∈-Elem′ a∈R))
```

Intersection for ListSetMap uses custom instance of the IntersectionWith parameter that calculates the intersection of the ListSet₁ associated with equal keys:

```
fR : (e₁ e₂ : Elem₀) → proj₁ e₁ ≈A proj₁ e₂ → Maybe (Σ [ e₃ : Elem₀ ] proj₁ e₁ ≈A proj₁ e₃)
fR (a₁, bs₁) (a₂, bs₂) a₁≈a₂ with SetB.intersection bs₁ bs₂
... | nothing = nothing
... | just bs  = just ((a₂, bs), a₁≈a₂)
```

As mentioned above, IntersectionWith is also used for the implementation of relation composition, more precisely for the implementation of domain restriction, which is in turn used to define composition. Domain restriction takes a list as of A elements and a relation R from A to B, and filters from R only the part that starts from elements in as:

```
dres₁ : ℙA₁ → ListSetMap → ListSetMap
dres₁ (_, as) (_, R) = ImgIntersection.intersection as R
```

The parameter function for IntersectionWith again just passes on the second argument:

```
fl : (a₁ : A₀) (a₂bs : Elem₀) → a₁ ≈A proj₁ a₂bs → Maybe (Σ [ p : Elem₀ ] a₁ ≈A proj₁ p)
fl a₁ a₂bs a₁≈a₂ = just (a₂bs, a₁≈a₂)
private module ImgM = SepElemTypes.M3 A A₀ idF Elem₀ proj₁ Elem₀ proj₁
module ImgIntersection = ImgM.IntersectionWith fl
```

Via programming at the level of SULists and proofs largely over the _∈_ semantics, we obtain a USLCCZ with StrictTotalOrder objects and ListSetMap morphisms. From this, another USLCCZ FinLSM is obtained via retraction to natural numbers as objects via the function Fin.strictTotalOrder which associates

with each natural number n : ℕ the standard strict total order of the set
Fin ņ = {0, ..., (n − 1)}. We proved that in FinLSM, the object suc n can be
represented as direct sum of n and the unit object 1; this enables us to obtain
a proven-correct implementation of Kleene star via the use of UnitSumStarOp of
Sect. 5.4 without further programming at the level of SULists.

5   Deriving a Default Implementation of Kleene Star

Equivalence closure of a relation R can be obtained from the reflexive-transitive
closure by applying it to the symmetric closure, (R ⊔ R ˘). Therefore, Kleene
categories with converse are the minimal setting required for calculating equivalence closure, and the main challenge in implementing these lies in creating a
verified implementation of reflexive-transitive closure.

In this section, we first present an overview of our formalisation of a categoric
version of Kleene algebras. In Sect. 5.3, we then make the standard construction
of Kleene star for 2×2-matrices more generally re-usable by reformulating it in
terms of a direct sum. Direct sums are the relation-algebraic version of the coproducts of category theory, and model disjoint union; we quickly present our direct sum formalisation in Sect. 5.2. Similarly, a unit object is a relation-algebraic
concept that characterises the terminal objects of the mapping category; since
defining a star operator on unit objects is trivial (Sect. 5.4), we obtain a default
implementation of Kleene star for FinLSM by natural induction in Sect. 5.5.

5.1   Kleene Categories

For the one-object case, our axiomatisation of Kleene categories is equivalent
to Kozen's Kleene algebra axiomatisation [Koz94]. We choose a variant with a
single equation as recursive definition axiom, and we start with axiomatising not
the star operator, but the relationship between a single relation R and the result
R^* of applying what will be the star operator to it (all in the context of an upper
semilattice category, that is, a locally ordered category where all homsets have
binary joins, and composition distributes over binary joins from both sides).

```
record IsStar {A : Obj} (R R* : Mor A A) : Set (i ⊔ j ⊔ k₁ ⊔ k₂) where
   field *-recDef  : R* ≈ Id ⊔ R ⊔ R* ⨾ R*
         *-leftInd  : {B : Obj} {S : Mor A B} → R ⨾ S ⊑ S → R* ⨾ S ⊑ S
         *-rightInd : {B : Obj} {Q : Mor B A} → Q ⨾ R ⊑ Q → Q ⨾ R* ⊑ Q
```

Among the derived properties, we show also Kozen's recursive definition axioms
and the alternative shapes of the induction axioms. The proofs are presented
in the calculational style, using the mixfix operators ≈-begin_ , ⊑-begin_ , _≈⟨ _ ⟩_
and _□ which are variants of the calculational reasoning operators provided by
the standard library, equipped with two additional implicit object parameters
(similar to _≈_ etc.) to enable calculational reasoning in homsets without having
to specify the homset explicitly.

```
*-recDef₁⊑ : Id ⊔ R ⨾ R* ⊑ R*
*-recDef₁⊑ = ⊑-begin
  Id ⊔ R ⨾ R*        ⊑⟨ ⊔-monotone₂ (⨾-monotone₁ *-increases ⟨⊑⊑⟩ ⊔-upper₂) ⟩
  Id ⊔ R ⊔ R* ⨾ R*  ≈⟨ ≈-sym *-recDef ⟩
  R*                □

*-recDef₁ : R* ≈ Id ⊔ R ⨾ R*
*-recDef₁ = ⊑-antisym
  (⊑-begin R*        ≈⟨ ≈-sym rightId ⟩
            R* ⨾ Id ⊑⟨ *-leftInd' (⊑-begin
                        Id ⊔ R ⨾ (Id ⊔ R ⨾ R*)
                        ⊑⟨ ⊔-monotone₂ (⨾-monotone₂ *-recDef₁⊑) ⟩
                        Id ⊔ R ⨾ R*                              □)⟩
          Id ⊔ R ⨾ R*                                          □)
  *-recDef₁⊑

*-leftInd' : {B : Obj} {P S : Mor A B} → P ⊔ R ⨾ S ⊑ S → R* ⨾ P ⊑ S
*-leftInd' {_} {P} {S} P⊔R⨾S⊑S = ⊑-begin
  R* ⨾ P ⊑⟨ ⨾-monotone₁ (⊔-upper₁ ⟨⊑⊑⟩ P⊔R⨾S⊑S) ⟩ R* ⨾ S
          ⊑⟨ *-leftInd (⊔-upper₂ ⟨⊑⊑⟩ P⊔R⨾S⊑S) ⟩    S                □
```

To prove the reverse implications, we define alternative IsStar constructors; including one that derives *-recDef from *-recDef₁.

A local star operator comes with the correctness proof of the operator itself:

```
record LocalStarOp (A : Obj) : Set (i ⊔ j ⊔ k₁ ⊔ k₂) where
  field _* : Mor A A → Mor A A
        isStar : (R : Mor A A) → IsStar {A} R (R *)
```

A Kleene category, equivalent to a typed Kleene algebra of [Koz98], is then an upper semilattice category with zero morphisms that has a local star operator on every object.

5.2 Direct Sums

In the context of an upper semilattice category with zero morphisms and converse, a direct sum for objects A and B is a co-span $A \xrightarrow{\iota} S \xleftarrow{\kappa} B$ for which the following properties hold:

```
record IsDirectSum {A B S : Obj} (ι : Mor A S) (κ : Mor B S) : Set (i ⊔ j ⊔ k₁) where
  field commutes   : ι ⨾ κ ˘ ≈ ⊥ {A} {B}
        jointId    : ι ˘ ⨾ ι ⊔ κ ˘ ⨾ κ ≈ Id {S}
        leftKernel : ι ⨾ ι ˘ ≈ Id {A}
        rightKernel : κ ⨾ κ ˘ ≈ Id {B}
```

A choice of direct sums makes the injection names polymorphic:

```
record HasDirectSum-L : Set (i ⊔ j ⊔ k₁ ⊔ k₂) where
  field _⊞_ : Obj → Obj → Obj
        ι : {A B : Obj} → Mor A (A ⊞ B)
        κ : {A B : Obj} → Mor B (A ⊞ B)
        dirSum : (A B : Obj) → IsDirectSum {A} {B} {A ⊞ B} ι κ
```

Direct sums are also categorical coproducts and products; we introduce the following notations for the associated universal morphisms:

$$\underline{}_{\mathfrak{D}}\underline{} \; : \; \{D : Obj\} \, (F : Mor\ A\ D) \, (G : Mor\ B\ D) \to Mor\ (A \boxplus B)\ D$$
$$F \mathfrak{D} G = \iota^{\smile} \, \mathring{}\, F \sqcup \kappa^{\smile} \, \mathring{}\, G$$

$$\underline{}_{\mathfrak{C}}\underline{} \; : \; \{Z : Obj\} \, (F : Mor\ Z\ A) \, (G : Mor\ Z\ B) \to Mor\ Z\ (A \boxplus B)$$
$$F \mathfrak{C} G = F \, \mathring{}\, \iota \sqcup G \, \mathring{}\, \kappa$$

Two useful laws relating these are the following (stated without full declarations):

$$\mathfrak{C}\text{-}\mathring{}\text{-}\mathfrak{D} \quad : \ldots \to (F_1 \mathfrak{C} F_2) \, \mathring{}\, (G_1 \mathfrak{D} G_2) \approx (F_1 \, \mathring{}\, G_1) \sqcup (F_2 \, \mathring{}\, G_2)$$

$$\mathfrak{C}\text{-}\mathfrak{D} \quad : \ldots \to (F_1 \mathfrak{C} F_2) \mathfrak{D} (G_1 \mathfrak{C} G_2) \approx (F_1 \mathfrak{D} G_1) \mathfrak{C} (F_2 \mathfrak{D} G_2)$$

5.3   Deriving the Star Operator for Direct Sums

Given a direct sum for A and B in a upper semilattice category with zero morphisms and converse, and given local star operators on A and B, we can derive a local star operator on the sum object A⊞B using what is usually considered as a matrix construction, see e.g. [Koz94]. (We do not assume a choice of sums here; "A⊞B" contains no spaces and is therefore a single variable name in Agda.) For considering the matrix $E = \begin{pmatrix} a & b \\ c & d \end{pmatrix}$ as a morphism on the direct sum A⊞B, we use the following setup:

module Square (a : Mor A A) (b : Mor A B) (c : Mor B A) (d : Mor B B) **where**
    E : Mor A⊞B A⊞B
    E = (a ᴅ c) ⋲ (b ᴅ d)

Then we define:

f = a ⊔ b ⸴ d* ⸴ c   h = d* ⸴ c ⸴ f*   E* : Mor A⊞B A⊞B
g = f* ⸴ b ⸴ d*   k = d* ⊔ d* ⸴ c ⸴ g   E* = (f* ᴅ h) ⋲ (g ᴅ k)

where d* = d *B and f* = f *A are defined using the local star operators.

Our proof of IsStar E E* corresponds directly to [Koz94, Lemma 3.2] about the matrix construction; however, by setting the construction in the context of an arbitrary direct sum, we achieve maximal re-usability. To give a flavour of the readability and writability of calculational proofs in Agda, we show, in Appendix A, the proof of *-recDef$_1$, which essentially uses ᴅ-monotone and ⋲-monotone to perform the same splitting as in [Koz94, Lemma 3.2] of the whole proof into the four subproofs for the matrix coefficients. A different formalisation of this proof, on the level of matrices like Kozen's, has been performed by Braibant and Pous [BP12], and is used there for the induction step of obtaining a definition of Kleene star on arbitrary square matrices.

In the case of FinLSM, we are not concerned with general matrices, and can employ our additional knowledge about the object 1 to obtain a simplified version of the sum star calculation.

5.4 Unit Objects Simplify the Direct Sum Star Operator

According to [FS90, 2.15], an object U in an allegory is a *partial unit* if $\mathsf{Id}\ \{U\}$ is a top morphism. The object $\mathcal{U}$ is a *unit* if, further, every object is the source of a total morphism targeted at U.

record IsUnit $(U : \mathsf{Obj}) : \mathsf{Set}\ (i \sqcup j \sqcup k_2)$ **where**
    field Id-isTop   : isTop $(\mathsf{Id}\ \{U\})$
            toUnit   : $\{A : \mathsf{Obj}\} \to \mathsf{Mor}\ A\ U$
            toUnit-isTotal : $\{A : \mathsf{Obj}\} \to$ isTotal $(\mathsf{toUnit}\ \{A\})$

On a unit object U, using $R^* \approx \mathsf{Id}$ for every $R : U \to U$ we can define:

UnitStarOp : $\{U : \mathsf{Obj}\} \to$ IsUnit $U \to$ LocalStarOp U

If, in the context of Sect. 5.3, A is a unit object, we have $f^* \approx \mathsf{Id}$, which allows us to simplify E^* to $E^{*\prime}$, with:

$$h' = d^* \,\fatsemi\, c \qquad\qquad E^{*\prime} : \mathsf{Mor}\ A \boxplus B\ A \boxplus B$$
$$g' = b \,\fatsemi\, d^* \qquad k' = d^* \sqcup h' \,\fatsemi\, g' \qquad E^{*\prime} = (\mathsf{Id} \ni h') \,\epsilon\, (g' \ni k')$$

As result, for a direct sum $A \boxplus B$ of A and B where A is a unit, we only need a local star operator on B to derive one on $A \boxplus B$:

UnitSumStarOp : IsUnit $A \to$ LocalStarOp $B \to$ LocalStarOp $A \boxplus B$

5.5 Simple Recursive Star Implementation on **Fin** n

Producing a local star operator on initial objects is even easier than on units; from that together with the direct sum representation of suc n mentioned at the end of Sect. 4 and with the specialised star operation from above we obtain local star operators on each object of FinLSM by primitive recursion over $\mathbb{N}$:

FinLSMLocalStarOp : $(n : \mathbb{N}) \to$ LocalStarOp FinLSMUSLCategory n
FinLSMLocalStarOp zero = InitialStarOp FinLSMUSLCategory FinLSMInitial
FinLSMLocalStarOp (suc n) = UnitSumStarOp FinLSMUnit (FinLSMLocalStarOp n)

6 Conclusion

We use the relation category FinLSM to satisfy Requirement 3 of the introduction, and for Requirement 4 we have proven isomorphism between its mapping category and the category we use for Requirement 2, with natural numbers as objects, and Mor m n = Vec (Fin n) m. With this progress in our Agda formalisations of both the theory of relation-algebraic categories and the implementation of verified container libraries, fully certified graph transformation now appears realistically reachable in the near future.

Although the calculation of R^* via the construction of Sect. 5.4 is a usable default implementation, it will typically not be a very efficient implementation.

However, that construction is still useful as a framework for more indirect correctness proofs; we conjecture that UnitSumStarOp applied to the direct sum Fin 1 ⊞ Fin n can be used to derive a purely functional version of Warshall's algorithm very similar to Berghammer's [Ber11]. This is just one example of many opportunities worth exploring for replacing implementations that arise directly from the theory with efficient implementations that connect intelligently with the theory so that correctness proofs do not become a huge burden.

References

[Ber11]   Berghammer, R.: A Functional, Successor List Based Version of Warshall's Algorithm with Applications. In: [ds11], pp. 109–124
[BP12]   Braibant, T., Pous, D.: Deciding Kleene Algebras in Coq. Logical Methods in Computer Science 8, 16 (2012)
[CMR⁺97]   Corradini, A., Montanari, U., Rossi, F., Ehrig, H., Heckel, R., Löwe, M.: Algebraic Approaches to Graph Transformation, Part I: Basic Concepts and Double Pushout Approach. In: Rozenberg, G. (ed.) Handbook of Graph Grammars and Computing by Graph Transformation. Foundations, vol. 1, ch. 3, pp. 163–245. World Scientific, Singapore (1997)
[dS11]   de Swart, H. (ed.): RAMICS 2011. LNCS, vol. 6663. Springer, Heidelberg (2011)
[EEPT06]   Ehrig, H., Ehrig, K., Prange, U., Taentzer, G.: Fundamentals of Algebraic Graph Transformation. Springer (2006)
[EPPH06]   Ehrig, H., Padberg, J., Prange, U., Habel, A.: Adhesive High-Level Replacement Systems: A New Categorical Framework for Graph Transformation. Fund. Inform. 74, 1–29 (2006)
[FS90]   Freyd, P.J., Scedrov, A.: Categories, Allegories, North-Holland Mathematical Library, vol. 39. North-Holland, Amsterdam (1990)
[Kah08]   Kahl, W.: Relational Semigroupoids: Abstract Relation-Algebraic Interfaces for Finite Relations between Infinite Types. J. Logic and Algebraic Programming 76, 60–89 (2008)
[Kah11a]   Kahl, W.: Collagories: Relation-Algebraic Reasoning for Gluing Constructions. J. Logic and Algebraic Programming 80, 297–338 (2011)
[Kah11b]   Kahl, W.: Dependently-Typed Formalisation of Relation-Algebraic Abstractions. In: [ds11], pp. 230–247
[Kaw90]   Kawahara, Y.: Pushout-Complements and Basic Concepts of Grammars in Toposes. Theoretical Computer Science 77, 267–289 (1990)
[Koz94]   Kozen, D.: A Completeness Theorem for Kleene Algebras and the Algebra of Regular Events. Inform. and Comput. 110, 366–390 (1994)
[Koz98]   Kozen, D.: Typed Kleene Algebra. Technical Report 98-1669, Computer Science Department, Cornell University (1998)
[MKJ09]   Mu, S.-C., Ko, H.-S., Jansson, P.: Algebra of Programming in Agda: Dependent Types for Relational Program Derivation. J. Functional Programming 19, 545–579 (2009) See also AoPA at, http://www.iis.sinica.edu.tw/~scm/2008/aopa/
[Nor07]   Norell, U.: Towards a Practical Programming Language Based on Dependent Type Theory. PhD thesis, Department of Computer Science and Engineering, Chalmers University of Technology (2007)

A   Proof of E^*-recDef$_1$: $Id \sqcup E \, {}_9^\circ \, E^* \sqsubseteq E^*$

E^*-recDef$_1$: $Id \sqcup E \, {}_9 \, E^* \sqsubseteq E^*$
E^*-recDef$_1$ = $\sqsubseteq$-begin
  $Id \sqcup E \, {}_9 \, E^*$
$\approx\check{}$($\sqcup$-cong (isIdentity-$\approx$Id Id$\boxplus$-isIdentity) ($\vdash$-cong$_1$ $\mathbb{E}$-$\ominus$))
  $(\iota \, \ominus \, \kappa) \sqcup ((a \oplus b) \, \ominus \, (c \oplus d)) \, {}_9 \, E^*$
$\approx$($\sqcup$-cong$_2$ $\ominus$-$_9$)
  $(\iota \, \ominus \, \kappa) \sqcup ((a \oplus b) \, {}_9 \, E^* \, \ominus \, (c \oplus d) \, {}_9 \, E^*)$
$\approx$($\ominus$-$\sqcup$-$\ominus$)
  $(\iota \sqcup (a \oplus b) \, {}_9 \, E^*) \, \ominus \, (\kappa \sqcup (c \oplus d) \, {}_9 \, E^*)$
$\sqsubseteq$($\ominus$-monotone
  ($\sqsubseteq$-begin
    $\iota \sqcup (a \oplus b) \, {}_9 \, E^*$
  $\approx$($\sqcup$-cong to-$\mathbb{E}$ ($\vdash$-$\mathbb{E}$ ($\approx\approx$) $\mathbb{E}$-cong $\mathbb{E}$-$\vdash$-$\ominus$ $\mathbb{E}$-$\vdash$-$\ominus$))
    $(\iota \, {}_9 \, \iota\check{} \, \mathbb{E} \, \iota \, {}_9 \, \kappa\check{}) \sqcup ((a \, {}_9 \, f^* \sqcup b \, {}_9 \, h) \, \mathbb{E} \, (a \, {}_9 \, g \sqcup b \, {}_9 \, k))$
  $\approx$($\mathbb{E}$-$\sqcup$-$\mathbb{E}$)
    $(\iota \, {}_9 \, \iota\check{} \sqcup a \, {}_9 \, f^* \sqcup b \, {}_9 \, h) \, \mathbb{E} \, (\iota \, {}_9 \, \kappa\check{} \sqcup a \, {}_9 \, g \sqcup b \, {}_9 \, k)$
  $\sqsubseteq$($\mathbb{E}$-monotone
    ($\sqsubseteq$-begin
      $\iota \, {}_9 \, \iota\check{} \sqcup a \, {}_9 \, f^* \sqcup b \, {}_9 \, d^* \, {}_9 \, c \, {}_9 \, f^*$
    $\approx$($\sqcup$-cong   (isIdentity-$\approx$Id leftKernel)
                    ($\sqcup$-cong$_2$ $\vdash$-assoc$_{3+1}$ ($\approx\check{}\approx\check{}$) $\vdash$-$\sqcup$-distribL))
      $Id \sqcup f \, {}_9 \, f^*$
    $\sqsubseteq$(*-recDef$_1\sqsubseteq$ StarA)
      f^*
    $\Box$)
    ($\sqsubseteq$-begin
      $\iota \, {}_9 \, \kappa\check{} \sqcup a \, {}_9 \, g \sqcup b \, {}_9 \, (d^* \sqcup d^* \, {}_9 \, c \, {}_9 \, g)$
    $\sqsubseteq$($\sqcup$-universal (commutes$\check{}$ ($\approx\sqsubseteq$) $\perp$-$\sqsubseteq$) ($\sqcup$-universal
      ($\sqsubseteq$-begin
        $a \, {}_9 \, f^* \, {}_9 \, b \, {}_9 \, d^*$
      $\sqsubseteq$($\vdash$-assocL ($\approx\sqsubseteq$) $\vdash$-monotone$_1$
          ($\vdash$-monotone$_1$ $\sqcup$-upper$_1$ ($\sqsubseteq\sqsubseteq$) *-stepL StarA))
        $f^* \, {}_9 \, b \, {}_9 \, d^*$
      $\Box$)
      ($\sqsubseteq$-begin
        $b \, {}_9 \, (d^* \sqcup d^* \, {}_9 \, c \, {}_9 \, g)$
      $\approx$($\vdash$-$\sqcup$-distribR)
        $b \, {}_9 \, d^* \sqcup b \, {}_9 \, d^* \, {}_9 \, c \, {}_9 \, g$
      $\sqsubseteq$($\sqcup$-universal
        ($\sqsubseteq$-begin
          $b \, {}_9 \, d^*$
        $\sqsubseteq$(proj$_1$ (*-isSuperidentity StarA))
          $f^* \, {}_9 \, b \, {}_9 \, d^*$
        $\Box$)
        ($\sqsubseteq$-begin
          $b \, {}_9 \, d^* \, {}_9 \, c \, {}_9 \, g$
        $\sqsubseteq$($\vdash$-assoc$_{3+1}$ ($\approx\check{}\sqsubseteq$) $\vdash$-monotone$_1$ $\sqcup$-upper$_2$)
          $f \, {}_9 \, f^* \, {}_9 \, b \, {}_9 \, d^*$
        $\sqsubseteq$($\vdash$-assocL ($\approx\sqsubseteq$) $\vdash$-monotone$_1$ (*-stepL StarA))
          $f^* \, {}_9 \, b \, {}_9 \, d^*$
        $\Box$)
)
      $\Box$)
)
                                    g
      $\Box$))
)
                                  g
    $\Box$)
)
  $f^* \, \mathbb{E} \, g$
$\Box$)

($\sqsubseteq$-begin
  $\kappa \sqcup (c \oplus d) \, {}_9 \, E^*$
$\approx$($\sqcup$-cong to-$\mathbb{E}$ ($\vdash$-$\mathbb{E}$ ($\approx\approx$) $\mathbb{E}$-cong $\mathbb{E}$-$\vdash$-$\ominus$ $\mathbb{E}$-$\vdash$-$\ominus$))
  $(\kappa \, {}_9 \, \iota\check{} \, \mathbb{E} \, \kappa \, {}_9 \, \kappa\check{}) \sqcup ((c \, {}_9 \, f^* \sqcup d \, {}_9 \, h) \, \mathbb{E} \, (c \, {}_9 \, g \sqcup d \, {}_9 \, k))$
$\approx$($\mathbb{E}$-$\sqcup$-$\mathbb{E}$)
  $(\kappa \, {}_9 \, \iota\check{} \sqcup c \, {}_9 \, f^* \sqcup d \, {}_9 \, h) \, \mathbb{E} \, (\kappa \, {}_9 \, \kappa\check{} \sqcup c \, {}_9 \, g \sqcup d \, {}_9 \, k)$
$\sqsubseteq$($\mathbb{E}$-monotone
  ($\sqsubseteq$-begin
    $\kappa \, {}_9 \, \iota\check{} \sqcup c \, {}_9 \, f^* \sqcup d \, {}_9 \, d^* \, {}_9 \, c \, {}_9 \, f^*$
  $\sqsubseteq$($\sqcup$-universal (commutes$\check{}$ ($\approx\sqsubseteq$) $\perp$-$\sqsubseteq$)
    ($\sqcup$-universal
      (proj$_1$ (*-isSuperidentity StarB))
      ($\vdash$-assoc$_1$ ($\approx\sqsubseteq$)
      $\vdash$-monotone$_1$ (*-stepL StarB)))
)
    $d^* \, {}_9 \, c \, {}_9 \, f^*$
  $\Box$)
  ($\sqsubseteq$-begin
    $\kappa \, {}_9 \, \kappa\check{} \sqcup c \, {}_9 \, g \sqcup d \, {}_9 \, (d^* \sqcup d^* \, {}_9 \, c \, {}_9 \, g)$
  $\sqsubseteq$($\vdash$-assocL ($\approx\sqsubseteq$) $\sqcup$-universal
    ($\sqcup$-monotone
      (isIdentity-$\approx$Id rightKernel
        ($\approx\sqsubseteq$) *-isReflexive StarB)
        (proj$_1$ (*-isSuperidentity StarB)))
      ($\vdash$-$\sqcup$-distribR ($\approx\sqsubseteq$) $\sqcup$-monotone
        (*-stepL StarB)
        ($\vdash$-assocL ($\approx\sqsubseteq$)
        $\vdash$-monotone$_1$ (*-stepL StarB)))
)
    $d^* \sqcup d^* \, {}_9 \, c \, {}_9 \, g$
  $\Box$)
)
                                  $h \, \mathbb{E} \, k$
  $\Box$)

)
                              $(f^* \, \mathbb{E} \, g) \, \ominus \, (h \, \mathbb{E} \, k)$
                              $\approx$($\mathbb{E}$-$\ominus$)
                                E^*
                              $\Box$

Deciding Regular Expressions (In-)Equivalence in Coq[*]

Nelma Moreira[1], David Pereira[1,**], and Simão Melo de Sousa[1]

[1] DCC-FC – University of Porto
Rua do Campo Alegre 1021, 4169-007, Porto, Portugal
`nam@dcc.fc.up.pt, dpereira@ncc.up.pt`
[2] LIACC & DI – University of Beira Interior
Rua Marquês d'Ávila e Bolama, 6201-001, Covilhã, Portugal
`desousa@di.ubi.pt`

Abstract. This work presents a mechanically verified implementation of an algorithm for deciding regular expression (in-)equivalence within the Coq proof assistant. This algorithm decides regular expression equivalence through an iterated process of testing the equivalence of their partial derivatives and also does not construct the underlying automata. Our implementation has a refutation step that improves the general efficiency of the decision procedure by enforcing the in-equivalence of regular expressions at early stages of computation. Recent theoretical and experimental research provide evidence that this method is, on average, more efficient than the classical methods based in automata. We present some performance tests and comparisons with similar approaches.

1 Introduction

Recently, much attention has been given to the mechanization of Kleene algebra (KA) within proof assistants. J.-C. Filliâtre [1] provided a first formalisation of the Kleene theorem for regular languages [2] within the Coq proof assistant [3]. Höfner and Struth [4] investigated the automated reasoning in variants of Kleene algebras with Prover9 and Mace4 [5]. Pereira and Moreira [6] implemented in Coq an abstract specification of Kleene algebra with tests (KAT) [7] and the proofs that propositional Hoare logic deduction rules are theorems of KAT. An obvious follow up of that work was to implement a certified procedure for deciding equivalence of KA terms, *i.e.*, regular expressions. A first step was the proof of the correctness of the partial derivative automaton construction from a regular expression presented in [8]. In this paper, our goal is to mechanically verify a decision procedure based on partial derivatives proposed by Almeida *et*

---

[*] This work was partially funded by the European Regional Development Fund through the programme COMPETE and by the Portuguese Government through the FCT under the projects PEst-OE/EEI/UI0027/2011, PEst-C/MAT/UI0144/2011, and CANTE-PTDC/EIA-CCO/101904/2008.
[**] David Pereira is funded by FCT grant SFRH/BD/33233/2007.

W. Kahl and T.G. Griffin (Eds.): RAMiCS 2012, LNCS 7560, pp. 98–113, 2012.

al. [9] that is a functional variant of the rewrite system proposed by Antimirov and Mosses [10]. This procedure decides regular expression equivalence through an iterated process of testing the equivalence of their partial derivatives.

Similar approaches based on the computation of a bisimulation between the two regular expressions were used recently. In 1971, Hopcroft and Karp [11] presented an almost linear algorithm for equivalence of two deterministic finite automata (DFA). By transforming regular expressions into equivalent DFAs, Hopcroft and Karp method can be used for regular expressions equivalence. A comparison of that method with the method here proposed is discussed by Almeida et. al [12,13]. There it is conjectured that a direct method should perform better on average, and that is corroborated by theoretical studies based on analytic combinatorics [14]. Hopcroft and Karp method was used by Braibant and Pous [15] to formally verify Kozen's proof of the completeness of Kleene algebra [16] in Coq. Although the relative inefficiency of the method chosen, and as we will note in Section 4, it seems the more competitive (and most general) implementation available currently in Coq.

Independently of the work here presented, Coquand and Siles [17] mechanically verified an algorithm for deciding regular expression equivalence based on Brzozowski's derivatives [18] and an inductive definition of finite sets called Kuratowski-finite sets. Also based on Brzozowski's derivatives, Krauss and Nipkow [19] provide an elegant and concise formalisation of Rutten's co-algebraic approach of regular expression equivalence [20] in the Isabelle proof assistant [21], but they do not address the termination of the formalized decision procedure. Vladimir Komendantsky provides a novel functional construction of the *partial derivative automaton* [22], and also made contributions [23] to the mechanization of concepts related to the Mirkin's construction [24] of that automata. More recently, Andrea Asperti formalized a decision procedure for the equivalence of *pointed regular expressions* [25], that is both compact and efficient.

Besides avoiding the need of building DFAs, our use of partial derivatives avoids also the necessary normalisation of regular expressions modulo *ACI* (i.e associativity, idempotence and commutativity of union) in order to ensure the finiteness of Brzozowski's derivatives. Like in other approaches [15], our method also includes a refutation step that improves the detection of inequivalent regular expressions. One of our goals is to use the procedure as a way to automate the process of reasoning about programs encoded as KAT terms in a certified framework. A first step towards this goal is reported in [26].

Although the algorithm we have chosen to verify seems straightforward, the process of its mechanical verification in a theorem prover based on a type theory such as the one behind the Coq proof assistant raises several issues which are quite different from an usual implementation in standard programming languages. The Coq proof assistant allows users to specify and implement programs, and also to prove that the implemented programs are compliant with their specification. In this sense, the first task is the effort of formalizing the underlying algebraic theory. Afterwards, and in order to encode the decision procedure, we have to provide a formal proof of its termination since our procedure is a general recursive one, whereas Coq's type system accepts only provable terminating functions. Finally, a formal proof must be provided in order to ensure that the functional behavior of the implemented procedure is correct wrt. regular expression (in-)equivalence. Moreover, the encoding effort must be conducted with care in order to obtain a solution that is able to compute inside Coq with a reasonable performance.

2   Some Basic Notions of Regular Languages

This section presents some basic notions of regular languages. These definitions can be found on standard books such as Hopcroft *et al.* [27], and their formalisation in the Coq proof assistant are presented by Almeida *et al.* [8].

2.1   Alphabets, Words, Languages and Regular Expressions

Let $\Sigma = \{a_1, a_2, \ldots, a_n\}$ be an *alphabet* (non-empty set of symbols). A *word* w over Σ is any finite sequence of symbols. The *empty word* is denoted by ε and the *concatenation* of two words w_1 and w_2 is the word $w = w_1 w_2$. Let $\Sigma^\star$ be the set of all words over Σ. A *language* over Σ is a subset of $\Sigma^\star$. If L_1 and L_2 are two languages, then $L_1 L_2 = \{w_1 w_2 \mid w_1 \in L_1, w_2 \in L_2\}$. The *power* of a language is inductively defined by $L^0 = \{\varepsilon\}$ and $L^n = L L^{n-1}$, with $n \geq 1$. The *Kleene star* $L^\star$ of a language L is $\cup_{n \geq 0} L^n$. Given a word $w \in \Sigma^\star$, the *(left-)quotient* of L by the word w is the language $w^{-1}(L) = \{v \mid wv \in L\}$.

A *regular expression* (*re*) α over Σ represents a *regular language* $\mathcal{L}(\alpha) \subseteq \Sigma^\star$ and is inductively defined by: $\emptyset$ is a *re* and $\mathcal{L}(\emptyset) = \emptyset$; ε is a *re* and $\mathcal{L}(\varepsilon) = \{\epsilon\}$; $\forall a \in \Sigma$, a is a *re* and $\mathcal{L}(a) = \{a\}$; if α and β are *re*'s, $(\alpha + \beta)$, $(\alpha\beta)$ and $(\alpha)^\star$ are *re*'s, respectively with $\mathcal{L}(\alpha + \beta) = \mathcal{L}(\alpha) \cup \mathcal{L}(\beta)$, $\mathcal{L}(\alpha\beta) = \mathcal{L}(\alpha)\mathcal{L}(\beta)$ and $\mathcal{L}(\alpha^\star) = \mathcal{L}(\alpha)^\star$. If Γ is a set of *re*'s, then $\mathcal{L}(\Gamma) = \cup_{\alpha \in \Gamma} \mathcal{L}(\alpha)$. The *alphabetic size* of a *re* α is the number of symbols of the alphabet in α and is denoted by $|\alpha|_\Sigma$. The *empty word property* (*ewp* for short) of a *re* α is denoted by $\varepsilon(\alpha)$ and is defined by $\varepsilon(\alpha) = \varepsilon$ if $\varepsilon \in \mathcal{L}(\alpha)$ and by $\varepsilon(\alpha) = \emptyset$, otherwise. If $\varepsilon(\alpha) = \varepsilon(\beta)$ we say that α and β *have the same ewp*. Given a set of *re*'s Γ we define $\varepsilon(\Gamma) = \varepsilon$ if there exists a *re* $\alpha \in \Gamma$ such that $\varepsilon(\alpha) = \varepsilon$ and $\varepsilon(\Gamma) = \emptyset$, otherwise. Two *re*'s α and β are *equivalent* if they represent the same language, that is, if $\mathcal{L}(\alpha) = \mathcal{L}(\beta)$, and we write $\alpha \sim \beta$.

2.2   Partial Derivatives

The notion of *derivative* of a *re* was introduced by Brzozowski [18]. Antimirov [10] extended this notion to the one of set of *partial derivatives*, which corresponds to a finite set representation of Brzozowski's derivatives.

Let α be a *re* and let $a \in \Sigma$. The *set* $\partial_a(\alpha)$ *of partial derivatives* of the *re* w.r.t. the symbol a is inductively defined as follows:

$$\partial_a(\emptyset) = \emptyset \qquad\qquad \partial_a(\alpha + \beta) = \partial_a(\alpha) \cup \partial_a(\beta)$$

$$\partial_a(\varepsilon) = \emptyset \qquad\qquad \partial_a(\alpha\beta) \;\; = \begin{cases} \partial_a(\alpha)\beta \cup \partial_a(\beta) & \text{if } \varepsilon(\alpha) = \varepsilon \\ \partial_a(\alpha)\beta & \text{otherwise} \end{cases}$$

$$\partial_a(b) = \begin{cases} \{\varepsilon\} & \text{if } a \equiv b \\ \emptyset & \text{otherwise} \end{cases} \qquad \partial_a(\alpha^\star) \;\; = \partial_a(\alpha)\alpha^\star,$$

where $\Gamma\beta = \{\alpha\beta \mid \alpha \in \Gamma\}$ if $\beta \neq \emptyset$ and $\beta \neq \varepsilon$, and $\Gamma\emptyset = \emptyset$ and $\Gamma\varepsilon = \Gamma$ otherwise (in the same way we define $\beta\Gamma$). Moreover one has

$$\mathcal{L}(\partial_a(\alpha)) = a^{-1}(\mathcal{L}(\alpha)). \tag{1}$$

The definition of set of partial derivatives is extended to sets of re's and to words. Given a re α, a symbol $a \in \Sigma$, a word $w \in \Sigma^\star$, and a set of re's Γ, we define $\partial_a(\Gamma) = \cup_{\alpha \in \Gamma} \partial_a(\alpha)$, $\partial_\varepsilon(\alpha) = \{\alpha\}$, and $\partial_{wa} = \partial_a(\partial_w(\alpha))$. Equation (1) can be extended to words $w \in \Sigma^\star$. The *set of partial derivatives* of a re α is defined by $PD(\alpha) = \cup_{w \in \Sigma^\star}(\partial_w(\alpha))$. This set is always finite and its cardinality is bounded by $|\alpha|_\Sigma + 1$.

Champarnaud and Ziadi show in [28] that partial derivatives and Mirkin's prebases [24] lead to identical constructions. Let $\pi(\alpha)$ be a function inductively defined as follows:

$$\begin{aligned} \pi(\emptyset) &= \emptyset & \pi(\alpha + \beta) &= \pi(\alpha) \cup \pi(\beta) \\ \pi(\varepsilon) &= \emptyset & \pi(\alpha\beta) &= \pi(\alpha)\beta \cup \pi(\beta) \\ \pi(a) &= \{\varepsilon\} & \pi(\alpha^\star) &= \pi(\alpha)\alpha^\star. \end{aligned} \tag{2}$$

In his original paper, Mirkin proved that $|\pi(\alpha)| \leq |\alpha|_\Sigma$, while Champarnaud and Ziadi established that $PD(\alpha) = \{\alpha\} \cup \pi(\alpha)$. These properties were proven correct in Coq by Almeida *et al.* [8] and will be used to prove the termination of the decision procedure described in this paper.

An important property of partial derivatives is that given a re α we have

$$\alpha \sim \varepsilon(\alpha) + \sum_{a \in \Sigma} a\partial_a(\alpha) \tag{3}$$

and so, checking if $\alpha \sim \beta$ can be reformulated as

$$\varepsilon(\alpha) + \sum_{a \in \Sigma} a\partial_a(\alpha) \sim \varepsilon(\beta) + \sum_{a \in \Sigma} a\partial_a(\beta). \tag{4}$$

This will be an essential ingredient to our decision method because deciding if $\alpha \sim \beta$ is tantamount to check if $\varepsilon(\alpha) = \varepsilon(\beta)$ and if $\partial_a(\alpha) \sim \partial_a(\beta)$, for each $a \in \Sigma$. We also note that testing if a word $w \in \Sigma^\star$ belongs to $\mathcal{L}(\alpha)$ can be reduced to the purely syntactical operation of checking if

$$\varepsilon(\partial_w(\alpha)) = \varepsilon. \tag{5}$$

By (4) and (5) we have that

$$(\forall w \in \Sigma^\star, \varepsilon(\partial_w(\alpha)) = \varepsilon(\partial_w(\beta))) \leftrightarrow \alpha \sim \beta. \tag{6}$$

3   The Decision Procedure

In this section we describe the implementation in Coq of a procedure for deciding the equivalence of re's based on partial derivatives. First we give the informal description of the procedure and afterwards we present the technical details of its implementation in Coq's type theory. The Coq development presented in this paper is available online in [29].

3.1   Informal Description

The procedure for deciding the equivalence of re's, which we call EQUIVP, is presented in Fig.1. Given two re's α and β this procedure corresponds to the iterated process of deciding the equivalence of their derivatives, in the way noted in equation (4). The procedure EQUIVP works over pairs of re's (Γ, Δ) such that $\Gamma = \partial_w(\alpha)$ and $\Delta = \partial_w(\beta)$, for some word $w \in \Sigma^\star$. The notion of set of partial derivatives can also be extended to these pairs that we refer from now on by $derivatives$. To check if $\alpha \sim \beta$ it is enough to test the ewp's of the derivatives, i.e., if (Γ, Δ) verify the condition $\varepsilon(\Gamma) = \varepsilon(\Delta)$.

Algorithm 1. The procedure EQUIVP.

Require: $S = \{(\{\alpha\}, \{\beta\})\}$, $H = \emptyset$
Ensure: true or false

```
 1: procedure EQUIVP(S, H)
 2:     while S ≠ ∅ do
 3:         (Γ, Δ) ← POP(S)
 4:         if ε(Γ) ≠ ε(Δ) then
 5:             return false
 6:         end if
 7:         H ← H ∪ {(Γ, Δ)}
 8:         for a ∈ Σ do
 9:             (Λ, Θ) ← ∂ₐ(Γ, Δ)
10:             if (Λ, Θ) ∉ H then
11:                 S ← S ∪ {(Λ, Θ)}
12:             end if
13:         end for
14:     end while
15: return true
16: end procedure
```

Two finite sets of derivatives are required for implementing EQUIVP: a set H that serves as an accumulator for the derivatives already processed by the procedure, and a set S which serves as a working set that gathers new derivatives yet to be processed. The set H ensures the termination of EQUIVP due to the finiteness of the number of derivatives.

When EQUIVP terminates, either the set H of all the derivatives of α and β has been computed, or a counter-example (Γ, Δ) has been found, i.e., $\varepsilon(\Gamma) \neq \varepsilon(\Delta)$. By equation (6), in the first case we conclude that $\alpha \sim \beta$ and, in the second case we conclude that $\alpha \not\sim \beta$. The correctness of this method can be found in Almeida *et al.* [9,12]. As an illustration of how EQUIVP computes, we present below two small examples of its execution, the first considering the equivalence of two *re*'s, and the second considering the in-equivalence of two *re*'s.

Example 1. Suppose we want to prove that $\alpha = (ab)^*a$ and $\beta = a(ba)^*$ are equivalent. Considering $s_0 = (\{(ab)^*a\}, \{a(ba)^*\})$, it is enough to show that

$$\text{EQUIVP}(\{s_0\}, \emptyset) = \texttt{true}.$$

The computation of EQUIVP is for these particular α and β involves the construction of the new derivatives $s_1 = (\{1, b(ab)^*a\}, \{(ba)^*\})$ and $s_2 = (\emptyset, \emptyset)$. We can trace the computation by the following table

i	S_i	H_i	drvs.
0	$\{s_0\}$	$\emptyset$	$\partial_a(s_0) = s_1, \partial_b(s_0) = s_2$
1	$\{s_1, s_2\}$	$\{s_0\}$	$\partial_a(s_1) = s_2, \partial_b(s_1) = s_0$
2	$\{s_2\}$	$\{s_0, s_1\}$	$\partial_a(s_2) = s_2, \partial_b(s_2) = s_2$
3	$\emptyset$	$\{s_0, s_1, s_2\}$	true

where i is the iteration number, and S_i and H_i are the arguments of EQUIVP in that same iteration. The trace terminates with $S_2 = \emptyset$ and thus we can conclude that $\alpha \sim \beta$.

Example 2. Suppose we want to check if $\alpha = b^*a$ and $\beta = b^*ba$ are not equivalent. Considering $s_0 = (\{b^*a\}, \{b^*ba\})$, to prove so it is enough to check if

$$\text{EQUIVP}(\{s_0\}, \emptyset) = \texttt{false}.$$

In this case, the computation of EQUIVP creates the new derivatives , $s_1 = (\{1\}, \emptyset)$ and $s_2 = (\{b^*a\}, \{a, b^*ba\})$, and takes two iterations to halt and return `false`. The counter example found is the pair s_1, as it is easy to see in the trace of computation presented in the table below.

i	S_i	H_i	drvs.
0	$\{s_0\}$	$\emptyset$	$\partial_a(s_0) = s_1, \partial_b(s_2) = s_2$
1	$\{s_1, s_2\}$	$\{s_0\}$	$\varepsilon(s_1) \rightarrow \texttt{false}$

3.2 Implementation in Coq

In this section we describe the mechanically verified formalisation of EQUIVP in the Coq proof assistant and show its termination and correctness.

The Coq Proof Assistant

The Coq proof assistant is an implementation of the Calculus of Inductive Constructions (CIC) [30], a typed λ-calculus that features polymorphism, dependent types and very expressive (co-)inductive types. Coq provides users with the means to define data-structures and functions, as in standard functional languages, and also allows to define specifications and to build proofs in the same language, if we consider the underlying λ-calculus as an higher-order logic. In CIC, every term has a type and also every type has its own type, called *sort*. The universe of sorts in Coq is defined as the set $\{\mathsf{Prop}, \mathsf{Set}, \mathsf{Type}(i) \mid i \in \mathbb{N}\}$, where Prop is the type of propositions and Set is the type of program specifications. Both Prop and Set are of type Type(0). This distinction between the type of propositions and the type of program specifications permits Coq to provide a mechanism that extracts functional programs directly from Coq scripts, by ignoring all the propositions and extracting only the computationally meaningful definitions. More details on the way certified program development and proof construction are carried out can be found in [3].

In the formalisation below we use the libraries and certified programming and proving mechanisms provided by the Coq official distribution [31]. Our implementation is not axiom-free, as it depends on set extensionality, but which does not interfere with the consistency of the development. We also use a specific library (which is not in Coq's standard library) to deal with finite sets: in this case we use Stephane Lescuyer's Containers library [32], which is a re-implementation of Coq's finite sets library using *typeclasses*. This library eases the implementation of functions that deal with finite sets and also provides facilities to handle ordered types. In particular, using this library we obtain the type of finite sets of a finite set for free. These properties revealed themselves quite handy for our development which is based mostly on sets, and sets of sets of *re*'s (and extensions).

Certified Pairs of Derivatives

The main data structures underlying the implementation of EQUIVP are pairs of sets of *re*'s and sets of these pairs. Each pair (Γ, Δ) corresponds to a word derivative $(\partial_w(\alpha), \partial_w(\beta))$, where $w \in \Sigma^\star$ and α and β are the *re*'s being tested by EQUIVP. The pairs (Γ, Δ) are encoded by the type Drv α β, presented in Fig.1. This is a *dependent record* built from three parameters: a pair of sets of *re*'s dp that corresponds to the actual pair (Γ, Δ), a word w, and a proof term cw that certifies that $(\Gamma, \Delta) = (\partial_w(\alpha), \partial_w(\beta))$. The dependency of Drv α β comes from cw, which is a proof depending on the values of the *re*'s α and β, and on the word parameter w. This dependency ensures, at compilation time, that EQUIVP will only accept as input pairs of *re*'s that correspond to derivatives of α and β.

The type Drv α β provides also an easy way to relate the computation of EQUIVP and the equivalence of α and β: if H is the set returned by EQUIVP, then the equation (6) is tantamount to check the *ewp* of the elements of H.

```
Record Drv (α β:re) := mkDrv {
  dp :> set re * set re ;
  w  : word ;
  cw : dp === (∂ᵥ(α),∂ᵥ(β))  (* "===" refers to finite set equivalence *)
}.

Program Definition Drv_1st (α β:re) : Drv α β.
refine(Build_Drv ({r1},{r2}) nil _).
(* Now comes the proof that ({α},{β}) = (∂ₑ(α),∂ₑ(β)) *)
abstract(unfold wpdrvp;simpl;constructor;
         unfold wpdrv_set;simpl;normalize_notations;auto).
Defined.

Definition Drv_pdrv(α β:re)(x:Drv α β)(a:A) : Drv α β.
refine(match x with
       | mkDrv α β K w P => mkDrv α β (pdrvp K a) (w++[a]) _
       end).
(* Now comes the proof that ∂ₐ(∂_w(α),∂_w(β)) = (∂_wa(α),∂_wa(β)) *)
abstract(unfold pdrvp;inversion_clear P;simpl in *;
         constructor;normalize_notations;simpl;
         [rewrite H|rewrite H0];rewrite wpdrv_set_app;
         unfold wpdrv_set;simpl;reflexivity).
Defined.

Definition Drv_pdrv_set(s:Drv α β)(sig:set A) : set (Drv α β) :=
  fold (fun x:A => add (Drv_pdrv s x)) sig ∅.

Definition Drv_wpdrv (α β:re)(w:word) : Drv α β.
refine(mkDrv α β (∂_w(α),∂_w(β)) w _).
(* Now comes the proof that (∂_w(α),∂_w(β)) = (∂_w(α),∂_w(β)) *)
abstract(reflexivity).
Defined.

Definition c_of_rep(x:set re * set re) :=
  Bool.eqb (c_of_re_set (fst x)) (c_of_re_set (snd x)).

Definition c_of_Drv(x:Drv α β) := c_of_rep (dp x).

Definition c_of_Drv_set (s:set (Drv α β)) : bool :=
  fold (fun x => andb (c_of_Drv x)) s true.
```

Fig. 1. The type Drv and the extension of derivatives and *ewp* functions

Furthermore, this type allows to keep the set of words from which the set of derivatives of α and β has been obtained. For that it is enough to apply the projection w to each pair $(\Gamma, \Delta) \in H$.

The notions of derivative and of *ewp* are extended to the type Drv α β as implemented by the functions Drv_pdrv and c_of_Drv, and to sets of terms Drv α β by the functions Drv_pdrv_set and c_of_Drv_set, respectively. Note that part of the implementation of these functions is done by explicitly building proof terms using Coq's tactical language. In order to improve the performance of the computation of these functions we have wrapped the corresponding proofs in the abstract tactic, which defines these proofs as external lemmas and, as a consequence, replaces the explicit computation of the proof terms by a function call to the corresponding external lemma.

Computation of New Derivatives

The *while-loop* of EQUIVP describes the process of testing the equivalence of the derivatives of α and β. In each iteration, new derivatives (Γ, Δ) are computed until either the set S becomes empty, or a pair (Γ, Δ) such that $\varepsilon(\Gamma) \neq \varepsilon(\Delta)$ is found. This is precisely what the function `step` presented in Fig.2 does (which corresponds to the *for-loop* from line 8 to line 12 of EQUIVP's pseudocode).

```
Definition Drv_pdrv_set_filtered(x:Drv α β)(H:set (Drv α β))
 (sig:set A) : set (Drv α β) :=
 filter (fun y => negb (y ∈ H)) (Drv_pdrv_set x sig).

Inductive step_case (α β:re) : Type :=
|proceed    : step_case α β
|termtrue   : set (Drv α β) → step_case α β
|termfalse : Drv α β → step_case α β.

Definition step (H S:set (Drv α β))(sig:set A) :
 ((set (Drv αβ) * set (Drv α β)) * step_case α β) :=
 match choose s with
 |None => ((H,S),termtrue α β H)
 |Some (Γ,Δ) =>
   if c_of_Drv _ _ (Γ,Δ) then
    let H' := add (Γ,Δ) H in
     let S' := remove (Γ,Δ) S in
      let ns := Drv_pdrv_set_filtered α β (Γ,Δ) H' sig in
       ((H',ns ∪ S'),proceed α β)
   else
    ((H,S),termfalse α β (Γ,Δ))
 end.
```

Fig. 2. The function `step`

The `step` function proceeds as follows: it obtains a pair (Γ, Δ) from the set S, generates new derivatives $(\Lambda, \Theta) = (\partial_a(\Gamma), \partial_a(\Delta))$ by a symbol $a \in \Sigma$, and adds to S all the (Λ, Θ) that are not elements of $\{(\Gamma, \Delta)\} \cup H$. This is implemented by `Drv_pdrv_set_filtered` which prevents the whole process from entering potential infinite loops since each derivative is considered only once during the execution of EQUIVP and the overall number of derivatives is finite. The return type of `step` is

$$((\text{set } (\text{Drv } \alpha \ \beta) \ * \ \text{set } (\text{Drv } \alpha \ \beta)) \ * \ \text{step\_case}).$$

The first component corresponds to the pair (H,S), constructed as described above. The second component is a term of type `step_case` which guides the iterative process of computing the equivalence of the derivatives of α and β: if it is the term `proceed`, then the iterative process should continue; if it is a term `termtrue` H then the process should terminate and H contains the set of all the derivatives of α and β. Finally, if it is a term `termfalse` (Γ, Δ), then the process should terminate. The pair (Γ, Δ) is a witness that $\alpha \not\sim \beta$, since $\varepsilon(\Gamma) \neq \varepsilon(\Delta)$.

Implementation and Termination of EQUIVP

The formalisation of EQUIVP in Coq is presented in Fig.4, and corresponds to the function equivP. Its main component is the function iterate which is responsible for the iterative process of calculating the derivatives of α and β, or to find a witness that $\alpha \not\sim \beta$ if that is the case. The function iterate executes recursively until step returns either a term termtrue H, or returns a term termfalse (Γ, Δ). Depending on the result of step, the function iterate returns a term of type term_cases, which can be the term Ok H indicating that $\alpha \sim \beta$, or the term NotOk (Γ, Δ) indicating that $\alpha \not\sim \beta$, respectively.

A peculiarity of the Coq proof assistant is that it only accepts terminating functions, and more precisely, it only accepts *structurally decreasing* functions. Nevertheless, *provably terminating functions* can be expressed via encoding into structural recursive functions. The Function [33] command helps users to define such functions which are not structurally decreasing along with an evidence of its termination, as an illustration of the *certified programming paradigm* that Coq promotes. In the case of iterate such evidence is given by the proof that its recursive calls follow a *well-founded relation*.

The decreasing measure (of the recursive calls) for iterate is defined as follows: in each recursive call the cardinal of the accumulator set H increases by one element due to the computation of step. This increase of H can occur only less than $2^{(|\alpha|_\Sigma+1)} \times 2^{(|\beta|_\Sigma+1)} + 1$ times, due to the upper bounds of the cardinalities of $PD(\alpha)$ and of $PD(\beta)$. Therefore, in each recursive call of iterate, if step $H\ S\ \_ = (H', \_, \_)$ then the following condition holds:

$$(2^{(|\alpha|_\Sigma+1)} \times 2^{(|\beta|_\Sigma+1)} + 1) - |H'| < (2^{(|\alpha|_\Sigma+1)} \times 2^{(|\beta|_\Sigma+1)} + 1) - |H| \quad (7)$$

The relation LLim presented in Fig.3 defines the decreasing measure imposed by equation (7). Furthermore, the definition of iterate requires an argument of type DP α β which determines that the sets H and S are invariably disjoint along the computation of iterate which is required to ensure that the set H is always increased by one element at each recursive call.

Besides the requirement of defining LLim to formalise iterate, we had to deal with two implementation details: first, we have used the type N which is a binary representation of natural numbers provided by Coq's standard library, instead of the type nat so that the computation of MAX becomes feasible for large natural numbers. The second detail is related to the computation over terms representing well founded relations: instead of using the proof LLim_wf directly in iterate, we follow a technique proposed by Bruno Barras that uses the proof returned by the call to the function guard, that lazily adds 2^n constructors Acc_intro in front of LLim_wf so, that the actual proof is never reached in practice, while maintaining the same logical meaning. This technique avoids normalisation of well founded relation proofs which is usually highly complex and may take too much time to compute.

Finally, the function equivP is defined as a call to equivP_aux with the correct input, *i.e.*, with the accumulator set $H = \emptyset$ and with the working set

```
Definition lim_cardN (z:N) : relation (set A) :=
  fun x y:set A => nat_of_N z - (cardinal x) < nat_of_N z - (cardinal y).

Lemma lim_cardN_wf : ∀ z, well_founded (lim_cardN z).

Section WfIterate.
  Variables α β : re.

  Definition MAX_fst := |α|_Σ + 1.
  Definition MAX_snd := |β|_Σ + 1.

  Definition MAX := (2^MAX_fst × 2^MAX_snd) + 1.
  Definition LLim := lim_cardN (Drv α β) MAX.

  Theorem LLim_wf : well_founded LLim.

  Fixpoint guard (n : nat)(wfp : well_founded (LLim)) : well_founded (LLim):=
    match n with
    |0 => wfp
    |S m => fun x => Acc_intro x (fun y _ => guard m (guard m wfp) y)
    end.

End WfIterate.
```

Fig. 3. The decreasing measure of `iterate`

$S = \{(\{\alpha\}, \{\beta\})\}$. The function `equivP_aux` is a wrapper that pattern matches over the term of type `term_cases` returned by `iterate` and returns the corresponding Boolean value.

Correctness and Completeness

To prove the correctness of `equivP` we must prove that, if `equivP` returns `true`, then `iterate` generates all the derivatives and prove that all these derivatives agree on the *ewp* of its components. To prove that all derivatives are computed, it is enough to ensure that the `step` function returns a new accumulator set H' such that:

$$\text{step } H \, S \, sig = (H', S', \_) \rightarrow \forall(\Gamma, \Delta) \in H', \forall a \in \Sigma, \partial_a(\Gamma, \Delta) \in (H' \cup S') \quad (8)$$

The predicate `invP` and the lemma `invP_step` presented in Fig.5 prove this property. This means that, in each recursive call to `iterate`, the sets H and S hold all the derivatives of the elements in H. At some point of the execution, by the finiteness of the number of derivatives, H will contain all such derivatives and S will eventually become empty. Lemma `invP_iterate` proves this fact by a proof by functional induction over the structure of `iterate`. From lemma `invP_equivP` we can prove that

$$\forall w \in \Sigma^\star, (\partial_w(\alpha), \partial_w(\beta)) \in \text{equivP } \{(\{\alpha\}, \{\beta\})\} \, \emptyset \, \Sigma \quad (9)$$

by induction on the length of the word w and using the invariants presented above.

To finish the correctness proof of `equivP` one needs to make sure that all the derivatives (Γ, Δ) verify the condition $\varepsilon(\Gamma) = \varepsilon(\Delta)$. For that, we have defined

```
Inductive term_cases α β : Type :=
|OK : set (Drv α β) → term_cases α β
|NotOk : Drv α β → term_cases α β.

Inductive DP (α β:re)(H S: set (Drv α β)) : Prop :=
| is_dp : H ∩ S = ∅ → c_of_Drv_set α β H = true → DP α β H S.

Lemma DP_upd : ∀ (h s : set (Drv α β)) (sig : set A), DP α β h s →
   DP α β (fst (fst (step α β h s sig))) (snd (fst (step α β h s sig))).

Lemma DP_wf : ∀ (h s : set (Drv r1 r2)) (sig : set A),
      DP _ _ h s → snd (StepFast' _ _ h s sig) = Process' _ _ →
      LLim _ _ (fst (fst (StepFast' _ _ h s sig))) h.

Function iterate(α β:re)(H S:set (Drv α β))(sig:set A)(D:DP α β h s)
  {wf (LLim α β) H}: term_cases α β :=
   let ((H',S',next) := step H S in
    match next with
    |termfalse x => NotOk α β x
    |termtrue  h => Ok α β h
    |progress    => iterate α β H' S' sig (DP_upd α β H S sig D)
   end.
Proof.
  (* Now comes the proof that LLim is a decreasing measure for iterate *)
  abstract(apply DP_wf).
  (* Now comes the proof that LLim is a well founded relation. *)
  exact(guard r1 r2 100 (LLim_wf r1 r2)).
Defined.

Definition equivP_aux(α β:re)(H S:set(Drv α β))(sig:set A)(D:DP α β H S):=
  let H' := iterate α β H S sig D in
   match H' with
   | Ok _    => true
   | NotOk _ => false
  end.

Definition mkDP_ini : DP α β ∅ {Drv_1st α β}.
(* Now comes the proof that {({α},{β})}∩∅ = ∅ and that ε(∅) = true *)
abstract(constructor;[split;intros;try(inversion H)|vm_compute];reflexivity).
Defined.

Definition equivP (α β:re)(sig:set A) :=
  equivP_aux α β ∅ {Drv_1st α β} sig (mkDP_ini α β).
```

Fig. 4. Implementation of equivP

the predicate invP_final which strengthens the predicate invP by imposing that the previous property is verified. The predicate invP_final is proved to be an invariant of equivP and this implies re equivalence by equation (6), as stated by theorem invP_final_eq_lang.

For the completeness, it is enough to reason by contradiction: assuming that $\alpha \sim \beta$ then it must be true that $\forall w \in \Sigma^\star$, $\varepsilon(\partial_w(\alpha)) = \varepsilon(\partial_w(\beta))$ which implies that iterate may not return a set of pairs that contain a pair (Γ, Δ) such that $\varepsilon(\Gamma) \neq \varepsilon(\Delta)$ and so, equivP must always answer true.

Using the lemmas equivP_correct and equivP_correct_dual of Fig.5 a tactic was developed to prove automatically the (in)equivalence of any two re's α and β. This tactic works by reducing the logical proof of the (in)equivalence of re's into a Boolean equality involving the computation of equivP. After effectively computing equivP into a Boolean constant, the rest of the proof amounts

```
Definition invP (α β:re)(H S:set (Drv α β))(sig:set A) :=
∀ x, x ∈ H → ∀a, a ∈ sig → (Drv_pdrv α β x a) ∈ (H∪S).

Lemma invP_step : ∀ H S sig,
  invP H S sig → invP (fst (fst (step α β H S sig)))
                      (snd (fst (step α β H S sig))) sig.

Lemma invP_iterate : ∀ H S sig D,
  invP H S sig → invP (iterate α β H S sig D) ∅ sig.

Lemma invP_equivP :
  invP (equivP α β Σ) ∅ Σ.

Definition invP_final (α β:re)(H S:set (Drv α β))(sig:set A) :=
  (Drv_1st α β) ∈ (H∪S) /\
  (∀ x, x ∈ (H∪S) → c_of_Drv α β x = true) /\ invP α β H S sig.

Lemma invP_final_eq_lang :
  invP_final α β (equivP α β Σ) ∅ Σ → α∼β.

Theorem equivP_correct : ∀ α β, equivP α β sigma = true → α∼β.
Theorem equivP_complete : ∀ α β, α∼β → equivP α β sigma = true.
Theorem equivP_correct_dual : ∀ α β, equivP α β sigma = false → α≁β.
Theorem equivP_complete_dual : ∀ α β, α≁β → equivP α β sigma = false.
```

Fig. 5. Invariants of step and iterate

at applying the reflexivity of Coq's primitive equality. Note that this tactic is also able to solve re containment due to the equivalence $\alpha \leq \beta \leftrightarrow \alpha + \beta \sim \beta$.

4   Performance

Although the main goal of our development was to provide a certified evidence that the decision algorithm suggested by Almeida $et.\ al.$ is correct, it is of obvious interest to infer the level of usability of equivP (and corresponding tactic) for conducting proofs involving re's (in-)equivalence within the Coq proof assistant. We have carried out two types of performance evaluation of the decision procedure. The first evaluation consisted in experiments[1] with the tactic developed over data sets of 10000 pairs of uniform-randomly generated re's for a given size and using the FAdo tool [34] so that the results are statistically relevant. Some results are presented in the table below. The value n is the size of the syntactic tree[2] of each re's generated. The value of k is the number of symbols of the alphabet. The columns eq and $ineq$ are the average time (in seconds) spent to decide equivalence and inequivalence of two re's, respectively. The column $iter$ is the average number of recursive calls needed for equivP to terminate. The

---

[1] The experiments were conducted on a Virtual Box environment with six cores and 8 Gb of RAM, using coq-8.3pl4. The virtual environment executes on a dual six-core processor AMD Opteron(tm) 2435 processor with 2.60 GHz, and with 16 Gb of RAM.

[2] This corresponds to the sum of the number of constants, symbols and regular operators of the re.

equivalence tests were performed by comparing a *re* with itself, whereas the in-equivalence tests were performed by comparing two consecutive *re*'s randomly generated.

k	$n = 25$				$n = 50$				$n = 100$			
	eq	*iter*	*ineq*	*iter*	*eq*	*iter*	*ineq*	*iter*	*eq*	*iter*	*ineq*	*iter*
10	0.142	9.137	0.025	1.452	0.406	16.746	0.033	1.465	1.568	34.834	0.047	1.510
20	0.152	9.136	0.041	1.860	0.446	16.124	0.060	1.795	1.028	30.733	0.081	1.919
30	0.163	9.104	0.052	2.060	0.499	15.857	0.080	2.074	1.142	29.713	0.112	2.107
40	0.162	9.102	0.056	2.200	0.456	15.717	0.105	2.178	0.972	29.152	0.148	2.266
50	0.158	9.508	0.065	2.392	0.568	15.693	0.125	2.272	1.182	28.879	0.170	2.374

In the second evaluation[3] we have compared the performance of our development with the developments [15,19,17,25]. The results are presented below. The equivalence test $A(n, m, o)$ is $(a^o)a^\star + (a^n + a^m)^\star \sim (a^n + a^m)^\star$ and the equivalence test $B(n)$ is $(\epsilon + a + aa + \ldots + a^{n-1})(a^n)^\star \sim a^\star$. The test $C(n)$ is the equivalence $\alpha_n \sim \alpha_n$, for $\alpha_n = (a + b)^\star (a(a + b)^n)$. The entries "–" and "$\geq 600$" refer, respectively, to tests that were not performed and tests which took more than 10 minutes to finish[4].

	$A(n, m, o)$			$B(n)$			$C(n)$		
	$(4, 5, 12)$	$(5, 6, 20)$	$(5, 7, 24)$	18	100	500	5	10	15
equivP	0.15	0.20	0.29	0.05	1.30	37.06	1.31	77.61	≥ 600
[15]	0.01	0.01	0.02	0.03	1.33	50.53	0.04	3.06	15.26
[19]	2.78	2.94	2.88	2.94	3.37	158.94	2.87	≥ 600	≥ 600
[17]	8.73	46.70	102.69	98.84	≥ 600	≥ 600	28.97	≥ 600	≥ 600
[25]	0.24	0.43	0.57	0.80	–	–	–	–	–

The development of Braibant and Pous[5] is globally the more efficient in the tests we have selected, thanks to their *reification* mechanism and efficient representation of automata. equivP is able to outperform it only for larger members of the family $B(n)$. When compared to the other formalizations, equivP exhibits better performance, which suggests that algorithms based on partial derivatives should be considered wrt. to other approaches.

5   Concluding Remarks and Applications

In this paper we have described the formalisation, within the Coq proof assistant, of the procedure EQUIVP for deciding *re* equivalence based in partial derivatives. This procedure has the advantage of not requiring the normalisation modulo

---

[3] These tests were performed on a Macbook Pro 13", with a 2.53 GHz Inter Core 2 Duo, with 4 GB 1067 MHz DD3 of RAM memory, using coq-8.3pl4.

[4] The times for [25] are the ones given there, and were obtained on a Pentium M Processor 750 1.86GHz with 1GB of RAM.

[5] The version of [15] is the one available at http://coq.inria.fr/pylons/pylons/contribs/view/ATBR/v8.3, and was tested with coq-8.3pl4.

ACI of *re*'s in order to prove its termination. We presented some performance tests and comparisons with similar approaches that suggest the acceptable behavior of our decision procedure. However, there is space for improvement of its performance and more throughout comparisons with the other developments should take place. The purpose of this research is part of a broader project where the equivalence of Kleene algebra with tests (KAT) terms is used to reason about the partial correctness of programs [35]. The development in [26] is a mechanization of KAT in the Coq proof assistant containing the extension of the decision procedure here presented for KAT terms (in-)equivalence.

Acknowledgments. We thank the anonymous referees for their constructive comments and criticisms, from which this paper has clearly benefited.

References

1. Filliâtre, J.C.: Finite Automata Theory in Coq: A constructive proof of Kleene's theorem. Research Report 97–04, LIP - ENS Lyon (February 1997)
2. Kleene, S.: In: Shannon, C., McCarthy, J. (eds.) Representation of Events in Nerve Nets and Finite Automata, pp. 3–42. Princeton University Press
3. Bertot, Y., Castéran, P.: Interactive Theorem Proving and Program Development. Coq'Art: The Calculus of Inductive Constructions. Texts in Theoretical Computer Science. Springer (2004)
4. Höfner, P., Struth, G.: Automated Reasoning in Kleene Algebra. In: Pfenning, F. (ed.) CADE 2007. LNCS (LNAI), vol. 4603, pp. 279–294. Springer, Heidelberg (2007)
5. McCune, W.: Prover9 and Mace4, http://www.cs.unm.edu/smccune/mace4 (access date: October 1, 2011)
6. Moreira, N., Pereira, D.: KAT and PHL in Coq. CSIS 05 (02) (December 2008) ISSN: 1820-0214
7. Kozen, D.: Kleene algebra with tests. Transactions on Programming Languages and Systems 19(3), 427–443 (1997)
8. Almeida, J.B., Moreira, N., Pereira, D., Melo de Sousa, S.: Partial Derivative Automata Formalized in Coq. In: Domaratzki, M., Salomaa, K. (eds.) CIAA 2010. LNCS, vol. 6482, pp. 59–68. Springer, Heidelberg (2011)
9. Almeida, M., Moreira, N., Reis, R.: Antimirov and Mosses's rewrite system revisited. Int. J. Found. Comput. Sci. 20(4), 669–684 (2009)
10. Antimirov, V.M., Mosses, P.D.: Rewriting extended regular expressions. In: Rozenberg, G., Salomaa, A. (eds.) DLT, pp. 195–209. World Scientific (1994)
11. Hopcroft, J., Karp, R.M.: A linear algorithm for testing equivalence of finite automata. Technical Report TR 71-114, University of California, Berkeley, California (1971)
12. Almeida, M., Moreira, N., Reis, R.: Testing regular languages equivalence. JALC 15(1/2), 7–25 (2010)
13. Almeida, M.: Equivalence of regular languages: an algorithmic approach and complexity analysis. PhD thesis, FCUP (2011), http://www.dcc.fc.up.pt/~mfa/thesis.pdf
14. Broda, S., Machiavelo, A., Moreira, N., Reis, R.: The Average Transition Complexity of Glushkov and Partial Derivative Automata. In: Mauri, G., Leporati, A. (eds.) DLT 2011. LNCS, vol. 6795, pp. 93–104. Springer, Heidelberg (2011)

15. Braibant, T., Pous, D.: An Efficient Coq Tactic for Deciding Kleene Algebras. In: Kaufmann, M., Paulson, L.C. (eds.) ITP 2010. LNCS, vol. 6172, pp. 163–178. Springer, Heidelberg (2010)
16. Kozen, D.: A completeness theorem for Kleene algebras and the algebra of regular events. Infor. and Comput. 110(2), 366–390 (1994)
17. Coquand, T., Siles, V.: A Decision Procedure for Regular Expression Equivalence in Type Theory. In: Jouannaud, J.-P., Shao, Z. (eds.) CPP 2011. LNCS, vol. 7086, pp. 119–134. Springer, Heidelberg (2011)
18. Brzozowski, J.A.: Derivatives of regular expressions. JACM 11(4), 481–494 (1964)
19. Krauss, A., Nipkow, T.: Proof pearl: Regular expression equivalence and relation algebra. J. Autom. Reasoning 49(1), 95–106 (2012)
20. Rutten, J.J.M.M.: Automata and Coinduction (An Exercise in Coalgebra). In: Sangiorgi, D., de Simone, R. (eds.) CONCUR 1998. LNCS, vol. 1466, pp. 194–218. Springer, Heidelberg (1998)
21. Nipkow, T., Paulson, L.C., Wenzel, M.: Isabelle/HOL. LNCS, vol. 2283. Springer, Heidelberg (2002)
22. Komendantsky, V.: Reflexive toolbox for regular expression matching: verification of functional programs in Coq+Ssreflect. In: Claessen, K., Swamy, N. (eds.) PLPV, pp. 61–70. ACM (2012)
23. Komendantsky, V.: Computable partial derivatives of regular expressions (2011), http://www.cs.st-andrews.ac.uk/~vk/papers.html (access date: July 1, 2011)
24. Mirkin, B.: An algorithm for constructing a base in a language of regular expressions. Engineering Cybernetics 5, 110–116 (1966)
25. Asperti, A.: A Compact Proof of Decidability for Regular Expression Equivalence. In: Beringer, L., Felty, A. (eds.) ITP 2012. LNCS, vol. 7406, pp. 283–298. Springer, Heidelberg (2012)
26. Moreira, N., Pereira, D., Melo de Sousa, S.: Deciding KAT terms equivalence in Coq. Technical Report DCC-2012-04, DCC-FC & LIACC, Universidade do Porto (2012)
27. Hopcroft, J., Motwani, R., Ullman, J.D.: Introduction to Automata Theory, Languages and Computation. Addison Wesley (2000)
28. Champarnaud, J.M., Ziadi, D.: From Mirkin's prebases to Antimirov's word partial derivatives. Fundam. Inform. 45(3), 195–205 (2001)
29. Moreira, N., Pereira, D., Melo de Sousa, S.: Coq library: PDCoq (Version 1.1 2012), http://www.liacc.up.pt/~kat/pdcoq/
30. Paulin-Mohring, C.: Inductive Definitions in the System Coq: Rules and Properties. In: Bezem, M., Groote, J.F. (eds.) TLCA 1993. LNCS, vol. 664, pp. 328–345. Springer, Heidelberg (1993)
31. The Coq Development Team: The Coq proof assistant, http://coq.inria.fr
32. Lescuyer, S.: First-class containers in Coq. Studia Informatica Universalis 9, 87–127 (2011)
33. Barthe, G., Courtieu, P.: Efficient Reasoning about Executable Specifications in Coq. In: Carreño, V.A., Muñoz, C., Tahar, S. (eds.) TPHOLs 2002. LNCS, vol. 2410, pp. 31–46. Springer, Heidelberg (2002)
34. Almeida, A., Almeida, M., Alves, J., Moreira, N., Reis, R.: FAdo and GUitar: Tools for Automata Manipulation and Visualization. In: Maneth, S. (ed.) CIAA 2009. LNCS, vol. 5642, pp. 65–74. Springer, Heidelberg (2009)
35. Kozen, D.: On Hoare logic and Kleene algebra with tests. ACM Transactions on Computational Logic (TOCL) 1(1), 60–76 (2000)

Simple Rectangle-Based Functional Programs for Computing Reflexive-Transitive Closures

Rudolf Berghammer and Sebastian Fischer

Institut für Informatik
Christian-Albrechts-Universität Kiel
Olshausenstraße 40, 24098 Kiel, Germany
{rub,sebf}@informatik.uni-kiel.de

Abstract. We show how to systematically derive simple purely functional algorithms for computing the reflexive-transitive closure of directed graphs. Directed graphs can be represented as binary relations and we develop our algorithms based on a relation-algebraic description of reflexive-transitive closures. This description employs the relation-algebraic notion of rectangles and instantiating the resulting algorithm with different kinds of rectangles leads to different algorithms for computing reflexive-transitive closures. Using data refinement, we then develop simple Haskell programs for two specific choices of rectangles and show that one of them has cubic runtime like an imperative implementation of Warshall's standard algorithm.

1 Introduction

The reflexive-transitive closure of a directed graph G is a directed graph with the same vertices as G that contains an edge from each vertex x to each vertex y if and only if y is reachable from x in G. We can represent any directed graph as a (binary) relation R that relates two vertices x and y if and only if there is an edge from x to y. The reflexive-transitive closure of a directed graph G corresponds to the reflexive-transitive closure R^* of the binary relation R which represents G, i.e., to the least reflexive and transitive relation that contains R.

Usually, the task of computing R^* is solved by a variant of Warshall's algorithm which represents R as a Boolean matrix where each entry that is 1 represents an edge of the graph. It first computes the transitive closure R^+ of R using Warshall's original method [30] and then obtains R^* from R^+ by putting all entries of the main diagonal of R^+ to 1. Representing the directed graph as a 2-dimensional Boolean array leads to a simple and efficient program with three nested loops for the computation of R^+ and a single loop for the subsequent treatment of the main diagonal, i.e., to a program of time complexity $O(n^3)$ with n as cardinality of the carrier set of R. However, in certain cases arrays are unfit for representing relations, especially if both R and R^* are of "medium density" or even sparse. Then a representation of relations by successor lists is much more economic. But such a representation destroys the simplicity and efficiency of the array implementation. Moreover, the traditional method of imperatively

W. Kahl and T.G. Griffin (Eds.): RAMiCS 2012, LNCS 7560, pp. 114–129, 2012.

updating an array is alien to the purely functional programming paradigm, which restricts the use of side effects.

There are sub-cubic algorithms for the computation of reflexive-transitive closures. Most of them are based upon sub-cubic algorithms for matrix multiplication, i.e., on Strassen's method [27] and its refinements. There is also a refinement of Warshall's algorithm (more precisely, Floyd's extension for the all-pairs shortest paths problem [15]) that computes the transitive closure of a relation in time $O(n^{2.5})$ [16]. But all these algorithms pay for their exponents by an intricacy that, particularly with regard to practice, makes it difficult to implement them correctly (w.r.t. the input/output behaviour as well as the theoretical runtime bound) in a conventional programming language like C, Java or Haskell. The complexity of the algorithms also leads in the O-estimation to such large constant coefficients that, again concerning practice, results are computed faster normally only in the case of very dense relations or very large carrier sets.

To get on with the results of [7], in this paper we present an alternative method for computing reflexive-transitive closures. Using relation algebra as the methodological tool, we develop a generic algorithm for computing R^*, that is based on the decomposition of R by so-called rectangles, and provide some instantiations. Each concrete algorithm – we now speak of a program – is determined by a specific choice of a rectangle that is removed from R. Due to its generality, our generic algorithm not only allows simple and efficient imperative array-based implementations, but also simple and efficient purely functional ones, which base upon a very simple representation of relations via successor lists. As in [7], also in this paper we focus on functional programming and Haskell as programming language. Its remainder is organized as follows. Section 2 is devoted to the relation-algebraic preliminaries. In Section 3, we first prove a property about reflexive-transitive closures and rectangles that is the base of the generic algorithm and then present the algorithm itself and its instantiations in the second part of the section. Section 4 shows how the instantiations can be implemented in the functional programming language Haskell using data refinement. In particular, we will present a simple Haskell program that uses a list representation of relations and computes reflexive-transitive closures in cubic runtime, i.e., has the same time complexity as the aforementioned traditional imperative method. We think that this is a valuable single result, but we also believe that the overall method we will apply in the same way as in [7] is the more valuable contribution of the paper. Section 5 reviews related work and contains concluding remarks.

2   Relation-Algebraic Preliminaries

In this paper, we denote the set (or type) of all relations with source X and target Y (both are assumed to be non-empty) by $[X \leftrightarrow Y]$ and write $R : X \leftrightarrow Y$ instead of $R \in [X \leftrightarrow Y]$. If the carrier sets X and Y of R are finite, we may consider R as a Boolean matrix with $|X|$ rows and $|Y|$ columns. Since this interpretation is well suited for many purposes, we will often use matrix notation and terminology in this paper. In particular, we talk about rows, columns and entries of relations.

We assume the reader to be familiar with the basic operations on relations, viz. R^T (transposition), $\overline{R}$ (complement), $R \cup S$ (join), $R \cap S$ (meet), $R; S$ (composition), the predicates indicating $R \subseteq S$ (inclusion) and $R = S$ (equality), and the special relations O (empty relation), L (universal relation) and I (identity relation). Furthermore, we assume the reader to know the most fundamental laws of relation algebra like $R^{\mathsf{T}^\mathsf{T}} = R$, $(R; S)^\mathsf{T} = S^\mathsf{T}; R^\mathsf{T}$, $R; (S \cup T) = R; S \cup R; T$, and the *Schröder rule* stating the equivalence of $Q; R \subseteq S$, $Q^\mathsf{T}; \overline{S} \subseteq \overline{R}$ and $\overline{S}; R^\mathsf{T} \subseteq \overline{Q}$. We will also use the relation-algebraic specifications of the following properties: reflexivity $\mathsf{I} \subseteq R$, transitivity $R; R \subseteq R$, injectivity $R; R^\mathsf{T} \subseteq \mathsf{I}$ and surjectivity $\mathsf{L}; R = \mathsf{L}$. For more details, concerning the algebraic treatment of relations, we refer to [24,28].

A *vector* is a relation v which satisfies the equation $v = v; \mathsf{L}$ and a *point* is an injective and surjective vector. If p is a point, then we have $p \neq \mathsf{O}$, $\overline{R; p} = \overline{R}; p$, and that $R \subseteq S; p$ is equivalent to $R; p^\mathsf{T} \subseteq S$; see [24]. For vectors the targets are irrelevant. Therefore, we consider in the following mostly vectors $v : X \leftrightarrow \mathbf{1}$ with a specific singleton set $\mathbf{1} = \{\bot\}$ as target and omit in such cases the second element $\bot$ in a pair, i.e., write $x \in v$ instead of $(x, \bot) \in v$. A vector $v : X \leftrightarrow \mathbf{1}$ can be considered as a Boolean matrix with $|X|$ rows and exactly one column, i.e., as a Boolean column vector, and *represents* the subset $\{x \in X \mid x \in v\}$ of X. In the Boolean matrix model a point from $[X \leftrightarrow \mathbf{1}]$ is a Boolean column vector in which exactly one entry is 1. This means that points represent singleton sets, or elements if we identify a singleton set with the only element it contains. Later we will use that if the point $p : X \leftrightarrow \mathbf{1}$ represents $x_1 \in X$ and the point $q : X \leftrightarrow \mathbf{1}$ represents $x_2 \in X$, then $(y, z) \in p; q^\mathsf{T}$ is equivalent to $x_1 = y$ and $x_2 = z$.

3 Rectangles and Reflexive-Transitive Closures

Given a relation $R : X \leftrightarrow X$, its *reflexive-transitive closure* $R^* : X \leftrightarrow X$ is defined as the least reflexive and transitive relation that contains R and its *transitive closure* $R^+ : X \leftrightarrow X$ is defined as the least transitive relation that contains R. It is well-known that R^* and R^+ can also be specified via least fixed point constructions. The least fixed point of $\tau_R : [X \leftrightarrow X] \to [X \leftrightarrow X]$, defined by $\tau_R(Q) = \mathsf{I} \cup R; Q$, is R^*, and the least fixed point of $\sigma_R : [X \leftrightarrow X] \to [X \leftrightarrow X]$, defined by $\sigma_R(Q) = R \cup R; Q$, is R^+. From these specifications, we obtain by fixed point considerations the equations

$$\mathsf{O}^* = \mathsf{I} \qquad R^* = \mathsf{I} \cup R^+ \qquad (R \cup S)^* = R^*; (S; R^*)^* \qquad (1)$$

and that R is transitive if and only if $R = R^+$. The rightmost equation of (1) also is called the *star-decomposition* rule. It has a nice graph-theoretic interpretation. If in a directed graph the edges are coloured with two colours, say r and s, then each path can be decomposed into a (possibly empty) initial part p_0 with r-edges only, followed by paths $p_1, \ldots, p_m$, where $m = 0$ is possible and each p_i starts with an s-edge and $n \geq 0$ subsequent r-edges.

3.1 A Generic Algorithm for Reflexive-Transitive Closures

Now, assume two relations $R : X \leftrightarrow X$ and $S : X \leftrightarrow X$ of the same type. Then we have the following decomposition property for reflexive-transitive closures that is decisive for the remainder of the paper.

$$S; \mathsf{L}; S \subseteq S \implies (R \cup S)^* = R^* \cup R^*; S; R^* \tag{2}$$

A proof of (2) starts with the following calculation that applies the left-hand side (the assumption on S) in the last step.

$$S; R^*; S; R^* \subseteq S; \mathsf{L}; S; R^* \subseteq S; R^*$$

This transitivity of $S; R^*$ yields $(S; R^*)^+ = S; R^*$. We can prove the right-hand side of (2) now as follows.

$$
\begin{aligned}
(R \cup S)^* &= R^*; (S; R^*)^* & \text{star-decomposition rule of (1)} \\
&= R^*; (\mathsf{I} \cup (S; R^*)^+) & \text{second equation of (1)} \\
&= R^*; (\mathsf{I} \cup S; R^*) & S; R^* \text{ is transitive} \\
&= R^* \cup R^*; S; R^*, & \text{distributivity}
\end{aligned}
$$

Of course, (2) can also be proved in the traditional way based upon paths and reachability in directed graphs. We believe that relation algebra provides a much greater level of formality and, hence, drastically reduces the danger of making errors. It additionally enables the application of automatic theorem proving; see [6] for examples.

A relation $S : X \leftrightarrow X$ such that $S; \mathsf{L}; S \subseteq S$ is called a *rectangle*. In set-theoretic notation this means that it is of the form $A \times B$, where A and B are subsets of X, and in Boolean matrix terminology this means that a permutation of rows and columns transforms S into a form such that the 1-entries form a rectangular block. If we assume a function *rectangle* to be at hand that yields for a given non-empty relation R a non-empty rectangle S such that $S \subseteq R$, then a combination of the right-hand side of (2) with the equations $R = (R \cap \overline{S}) \cup S$ and $\mathsf{O}^* = \mathsf{I}$ leads to the following recursive algorithm for computing R^* that uses a let-clause to avoid duplicating the computation of $(R \cap \overline{S})^*$.

$$
\begin{aligned}
rtc(R) = \ &\texttt{if } R = \mathsf{O} \texttt{ then } \mathsf{I} \\
&\texttt{else let } S = rectangle(R) \\
&\qquad\quad C = rtc(R \cap \overline{S}) \\
&\texttt{in } C \cup C; S; C
\end{aligned}
\tag{3}
$$

If the input is a relation on a finite set, then this algorithm terminates since the arguments of the recursive calls become strictly smaller untill the termination condition is fulfilled.

3.2 Some Instantiations

For giving concrete code for the choice of the rectangle S in algorithm (3), assume $R : X \leftrightarrow X$ to be a non-empty relation. Then a first possibility to obtain a non-empty rectangle inside of R is as follows. Select a point $p : X \leftrightarrow \mathbf{1}$ such that

$p \subseteq R; \mathsf{L}$. After that, select a point $q : X \leftrightarrow \mathbf{1}$ such that $q \subseteq R^\mathsf{T}; p$. That $R^\mathsf{T}; p$ is non-empty, i.e., contains a point, can be shown as follows. Assume $R^\mathsf{T}; p = \mathsf{O}$. Then this is equivalent to $R; \mathsf{L} \subseteq \overline{p}$ because of the Schröder rule, so that we have $p \subseteq \overline{R; \mathsf{L}}$. Together with $p \subseteq R; \mathsf{L}$ this yields the contradiction $p = \mathsf{O}$. By means of the points p and q, finally, define S as $p; q^\mathsf{T} : X \leftrightarrow X$.

Using graph-theoretic terminology, the point p represents a vertex x of the directed graph $G = (X, R)$ with set X of vertices and set R of edges that possesses at least one successor, the vector $R^\mathsf{T}; p$ represents the successor set $succ_R(x) := \{y \in X \mid (x, y) \in R\}$ of x and the point q represents an element of $succ_R(x)$, that is, a single successor of x. Here is the correctness proof for the above procedure. From the vector property of p we get

$$S; \mathsf{L}; S = p; q^\mathsf{T}; \mathsf{L}; p; q^\mathsf{T} \subseteq p; \mathsf{L}; q^\mathsf{T} = p; q^\mathsf{T} = S, \;$$

i.e., that the relation S indeed is a rectangle. Non-emptiness of S follows from the fact that p and q are points, since the assumption $S = p; q^\mathsf{T} = \mathsf{O}$ is equivalent to the inclusion $p \subseteq \mathsf{O}; q$ (see Section 2), i.e., contradicts the point property $p \neq \mathsf{O}$. Finally, if we combine the assumption $q \subseteq R^\mathsf{T}; p$ with the injectivity of the point p, this yields that S is contained in R by the calculation

$$S = p; q^\mathsf{T} \subseteq p; (R^\mathsf{T}; p)^\mathsf{T} = p; p^\mathsf{T}; R^{\mathsf{T}^\mathsf{T}} \subseteq R.$$

Let us assume a pre-defined operation *point* for the selection of a point from a non-empty vector to be available as, for instance, in the programming language of the Kiel relation algebra tool RELVIEW [4]. Then the above approach leads to the following refinement of algorithm (3).

$$
\begin{aligned}
rtc(R) = \text{ if } R &= \mathsf{O} \text{ then } \mathsf{I} \\
\text{else let } p &= point(R; \mathsf{L}) \\
q &= point(R^\mathsf{T}; p) \\
S &= p; q^\mathsf{T} \\
C &= rtc(R \cap \overline{S}) \\
\text{in } C &\cup C; S; C
\end{aligned}
\qquad (4)
$$

Because p and q are selected as points, the rectangle $S := p; q^\mathsf{T}$ is an atom of the ordered set $([X \leftrightarrow X], \subseteq)$. The latter means that it contains exactly one pair, or, in Boolean matrix terminology, exactly one entry is 1. As direct product, S is of the form $\{x\} \times \{y\}$, with $x \in X$ represented by p and $y \in X$ represented by q. So, the (proper) subrelation $R \cap \overline{S}$ of R is obtained by taking off from R the single pair (x, y). Due to this property, program (4) leads to $|R|$ recursive calls.

Obviously, an operation *atom* such that $atom(R)$ is an atom contained in R would simplify the first possibility for obtaining a rectangle. We renounce its use for two reasons. First, it would require additional effort to axiomize *atom*. Secondly, a use of *atom* would obfuscate the general principle of obtaining from the point p step-wisely more appropriate rectangles.

For presenting a second possibility to obtain a rectangle inside of the given relation $R : X \leftrightarrow X$, we assume again that $p : X \leftrightarrow \mathbf{1}$ is a point contained in the vector $R; \mathsf{L}$. If we do not select a single point q from $R^{\mathsf{T}}; p$ as above, but take the entire vector $R^{\mathsf{T}}; p$ instead, then in combination with

$$p; \left(R^{\mathsf{T}}; p\right)^{\mathsf{T}} = p; p^{\mathsf{T}}; R^{\mathsf{T}^{\mathsf{T}}} = p; p^{\mathsf{T}}; R$$

this choice yields the following instantiation of algorithm (3). In graph-theoretic terminology, this means that instead of removing a single edge in each step we remove all edges that originate from a specific vertex.

$$
\begin{aligned}
rtc(R) = \ &\textbf{if } R = \mathsf{O} \textbf{ then } \mathsf{I} \\
&\textbf{else let } p = point(R; \mathsf{L}) \\
&\qquad\quad S = p; p^{\mathsf{T}}; R \\
&\qquad\quad C = rtc(R \cap \overline{S}) \\
&\textbf{in } C \cup C; S; C
\end{aligned}
\qquad (5)
$$

Concerning the correctness of program (5), the proof of the rectangle property of $S := p; p^{\mathsf{T}}; R$ via the calculation

$$S; \mathsf{L}; S = p; p^{\mathsf{T}}; R; \mathsf{L}; p; p^{\mathsf{T}}; R \subseteq p; \mathsf{L}; p^{\mathsf{T}}; R = p; p^{\mathsf{T}}; R = S$$

uses (in the third step) again that p is a vector. To show non-emptiness of the rectangle S, we assume $S = \mathsf{O}$. Starting with this, we get

$$
\begin{aligned}
p; p^{\mathsf{T}}; R = \mathsf{O} &\iff (p; p^{\mathsf{T}})^{\mathsf{T}}; \mathsf{L} \subseteq \overline{R} && \text{Schröder rule} \\
&\iff p; p^{\mathsf{T}}; \mathsf{L} \subseteq \overline{R} && \\
&\iff p; \mathsf{L} \subseteq \overline{R} && p \text{ surjective} \\
&\iff R; \mathsf{L} \subseteq \overline{p} && \text{Schröder rule, } \mathsf{L} = \mathsf{L}^{\mathsf{T}} \\
&\iff p \subseteq \overline{R; \mathsf{L}}
\end{aligned}
$$

so that, together with $p \subseteq R; \mathsf{L}$, we arrive at the contradiction $p = \mathsf{O}$. Finally, $S = p; p^{\mathsf{T}}; R \subseteq R$ immediately follows from the injectivity of p.

Assume that the point $p : X \leftrightarrow \mathbf{1}$ represents the element $x \in X$. As a direct product then the rectangle $S = p; p^{\mathsf{T}}; R$ is of the form $\{x\} \times X$. Using matrix terminology, S is obtained from R by, as [24] calls it, "singling out" the x-row of R. This means that the x-row of S coincides with the x-row of R and all other entries of S are 0. Hence, $R \cap \overline{S}$ is obtained from R by "zeroing out" the x-row. As a consequence, program (5) leads to at most $|X|$ recursive calls.

It should be mentioned that in graph-theoretic terminology p represents a vertex x and zeroing out the x-row of R means to take off from the directed graph $G = (X, R)$ all edges starting from x.

Here is an example for the zeroing-out technique. The leftmost of the following four pictures depicts a relation R on the set $X := \{1, \dots, 8\}$ as a Boolean 8×8 matrix. The remaining three pictures show a vector p that represents the element 4 of X, then the rectangle $S = p; p^{\mathsf{T}}; R$ and finally, the relation $R \cap \overline{S}$. All pictures are produced with the help of RELVIEW. A grey square denotes a 1-entry and a white square denotes a 0-entry.

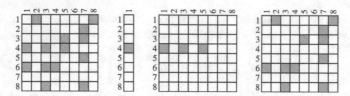

It is obvious that in program (5) the rectangle $p; p^\mathsf{T}; R$ can be replaced by the rectangle $R; p; p^\mathsf{T}$ if additionally the selection of p is done by $p = point(R^\mathsf{T}; \mathsf{L})$. This rectangle singles out the x-column of R. Its use again leads to an instantiation of algorithm (3) with at most $|X|$ recursive calls.

A further instantiation of algorithm (3) can be obtained if we select a *maximal rectangle* contained in the input relation $R : X \leftrightarrow X$. As shown in [26], every point $p : X \leftrightarrow X$ that is contained in $R; \mathsf{L}$ gives rise to a specific maximal rectangle inside R started horizontally by

$$S := \overline{R; R^\mathsf{T}}; p; p^\mathsf{T}; R = \overline{R; R^\mathsf{T}; p}; (R^\mathsf{T}; p)^\mathsf{T}.$$

Using this rectangle and a variable v within the let-binding to avoid the two-fold evaluation of $R^\mathsf{T}; p$, we obtain the following program for computing R^*.

$$
\begin{aligned}
rtc(R) = \text{ if } & R = \mathsf{O} \text{ then } \mathsf{I} \\
\text{else let } & p = point(R; \mathsf{L}) \\
& v = R^\mathsf{T}; p \\
& S = \overline{R}; v; v^\mathsf{T} \\
& C = rtc(R \cap \overline{S}) \\
\text{in } & C \cup C; S; C
\end{aligned}
\tag{6}
$$

With the help of the Schröder rule we get $\mathsf{I} \subseteq \overline{R; R^\mathsf{T}}$ from $\mathsf{I}; R \subseteq R$, so that the rectangle used in program (5) is contained in that used in program (6). As a consequence, the number of recursive calls of program (6) is bounded by the number of recursive calls of program (5). As experiments with RELVIEW have shown, program (6) frequently leads to fewer recursive calls than program (5) in practice. This is due to the fact that, with $x \in X$ represented by $p : X \leftrightarrow \mathbf{1}$, the step from R to $R \cap \overline{S}$, where $S = \overline{R; R^\mathsf{T}}; p; p^\mathsf{T}; R$, not only zeroes out the x-row of R but, at the same time, all rows of R identical to it. In graph-theoretic terminology this "parallel zeroing out" removes all edges from $G = (X, R)$ that start from a vertex $y \in X$ such that $succ_R(y) = succ_R(x)$.

If the above point p is contained in $R^\mathsf{T}; \mathsf{L}$ instead of $R; \mathsf{L}$, again due to [26] this gives rise to another specific maximal rectangle inside R, but now started vertically. This rectangle is given by $S := R; p; p^\mathsf{T}; R^\mathsf{T}; \overline{R}$ and leads to a variant which is on a par with program (6).

4   Implementations in Haskell

The aim of Section 4.1 is to develop by means of the method of [7] systematically an implementation of program (5) of Section 3.2 that uses linear lists of successor sets for representing relations. To simplify the development, we assume that the finite carrier set X of the input relation $R : X \leftrightarrow X$ consists of the natural numbers $0, 1, \ldots, m$. In Section 4.2 we then translate (using again data refinement) the result of Section 4.1 into the functional programming language Haskell and show that the resulting program runs in cubic time. We assume readers to be familiar with Haskell. Otherwise, they may consult one of the well-known textbooks about it, for example [9,29]. Possible approaches for programs (4) and (6) are discussed in Section 4.3.

4.1   From Relation Algebra to Linear Lists

To obtain a version of program (5) that works on linear lists, we represent in a first step relations of type $[X \leftrightarrow X]$ by functions from X to its powerset 2^X which map each element to its successor set. Doing so, the empty relation O is represented by $\lambda x.\emptyset$ and the identity relation I by $\lambda x.\{x\}$, where we denote anonymous functions as λ-terms.

Now, let the relations R and C of program (5) be represented by the functions $r : X \to 2^X$ and $c : X \to 2^X$, respectively. Then the selection of a point p such that $p \subseteq R; \mathsf{L}$ corresponds to the selection of an element n from X such that $r(n) \neq \emptyset$. This element may be assumed to be represented by p. Based on r, c and n, the function representing $R \cap \overline{p; p^\mathsf{T}}; R$ maps all $x \neq n$ to $r(x)$ and n to $\emptyset$. Usually this function is denoted as $r[n \leftarrow \emptyset]$. For the relation $C \cup C; p; p^\mathsf{T}; R; C$ we get $\lambda x.\text{if } n \in c(x) \text{ then } c(x) \cup ks \text{ else } c(x)$ as representing function, where ks abbreviates the set $\bigcup\{c(k) \mid k \in r(n)\}$. This follows from the subsequent calculation in which x and y are arbitrary elements from the set X and in the existentially quantified formulae i, j, k range over X.

$$
\begin{aligned}
&\quad (x, y) \in C \cup C; p; p^\mathsf{1}; R; C \\
&\Longleftrightarrow (x, y) \in C \vee (x, y) \in C; p; p^\mathsf{T}; R; C \\
&\Longleftrightarrow (x, y) \in C \vee \exists i : (x, i) \in C \wedge \exists j : (i, j) \in p; p^\mathsf{T} \wedge (j, y) \in R; C \\
&\Longleftrightarrow (x, y) \in C \vee \exists i : (x, i) \in C \wedge \exists j : i = n \wedge j = n \wedge (j, y) \in R; C \\
&\Longleftrightarrow (x, y) \in C \vee ((x, n) \in C \wedge (n, y) \in R; C) \\
&\Longleftrightarrow (x, y) \in C \vee ((x, n) \in C \wedge \exists k : (n, k) \in R \wedge (k, y) \in C) \\
&\Longleftrightarrow y \in c(x) \vee \text{if } n \in c(x) \text{ then } \exists k : k \in r(n) \wedge y \in c(k) \text{ else } \textit{false} \\
&\Longleftrightarrow y \in c(x) \vee \text{if } n \in c(x) \text{ then } y \in \bigcup\{c(k) \mid k \in r(n)\} \text{ else } y \in \emptyset \\
&\Longleftrightarrow y \in c(x) \vee y \in \text{if } n \in c(x) \text{ then } ks \text{ else } \emptyset \\
&\Longleftrightarrow y \in \text{if } n \in c(x) \text{ then } c(x) \cup ks \text{ else } c(x)
\end{aligned}
$$

This calculation only uses some well-known correspondences between logical and relation-algebraic respectively set-theoretic constructions and the definition of the set ks. The conditional is introduced only to enhance readability and to prepare a later translation into Haskell.

If we replace in program (5) all relations by their representing functions and the selection of the point p by that of the element n from $\{x \in X \mid r(x) \neq \emptyset\}$ via *elem*, we arrive at the following functional program; the function modification $r[n \leftarrow \emptyset]$ behaves like r with the exception that it maps n to the empty set. Although constructions are used which do not exist in conventional programming languages (equality test on functions, non-deterministic choice, function modification, arbitrary set union), we speak of a program since their realizations are rather trivial due to the finiteness of X.

$$rtc(r) = \text{if } r = \lambda x.\emptyset \text{ then } \lambda x.\{x\}$$
$$\text{else let } n = elem(\{x \in X \mid r(x) \neq \emptyset\})$$
$$c = rtc(r[n \leftarrow \emptyset])$$
$$ks = \bigcup\{c(k) \mid k \in r(n)\}$$
$$\text{in } \lambda x.\text{if } n \in c(x) \text{ then } c(x) \cup ks \text{ else } c(x)$$

In terms of directed graphs G this algorithm describes the following fact. If the directed graph G' is obtained from G by removing all outgoing edges of a certain vertex n of G and x is a G'-ancestor of n, then y is a G-descendant of x if and only if it is a G'-descendant of x or a G'-descendant of a G-successor of n.

In the next steps, we represent functions by linear lists. We start with the functions which appear as results of *rtc*, i.e., $c : X \to 2^X$ and the anonymous ones $\lambda x.\{x\}$ and $\lambda x.\text{if } n \in c(x) \text{ then } c(x) \cup ks \text{ else } c(x)$. By assumption, the set X consists of the natural numbers $0, 1, \ldots, m$. Hence, it seems to be obvious to represent a function from X to 2^X by a linear list of length $m + 1$ such that for all x with $0 \leq x \leq m$ the x-th component of the list equals the result of the application of the function to x, i.e., is a subset of X. In the case of $\lambda x.\{x\}$ this yields the linear list $[\{0\}, \ldots, \{m\}]$ of singleton sets. Using a notation similar to Haskell's list comprehension, we denote it by $[\{x\} \mid x \in [0..m]]$. For the anonymous function

$$\lambda x.\text{if } n \in c(x) \text{ then } c(x) \cup ks \text{ else } c(x)$$

we obtain, again using the list comprehension notation just introduced, the list representation

$$[\text{if } n \in cs!!x \text{ then } cs!!x \cup ks \text{ else } cs!!x \mid x \in [0..m]], \tag{7}$$

where cs is the list representation of c and, as in Haskell, $cs!!x$ denotes the x-th component of the linear list cs. The list cs consists exactly of the lists $cs!!x$, where x ranges over $0, 1, \ldots, m$. Hence, instead of ranging in the list comprehension (7) over all these indices, we alternatively can range over all components ms of the linear list cs, i.e., can replace in (7) the formula $x \in [0..m]$ by $ms \in cs$, if simultaneously the expression $cs!!x$ is replaced by ms.

Moving from the output functions to the list representations we have transformed the above program into the following one.

$$rtc(r) = \text{if } r = \lambda x.\emptyset \text{ then } [\{x\} \mid x \in [0..m]]$$
$$\text{else let } n = elem(\{x \in X \mid r(x) \neq \emptyset\})$$
$$cs = rtc(r[n \leftarrow \emptyset])$$
$$ks = \bigcup\{cs!!k \mid k \in r(n)\}$$
$$\text{in } [\text{if } n \in ms \text{ then } ms \cup ks \text{ else } ms \mid ms \in cs]$$

It remains to give list representations of the input functions r and $r[n \leftarrow \emptyset]$ of this version of rtc. Here the approach used for the result functions of rtc seems not to be reasonable. In the case of $\lambda x.\emptyset$ such a representation leads to the task of testing whether all components of a linear list of sets are $\emptyset$. Also the selection of an n such that $0 \leq n \leq m$ and the n-th component of the list representation of r is not $\emptyset$ requires some effort. The same holds for the computation of the list representation of $r[n \leftarrow \emptyset]$ from that of r.

If, however, the input function r of rtc is represented as a linear list rs of pairs (x, xs), where x ranges over all numbers from 0 to m for which $r(x) \neq \emptyset$ and xs is a set that equals $r(x)$, and the function $r[n \leftarrow \emptyset]$ is represented in the same way, then the solutions of the just mentioned three tasks are rather simple. Testing $r = \lambda x.\emptyset$ reduces to the list-emptiness test $rs = []$. An n with $0 \leq n \leq m$ and $r(n) \neq \emptyset$ is given by the first component of the head of the linear list rs. Because of the relationship between the two components of each list element, with this choice of n the list representation of the function $r[n \leftarrow \emptyset]$ is obtained from that of r by simply removing the head of rs and, furthermore, the set $r(n)$ in $\bigcup\{cs!!k \mid k \in r(n)\}$ equals the second component of the head of rs. If we use pattern matching in the let-clause and the well-known list-operations $head$ and $tail$, we obtain the following program.

$$
rtc(rs) = \text{if } rs = [] \text{ then } [\{x\} \mid x \in [0..m]]
$$
$$
\text{else let } (n, ns) = head(rs)
$$
$$
cs = rtc(tail(rs))
$$
$$
ks = \bigcup\{cs!!k \mid k \in ns\}
$$
$$
\text{in } [\text{if } n \in ms \text{ then } ms \cup ks \text{ else } ms \mid ms \in cs]
$$

This program can also be written in the following form. It uses a non-recursive auxiliary function $step$ that performs, for the head of rs and the result of the recursive call of rtc as inputs, the essential computations of the recursion.

$$
step((n, ns), cs) = \text{let } ks = \bigcup\{cs!!k \mid k \in ns\}
$$
$$
\text{in } [\text{if } n \in ms \text{ then } ms \cup ks \text{ else } ms \mid ms \in cs]
$$
$$
rtc(rs) = \text{if } rs = [] \text{ then } [\{x\} \mid x \in [0..m]]
$$
$$
\text{else } step(head(rs), rtc(tail(rs)))
$$

(8)

Obviously, an execution of this program leads to at most $m + 1 = |X|$ recursive calls of rtc. Since the generation of the linear list $[\{x\} \mid x \in [0..m]]$ can be done in time $O(m)$ and $head$ and $tail$ require constant time, the total runtime of program (8) depends on the time complexity of the auxiliary function $step$.

4.2 A Haskell Program for Reflexive-Transitive Closures

For the following, we assume the largest number m of the set X to be at hand as Haskell constant m of type Int. To obtain an implementation of program (8) in Haskell, the main task is to represent subsets of X in Haskell and to formulate a Haskell function that computes the union of two sets. From the latter, a simple recursion (that we will realize by a pre-defined higher-order Haskell function) then at once leads to the computation of the set $\bigcup\{cs!!k \mid k \in ns\}$.

An obvious representation of subsets of X in Haskell is given by linear lists over X *without multiple occurrences of elements*. Then $\emptyset$ is implemented by [] and set-membership by the pre-defined Haskell operation `elem`. A straightforward implementation of set union requires quadratic time. But we can do better using that the set $X = \{0, 1, \ldots, m\}$ is linearly ordered. If we additionally demand that *all lists that occur as representations of successor sets are increasingly sorted*, then an obvious implementation of set union is given by the following Haskell function. It needs linear time to merge two sorted lists into a sorted one and to remove at the same time all multiple occurrences of elements.

```
merge :: [Int] -> [Int] -> [Int]

merge []     ys              = ys
merge xs     []              = xs
merge (x:xs) (y:ys) | x == y = x : merge xs ys
                    | x <  y = x : merge xs (y:ys)
                    | x >  y = y : merge (x:xs) ys
```

Now, assume the set ns of the expression $\bigcup\{cs!!k \mid k \in ns\}$ of program (8) to be given as Haskell list `ns` and let `cs` and `k` be the Haskell counterparts of cs and k. Then the Haskell list for $\bigcup\{cs!!k \mid k \in ns\}$ is obtained by merging all sorted lists `cs!!k`, where k ranges over the elements of `ns`. Such a repeated application of `merge` corresponds to a fold over a list containing all lists `cs!!k`, with the empty list as initial value. If we use right-fold, in Haskell realized by the higher-order function `foldr`, we get `foldr merge [] [cs!!k | k <- ns]` as list representation of $\bigcup\{cs!!k \mid k \in ns\}$ and, thus, the following Haskell implementation of *step*.

```
step :: (Int,[Int]) -> [[Int]]  -> [[Int]]
step (n,ns) cs =
  let ks = foldr merge [] [cs!!k | k <- ns]
  in  [if elem n ms then merge ms ks else ms | ms <- cs]
```

Tests have shown that this function becomes slightly faster if the pre-defined operation `elem` is replaced by a user-defined one that takes advantage of the fact that all successor lists are sorted. Since `ns` is sorted, also `[cs!!k | k <- ns]` can be done in linear time by a user-defined function.

In view of the main function rtc of (8), our Haskell implementation of successor sets leads to `[[x] | x <- [0..m]]` as Haskell code for $[\{x\} \mid x \in [0..m]]$. Also rtc can be seen as a right-fold over the Haskell counterpart `rs` of rs that realizes the repeated application of `step`. The initial value is `[[x] | x <- [0..m]]`. So, we arrive at the following Haskell implementation.

```
rtc :: [(Int,[Int])] -> [[Int]]
rtc rs =
  foldr step [[x] | x <- [0..m]] rs
```

For estimating the runtime of step, assume that the length of the input lists ns, cs and of all components of cs is at most $m + 1$ and that the linear list ns and all components of the linear list cs are sorted. Then the computation of ks can be done in time $O(m^2)$ since this time suffices to generate [cs!!k | k <- ns] as well as to merge its components to a single list. Notice that during the merging process each argument of merge has at most length $m + 1$, the same also holds for the result and merge merges sorted lists in linear runtime. Also the list comprehension of step can be evaluated in $O(m^2)$ steps. This follows again from the assumptions on the input lists since they imply that each of the at most $m + 1$ membership-tests and mergings can be done in runtime $O(m)$. (The length of the linear list ks is at most $m + 1$).

If the Haskell representation rs of the original input relation R satisfies the requirements stated at the beginning of this section, i.e., each successor list is sorted and without multiple occurrences of elements, then all arguments of the calls of step that occur during the evaluation of rtc rs satisfy the assumptions of the runtime estimation of step. Together with the remark made at the end of Section 4.1, we get the claimed runtime $O(m^3)$.

Slightly dissatisfying is that the above Haskell function rtc depends on the constant m and works with two different implementations of relations. But it is not difficult to change it in such a way that it is independent of m and also the input rs is a linear list of type [[Int]] that contains as x-th component the (possibly empty) successor list of x. Here is the result, where the Haskell function step is as declared above.

```
rtc :: [[Int]] -> [[Int]]
rtc rs =
    let xs = [0..length rs - 1]
    in  foldr step [[n] | n <- xs] (zip xs rs)
```

Since the length of rs is $m + 1$, the linear list xs equals [0..m]. Using the pre-defined operation zip that takes two linear lists and returns the linear list of corresponding pairs, the list rs of successor lists is transformed into a list of pairs (x, xs), where x ranges over $X = \{0, 1, \ldots, m\}$ and xs is the successor list of x. This is almost the same representation as used in Section 4.1 for the output of rtc; compared with Section 4.1 the only difference is that now a pair of the linear list rs may contain an empty second component. Such empty components do not affect correctness since a call step (n,[]) cs yields cs as its result.

The relation-algebraic background of this fact is as follows: If $succ_R(x) = \emptyset$, then x is represented by a point p such that $p \sqsubseteq \overline{R; \mathsf{L}}$. From this inclusion we get $p^\mathsf{T}; R = \mathsf{O}$; cf. Section 3.2. Hence, we have $(R \cap p; p^\mathsf{T}; R)^* = R^*$. The only reason for the requirement $rectangle(R) \neq \mathsf{O}$ in algorithm (3) is to ensure termination.

The just described version of rtc has the same time complexity as the original one even if (zip xs rs) is replaced by (zip xs (map (nub.sort) rs)), where nub removes duplicate elements from a list, sort is the pre-defined sorting function on lists, " . " denotes function composition, and map is the higher-order function that applies a function to each component of a list. Hence, we can provide an implementation of `rtc` that does not rely on the successor lists to be strictly increasing without sacrificing time complexity.

4.3 How to Deal with Single Pairs and Maximal Rectangles

The program (4) of Section 3.2 can also be turned into a Haskell program. This time we consider the input relation R as a set of pairs, i.e., as a set-theoretic relation, and only the output as a function. (In set theory, a relation is untyped and specified as set of pairs. Strictly speaking, the relations we have introduced in Section 2 are typed and consist of a source, a target and a set of pairs.) Let r be the set of pairs contained in the relation $R : X \leftrightarrow X$, the function $c : X \to 2^X$ implement the relation $C : X \leftrightarrow X$ and $n, k \in X$ be the elements of X represented by the points $p : X \leftrightarrow \mathbf{1}$ and $q : X \leftrightarrow \mathbf{1}$, respectively. Then obviously $r \setminus \{(n, k)\}$ is the set of pairs of $R \cap \overline{p; q^\mathsf{T}}$. Furthermore, a calculation similar to that of Section 4.1 yields the equivalence

$$(x, y) \in C \cup C; p; q^\mathsf{T}; C \iff y \in \text{if } n \in c(x) \text{ then } c(x) \cup c(k) \text{ else } c(x)$$

for all $x, y \in X$. From this equivalence we obtain the anonymous function $\lambda x.\text{if } n \in c(x) \text{ then } c(x) \cup c(k) \text{ else } c(x)$ as implementation of the expression $C \cup C; p; q^\mathsf{T}; C$. Altogether, we get the following version of program (4).

$$\begin{aligned}
rtc(r) = \ &\text{if } r = \emptyset \text{ then } \lambda x.\{x\} \\
&\text{else let } (n, k) = elem(r) \\
&\qquad\quad c = rtc(r \setminus \{(n, k)\}) \\
&\quad \text{in } \lambda x.\text{if } n \in c(x) \text{ then } c(x) \cup c(k) \text{ else } c(x)
\end{aligned}$$

In order to translate this function into Haskell code, we represent the set r of pairs as linear list rs of pairs. We implement the function c by a linear list cs of sets such that (as in Section 4.1) $cs!!x$ equals $c(x)$. The comparison $r = \emptyset$ can be implemented in Haskell by checking whether rs is the empty list. If rs is non-empty, then we implement $elem(r)$ as $head(r)$ and $r \setminus \{(n, k)\}$ as $tail(r)$. For the function $\lambda x.\text{if } n \in c(x) \text{ then } c(x) \cup c(k) \text{ else } c(x)$ a little reflection yields the list implementation [if $n \in ms$ then $ms \cup cs!!k$ else $ms \mid ms \in cs$]. If successor sets are implemented as in Section 4.2, then the just sketched list version succinctly can be formulated in Haskell by means of pattern matching. The result is then as follows.

```
rtc :: [(Int,Int)] -> [[Int]]
rtc [] = [[x] | x <- [0..m]]
rtc ((n,k):rs) =
   let cs = rtc rs
   in [if elem n ms then merge ms (cs!!k) else ms | ms <- cs]
```

An analysis similar to that of Section 4.2 shows that the time complexity of this program is $O(|R| m^2)$, i.e. $O(m^4)$. It remains unchanged if additionally a small Haskell program is used that transforms lists of successor lists into the corresponding lists of pairs, i.e., adapts the result type of `rtc` to its input type.

We can also obtain a Haskell version of program (6) along the lines of Sections 4.1 and 4.2. But the result is much more complicated as the Haskell programs we have developed so far. To give an impression for the difficulties that occur, we consider the computation of the argument of the recursive call of rtc from the input list rs. Assume that $n \in X$ is represented by the point p. The maximal rectangle $\overline{R}; R^{\mathsf{T}}; p; p^{\mathsf{T}}; R$ in R cannot be removed by a simple call such as $tail(rs)$. Instead, we have to compute a linear list that consists of all those pairs (x, xs) from rs such that xs is not equal to the second component of $head(rs)$. In most cases the intricacy of the final program (which we do not want to present here) annihilates the advantage of fewer recursive calls.

5   Related Work and Concluding Remarks

Since directed graphs are nothing else than relations on sets of vertices, computing the reflexive-transitive closure of a relation can be seen as a problem of graph-theory. Therefore, related to our work are all the approaches to program graph algorithms in a functional programming language. In the meantime functional graph algorithms have a long tradition. Here we only want to mention some papers that deal with different aspects. In [1] the transformational programming technique is applied to derive certain functional reachability algorithms. The papers [20,21] deal with the specification and functional computation of the depth-first search forest and present some classical applications (topological sorting, testing for cycles, strongly connected components) in the functional style. To achieve a linear runtime, monads are used to mimic the imperative marking technique. In [12] pre-defined operators on graphs are introduced to program functional graph algorithms. The realization of a program consists in the selection of an appropriate operator together with fitting parameter functions and data structures. Instead of regarding graphs as monolithic data as all the just mentioned papers do, in [13,14] graphs are inductively generated. This approach allows to write many graph algorithms in the typical pattern-based functional style. Also in [17] an inductive definition of graphs is given, but restricted to DAGs (directed acyclic graphs) and without discussing an implementation.

Since many years also relation algebra, the methodological tool we have used in this paper, and some variations are used in algorithmics. The papers can be discriminated by the way how the manipulation of relations is separated from the algorithmic principles involved. Some of them consider programs as well as the data which are manipulated by them as relations such that a separation is not given. Typical examples are [10,23]. But in most cases the separation is very clear since relations are only used as data and manipulated with a conventional programming language. Concerning the programming style, the imperative one is predominant; see [2,3,8,22] for typical examples. Papers on functional programming with relations seem to be rare. Besides [7], we want to refer to [11], where relations are used to develop a Haskell program for computing maximum bipartite matchings, and [19,25], where Haskell libraries for relation algebra are presented. Using such libraries, each of the programs (4), (5) and (6) immediately can be written in Haskell but, compared with our programs from Sections 4.2 and 4.3, with loss of efficiency. In [5] it is shown how relation algebra and the features of the functional-logic extension Curry (see [18]) of Haskell can be employed to solve problems in a very high-level declarative style.

Having a Haskell library for relation algebra at hand, it is obvious that also the algorithm scheme (3) immediately can be translated into Haskell, even into a generic program parameterized with the function *rectangle*. But presently we do not see a possibility to provide for this (admittedly much more elegant) solution a successor list implementation that possesses cubic runtime if *rectangle* singles out a row (or column). Even Boolean arrays lead to a worser algorithm if the computation of $S := p; p^\mathsf{T}; R$ and its use in $C \cup C; S; C$ are separated. In this case for computing $C \cup C; S; C$ from S and C cubic runtime is necessary, whereas for given p and C the clever use of the brackets in $C \cup (C; p); ((p^\mathsf{T}; R); C)$ leads to quadratic runtime.

Each relation can be specified as a union of disjoint rectangles. This is closely related to the inductive generation of graphs in [13,14] since there in each step of the generation process a graph g is extended by a new vertex x and edges connecting x with vertices that are already in g. The insertion of the edges corresponds to a union with two rectangles, one consisting of the edges with source x and one consisting of the edges with sink x. It would be interesting to explore how the inductive graph view can be transferred to relations to make algorithms on relations and relational structures (like graphs, orders, Petri nets, games) more amenable for the typical functional style.

Acknowledgement. We want to thank B. Braßel, J. Christiansen, F. Huch, and the anonymous referees for valuable hints.

References

1. Berghammer, R., Ehler, H., Zierer, H.: Development of Graph Algorithms by Program Transformation. In: Göttler, H., Schneider, H.-J. (eds.) WG 1987. LNCS, vol. 314, pp. 206–218. Springer, Heidelberg (1988)
2. Berghammer, R., Hoffmann, T.: Deriving relational programs for computing kernels by reconstructing a proof of Richardson's theorem. Sci. of Comput. Progr. 38, 1–25 (2000)
3. Berghammer, R., Hoffmann, T.: Relational depth-first-search with applications. Inform. Sci. 139, 167–186 (2001)
4. Berghammer, R., Neumann, F.: RelView – An OBDD-Based Computer Algebra System for Relations. In: Ganzha, V.G., Mayr, E.W., Vorozhtsov, E.V. (eds.) CASC 2005. LNCS, vol. 3718, pp. 40–51. Springer, Heidelberg (2005)
5. Berghammer, R., Fischer, S.: Implementing relational specifications in a constraint functional language. Electr. Notes on Theor. Comput. Sci. 177, 169–183 (2007)
6. Berghammer, R., Struth, G.: On Automated Program Construction and Verification. In: Bolduc, C., Desharnais, J., Ktari, B. (eds.) MPC 2010. LNCS, vol. 6120, pp. 22–41. Springer, Heidelberg (2010)
7. Berghammer, R.: A Functional, Successor List Based Version of Warshall's Algorithm with Applications. In: de Swart, H. (ed.) RAMICS 2011. LNCS, vol. 6663, pp. 109–124. Springer, Heidelberg (2011)

8. Bird, R., Ravelo, J.: On computing representatives. Inf. Proc. Lett. 63, 1–7 (1997)
9. Bird, R.: Introduction to functional programming using Haskell, 2nd edn. Prentice Hall (1998)
10. Brunn, T., Möller, B., Russling, M.: Layered Graph Traversals and Hamiltonian Path Problems - An Algebraic Approach. In: Jeuring, J. (ed.) MPC 1998. LNCS, vol. 1422, pp. 96–121. Springer, Heidelberg (1998)
11. Danilenko, N.: Using Relations to Develop a Haskell Program for Computing Maximum Bipartite Matchings. In: Kahl, W., Griffin, T.G. (eds.) RAMiCS 2012. LNCS, vol. 7560, pp. 130–145. Springer, Heidelberg (2012)
12. Erwig, M.: Graph Algorithms = Iteration + Data Structures? The Structure of Graph Algoritms and a Corresponding Style of Programming. In: Mayr, E.W. (ed.) WG 1992. LNCS, vol. 657, pp. 277–292. Springer, Heidelberg (1993)
13. Erwig, M.: Functional programming with graphs. ACM SIGPLAN Notices 32, 52–65 (1997)
14. Erwig, M.: Inductive graphs and functional graph algorithms. J. of Funct. Progr. 11, 467–492 (2001)
15. Floyd, R.W.: Algorithm 97 (Shortest path). Comm. of the ACM 5, 345 (1962)
16. Fredmann, M.L.: New bounds on the complexity of the shortest path problem. SIAM J. on Comput. 5, 83–89 (1976)
17. Gibbons, J.: An Initial Algebra Approach to Directed Graphs. In: Möller, B. (ed.) MPC 1995. LNCS, vol. 947, pp. 282–303. Springer, Heidelberg (1995)
18. Hanus, M.: The integration of functions into logic programming: From theory to practice. J. of Logic Progr. 19&20, 583–628 (1994)
19. Kahl, W.: Semigroupoid Interfaces for Relation-Algebraic Programming in Haskell. In: Schmidt, R.A. (ed.) RelMiCS/AKA 2006. LNCS, vol. 4136, pp. 235–250. Springer, Heidelberg (2006)
20. King, D.J., Launchbury, J.: Structuring depth-first search algorithms in Haskell. In: Proc. ACM Symposium on Principles of Programming, pp. 344–356. ACM Press (1995)
21. Launchbury, J.: Graph Algorithms with a Functional Flavour. In: Jeuring, J., Meijer, E. (eds.) AFP 1995. LNCS, vol. 925, pp. 308–331. Springer, Heidelberg (1995)
22. Ravelo, J.: Two graph algorithms derived. Acta Informat. 36, 489–510 (1999)
23. Russling, M.: Deriving a class of layer-oriented graph algorithms. Sci. of Comput. Progr. 26, 117–132 (1996)
24. Schmidt, G., Ströhlein, T.: Relations and graphs. Discrete Mathematics for Computer Scientists. EATCS Monographs on Theoretical Computer Science. Springer (1993)
25. Schmidt, G.: A proposal for a multilevel relational reference language. J. of Relat. Meth. in Comput. Sci. 1, 314–338 (2004)
26. Schmidt, G.: Rectangles, Fringes, and Inverses. In: Berghammer, R., Möller, B., Struth, G. (eds.) RelMiCS/AKA 2008. LNCS, vol. 4988, pp. 352–366. Springer, Heidelberg (2008)
27. Strassen, V.: Gaussian elimination is not optimal. Num. Math. 13, 354–356 (1969)
28. Tarski, A.: On the calculus of relations. J. of Symb. Logic 6, 73–89 (1941)
29. Thomson, S.: Haskell – The craft of functional programming. Addison-Wesley (1999)
30. Warshall, S.: A theorem on Boolean matrices. J. of the ACM 9, 11–12 (1962)

Using Relations to Develop a Haskell Program for Computing Maximum Bipartite Matchings

Nikita Danilenko

Institut für Informatik, Christian-Albrechts-Universität Kiel
Olshausenstraße 40, D-24098 Kiel
nda@informatik.uni-kiel.de

Abstract. We show how to develop a purely functional algorithm that computes maximum matchings in bipartite graphs by using relation algebra. Our algorithm is based upon the representation of graphs by lists of successor lists and a generalisation to specific container types is discussed. The algorithm itself can be implemented in HASKELL and we will provide a complete implementation using the successor list model.

1 Introduction

Maximum matchings in bipartite graphs are a natural and well studied problem. When given a bipartite graph, that is a graph that can be partitioned into two sets of vertices, such that all edges of the graph run between these two sets, one can interpret the two vertex sets as "employees" and "jobs" and the edges as "can work as" conditions. A maximum matching then is an assignment of a maximum of employees to exactly one job, such that a job is assigned to at most one employee.

A solution for this problem can be obtained from a lemma by Berge (see [2]). The lemma yields a simple algorithm that computes a maximum matching in a bipartite graphs in $\mathcal{O}(|V| \cdot |E|)$ steps. In [10] a more efficient algorithm is given that is based upon Berge's lemma, but has an $\mathcal{O}(\sqrt{|V|} \cdot |E|)$ worst case time bound. A further improvement that is based upon fast matrix multiplication is presented in [17] and has a complexity of less than $\mathcal{O}(|V|^{2.38})$. These algorithms are usually implemented in an imperative language, but to our knowledge not in a functional programming language like Haskell.

A number of graph algorithms has been implemented in Haskell in [9] based upon an inductive definition of graphs that allows typical functional programming techniques like pattern matching (this is accomplished via active patterns theoretically and implemented via an explicit decomposition function). The representation of graphs by adjacency lists has beed used in [12] (and in a modified version in [13]) and more recently in [4]. A great advantage of this implementation is the fact that many graph-theoretic statements can be written in a very concise manner which is closely related to the underlying mathematical description. In [20] some graph representations (including ours) are given and a number of algorithms is developed specifically to solve several graph-theoretic problems.

W. Kahl and T.G. Griffin (Eds.): RAMiCS 2012, LNCS 7560, pp. 130–145, 2012.
© Springer-Verlag Berlin Heidelberg 2012

In [12] some graph-theoretic problems are solved using depth-first search. The implementation is based upon monadic effects that allow constant time array updates by mimicking side effects. The approach taken in [11] defines efficient array operations on a low level (in C) and encapsulates the efficient arrays in a monad similar to the state monad. The algorithms presented in [11] are based upon the representation of graphs by such an efficient array containing successor lists and implemented closely to the imperative versions. More recently the stateful, monadic approach has been used in [15] with a focus on parallelisation and providing some examples (e. g. the single source shortest path algorithm). Graph algorithm derivation based on general schemes has been studied in [21]. These schemes allow to derive several graph algorithms using concise specifications, that employ algebraic closures, that are then deconstructed by supplied functions.

Our focus will be not on efficiency, but on modularity and reusability. Many graph implementations in both imperative and functional languages provide skeletons for certain computations, but require rewriting when the problem changes even slightly. For instance breadth-first search can be used to obtain a list of reachable vertices. It is just as simple to write a version that actually provides some path to a reached vertex, but these two algorithms are usually presented by rewriting the first algorithm to fit the new requirements. Also our methods are closely related to the mathematical foundation of relation algebra. This approach allows some modularity in data structures so that graphs can be treated as ADTs (c.f. [20,7]).

We assume the reader to be familiar with basic Haskell (see [16,19] for introduction and [1] for more details). We will implement the algorithm obtained from Berge's result using the classical view on graphs as a list of adjacency lists and discuss an improvement in the data types. The algorithm obtained by our approach has a cubic running time w.r.t. the number of vertices.

2   Basic Setting and Implementation

We will represent the edges of a graph by pairs. This approach is close to relation algebra and allows directed graphs as well. Let $G = (V, E)$ be an undirected graph. Then have $E \subseteq V \times V$ and since G is undirected we have that $(x, y) \in E$ if and only if $(y, x) \in E$ for all $x, y \in V$. To represent graphs in Haskell we will use an adjacency list model based on vertices and vertex sets.

```
type Vertex    = Int
type VertexSet = [Vertex]
```

We will use the implicitly maintained assumption that all vertex sets are increasingly sorted and do not contain multiple occurrences of the same vertex.

Then the following implementation of a graph

```
type Graph = [(Vertex, VertexSet)]
```

will suffice for our purposes. This implementation is based upon the one given in [4]. The only difference is the fact that the adjacency lists are labelled with the vertex they belong to. This will allow some simplifications as we will see later. Since we consider undirected graphs only, we will assume (and maintain) the condition, that all given graphs are symmetric. We will also assume (and maintain) the condition that a graph on n vertices has the vertices $\{0, \ldots, n-1\}$ and is represented by a list g of exactly n pairs such that map fst g = [0..n-1]. The reason for this technicality is that for our purposes it is necessary to have a function that provides the vertices of a graph.

We require some functions on lists that will allow us to express certain set operations. Many of these functions are given in the existing package [22], but are not be general enough for our purpose. We will use the names provided in [22] nevertheless. In the following we will omit the definition of the general functions (they are provided in Appendix A).

The union of sets can be implemented by supplying some comparison function and an operation to perform in case of equality. The function then works as follows: it passes through both lists simultaneously and compares the encountered elements. If one of the two elements is smaller, it is added to the output list. Otherwise the supplied operation is applied to both elements and the result is added to the output list.

```
unionByWith :: (a → a → Ordering) → (a → a → a) → [a] → [a] → [a]
```

A simple instance hereof is the so-called left-biased union, which adds the left occurrence of "equal" elements to the result list. This can be defined as follows.

```
unionBy :: (a → a → Ordering) → [a] → [a] → [a]
unionBy cmp = unionByWith cmp const
```

Since const x y = x this instance of the general union function in fact places only the left occurrence of the "equal" elements in the result list.

When the type a belongs to the type class Ord, we can define an overloaded version that uses the function compare provided by the class context Ord to compare two elements.

```
(∪) :: Ord a ⇒ [a] → [a] → [a]
(∪) = unionBy compare
```

Likewise the symmetric difference (and its overloaded version) is given by

```
xunionBy :: (a → a → Ordering) → [a] → [a] → [a]

xunion :: Ord a ⇒ [a] → [a] → [a]
xunion = xunionBy compare
```

where the name xunionBy is a mnemonic for "exclusive union".

Similarly we obtain the difference of sets.

```
(\) :: Ord a ⇒ [a] → [a] → [a]
```

As for the intersection let us note that intersecting indexed sets can be viewed as a specific instance of zipping these sets, which gives a more general type. To that end we would like to introduce the following function.

```
isectByWith :: (a → b → Ordering) → (a → b → c) → [a] → [b] → [c]
```

The merging technique employed here is basically the same one as seen before for our function unionByWith. Since an element is contained in the intersection of two lists if and only if it is contained in both lists, the supplied operation has a more general type.

We proceed to define the left-biased intersection that is defined similarly to the previously implemented function unionBy.

```
isectBy :: (a → b → Ordering) → [a] → [b] → [a]
isectBy cmp = isectByWith cmp const
```

We will also need the following intersection.

```
(∩₁) :: Ord a ⇒ [(a, b)] → [a] → [(a, b)]
(∩₁) = isectBy (compare ∘ fst)
```

This function compares the first components of the elements of the first list with the elements of the second list by using the provided function compare.

3   Relation-Algebraic Preliminaries

Let V be some set, $R, S \subseteq V \times V$ and $s \subseteq V$. We will use the well-known fact that graphs can be viewed as relations. Also since set operations with graphs on V are defined by the respective operation on the edge sets, we will not distinguish between graphs and relations.

The composition of a set and a relation (frequently also called relational image) is defined as

$$s \cdot R := \{y \in V \mid \exists x \in s : (x, y) \in R\} = \bigcup_{x \in s} \{y \in V \mid (x, y) \in R\}.$$

Graph-theoretically the second equality gives us that $s \cdot R$ is precisely the set of all R-successors of the vertices in s. Using this property we can easily implement this function using two previously defined merging functions – to get all the necessary successor lists we can intersect the graph with the set, then drop the vertex numbers and unify the collection of the obtained lists.

```
(⊙) :: VertexSet → Graph → VertexSet
s ⊙ g = foldr (∪) [] (map snd (g ∩₁ s))
```

We assume the reader to be familiar with the following operations on relations and relational constants: $R \cup S$ (union), $R \oplus S$ (symmetric difference), $R \; S$ (relational composition)[1], I (identity relation), $O := \emptyset$ (empty relation) and R^* (reflexive transitive closure). When mixing relational operations, $\cdot$ has the strongest binding and $\cup$ the weakest. We will need the fact that $\cdot$ is associative (in particular: $s \cdot (R \cdot S) = (s \cdot R) \cdot S$) and that I is a neutral element w.r.t.

---

[1] We overload $\cdot$, since sets can be viewed as relations by associating s with $\{x_0\} \times s$ for some fixed $x_0 \in V$. Then both versions of $\cdot$ provide the same associated result. This way the set s is identified with the row vector that represents s.

composition. For abbreviation purposes we will write $R \setminus S := R \cap \overline{S}$. We will need the well-known rule

$$(R \cdot S)^* \cdot R = R \cdot (S \cdot R)^*. \tag{1}$$

To implement the necessary functions on graphs, we can use the functions presented in Section 2. The union of two graphs is implemented as follows.

```
(⊔) :: Graph → Graph → Graph
(⊔) = unionByWith (comparing fst) (λ(i, xs) (_, ys) → (i, xs ∪ ys))
```

Here the function `comparing` is an auxiliary function from the Haskell module *Data.Ord*. It takes a function and two arguments, applies the function to each of the two arguments and compares the results.

We implement the symmetric difference of graphs as follows.

```
(⊕) :: Graph → Graph → Graph
(⊕) = unionByWith (comparing fst) (λ(i, xs) (_, ys) → (i, xunion xs ys))
```

Note, that the "outer operation" has to be a union and not a symmetric difference[2]. This definition allows us to apply it to a graph that meets our internal requirements and another graph where some indices are missing and still obtain a graph as required.

4   Problem Description and Theoretical Solution

From now on let $G = (V, E)$ be a finite, undirected graph. The graph G is called bipartite if and only if there are $V_1, V_2 \subseteq V$ such that $V_1 \cup V_2 = V$, $V_1 \cap V_2 = \emptyset$ and $E \subseteq (V_1 \times V_2) \cup (V_2 \times V_1)$. From a theorem by König (see [6], Proposition 0.6.1) we obtain that a graph is bipartite if and only if it doesn't contain odd cycles. This property can be described relationally by the condition $E \cdot (E \cdot E)^* \subseteq \overline{\mathsf{I}}$.[3] For the remainder of this text we will assume that G is bipartite.

Definition 1 (Matching)
*An $M \subseteq E$ is called **matching** if and only if $M = M^\top$ and $M \cdot M \subseteq \mathsf{I}$. A matching is called **maximum**, iff it is a maximal element of all matchings (of G) w.r.t. cardinality.*

The following theorem by Berge (see [2]) is the key to the matching algorithm.

Theorem 1 (Characterization of maximum matchings, Berge)
Let $M \subseteq E$ be a matching. Let $\oplus$ denote the symmetric difference. For a path p we denote the set of the edges along p by $E(p)$. A path is called M-augmenting iff it starts and ends in a vertex that is not contained in some edge of M and alternates between edges of $E \setminus M$ and those of M. Then we have

---

[2] Otherwise the graph `g = [(0, [0])]` would give `g ⊕ g = []`, but in our representation `[]` is the empty graph on no vertices, whereas `g ⊕ g` should be `[(0, [])]`, which is the empty graph on the vertex set `[0]`.

[3] This condition can be checked efficiently, but our implementation of the matching algorithm will assume the given graph to be bipartite.

1. *If there are no M-augmenting paths in G, then M is a maximum matching.*
2. *If there is an M-augmenting path p in G, then $M \oplus E(p)$ is a larger matching.*

This theorem yields a well-known algorithm for finding maximum matchings: begin with any matching (e. g. $M := \emptyset$), find an augmenting path and enlarge the matching according to Theorem 1 until there are no augmenting paths.

5   Augmenting Paths and Reachability

Now let's have a look at the necessary components of the previous theorem. There are two specific objects we need to deal with – the vertices that are not contained in some edge of M and alternating paths. Let us begin with the former.

Definition 2 (Uncovered vertices)
We define: $\mathrm{uncovered}(M) := \overline{V \cdot M} = \{x \in V \mid \{y \in V \mid (x, y) \in M\} = \emptyset\}$.

For symmetric graphs the following function returns exactly those vertices that have no neighbours (i.e. successors, because of symmetry) in the graph.

```
uncovered :: Graph → VertexSet
uncovered = map fst o filter (null o snd)
```

The fact that the adjacency lists carry their vertices is convenient in this definition.

Lemma 1 (Existence of augmenting paths)
In a bipartite graph there is an M–augmenting path if and only if

$$\mathrm{uncovered}(M) \cdot ((E \setminus M) \cdot M)^* \cdot (E \setminus M) \cap \mathrm{uncovered}(M) \neq 0.$$

The proof is given in Appendix B. Implementing this characterisation would result in a quite strict program, since we would need to compute the reachability of some vertex set and then apply another multiplication. We can do better by using rule (1). This results in the following trivial corollary.

Corollary 1 (Existence of augmenting paths II)
In a bipartite graph there is an M–augmenting path if and only if

$$\mathrm{uncovered}(M) \cdot (E \setminus M) \cdot (M \cdot (E \setminus M))^* \cap \mathrm{uncovered}(M) \neq 0. \qquad (2)$$

By abstracting the components of this description we have some vertex set v composed with some relation R^* and then intersected with another vertex set w, so that we are interested in computing $v \cdot R^* \cap w$.

Since $v \cdot R^*$ describes reachability from the vertices in v along edges in R we can use breadth-first search to obtain the reachable vertices. In [3] a relational specification of a reachability algorithm based upon a breadth-first search is given and can be translated immediately to our setting. It is based upon the computation of certain pairwise disjoint sets s_i for all $i \in \{0, \ldots, |V| - 1\}$ such that s_i contains those vertices reachable from v by walking along i edges, but not less. These sets can be obtained by defining

$$s_0 := v \text{ and } s_i := s_{i-1} \cdot R \cap \bigcap_{j=0}^{i-1} \overline{s_j} \text{ for all } i \in \{1, \ldots, |V| - 1\}. \tag{3}$$

This results in $v \cdot R^* = \bigcup \{s_i \mid i \in \{0, \ldots, |V| - 1\}\}$ and thus

$$v \cdot R^* \cap w = \left(\bigcup_{i=0}^{|V|-1} s_i \right) \cap w = \bigcup_{i=0}^{|V|-1} (s_i \cap w). \tag{4}$$

This distributive regrouping allows to check the condition $v \cdot R^* \cap w$ in a more non-strict (hence more Haskell-like) manner, since we can immediately stop computing further steps after we have found some step s_i such that $s_i \cap w \neq \emptyset$.

Merely checking, whether an augmenting path exists is insufficient in our case, since we are interested in finding an actual augmenting path if such a path exists. To accomplish that we can apply a technique similar to the imperative solution by introducing "markings". These markings are an additional information carried through the relational computation. Let us therefore define

```
type VertexI a = (Vertex, a)
type VertexISet a = [VertexI a]
```

where we again assume (and maintain) the condition that the vertices in the marked vertex set are increasingly sorted by their vertex values and no vertex value appears more than once. The additional I in the name is a mnemonic for "information". A similar approach is taken in [7].

Since we are interested in a path, we can mark vertices with a list of their predecessors during the search. As we will compute the sets from (3) successively, the only possible place to embed this marking function is the operation · that shall be rewritten to this end.

In the context of a vertex set composed with a relation this operation is precisely a multiplication of a Boolean (row) vector with a Boolean matrix. Then first of all any row i of the matrix is multiplied (in the sense of scalar multiplication) with the respective entry of the vector, thus obtaining a collection of vectors and then unifying (i.e. adding in the Boolean context) these vectors. For Boolean values the scalar multiplication is obviously trivial, since multiplying by True yields the whole row, whereas multiplication with False is the function that always returns an empty list.

To establish a marking we simply need a modification of the scalar multiplication: when a marked vertex (i, xs) encounters a successor j we obtain a marked vertex $(j, i : xs)$ which is easily computed as $\lambda(\texttt{i, xs}) \texttt{ j} \rightarrow (\texttt{j, i : xs})$. Now we can derive a scalar-multiplication-like function from this definition:

```
sMult :: VertexI [Vertex] → (Vertex, VertexSet) → VertexISet [Vertex]
sMult (v, ps) (_, vs) = map (λx → (x, v : ps)) vs
```

Using this scalar multiplication we can obtain our desired multiplication in an intersection-like manner as we did for (⊙).

```
(⊙[]) :: VertexISet [Vertex] → Graph → VertexISet [Vertex]
mv ⊙[] g = foldr (unionBy cmp) [] (isectByWith cmp sMult mv g)
    where cmp = comparing fst
```

Please note, that the relational context is maintained in the first component and thus the successors computed by (⊙) are precisely the ones computed by (⊙[]).

While this new multiplication is very similar to the one obtained by extending the operations of the Kleene algebra of paths to a vector-matrix level, it is actually conceptually different. The main difference is that we do not collect all paths, but use the left-biased union to obtain a single path. Also, the scalar multiplication sMult is inhomogenous – the "scalar" is a vertex labelled with a path, while the vector contains vertices only. In a mathematical setting letters of an alphabet are usually identified with one-letter words, but using this identification in our implementation would require the unnecessary transformation of a successor list to a list of singleton lists. A technical difference is that paths are constructed in the reversed order, since prepending an element to a list has constant time complexity in Haskell, while appending a singleton list to another list is linear in the size of the first list.

A final observation yields that the relation R in our previous abstraction is actually a product of two other relations. This way we will be computing $v \cdot (A \cdot B)$, which requires a computation of $A \cdot B$. In our case this would result in a mere technicality, but in the general case this could lead to an efficiency loss. Instead we use the associativity of $\cdot$ and compute $v \cdot (A \cdot B) = (v \cdot A) \cdot B$. This rearrangement is very convenient for our purpose since it allows us to mark our paths without any further modifications. In a similar fashion we can replace the two graphs by an arbitrary finite list of graphs and thus gain generality.

This results in the following version of a reachability algorithm which modifies the version presented in [3] to meet our requirements. Its inputs are a vertex set where each vertex is marked with a path and a list of graphs in the reversed order of traversal[4]. Its result is a list that consists of the marked versions of those s_i from (3), that are non-empty. Each vertex in such an s_i is labelled with a path (in the reversed order of traversal) that leads to this vertex from s_0, followed by the initial list specified in s_0.

```
reach :: VertexISet [Vertex] → [Graph] → [VertexISet [Vertex]]
reach mv []  = [mv]
reach mv gs  = reach' mv (vertices (head gs))
    where vertices = map fst
          reach' [] _ = []
          reach' v  w = let w' = w \ map fst v
                            v' = (foldr (flip (⊙[])) v gs) ∩₁ w'
                        in v : reach' v' w'
```

All paths constructed by this function are lists of vertices in the reversed order of traversal. The latter is irrelevant for our problem, since obviously in an undirected graph a reversed augmenting path is also an augmenting path. Note that

---

[4] This allows a right-fold.

any path obtained by reach is missing its last vertex, since the label consists of strict predecessors only.

Also we are not so much interested in a vertex path, but rather in the edges along the path. To that end we require some notion of an edge.

As mentioned before we now consider a directed edge (i, j) to be a graph, i. e. we associate (i, j) with the graph $(\{i, j\}, \{(i, j)\})$. Such a graph does not meet our internal requirements, but behaves similarly. Therefore we conceal this "inconsistency" with a new type.

```
type Edge = Graph
```

An undirected edge between vertices i and j is then a graph consisting of the adjacency lists (i, [j]) and (j, [i]). Since we would like the vertices in the first component to be sorted increasingly and without multiple occurrences, we can simply use our union function from before.

```
edge :: Vertex → Vertex → Edge
edge i j = [(i, [j])] ⊔ [(j, [i])]
```

This allows us to omit an explicit computation of the minimum of i and j as well as whether they differ.

Now we can define a function that finds a shortest path between two given vertex sets. First we compute the list of the reachability steps using our reachability algorithm from the first vertex set with an initial path, such that each vertex is labelled with some path that leads to this vertex[5]. Then we use the rearrangement from (4) and compute the intersections of the second vertex set with all steps. This can be done using our previously defined function ($\cap_1$) and map. The third step is to drop all empty intersections, which is accomplished by dropWhile null.

Now if there is a path between the given vertex sets, then any shortest path between these two sets is contained in the first element of the remaining list. If on the other hand there is no path between these sets, then the remaining list is empty. The function listToMaybe (from *Data.Maybe*) returns Nothing when applied to an empty list and Just x when applied to a list x:xs .

We can now use fmap to apply some function to the Maybe container and thus can deal with the non-Nothing case only. If we got Just l from the previous calculation, then l consists of all those vertices in the second vertex set that are reachable from the first vertex set by walking along a shortest path. Since we are interested in a single shortest path, we may simply take the first element of this list. This element itself is a vertex labelled with a path that leads to this vertex and thus some (i, ps). The complete path is then i : ps and the complete edge path is simply zipWith edge (i:ps) ps

```
shortest :: VertexISet [Vertex] → [Graph] → VertexSet → Maybe [Edge]
shortest v gs w =
    fmap finish (listToMaybe (dropWhile null (map ( ∩₁ w) (reach v gs))))
        where finish = (λ(i, ps) → zipWith edge (i:ps) ps) ∘ head
```

This function is the key to our implementation of the matching algorithm.

---

[5] This can be the empty path, but in our case we will already have a "first step".

6   The Bipartite Matching Algorithm

As noted before (see Section 4) the actual maximum matching algorithm consists of two parts – a loop (function) and a function that is applied as long as the loop condition is true.

Let us first deal with the latter function. This function checks whether an augmenting path exists and enlarges the current matching in the positive case. Since matchings are edge sets we will view them as graphs, too. The only edge sets that appear in the condition (2) are M and $E \setminus M$, but the actual edge set E is never used. Now suppose we have found an augmenting path p. Then by Theorem 1 we can enlarge the matching by computing $M \oplus E(p)$. Independent of whether another enlargement is possible or not we will need to check the condition (2) again with $M \oplus E(p)$ substituted for M. At first glance it seems that we will need E, since we have to compute $E \setminus (M \oplus E(p))$. For all sets A, B, C we have that $\overline{A \oplus B} = \overline{A} \oplus B$ and $A \cap (B \oplus C) = (A \cap B) \oplus (A \cap C)$ which in our case yields:

$$E \setminus (M \oplus E(p)) = E \cap \overline{M \oplus E(p)} = E \cap (\overline{M} \oplus E(p))$$
$$= (E \cap \overline{M}) \oplus (E \cap E(p)) = (E \setminus M) \oplus E(p) \ ,$$

since $E(p) \subseteq E$. Thus M and $E \setminus M$ are changed in the same manner.

We use this circumstance in our enlargement function by supplying two graphs, namely M and $E \setminus M$, and obtaining the updated versions of both. Since there might be no update applicable, we use a Maybe-wrapper.

```
enlargeMatching :: (Graph, Graph) → Maybe (Graph, Graph)
```

The implementation then is quite straight-forward – compute a shortest path using our previously defined function shortest from $v = $ uncovered$(M) \cdot (E \setminus M)$ through the graphs M and $E \setminus M$ to uncovered(M) and apply $\oplus$ to all edges along this path, where the latter is easily accomplished using a fold.

```
enlargeMatching (m, eNotM) =
    fmap augment (shortest (mark unc ⊙[] eNotM) [eNotM, m] unc)
      where unc     = uncovered m
            mark    = map (λi → (i, []))
            augment = foldr (λe (a, b) → (e ⊕ a, e ⊕ b)) (m, eNotM)
```

The local function mark is used to supply an initially empty path. The actual loop can be realised by the function maybe from *Data.Maybe*[6].

```
maximumMatching  :: Graph → Graph
maximumMatching g  = mmgo (emptyGraph g, g)
    where mmgo (m, eNotM) = maybe m mmgo (enlargeMatching (m, eNotM))
          emptyGraph      = map (λ(i, _) → (i, []))
```

---

[6] The function maybe takes a default value, a Maybe-value and a function f and applies f to x if the Maybe-value is Just x and returns the default value otherwise.

The presented algorithm incorporates a calculation that is usually implemented by hand in an imperative language. Instead of starting with the empty matching one can implement a simple greedy algorithm that computes a maximal matching (where a maximal matching is a maximal element of all matchings w.r.t. inclusion). The more sophisticated algorithm that we just implemented can then begin with this maximal matching. In our implementation we are always searching for shortest paths, thus if an edge can be added to the matching our matching enlargement function will do that and not search for a longer path. While a specifically tailored function that computes a maximal matching can be used to improve the constants in the big-O notation, the overall running time is not affected by this modification.

7   Discussion and Future Work

We have shown how to develop a functional algorithm for computing maximum matchings in a bipartite graph. This has been accomplished by maintaining a strong emphasis on the relational view on graphs. The presented implementation is very close to the mathematical foundation and theoretical modifications can be easily incorporated. Our enlargement function for instance computes `shortest (mark unc ⊙[] eNotM) [eNotM, m] unc` where the list is in the reversed order of traversal for technical reasons only. Omitting this technicality would result in `shortest (mark unc ⊙[] eNotM) [m, eNotM] unc` that bears a striking similarity with the purely relational formula (2).

Our approach lead to certain generalisations, that resulted in some immediate benefits. For instance the general approach to reachability, where the initial vertex set is already labelled with some paths allowed the somewhat technical decomposition during the path computation (first `mark (uncovered m) ⊙[] eNotM` and then a reachability search through two different graphs) to be implemented rather canonically since all we had to do was to begin the actual path search from the first step of the path instead of the zeroth step.

Also we obtained a more general view on vector-matrix multiplication by abstracting the scalar multiplication and the sum. The very same approach can be used for instance to find all paths from a vertex set. To do that we can define

```
type Paths = [[Vertex]]

allsum :: VertexISet Paths → VertexISet Paths → VertexISet Paths
allsum = unionByWith (comparing fst) (λ(i, xs) (_, ys) → (i, xs ++ ys))

sMult' :: VertexI Paths → (Vertex, VertexSet) → VertexISet Paths
sMult' (v, ps) (_, vs) = map (λx → (x, map (v :) ps)) vs

(⊙all) :: VertexISet Paths → Graph → VertexISet Paths
mv ⊙all g = foldr allsum [] (isectByWith cmp sMult' mv g)
    where cmp x y = compare (fst x) (fst y)
```

thus obtaining a multiplication that collects all paths in the reversed order of traversal. If we substitute `map (v :)` by `map (++[v])` in the above definition, we obtain the multiplication in the Kleene algebra of paths (which collects all paths in the correct order of traversal). Extending graphs to having labelled edges provides even more applications, while the actual approach to this multiplication can be modularised in a fashion that allows programmers to obtain such multiplications simply from specifying certain functions. These functions are often very descriptive in nature and occur naturally when reasoning algebraically. Our function sMult is an example of such a function. These multiplications can be easily incorporated into our reachability function, by abstracting the multiplication function used in each step thus obtaining

```
reachWith ::    (VertexISet a → Graph b → VertexISet a)
          →   VertexISet a → [Graph b] → [VertexISet a]
```

where Graph b represents an edge-labelled graph and the first argument of the function reachWith is the multiplication. With this function at hand we have reach = reachWith $(\odot_{[]})$. Note again, that the supplied multiplication can be inhomogenous in the same way as mentioned before. This inhomogenity can be useful in edge-labelled graphs for instance, when the task is to compute a shortest (w.r.t. the number of vertices) path and to ignore the labels encountered during the computation completely.

While these generalisations can be implemented in (some) imperative languages as well, functional languages provide particularly elegant solutions, since the use of functions as arguments (in higher-order functions) is one of their key aspects. This way many algorithms can be implemented in a similar fashion as our merging functions by abstracting several components and instantiating the same schemes with different parameters (c.f. Section 2, Appendix A, [21]).

The faster algorithm from [10] is based upon the computation of a maximal set of vertex-disjoint paths. Using the functions $(\odot_{all})$ and reachWith one can easily implement this technique. While this trivial modification leads to an efficiency loss[7], it still provides a useful multiplication and demonstrates the modularity of the components we developed. Also it makes use of the non-strictness of Haskell, since the first components of the successive steps from (3) are still computed with the same complexity and the computation that results in the list of all paths is threaded through the second component and evaluated on demand only.

The downside of our implementation is its theoretical running time, which is cubic w.r.t. the number of vertices. The enlargement function is applied at most $\mathcal{O}(|V|)$ times and contains two components with an $\mathcal{O}(|V|^2)$ complexity. The first of these is the modification of the matching and its complement along an augmenting path. Using tree-like structures with logarithmic access and updates we could redefine graphs to be trees that consist of trees and improve this component. The other component is our implementation of reach, whose inefficiency is not rooted in the use of lists, but in the types of the employed operations.

While we have presented a list based implementation of graphs and vertex sets, most of these can be easily replaced by container types that support fast merging operations. Assuming the same nomenclature on such containers, we

---

[7] The number of paths between two vertex sets is not linear in the number of edges.

need to replace the definition of an edge only, as it is the single position where we use the definition of a graph explicitly. For many indexed containers[8] "fast" means "linear in the size of both". Replacing the lists by such containers may then lead to noticeable performance gain, but the theoretical worst case complexity remains the same. This is due to the homogenous structure of these merging operations. This homogenity allows an immediate translation of the mathematical context into Haskell functions.

A slightly different approach would be to maintain natural inhomogenous, but more efficient operations. Suppose that we have some `Container` that allows the fast inhomogenous operations

```
fastUnion :: [(Int, a)] → Container a → Container a
fastIsect :: [(Int, a)] → Container a → [(Int, a)]
```

that are linear in the size of the first argument. Then these two operations allow to rewrite ⊙ efficiently, but only if an empty container is already supplied somehow. The initialisation of such containers may be costly (c.f. [12]) and thus we cannot simply create such a container in every step of `reach` to benefit from the better complexity.

While the latter function can be trivially obtained using standard Haskell arrays, since these allow constant time access, it is not possible to use them for the former function, as they do not provide efficient updates. To achieve efficient updates the Haskell libraries contain a type class for mutable arrays that reside in certain monads (the state monad or IO monad, for instance) and instances thereof. The monadic context is used to mimic side effects. This technique is presented in [14] and used in [12]. Another approach to fast arrays is given in [5]. The use of such containers at top level often leads to graph algorithms that are basically rewritten from imperative specifications[9]. We would like to employ these fast arrays on a low level only to provide a more efficient implementation of the set operations we used (for instance `foldr` (∪)) while still maintaining the same algebraic concept.

Our approach to graph operations allows a highly compositional programming style and provides immediate insights into abstraction possibilities. Despite the bad complexity we believe that the presented implementation is of theoretical value. It is another example (along with [4]) of a great bond between relational reasoning and functional programming (e.g. greedy specification, but less strict implementation). We use theoretical models to obtain practical results (e.g. marking) thus maintaining a very structured approach to graphs while meeting the requirements of practical algorithms.

Acknowledgements. I would like to thank Rudolf Berghammer for pointing out ideas for simplification and Jan Christiansen for valuable discussions and suggestions. I also wish to thank the anonymous reviewers of this paper for their much appreciated input.

---

[8] For instance `IntMap` from the Haskell module *Data.IntMap* which is based upon [18].

[9] This drawback is pointed out in [8] as well and the graph representation of [9] in fact allows to omit any explicit notion of "visited vertices".

References

1. The Haskell report, http://www.haskell.org/onlinereport/haskell2010
2. Berge, C.: Two theorems in graph theory. PNAS. National Academy of Sciences 43(9), 842–844 (1957)
3. Berghammer, R.: Ordnungen, Verbände und Relationen mit Anwendungen. Vieweg-Teubner (2008)
4. Berghammer, R.: A Functional, Successor List Based Version of Warshall's Algorithm with Applications. In: de Swart, H. (ed.) RAMICS 2011. LNCS, vol. 6663, pp. 109–124. Springer, Heidelberg (2011)
5. Chakravarty, M.M.T., Keller, G.: An Approach to Fast Arrays in Haskell. In: Jeuring, J., Jones, S.L.P. (eds.) AFP 2002. LNCS, vol. 2638, pp. 27–58. Springer, Heidelberg (2003)
6. Diestel, R.: Graphentheorie. Springer (2000)
7. Erwig, M.: The FGL package, http://hackage.haskell.org/package/fgl
8. Erwig, M.: Functional programming with graphs. In: Jones, S.L.P., Tofte, M., Berman, A.M. (eds.) ICFP, pp. 52–65. ACM (1997)
9. Erwig, M.: Inductive graphs and functional graph algorithms. J. Funct. Program. 11(5), 467–492 (2001)
10. Hopcroft, J.E., Karp, R.M.: An $n^{5/2}$ algorithm for maximum matchings in bipartite graphs. SIAM J. Comput. 2(4), 225–231 (1973)
11. Johnsson, T.: Efficient graph algorithms using lazy monolithic arrays. J. Funct. Program. 8(4), 323–333 (1998)
12. King, D.J., Launchbury, J.: Structuring depth-first search algorithms in Haskell. In: POPL, pp. 344–354 (1995)
13. Launchbury, J.: Graph Algorithms with a Functional Flavour. In: Jeuring, J., Meijer, E. (eds.) AFP 1995. LNCS, vol. 925, pp. 308–331. Springer, Heidelberg (1995)
14. Launchbury, J., Jones, S.L.P.: Lazy functional state threads. In: PLDI, pp. 24–35 (1994)
15. Lesniak, M.: Palovca: Describing and Executing Graph Algorithms in Haskell. In: Russo, C., Zhou, N.-F. (eds.) PADL 2012. LNCS, vol. 7149, pp. 153–167. Springer, Heidelberg (2012)
16. Lipovaca, M.: Learn You a Haskell for Great Good!. No Starch Press (2011), http://learnyouahaskell.com
17. Mucha, M., Sankowski, P.: Maximum matchings via Gaussian elimination. In: FOCS, pp. 248–255. IEEE Computer Society (2004)
18. Okasaki, C., Gill, A.: Fast mergeable integer maps. In: Workshop on ML, pp. 77–86 (1998)
19. O'Sullivan, B., Stewart, D., Goerzen, J.: Real World Haskell. O'Reilly (2009), http://book.realworldhaskell.org/read
20. Rabhi, F., Lapalme, G.: Algorithms: a functional programming approach, 2nd edn. Addison-Wesley (1999)
21. Russling, M.: Deriving General Schemes for Classes of Graph Algorithms. PhD thesis (1996)
22. Smith, L.: The data-ordlist package, http://hackage.haskell.org/package/data-ordlist

A   Implementation of Merging Functions on Lists

The following function generalises merging functions on lists.

```
merge ::     (a → b → Ordering) -- comparison function
         → ([b] → c)            -- action if the first list is empty
         → ([a] → c)            -- action if the second list is empty
         → (a → b → c → c)      -- operation in the EQ case
         → (a → b → c → c)      -- operation in the LT case
         → (a → b → c → c)      -- operation in the GT case
         → [a] → [b] → c
merge cmp lEmpty rEmpty eq lt gt = merge' where
    merge' [] ys = lEmpty ys
    merge' xs [] = rEmpty xs
    merge' l@(x:xs) m@(y:ys) = case cmp x y of
                                EQ → eq x y (merge' xs ys)
                                LT → lt x y (merge' xs  m)
                                GT → gt x y (merge' l  ys)
```

The following auxiliary functions will simplify the code:

```
consWith :: (a → b → c) → a → b → [c] → [c]
consWith op x y l = (op x y) : l

thirdArg :: a → b → c → c
thirdArg _ _ x = x
```

Using these functions we can now easily define the necessary operations.

```
unionByWith cmp op =
 merge cmp id id (consWith op) (λx _ l → x : l) (λ_ y l → y:l)
```

The function cmp is the comparison function. If one of the lists is empty, the other one is returned, hence we have rEmpty = id and lEmpty = id in the general scheme. If two elements are equal w.r.t. the supplied comparison function, the given operation is applied to these arguments and the result is added to the result list. Finally if one of the elements is found to be smaller than the other (again w.r.t. the comparison funcion) this element is added to the result list and the recursion continues as specified in the function merge. The remaining three functions are obtained in a similar fashion.

```
isectByWith cmp op =
 merge cmp (const []) (const []) (consWith op) thirdArg thirdArg

(\) = merge compare (const []) id thirdArg (λx _ l → x : l) thirdArg

xunionBy cmp =
 merge cmp id id thirdArg (λx _ l → x : l) (λ_ y l → y : l)
```

B   Proof of Lemma 1

Let $u := \text{uncovered}(M)$. First we note the following equivalencies:

$$u \cdot ((E \setminus M) \cdot M)^* \cdot (E \setminus M) \cap u \neq \emptyset$$
$$\iff \exists y \in u : y \in u \cdot ((E \setminus M) \cdot M)^* \cdot (E \setminus M)$$
$$\iff \exists y \in u : \exists x \in u : (x, y) \in ((E \setminus M) \cdot M)^* \cdot (E \setminus M)$$
$$\iff \exists x, y \in u : \exists z \in V : (x, z) \in ((E \setminus M) \cdot M)^* \wedge (z, y) \in E \setminus M. \quad (5)$$

We set $I_n := \{m \in \mathbb{N} \mid 1 \leq m \leq n\}$ for all $n \in \mathbb{N}$ (implying that $I_0 = \emptyset$).
"$\implies$": Suppose $p \in V^+$ is an M-augmenting path. Then there is a $k \in \mathbb{N}$ such that $p \in V^k$. Since p is alternating and $p_1, p_k \in u$ we have that k is even, so there is $t \in \mathbb{N}_{>0}$ such that $k = 2t$. It holds that $(p_{2t-1}, p_{2t}) \in E \setminus M$. For all $i \in I_{t-1}$ we have $(p_{2i-1}, p_{2i}) \in E \setminus M$ and $(p_{2i}, p_{2i+1}) \in M$, so $(p_{2i-1}, p_{2i+1}) \in (E \setminus M) \cdot M$. Thus $(p_{2i-1})_{i=1}^t$ is a path from p_1 to p_{2t-1} in the graph $(V, (E \setminus M) \cdot M)$ and this is equivalent to $(p_1, p_{2t-1}) \in ((E \setminus M) \cdot M)^*$. Thus (5) holds (with $x := p_1$, $y := p_{2t}$, $z := p_{2t-1}$), leading to the desired result.
"$\impliedby$": Now suppose that $u \cdot ((E \setminus M) \cdot M)^* \cdot (E \setminus M) \cap u \neq \emptyset$. Then by (5) there are $x, y \in u$ and $z \in V$ such that $(x, z) \in ((E \setminus M) \cdot M)^*$ and $(z, y) \in E \setminus M$. Since G is bipartite we have that:

$$(x, y) \in ((E \setminus M) \cdot M)^* \cdot (E \setminus M) \subseteq (E \cdot E)^* \cdot E \subseteq \bar{\mathbf{I}}$$

and thus $x \neq y$. If $z = x$ we have that $x, y \in u$ and $(x, y) \in E \setminus M$ and thus (x, y) is a trivial M-augmenting path. Now consider the case that $z \neq x$. Then the graph $(V, (E \setminus M) \cdot M)$ contains a path $p \in V^+$ from x to z and there is a $k \in \mathbb{N}_{\geq 2}$ such that $p \in V^k$. Now for all $i \in I_{k-1}$ we have that $(p_i, p_{i+1}) \in (E \setminus M) \cdot M$ and thus there is an $m_i \in V$ such that $(p_i, m_i) \in E \setminus M$ and $(m_i, p_{i+1}) \in M$. We choose such an intermediate vertex $m_i \in V$ for all $i \in I_{k-1}$. We will now show that the list $p' := (p_1, m_1, \ldots, m_{k-1}, p_k)$ is a path. Clearly each two consecutive vertices are connected by an edge that is contained in E and thus the above list is a walk from $x = p_1$ to $z = p_k$ in G. Let $i, j \in I_k$. Since p is a path, the condition $p_i = p_j$ implies $i = j$. Now suppose that $i, j \leq k - 1$ and $m_i = m_j$. Then we have that $(p_i, m_i) \in M$ and $(p_j, m_i) = (p_j, m_j) \in M$. Since M is a matching we obtain that $p_i = p_j$ and thus $i = j$. Finally assume w.l.o.g. that $i \leq k - 1$ and $m_i = p_j$. The assumption that $i = j$ yields the following contradiction:

$$(p_i, m_i) \in (E \setminus M) \cap \mathbf{I} \subseteq ((E \cdot E)^* \cdot E) \cap \mathbf{I} \subseteq \bar{\mathbf{I}} \cap \mathbf{I} = \emptyset.$$

Thus we have that $i \neq j$. If $i < j$ then we have $2i < 2j$, thus $2i \leq 2j - 1$ and since $2i$ is even and $2j - 1$ is odd we find that $2i < 2j - 1$. Now the walk $(p'_t)_{t=2i}^{2j-1}$ is a cycle (since $p'_{2i} = m_i$ and $p'_{2j-1} = p_j$) that contains an odd number of edges, which cannot occur in a bipartite graph. Similarly if $j < i$ we have that $2j - 1 < 2i$ and the walk $(p'_t)_{t=2j-1}^{2i}$ is an odd cycle. We have shown that p' is a path. Note that all vertices of p' except for $p'_1 = p_1 = x$ are not contained in u and that by $y \neq x$ and $y \in u$ the list $a := (p_1, m_1, \ldots, m_{k-1}, p_k, y)$ is a path. Its edges alternate between $E \setminus M$ and M and its first and last vertex are uncovered by M and thus a is an M-augmenting path. $\qquad \square$

Relations as Executable Specifications: Taming Partiality and Non-determinism Using Invariants

Nuno Macedo, Hugo Pacheco, and Alcino Cunha

HASLab — High Assurance Software Laboratory
INESC TEC & Universidade do Minho, Braga, Portugal
{nfmmacedo,hpacheco,alcino}@di.uminho.pt

Abstract. The calculus of relations has been widely used in program specification and reasoning. It is very tempting to use such specifications as running prototypes of the desired program, but, even considering finite domains, the inherent partiality and non-determinism of relations makes this impractical and highly inefficient. To tame partiality we prescribe the usage of invariants, represented by coreflexives, to characterize the exact domains and codomains of relational specifications. Such invariants can be used as pre-condition checkers to avoid runtime errors. Moreover, we show how such invariants can be used to narrow the non-deterministic execution of relational specifications, making it viable for a relevant class of problems. In particular, we show how the proposed techniques can be applied to execute specifications of bidirectional transformations, a domain where partiality and non-determinism are paramount.

1 Introduction

The *relational calculus* provides a more natural way to specify programs than purely functional formalisms: most so-called functions in computer science are actually *partial*, and *non-determinism* is many times an essential characteristic of the program. In particular, since its first axiomatization by Tarski, a *point-free* (PF) version of the calculus of relations has been used in a variety of areas of computer science [3,15,16] in order to specify and reason about programs, due to its high simplicity and ease of manipulation.

However, relational specifications are frequently not amenable for execution: with partiality the behavior of the program may become unpredictable and give rise to runtime errors, while non-determinism may produce infinite runs without returning a single valid value. For instance, consider the expression $(\text{id} \triangle \text{id})^\circ \circ (\text{length}^\circ \triangle \text{head}^\circ) : \text{Nat} \to [\text{Nat}]$, where head returns the first element of a list, length its length, and $^\circ$ and $\triangle$ are the converse and split of relations, respectively. Given a natural n, this expression calculates a list with length n, whose first element is also n. This is not a total relation as it is not defined for the value 0, since no list with length 0 could have the same 0 as its head. We resort to the converse from the relational calculus to generate these lists: head° generates all lists with the input value at its head, while length° generates all lists with the

W. Kahl and T.G. Griffin (Eds.): RAMiCS 2012, LNCS 7560, pp. 146–161, 2012.

given length; both these operations are total and non-deterministic. The expression $(\text{id} \triangle \text{id})^\circ$ is the converse of the duplication operation: it is a partial function that takes as input tuples with two copies of the same element, and returns such element. In an unbounded execution, length° and head° would evaluate freely until they both return the same list that could be consumed by $(\text{id} \triangle \text{id})^\circ$. Such execution may not even terminate, since, for instance, head° could be generating all possible lists by increasing length.

If we are able to determine exactly the domain (and range)[1] of an expression, such mechanism can be used to predict the behavior of partial expressions by being used as a pre-condition checker. In this case, we are able to calculate both the domain ($n \neq 0$) and the range ($\text{length } l = \text{head } l$, which also implies that l is not empty) of this expression. Moreover, these domains can also be propagated down the expression to the inner combinators, avoiding unnecessary computations. In this case, due to $(\text{id} \triangle \text{id})^\circ$, length° and head° must generate the same list, and this information can be used to narrow their executions. In particular, given an input n, we can either restrict the values generated by length° to those lists whose head is n or, dually, restrict head° to produce lists with length n. This will result in an efficient and complete (in the sense that all values will eventually be produced) non-deterministic evaluation.

In this paper, we propose a PF relational framework whose type system is enhanced with the introduction of *invariants* (represented by coreflexives), allowing the definition of more refined data-types, in order to address the abovementioned issues. To carry this development, a powerful and simple calculus of invariants based on the relational PF notation [16] is harnessed into a type-inference and type-checking algorithm that works for many practical examples. The inferred invariants are also used to optimize the execution of a relational expression, making them viable as running prototypes of the specified program.

Our framework proves to be particularly useful in the area of bidirectional transformations (BX), where partiality and non-determinism play an important role. In particular, we put it to use in the specification of *lenses* [5], one of the most successful BX approaches. Using invariants to precisely characterize the domain and range of a lens, we can safely extend the class of expressible transformations, namely by allowing unrestricted usage of duplication, a well-known problematic feature in such frameworks. Also, propagation of such invariants allows us to efficiently execute the non-deterministic update propagation function for a wider class of transformations than before [5,17].

Section 2 introduces the PF relational calculus that is at the core of our framework. Section 3 presents our optimizations on invariant calculation and non-deterministic evaluation. Section 4 shows how standard recursion patterns can also be supported. In Sect. 5 we apply our framework in the specification of BX, obtaining non-deterministic lenses enhanced with invariants. Finally, Sect. 6 discusses related work and Sect. 7 draws the final conclusions and points directions for future work.

---

[1] By domain and range we refer to the exact set of values which a relation consumes and produces, respectively.

$$\cdot \circ \cdot \; : (B \to C) \to (A \to B) \to (A \to C) \qquad \mathrm{id} \; : A \to A$$
$$\cdot \cap \cdot \; : (A \to B) \to (A \to B) \to (A \to B) \qquad \pi_1 : A \times B \to A$$
$$\cdot \cup \cdot \; : (A \to B) \to (A \to B) \to (A \to B) \qquad \pi_2 : A \times B \to B$$
$$\cdot \vartriangle \cdot \; : (A \to B) \to (A \to C) \to (A \to B \times C) \qquad i_1 \; : A \to (A + B)$$
$$\cdot \triangledown \cdot \; : (B \to A) \to (C \to A) \to (B + C \to A) \qquad i_2 \; : B \to (A + B)$$
$$\top \quad : A \to B \qquad\qquad\qquad\qquad\qquad\qquad ! \; : A \to 1$$
$$\bot \quad : A \to B \qquad\qquad\qquad\qquad\qquad\qquad \underline{\cdot} \; : B \to (A \to B)$$
$$\cdot^{\circ} \quad : (A \to B) \to (B \to A)$$

Fig. 1. PF relational combinators

2   Point-Free Relational Calculus

Relation algebra [3,16,19] is a key ingredient in the formalization of our framework. It generalizes the well-known PF functional calculus, allowing us to reason about partiality and non-determinism using a powerful set of algebraic laws.

2.1   Syntax and Semantics

A *relation* R is said to have type $A \to B$ if it is the subset of the Cartesian product $A \times B$. We write $b \, R \, a$ if the pair (a, b) is in R. Relations can be built using the combinators presented in Fig. 1. The key combinator is *composition*, that given $R : A \to B$ and $S : B \to C$ builds a relation $S \circ R : A \to C$, which is associative and has the *identity* relation $\mathrm{id} : A \to A$ as neutral element (we thus have a category of relations). Relations $R : A \to B$ and $S : A \to B$ can be combined using the standard *intersection* and *union* operators. Every relation $R : A \to B$ also possesses a well-defined *converse* $R^{\circ} : B \to A$. For any two types A and B, $\top : A \to B$ is the largest relation over those types (their Cartesian product) and $\bot : A \to B$ the smallest (the empty relation). A special case of $\top$ with final type 1 as range is denoted as $! : A \to 1$. For any value $b \in B$, the constant relation $\underline{b} : A \to B$ always returns b.

We also have *products* and *coproducts* (or *sums*). For any two relations $R : A \to B$ and $S : A \to C$, the *split* combinator is defined as $R \vartriangle S : A \to B \times C$. The left and right components of a pair can be projected with $\pi_1 : A \times B \to A$ and $\pi_2 : A \times B \to B$, respectively. Dually, for any two relations $R : B \to A$ and $S : C \to A$, the *either* combinator is defined as $R \triangledown S : B + C \to A$. Left and right tagged elements can be built with $i_1 : A \to A + B$ and $i_2 : B \to A + B$, respectively. Two derived combinators are the product and sum bifunctors, defined respectively as $R \times S = R \circ \pi_1 \vartriangle S \circ \pi_2$ and $R + S = i_1 \circ R \triangledown i_2 \circ S$. Some of the laws ruling the PF relational calculus are presented in an accompanying technical report [11].

The formal semantics of relational expressions as membership predicates is given in Fig. 2. Notice that, apart from composition, this semantics can be directly and efficiently executed. If we assume that all types are finite, composition could also be implemented but would obviously be very inefficient. An alternative semantics as non-deterministic functions (functions returning sets of values)

$$
\begin{array}{llll}
b \; [\![S \circ R]\!] \; a & = \exists \; c. \; b \; [\![S]\!] \; c \wedge c \; [\![R]\!] \; a & a' \; [\![\mathrm{id}]\!] \; a & = a \equiv a' \\
b \; [\![R \cap S]\!] \; a & = b \; [\![R]\!] \; a \wedge b \; [\![S]\!] \; a & a' \; [\![\pi_1]\!] \; (a, b) & = a \equiv a' \\
b \; [\![R \cup S]\!] \; a & = b \; [\![R]\!] \; a \vee b \; [\![S]\!] \; a & b' \; [\![\pi_2]\!] \; (a, b) & = b \equiv b' \\
(b, c) \; [\![R \vartriangle S]\!] \; a & = b \; [\![R]\!] \; a \wedge c \; [\![S]\!] \; a & (\mathrm{Left} \; a') \; [\![i_1]\!] \; a & = a \equiv b \\
a \; [\![R \triangledown S]\!] \; (\mathrm{Left} \; b) & = a \; [\![R]\!] \; b & (\mathrm{Left} \; a') \; [\![i_2]\!] \; b & = \mathrm{False} \\
a \; [\![R \triangledown S]\!] \; (\mathrm{Right} \; c) & = a \; [\![S]\!] \; c & (\mathrm{Right} \; b') \; [\![i_1]\!] \; a & = \mathrm{False} \\
a \; [\![R^\circ]\!] \; b & = b \; [\![R]\!] \; a & (\mathrm{Right} \; b') \; [\![i_2]\!] \; b & = a \equiv b \\
b \; [\![\bot]\!] \; a & = \mathrm{False} & 1 \; [\![!]\!] \; a & = \mathrm{True} \\
b \; [\![\top]\!] \; a & = \mathrm{True} & b' \; [\![\underline{b}]\!] \; a & = b \equiv b'
\end{array}
$$

Fig. 2. Semantics as predicates

$$
\begin{array}{llll}
[\![S \circ R]\!] \; a & = \{ b \mid c \leftarrow [\![R]\!] \; a, b \leftarrow [\![S]\!] \; c \} & [\![\mathrm{id}]\!] \; a & = \{ a \} \\
[\![R \cap S]\!] \; a & = \{ b \mid b \leftarrow [\![R]\!] \; a, b \; [\![S]\!] \; a \} & [\![\pi_1]\!] \; (a, b) & = \{ a \} \\
[\![R \cup S]\!] \; a & = [\![R]\!] \; a \; \cup \; [\![S]\!] \; a & [\![\pi_2]\!] \; (a, b) & = \{ b \} \\
[\![R \vartriangle S]\!] \; a & = \{ (b, c) \mid b \leftarrow [\![R]\!] \; a, c \leftarrow [\![S]\!] \; a \} & [\![i_1]\!] \; a & = \{ \mathrm{Left} \; a \} \\
[\![R \triangledown S]\!] \; (\mathrm{Left} \; b) & = [\![R]\!] \; b & [\![i_2]\!] \; b & = \{ \mathrm{Right} \; b \} \\
[\![R \triangledown S]\!] \; (\mathrm{Right} \; c) & = [\![S]\!] \; c & [\![!]\!] \; a & = \{ 1 \} \\
[\![R^\circ]\!] \; b & = \{ a \mid a \leftarrow A, a \; [\![R^\circ]\!] \; b \} & [\![\underline{b}]\!] \; a & = \{ b \} \\
[\![\top]\!] \; a & = B & [\![\bot]\!] \; a & = \{ \}
\end{array}
$$

Fig. 3. Semantics as non-deterministic functions

is more useful if we intend to execute relational specifications. It can trivially be defined by set comprehension as $[\![R\!:\!A \to B]\!] \; a = \{ b \mid b \; [\![R]\!] \; a, b \leftarrow B \}$, but such definition is highly inefficient and cannot be used in practice. Figure 3 presents an alternative optimized definition that avoids the exhaustive search over B for all combinators but the converse. Again, this semantics can be directly implemented, for example using the non-determinism monad in a functional language like Haskell. However, given a value a, even if we are only interested in just one of the results of $[\![R]\!] \; a$, there are still several concerns for efficiency (besides the converse) that make such definition impractical. For example, in the left-biased implementation of intersection we still need to iterate over all results of R until a suitable value that also satisfies S is found.

The *kernel* of a relation is defined as $\ker R = R^\circ \circ R$, while its counterpart, the *image*, is defined as $\mathrm{img} \; R = R \circ R^\circ$. A relation R is said to be *reflexive* if it is at least the identity ($\mathrm{id} \subseteq R$), and *coreflexive* if it is at most the identity ($R \subseteq \mathrm{id}$). Coreflexives will be denoted by upper-case Greek letters ($\Psi, \Phi, \Omega, ...$). Relations can be classified according to the properties of their kernel and image. A relation is said to be total or surjective if its kernel and image are reflexive, respectively, and injective or simple if its kernel or image are coreflexive, respectively. Functions arise as the particular class of relations that are total and simple. As a convention, the identifiers of relational expressions that happen to be simple will begin with a lower-case. So, while $R, S, T, ...$ are typical identifiers for relational expressions, $f, g, h, ...$ will denote simple relations (partial functions).

2.2   Predicates as Coreflexives

Coreflexives act as filters of data and can be used to model predicates (and thus invariants): values a for which $a \; \llbracket \Phi \rrbracket \; a$ satisfy the predicate Φ. We will often see them as sets and denote predicate satisfiability using just set membership $a \in \llbracket \Phi \rrbracket \equiv a \; \llbracket \Phi \rrbracket \; a$. Coreflexives have interesting algebraic properties that simplify their manipulation like, for example, $\Phi^\circ = \Phi$, $\Phi \circ \Phi = \Phi$, and $\Phi \circ \Psi = \Phi \cap \Psi$. Evaluation of coreflexives also reduces to membership test as $\llbracket \Phi \rrbracket \; a = \{ a \mid a \in \llbracket \Phi \rrbracket \}$, meaning that its evaluation is typically efficient. The only problematic case is again composition, but as we will see shortly most of the compositions appearing in coreflexives can be evaluated efficiently. In particular, composition of coreflexives is just a conjunction of predicates.

A predicate on products can always be specified by a relation between its elements. Any relation $R : A \to B$ can be lifted to a coreflexive $[R] : A \times B \to A \times B$ defined as $[R] = (\pi_2^\circ \circ R \circ \pi_1) \cap \mathsf{id}$. Another way to put it is to say that $[R]$ is the largest coreflexive Φ such that $\pi_2 \circ \Phi \subseteq R \circ \pi_1$, since $\Phi \subseteq [R] \Leftrightarrow \pi_2 \circ \Phi \circ \pi_1^\circ \subseteq R$.

From this we can derive many interesting properties of this combinator, such as $[\top] = \mathsf{id}$, $[\bot] = \bot$, the cancellation rules $\pi_1 \circ [R] = (\mathsf{id} \triangle R)^\circ$ and $\pi_2 \circ [R] = (R^\circ \triangle \mathsf{id})^\circ$, and $[\pi_2 \circ \Phi \circ \pi_1^\circ] = \Phi$ for any coreflexive on pairs Φ. For example, using this combinator we can trivially specify the predicate stating that both components of a pair are equal using the coreflexive $[\mathsf{id}] : A \times A \to A \times A$. Given coreflexives $\Phi : A \to A$ and $\Psi : B \to B$, their product is the coreflexive $\Phi \times \Psi : A \times B \to A \times B$ that holds for pairs whose left element satisfies Φ and whose right element satisfies Ψ. It can alternatively be specified as $\Phi \times \Psi = [\Psi \circ \top \circ \Phi]$.

Coreflexives on sums are considerably simpler, since predicates on sums can always be specified using the sum combinator. The coreflexive $\Phi + \Psi : A + B \to A + B$ holds for left values that satisfy Φ and for right values that satisfy Ψ.

Every coreflexive has a *complement* $\overline{\Phi} : A \to A$ such that $a \; \llbracket \Phi \rrbracket \; a \Leftrightarrow \neg (a \; \llbracket \Phi \rrbracket \; a)$. A useful combinator for coreflexives is the guard $\Phi? = (\Phi \triangledown \overline{\Phi})^\circ : A \to A + A$ that tags the input as a left or right value in a sum, depending on the result of testing Φ. Composed with an either, it allows the representation conditionals, i.e., $(R \triangledown S) \circ \Phi?$ applies R if the input is in Φ, and S otherwise.

In this paper, we will use coreflexives to specify the invariants that characterize the *domain* and *range* of a relation, thus type-inference will amount to calculating the domain/range of a relation, while type-checking will consist of a membership test on those coreflexives. Given a relation $R : A \to B$, its domain, denoted as $\delta R : A \to A$, is the coreflexive $\delta R = \mathsf{ker}\; R \cap \mathsf{id}$. Dually, its range, denoted as $\rho R : B \to B$, is the coreflexive $\rho R = \mathsf{img}\; R \cap \mathsf{id}$. If R is total, its kernel is larger than id and thus $\delta R = \mathsf{ker}\; R \cap \mathsf{id} = \mathsf{id}$, as expected, while if R is simple, its image is smaller than the identity and thus $\rho R = \mathsf{img}\; R \cap \mathsf{id} = \mathsf{img}\; R$. These definitions simplify in a similar way for surjective and injective relations.

A relation $R : A \to B$ that is only defined for inputs satisfying Φ and always produces outputs satisfying Ψ ($R \subseteq \Psi \circ \top \circ \Phi$) will be typed as $R : A_\Phi \to B_\Psi$, or just $R : \Phi \to \Psi$ if the underlying types are irrelevant or clear from the context.

$$\delta\text{id} = \text{id} \qquad \delta\Phi = \Phi \qquad \rho\text{id} = \text{id} \qquad \rho\Phi = \Phi$$
$$\delta\bot = \bot \qquad \delta\top = \text{id} \qquad \rho\bot = \bot \qquad \rho\top = \text{id}$$
$$\delta(R \cap S) = R^\circ \circ S \cap \text{id} \qquad \delta\pi_1 = \text{id} \qquad \rho(R \cap S) = R \circ S^\circ \cap \text{id} \qquad \rho\pi_1 = \text{id}$$
$$\delta(R \cup S) = \delta R \cup \delta S \qquad \delta\pi_2 = \text{id} \qquad \rho(R \cup S) = \rho R \cup \rho S \qquad \rho\pi_2 = \text{id}$$
$$\delta(R \vartriangle S) = \delta R \cap \delta S \qquad \delta i_1 = \text{id} \qquad \rho(R \vartriangle S) = [S \circ R^\circ] \qquad \rho i_1 = \text{id} + \bot$$
$$\delta(R \triangledown S) = \delta R + \delta S \qquad \delta i_2 = \text{id} \qquad \rho(R \triangledown S) = \rho R \cup \rho S \qquad \rho i_2 = \bot + \text{id}$$
$$\delta(R^\circ) = \rho R \qquad \delta! = \text{id} \qquad \rho\underline{b} = \underline{b} \cap \text{id} \qquad \rho! = \text{id}_1$$
$$\delta(\Phi?) = \text{id} \qquad \delta\underline{b} = \text{id} \qquad \rho(\Phi?) = \Phi + \overline{\Phi} \qquad \rho(R^\circ) = \delta R$$
$$\delta(R \circ S) = \delta(\delta R \circ S) \qquad \rho(R \circ S) = \rho(R \circ \rho S)$$

Fig. 4. Domain and range of PF combinators

3   Optimizations

In the previous section, we have shown how the domain and range of a relation can be specified. However, such specifications involve relational compositions that hinder their efficient execution as pre- and pos-condition checkers of a relation. In this section we will first show how the calculation of the domain and range can be optimized to yield expressions more amenable to execution. Then, we will show how we can take advantage of such domain and range expressions to optimize the semantics defined in Fig. 3.

3.1   Optimizing Domain and Range Calculation

For relational programs written using the PF combinators from Fig. 1, their respective domains and ranges can be defined by induction as presented in Figs. 4 and 5. To avoid infinite reductions in compositions, the laws of Fig. 5 should be prioritized. These laws detail how the domain and range of a combinator should be further restricted in presence of a coreflexive.

The expressions resulting from these definitions are more amenable for execution (and consequently, type-checking) than the default domain and range definitions because most of the compositions are eliminated. The remaining ones (except for the range of the split combinator) fall in the special case $R^\circ \circ U \circ S$, that, as shown in [15], can be evaluated deterministically as $a \ [\![R^\circ \circ U \circ S]\!] \ b = (R\ a) \ [\![U]\!] \ (S\ b)$ if R and S are functions. After applying the laws of Figs. 4 and 5, we further simplify the resulting expression using a rewrite system similar to one previously developed for the optimization of PF functional expressions [4,18]. Essentially, this rewrite system applies some of the PF laws [11] as unidirectional rewrite rules oriented from left to right. This simplification phase can further eliminate problematic compositions. If the final expression still contains some of those, our implementation can issue a warning informing that its usage as an invariant checker may not be feasible.

This rewrite system is also used to perform the equality test $\Phi = \bot$ that occurs in some of the definitions in Fig. 5. However, since such test may not be not decidable, i.e., the rewrite system may not be able to reduce into $\bot$ an

$$\delta(\Phi \circ \Psi) = \Phi \cap \Psi$$

$$\delta(\Phi \circ \mathsf{id}) = \Phi$$

$$\delta(\Phi \circ \top) = \begin{cases} \bot & \text{if } \Phi = \bot \\ (\top \circ \Phi \circ \top) \cap \mathsf{id} & \text{otherwise} \end{cases}$$

$$\delta(\Phi \circ \bot) = \bot$$

$$\delta(\Phi \circ R^\circ) = \rho(R \circ \Phi)$$

$$\delta(\Phi \circ (R \cup S)) = \delta(\Phi \circ R) \cup \delta(\Phi \circ S)$$

$$\delta(\Phi \circ (R \cap S)) = (R^\circ \circ \Phi \circ S) \cap \mathsf{id}$$

$$\delta(\Phi \circ \pi_1) = \Phi \times \mathsf{id}$$

$$\delta(\Phi \circ \pi_2) = \mathsf{id} \times \Phi$$

$$\delta([U] \circ (R \triangle S)) = (\delta S \circ R^\circ \circ U^\circ \circ S \circ \delta R) \cap \mathsf{id}$$

$$\delta((\Phi + \Psi) \circ i_1) = \Phi$$

$$\delta((\Phi + \Psi) \circ i_2) = \Psi$$

$$\delta(\Phi \circ (R \triangledown S)) = \delta(\Phi \circ R) + \delta(\Phi \circ S)$$

$$\delta(\Phi \circ \underline{b}) = \begin{cases} \mathsf{id} \text{ if } b \; [\![\Phi]\!] \; b \\ \bot \text{ otherwise} \end{cases}$$

$$\delta(\Phi \circ !) = \begin{cases} \mathsf{id} \text{ if } 1 \; [\![\Phi]\!] \; 1 \\ \bot \text{ otherwise} \end{cases}$$

$$\delta((\Phi + \Psi) \circ \Omega?) = (\Phi \cap \Omega) \cup (\Psi \cap \overline{\Omega})$$

$$\rho(\Psi \circ \Phi) = \Psi \cap \Phi$$

$$\rho(\mathsf{id} \circ \Phi) = \Phi$$

$$\rho(\top \circ \Phi) = \begin{cases} \bot & \text{if } \Phi = \bot \\ (\top \circ \Phi \circ \top) \cap \mathsf{id} & \text{otherwise} \end{cases}$$

$$\rho(\bot \circ \Phi) = \bot$$

$$\rho(R^\circ \circ \Phi) = \delta(\Phi \circ R)$$

$$\rho((R \cup S) \circ \Phi) = \rho(R \circ \Phi) \cup \rho(S \circ \Phi)$$

$$\rho((R \cap S) \circ \Phi) = (R \circ \Phi \circ S^\circ) \cap \mathsf{id}$$

$$\rho(\pi_1 \circ [U]) = \delta U$$

$$\rho(\pi_2 \circ [U]) = \rho U$$

$$\rho((R \triangle S) \circ \Phi) = [S \circ \Phi \circ R^\circ]$$

$$\rho(i_1 \circ \Phi) = \Phi + \bot$$

$$\rho(i_2 \circ \Phi) = \bot + \Phi$$

$$\rho((R \triangledown S) \circ (\Phi + \Psi)) = \rho(R \circ \Phi) \cup \rho(S \circ \Psi)$$

$$\rho(\underline{b} \circ \Phi) = \begin{cases} \bot & \text{if } \Phi = \bot \\ \underline{b} \circ \Phi \circ \underline{b}^\circ & \text{otherwise} \end{cases}$$

$$\rho(! \circ \Phi) = \begin{cases} \bot & \text{if } \Phi = \bot \\ ! \circ \Phi \circ !^\circ & \text{otherwise} \end{cases}$$

$$\rho(\Psi ? \circ \Phi) = (\Psi \cap \Phi) + (\overline{\Psi} \cap \Phi)$$

Fig. 5. Domain and range of compositions

expression that is semantically equivalent to $\bot$, if we cannot show that Φ is empty, the default definitions of range and domain are applied instead. Still, for some cases when we can prove that $\Phi \neq \bot$, the range of (for instance) $! \circ \Phi$ and $\top \circ \Phi$ can be further simplified to id.

3.2   Optimizing Non-deterministic Executions

The executable semantics of Fig. 3 can be optimized by propagating the domains and ranges of the outer expressions down to the inner expressions, in order to avoid the computation of intermediate values that are valid for sub-expressions but are not valid for the global expression. Figure 6 shows how this propagation can be performed (A and B denote the set of all elements of the respective type), where input values are assumed to have already passed the pre-condition test, i.e., for an evaluation $[\![R : \Phi \to \Psi]\!] \; a$, we assume that $a \in \Phi$. For instance, in the evaluation of $R \circ S$ we can narrow the evaluation of S to return only values in the domain of R (and vice-versa), thus avoiding generation of values not accepted by R; since the split $R \triangle S$ is only defined for values in the domain of both R and S, the domain invariant of each branch takes the domain of the other, in order to disregard invalid values during execution. The converse of expressions is presented in Fig. 7, where each case is analyzed individually to achieve better efficiency. We omit the evaluation of the converse of idempotent combinators (id, Ω, $\top$, $\bot$) and of combinators whose converse can be easily propagated ($R \circ S$, $R \cap S$, $R \cup S$) and thus can be executed by the definitions in Fig. 6. The proof of the semantic equivalence between the two versions, in the sense that $[\![R : \Phi \to \Psi]\!] = [\![\Psi \circ R \circ \Phi]\!]$, is given in [11].

$$[\![R \circ S : \Phi \to \Psi]\!]\, a = \{b \mid c \leftarrow [\![S : \Phi \to \delta R]\!]\, a, b \leftarrow [\![R : \rho S \to \Psi]\!]\, c\}$$
$$[\![R \cap S : \Phi \to \Psi]\!]\, a = \{b \mid b \leftarrow [\![R : \Phi \cap \delta S \to \rho(\Psi \circ S \circ \underline{a})]\!]\, a\}$$
$$[\![R \cup S : \Phi \to \Psi]\!]\, a = [\![R : \Phi \to \Psi]\!]\, a \cup [\![S : \Phi \to \Psi]\!]\, a$$
$$[\![R \vartriangle S : \Phi \to [U]]\!]\, a = \{(b, c) \mid b \leftarrow [\![R : \Phi \cap \delta S \to \rho(U^\circ \circ S \circ \underline{a})]\!]\, a,$$
$$c \leftarrow [\![S : \Phi \cap \delta R \to \rho(U \circ \underline{b})]\!]\, a\}$$

$$[\![\mathsf{id} : \Phi \to \Psi]\!]\, a = [\![\Psi]\!]\, a \qquad\qquad [\![\pi_1 : [U] \to \Psi]\!]\, (a, b) = [\![\Psi]\!]\, a$$
$$[\![\Omega : \Phi \to \Psi]\!]\, a = \{a' \mid a' \leftarrow [\![\Omega]\!]\, a, a' \,[\![\Psi]\!]\, a'\} \quad [\![\pi_2 : [U] \to \Psi]\!]\, (a, b) = [\![\Psi]\!]\, b$$
$$[\![i_1 : \Phi \to \Psi + \Omega]\!]\, a = \{\mathsf{Left}\, a' \mid a' \leftarrow [\![\Psi]\!]\, a\} \quad [\![\bot : \Phi \to \Psi]\!]\, a = \{\,\}$$
$$[\![i_2 : \Phi \to \Psi + \Omega]\!]\, b = \{\mathsf{Right}\, b' \mid b' \leftarrow [\![\Omega]\!]\, b\} \quad [\![\top : \Phi \to \Psi]\!]\, a = \{b \mid b \leftarrow B, b\,[\![\Psi]\!]\, b\}$$
$$[\![R \triangledown S : \Phi + \Omega \to \Psi]\!]\, (\mathsf{Left}\, b) = [\![R : \Phi \to \Psi]\!]\, b \quad [\![\underline{b} : \Phi \to \Psi]\!]\, a = [\![\Psi]\!]\, b$$
$$[\![R \triangledown S : \Phi + \Omega \to \Psi]\!]\, (\mathsf{Right}\, c) = [\![S : \Omega \to \Psi]\!]\, c \quad [\![! : \Phi \to \Psi]\!]\, a = [\![\Psi]\!]\, 1$$

Fig. 6. Optimized non-deterministic evaluation

$$[\![!^\circ : \Phi \to \Psi]\!]\, 1 = \{a \mid a \leftarrow A, a\,[\![\Psi]\!]\, a\} \qquad [\![\underline{b}^\circ : \Phi \to \Psi]\!]\, b = \{a \mid a \leftarrow A, a\,[\![\Psi]\!]\, a\}$$
$$[\![\pi_1^\circ : \Phi \to [U]]\!]\, a = \{(a, b) \mid b \leftarrow [\![U]\!]\, a\} \,[\![\pi_2^\circ : \Phi \to [U]]\!]\, b = \{(a, b) \mid a \leftarrow [\![U^\circ]\!]\, b\}$$
$$[\![i_1^\circ : \Phi + \bot \to \Psi]\!]\, (\mathsf{Left}\, a) = [\![\Psi]\!]\, a \qquad [\![i_2^\circ : \bot + \Phi \to \Psi]\!]\, (\mathsf{Right}\, b) = [\![\Psi]\!]\, b$$
$$[\![(R \vartriangle S)^\circ : [U] \to \Psi]\!]\, (b, c) = \{a \mid a \leftarrow [\![R^\circ : \rho U \to \rho(\Psi \circ S^\circ \circ \underline{c})]\!]\, b\}$$
$$[\![(R \triangledown S)^\circ : \Phi \to \Psi + \Omega]\!]\, a = [\![i_1 \circ R^\circ : \Phi \to \Psi]\!]\, a \cup [\![i_2 \circ S^\circ : \Phi \to \Omega]\!]\, a$$

Fig. 7. Optimized non-deterministic evaluation of converses

The evaluation of the primitive combinators is, for most cases, fairly obvious, since it consists in their standard definition, with a membership test for the desired invariant. Note however that all invariant tests occur at the primitives, meaning that infeasible values are not passed through higher-order combinators. Nevertheless, redundant values can still be generated, even if they produce a valid output. For instance, in the expression $\underline{b} \circ \top$, $\top$ will generate all possible values, even though they will all be transformed into the same value by $\underline{b}$. In many of such cases, the rewrite system already presented can be used to remove redundant value generation. In this case, $\underline{b} \circ \top$ would be reduced to $\underline{b}$.

The most interesting narrowing cases are those of the meet and the converse of split (which is itself a meet). For these cases, with the definition from Fig. 3, the R branch would execute independently of the invariants of S and its output would be tested in S. Naturally, the unconstrained evaluation of R can be very inefficient and may process and generate infeasible values that are not in the domain or range of S, respectively. Using invariants, we restrict R to the domain of S and constrain the values generated by R to only those that would also be produced by S. For instance, in the execution of the converse of the split $[\![(R \vartriangle S)^\circ]\!]\, (b, c)$, instead of having $[\![R^\circ]\!]\, b$ running freely, it is restricted to produce values that would also be produced by $[\![S^\circ]\!]\, c$, as specified by its post-condition $\rho(\Psi \circ S^\circ \circ \underline{c})$. Again, a right-biased implementation would be equivalent.

4   Recursive Relations with Invariants

In this section, we investigate the construction of expressions and the calculation of invariants for recursive types. Most user-defined data types can be defined as fixed points of regular functors. Given a base functor, the inductive type generated by its least fixed point will be denoted by μF. A regular functor is either the identity functor Id (denoting recursive invocation), the constant functor $\underline{A}$, the lifting of the sum $\oplus$ and product bifunctors $\otimes$, or the composition of functors $\odot$. For example, for lists we have $[A] = \mu \mathsf{L}$, where $\mathsf{L} = \underline{1} \oplus (\underline{A} \otimes \mathsf{Id})$, and for naturals $\mathbb{N} = \mu \mathsf{N}$, where $\mathsf{N} = \underline{1} \oplus \mathsf{Id}$. Associated with each data type μF we have also two unique functions $\mathsf{in}_F : F\ \mu F \to \mu F$ and $\mathsf{out}_F : \mu F \to F\ \mu F$, that are each other's inverse. The list constructors can be defined as $\mathsf{nil} = \mathsf{in_L} \circ i_1$ and $\mathsf{cons} = \mathsf{in_L} \circ i_2$, (thus $\mathsf{nil} \triangledown \mathsf{cons} = \mathsf{in_L}$), and for naturals as $\mathsf{zero} = \mathsf{in_N} \circ i_1$ and $\mathsf{succ} = \mathsf{in_N} \circ i_2$. They allow us to encode and inspect values of the given type, respectively. The application of out results on a one-level unfolding to a sums-of-products representation capable of being processed with PF combinators. For a functor F and a function $f : A \to B$, the functor mapping $F\ f : F\ A \to F\ B$ is a function that maps f over the instances of the type argument, and can be defined inductively over the structure of the functor. Since in and out are bijections, they are total and surjective, and for in (and dually out) we have that:

$$\delta(\Phi \circ \mathsf{in}) = \mathsf{out} \circ \Phi \circ \mathsf{in} \qquad\qquad \rho(\mathsf{in} \circ \Phi) = \mathsf{in} \circ \Phi \circ \mathsf{out}$$

Instead of defining expressions by general recursion, we resort to well-known recursion patterns, namely folds (catamorphisms) and unfolds (anamorphisms), that encode the recursion patterns of iteration and coiteration, respectively. The fold $(\!|R|\!)_F : \mu F \to A$ consumes values of a recursive type μF according to an algebra $R : F\ A \to A$, while the dual unfold $[\![S]\!]_F : A \to \mu F$ produces elements of a recursive type μF according to a coalgebra $S : A \to F\ A$, and are the unique relations that make the hereunder diagrams commute:

$$
\begin{array}{ccc}
\mu F \xrightarrow{\ \mathsf{out}_F\ } F\ \mu F & \qquad & \mu F \xleftarrow{\ \mathsf{in}_F\ } F\ \mu F \\
\left\downarrow{\scriptstyle (\!|R|\!)_F} \qquad\quad \downarrow{\scriptstyle F\ (\!|R|\!)_F}\right. & \quad {\scriptstyle [\![S]\!]_F}\uparrow \qquad\quad \uparrow {\scriptstyle F\ [\![S]\!]_F} \\
A \xleftarrow[\ R\ ] F\ A & \qquad & A \xrightarrow[\ S\ ] F\ A
\end{array}
$$

As expected, these recursion patterns preserve the simplicity of their argument algebras or coalgebras [2]. Forward and backward (converse) evaluation is not problematic, because we can proceed recursively by unfolding their definitions:

$$(\!|R|\!)_F = R \circ F\ (\!|R|\!)_F \circ \mathsf{out}_F \qquad\qquad [\![S]\!]_F = \mathsf{in}_F \circ F\ [\![S]\!]_F \circ S$$

The main problem, however, is the optimization of domain/range calculation for folds and unfolds due to the nonexistence of a normal form to express invariants over recursive types. For some simple cases, we can rely on the following laws [2]:

$$F \rho R \subseteq \delta R \Rightarrow \delta (\!(R)\!) = \mathsf{id} \qquad\qquad F\, \delta S \subseteq \rho S \Rightarrow \rho [\![S]\!] = \mathsf{id}$$

$$R : F\,\Phi \to \Phi \Rightarrow (\!(R)\!)_F : \mathsf{id} \to \Phi \qquad S : \Phi \to F\,\Phi \Rightarrow [\![S]\!]_F : \Phi \to \mathsf{id}$$

Focusing on the left column (the other is dual for unfolds), the first law states that a fold is total if the range of its algebra is contained in its own domain (in particular, total algebras yield total folds); the second law states a simple consistency condition needed to establish the range of a fold. Whenever these laws do not apply, we resort to the general definitions of domain and range presented in Sect. 2.2, and then apply the rewrite system briefly presented in Sect. 3.1, enriched with laws to handle recursive patterns, namely fusion[2]:

$$S \circ (\!(R)\!)_F = (\!(T)\!)_F \Leftarrow S \circ R = T \circ F\, S \qquad [\![S]\!]_F \circ R = [\![T]\!]_F \Leftarrow S \circ R = F\, R \circ T$$

Fusion laws transform the composition of a relation with a recursion pattern into a single recursion pattern. As explained before, we issue a warning if the rewrite system yields expressions whose evaluation may be problematic.

We now give some examples of recursive expressions that are already supported in our framework. We begin with $(\mathsf{id} \bigtriangleup \mathsf{id})^\circ \circ (\mathsf{length}^\circ \bigtriangleup \mathsf{head}^\circ)$, the example from the introduction, where $\mathsf{head} = \pi_1 \circ \mathsf{cons}^\circ$ and $\mathsf{length} = (\!(\mathsf{in_N} \circ (\mathsf{id} + \pi_2))\!)$. The domain of head is $\mathsf{in_L} \circ (\bot + \mathsf{id}) \circ \mathsf{out_L}$, meaning that the list cannot be empty, while $\mathsf{length} : \mathsf{id} \to \mathsf{id}$ by applying the above laws for folds, since its algebra has type $F\,\mathsf{id} \to \mathsf{id}$. The range of the whole expression can be computed as follows:

$\rho((\mathsf{id} \bigtriangleup \mathsf{id})^\circ \circ (\mathsf{head}^\circ \bigtriangleup \mathsf{length}^\circ))$
$\quad = \quad \{\text{-Range definition: Fig. 4 -}\}$
$\rho((\mathsf{id} \bigtriangleup \mathsf{id})^\circ \circ \rho(\mathsf{head}^\circ \bigtriangleup \mathsf{length}^\circ))$
$\quad = \quad \{\text{-Range definition: Fig. 4 -}\}$
$\rho((\mathsf{id} \bigtriangleup \mathsf{id})^\circ \circ [\mathsf{length}^\circ \circ \mathsf{head}])$
$\quad = \quad \{\text{-Range definition: Fig. 5 -}\}$
$\delta([\mathsf{length}^\circ \circ \mathsf{head}] \circ (\mathsf{id} \bigtriangleup \mathsf{id}))$
$\quad = \quad \{\text{-Domain definition: Fig. 5, Simplifications: PF Laws [11] -}\}$
$(\mathsf{head}^\circ \circ \mathsf{length}) \cap \mathsf{id}$

Since $l \in [\![(\mathsf{head}^\circ \circ \mathsf{length}) \cap \mathsf{id}]\!] \Leftrightarrow \mathsf{head}\, l \equiv \mathsf{length}\, n$, we have the expected invariant on the range. On the other hand, its domain is $\mathsf{in_L} \circ (\bot + \mathsf{id}) \circ \mathsf{out_L}$ (the proof can be found in [11]), and thus, $(\mathsf{id} \bigtriangleup \mathsf{id})^\circ \circ (\mathsf{length}^\circ \bigtriangleup \mathsf{head}^\circ) : \mathsf{in_L} \circ (\bot + \mathsf{id}) \circ \mathsf{out_L} \to (\mathsf{head}^\circ \circ \mathsf{length}) \cap \mathsf{id}$.

Another example of a catamorphism is the $\mathsf{unzip} : [A \times B] \to [A] \times [B]$ function, that splits a list of pairs into two lists with the left and right elements. Since the algebra of unzip is a total function $g = (\mathsf{nil} \bigtriangleup \mathsf{nil}) \triangledown ((\mathsf{cons} \circ \pi_1 \times \pi_1) \bigtriangleup (\mathsf{cons} \circ \pi_2 \times \pi_2))$, the domain of the catamorphism is id. As for its range, our rewrite system performs a calculation equivalent to the following:

---

[2] We implement the "guessing step" required for fusion using the technique from [18].

ρunzip

$\quad = \quad$ {-Definitions: range -}

$(\text{unzip} \circ \text{unzip}^\circ) \cap \text{id}$

$\quad = \quad$ {-Simplifications: unzip is simple, Liftify: range of unzip is a product -}

$[\pi_2 \circ \text{unzip} \circ \text{unzip}^\circ \circ \pi_1^\circ]$

$\quad = \quad$ {-Catamorphism fusion: $\pi_1 \circ g = \text{nil} \nabla (\text{cons} \circ (\pi_1 \times \text{id})) \circ F\ \pi_1$ -}

$[(\!|\text{nil} \nabla (\text{cons} \circ (\pi_1 \times \text{id}))|\!) \circ (\!|\text{nil} \nabla (\text{cons} \circ (\pi_2 \times \text{id}))|\!)^\circ]$

$\quad = \quad$ {-Definitions: map -}

$[(\text{map}\ \pi_1) \circ (\text{map}\ \pi_2)^\circ]$

$\quad = \quad$ {-Simplifications: map converse, map fusion (see below) -}

$[\text{map}\ (\pi_1 \circ \pi_2^\circ)]$

$\quad = \quad$ {-Simplifications: PF Laws [11] -}

$[\text{map}\ \top]$

Here, map $f = (\!|\text{in}_L \circ (\text{id} + (f \times \text{id}))|\!)$ is the mapping that applies f to all elements of a list, whose converse and fusion properties are defined as $(\text{map}\ f)^\circ = \text{map}\ f^\circ$ and $\text{map}\ f \circ \text{map}\ g = \text{map}\ (f \circ g)$. The resulting range $[\text{map}\ \top]$ means that unzip always produces lists with the same length but unrelated elements. Maps of coreflexives are themselves coreflexives, and represent a special shape of invariants over recursive types. For instance, the domain of in (and dually the range of out) over map invariants can be calculated as $\delta(\text{map}\ \Phi \circ \text{in}_L) = \text{id} + (\Phi \times \text{map}\ \Phi)$.

The reasoning about anamorphisms follows the same rationale and is omitted.

5 Application Scenario: Bidirectional Transformations

Lenses [5] are one of the most successful BX approaches. A lens, denoted by $S \rhd V$, is a bidirectional transformation between sources of type S and views of type V that comprises two functions: a forward transformation Get: $S \to V$ that abstracts a source into a view; and a backward transformation Put : $V \times S \to S$ that takes an updated view and the original source to return an updated source. A lens is *well-behaved* if it satisfies the round-tripping properties Get $\circ$ Put $\subseteq \pi_1$ (denoted *acceptability* or PUTGET) and Put $\circ$ (Get $\triangle$ id) $\subseteq$ id (denoted *stability* or GETPUT). A lens is also said to be *total* if Get and Put are total functions. Due to these laws, the Get of a total well-behaved lenses must be a surjective function (where any value of V must be the view of some source) and obviously total. For this reason, many interesting transformations (such as the split combinator) are not admissible as total well-behaved lenses since they are not surjective.

In fact, when designing a BX language there is a well-known tradeoff between the expressiveness allowed by its syntax and the robustness enforced by the totality and round-tripping laws. Some approaches [17,22] compromise the expressiveness; others ignore the totality requirement [12,20,21]; others maintain totality, but weaken the round-tripping laws [6]; some relax both totality and round-tripping laws [13,10,9,8]; finally, it also possible to avoid compromising the laws by developing a more refined type system, as proposed in the original lens framework [5]: in order to preserve totality, a powerful semantic type system with invariants was used to specify the exact domain and range of lenses, which allowed the definition of duplication and conditional combinators as total well-behaved lenses. Unfortunately, to retain decidability in the type system, the expressiveness was still restricted by forcing composed lenses to agree not only on types but also on invariants. For example, in such a scenario, duplication could be followed by a *merge* combinator that only accepts pairs with two equal values, but not by a generic projection that works for whatever pair.

Consider the composition of two transformations $f: A_\Phi \to B_\Psi$ and $g: B_\Gamma \to C_\Omega$, where Ψ is more restrictive then Γ, as depicted in the following diagram:

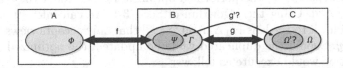

Since the range of f and the domain of g do not match, the Put of the composed lens would only be defined for Ω', the values in the range of g for which the values produced by Put_g are within the range of f. To support such generalized composition, we wil use the techniques proposed in this paper to: 1) perform invariant inference to discover the exact range Ω' of the (global) transformation; 2) specify a non-deterministic Put, whose optimization can be efficiently narrowed to the Put of a lens $g' : B_\Psi \unrhd C_{\Omega'}$ that only generates values in Ψ.

Using the relational calculus, it is quite simple to specify a generic non-deterministic Put that is the largest relation that satisfies the round-tripping properties. Any transformation (i.e., a simple relation) $f : A \to B$ can be lifted to a total, well-behaved non-deterministic lens $\lfloor f \rfloor : \delta f \unrhd \rho f$, with $\mathsf{Get}_f = f$ and $\mathsf{Put}_f = (\pi_2 \triangledown (f^\circ \circ \pi_1)) \circ [f^\circ]?$. This specification of Put_f trivially satisfies the round-tripping laws (a formal proof can be found in [11]), because it explicitly tests if the view was modified using the coreflexive $[f^\circ]$, as $(v, s) \in [\![[f^\circ]]\!] \Leftrightarrow v \equiv f\ s$. If so, it returns the original source; otherwise, it runs Get backwards to recover all possible sources that could have originated that view. Thus, Put_f is also the largest non-deterministic relation that keeps the lens well-behaved. Although trivial, the lens resulting from this lifting cannot be used in a practical BX framework. Of course, we could use the semantics of Fig. 2 to evaluate the invariants δf and ρf and perform type-checking, but as explained in Sect. 2.1, due to composition the resulting algorithm would be undecidable. Similarly for Put_f, we could use the semantics of Fig. 3 to perform evaluation. Even (reasonably) assuming that the user only wants a single updated source, and relying on lazy evaluation, the efficiency problem would be even worse, due to the central role played by the converse in the definition.

Both these problems can be handled by the optimizations presented in the previous sections. Our lens language allows any simple (or simplicity-preserving) PF combinator to be used to specify the forward transformation. Although unconstrained converse is not allowed (since it is not simple in general), we include the converses of the injections that are partial functions useful for "destructing sums". Thus, the domain and range of the transformations can be trivially calculated, and except particular ranges of splits, type checking is decidable.

As for the backward transformation, by applying the rules already presented in Fig. 6, the generic definition can be efficiently executed. Our language supports transformations including splits, conditionals, and converses of injections, that are not supported by most existing lens frameworks. In particular, the duplication operator $id \bigtriangleup id : A \to A \times A$ yields a lens $\lfloor id \bigtriangleup id \rfloor : id \trianglerighteq [id]$ (whose backward transformation only accepts pairs with equal components) that can be freely composed with other lenses irrespective of their invariants. Recursive expressions are also supported as they preserve the simplicity of their algebras.

To give an example of the performed optimizations, consider the transformation $f = \pi_1 \bigtriangleup id$. Using the algorithm of Sect. 3.1, we can infer its range and domain and lift it to the lens $\lfloor \pi_1 \bigtriangleup id \rfloor : id \trianglerighteq [\pi_1^\circ]$ that only accepts views $(x, (y, z))$ where $x \equiv y$. Should the duplicated value be updated, the optimized backward transformation would execute as follows:

$$\llbracket (\pi_2 \triangledown (f^\circ \circ \pi_1)) \circ [f^\circ] \, ? : [\pi_1^\circ] \times id \to id \rrbracket \, ((a, (a, y)), (x, y))$$
$$= \; \{\text{-Optimized semantics: } R \circ S \text{ (Fig. 6); Domain/Range (Fig. 4) -}\}$$
$$\{c \mid b \leftarrow \llbracket [f^\circ] \, ? : [\pi_1^\circ] \times id \to id + [\pi_1^\circ] \times id \rrbracket \, ((a, (a, y)), (x, y)),$$
$$\qquad c \leftarrow \llbracket \pi_2 \triangledown (f^\circ \circ \pi_1) : [f^\circ] + \overline{[f^\circ]} \to id \rrbracket \, b\}$$
$$= \; \{\text{-Optimized semantics: } \Phi?; \, ((a, (a, y)), (x, y)) \in \llbracket \overline{[f^\circ]} \rrbracket) \, \text{-}\}$$
$$\{c \mid c \leftarrow \llbracket (\pi_2 \triangledown (f^\circ \circ \pi_1)) : [f^\circ] + \overline{[f^\circ]} \to id \rrbracket \, (\mathsf{Right} \, ((a, (a, y)), (x, y)))\}$$
$$= \; \{\text{-Optimized semantics: } R \triangledown S \text{ (Fig. 6) -}\}$$
$$\{c \mid c \leftarrow \llbracket (f^\circ \circ \pi_1) : \overline{[f^\circ]} \to id \rrbracket \, ((a, (a, y)), (x, y))\}$$
$$= \; \{\text{-Optimized semantics: } R \circ S, \, \pi_1 \text{ (Fig. 6); Domain/Range (Fig. 4) -}\}$$
$$\{c \mid k \leftarrow \llbracket [\pi_1^\circ] \rrbracket \, (a, (a, y)), c \leftarrow \llbracket f^\circ : id \to id \rrbracket \, k\}$$
$$= \; \{\text{-}(a, (a, y)) \in \llbracket [\pi_1^\circ] \rrbracket \text{ -}\}$$
$$\{c \mid c \leftarrow \llbracket (\pi_1 \bigtriangleup id)^\circ : id \to id \rrbracket \, (a, (a, y))\}$$
$$= \; \{\text{-Optimized semantics: } (R \bigtriangleup S)^\circ \text{ (Fig. 6); Domain/Range (Fig. 4) -}\}$$
$$\{c \mid c \leftarrow \llbracket \pi_1^\circ : id \to \rho(a, y) \rrbracket \, a\}$$
$$= \; \{\text{-}\rho(a, y) = \rho\underline{a} \times \rho\underline{y} = [\rho\underline{a} \circ \top \circ \rho\underline{y}]; \text{ Optimized semantics: } \pi_1^\circ \text{ (Fig. 6) -}\}$$
$$\{(a, l) \mid l \leftarrow \llbracket \rho\underline{a} \circ \top \circ \rho\underline{y} \rrbracket \, a\}$$
$$= \; \{\text{-Simplifications: PF Laws [11]; Semantics: } R \circ S, \, \Phi \text{ (Fig. 3); } a \in \llbracket \rho\underline{a} \rrbracket \text{ -}\}$$
$$\{(a, y)\}$$

Note how the invariants only need to be evaluated for the primitives. Although the semantics of π_1° is non-deterministic, id forces a single result. Simplifications are applied to convert the invariant over pairs into the lifted form.

Recursive specifications can also be lifted to lenses. For instance, the transformation $tail \bigtriangleup length$ can be lifted to the lens $\lfloor tail \bigtriangleup length \rfloor : id \trianglerighteq \lfloor succ \circ length \rfloor$, whose backward transformation only accepts values such that $(l, n) \in \llbracket [succ \circ length] \rrbracket \Leftrightarrow length \, l + 1 \equiv n$. In this case, $\llbracket \mathsf{Put} \rrbracket \, (([2, 3], 3), [1, 2, 3]) = \{[1, 2, 3]\}$ since the view did not change, while $\llbracket \mathsf{Put} \rrbracket \, (([2, 0], 3), [1, 2, 3]) = \{[0, 2, 0], [1, 2, 0], [2, 2, 0], ...\}$, generating all possible lists with $[2, 0]$ as tail. Since $\lfloor unzip \rfloor : id \trianglerighteq [map \, \top]$ is an injective relation, its backward transformation is simple; therefore, even if the view lists are updated, Put_{unzip} always returns a single result that is the zip of the view pair.

6   Related Work

Although our calculus of invariants was inspired in [16], our typing rules impose a stronger restriction. In our case, a relation $R : \Phi \to \Psi$ is exactly defined only for values of Φ and only produces values in Ψ, while in [16] invariants represent pre- and post-conditions, i.e., $R \circ \Phi \subseteq \Psi \circ R$, meaning that there may exist values outside Φ for which R is defined but whose behavior is unpredictable. It follows that all typing rules of [16] are applicable to our framework.

Functional logic programming languages like Curry [7] focus on the non-deterministic evaluation of specifications written in a functional programming style. While such languages focus on the evaluation of the specifications, our approach provides a better understanding of the program and its behavior during executing, resorting to a calculus of invariants.

The universal resolving algorithm (URA) [1] has been developed to compute the inverses of functional programs. Like our evaluation algorithm, it is complete (it lazily enumerates all possible values) but not always terminating (since recursive types may admit infinitely many values). Nevertheless, unlike in URA, we are able to optimize expressions before evaluation using the relational calculus. This allows to cut many intermediate infeasible values, making value generation for most invariants much more efficient.

Regarding BX, our framework can be seen as a domain-specific language over inductive types similar to the one for lenses over generalized trees first developed by Foster *et al* [5]. They devise a complex set-based type system with invariants to precisely define the domains for which their combinators are well-behaved. However, combining lenses requires matching on invariants rather than on types, which is too restrictive. A dual approach is followed in [6], where composition requires matching on equivalence relations that relax the lens domains.

Our application of the relational PF calculus to BX builds up from [17,18], where we have developed a language of functional PF combinators allowing only surjective transformations over inductive types. In this paper, we extend such language to support typical non-surjective combinators such as splits and injections. Unlike the data abstraction approach from [22], our lens language allows arbitrary type constructors and deconstructors without extending the language with ad-hoc primitives and surjectivity tests.

Most BX approaches rely on more standard and decidable type systems, at the cost of a more limited expressiveness [17,22], by allowing partial transformations [12,20,21] or by assuming both partiality and weaker round-tripping laws [13,10,9,8]. More closely related to our approach, some frameworks derive the backward transformations by calculation, but are less expressive than ours. In [13], Put is derived by inverting *injective* forward transformations through algebraic reasoning, while [12] bidirectionalizes a restricted first-order language (namely, without duplication) based on a notion of view-update under constant complement. They also calculate an automata that matches the exact domain of the transformations, and acts similarly to our invariants. The lens language for graph transformations proposed in [8] processes view insertions using URA, exploring all possible right inverses for the forward transformation.

7   Conclusion

In this paper, we have presented mechanisms for the efficient execution of expressions in a PF relational language over data-types with invariants. By defining a careful semantics that uses invariants to narrow evaluation, we attain a viable non-deterministic implementation. In retrospect, our handling of product invariants in lifted form made the difference from previous approaches to domain and range calculation, and ended up being a key component of our framework.

In the context of BX, we identify an open problem in the composition of lenses with (explicit or implicit) invariants that is responsible for the latent partiality found in most practical BX frameworks. We have proposed to alleviate this problem by modeling lenses using the relational calculus and their particular domains using invariants. Applying our proposed non-deterministic calculus and semantics, we were able to implement an expressive PF BX language that supports duplication, conditional choice and recursion patterns, whose backward transformations emerge naturally from the lens laws.

Although we are already able to handle many interesting recursive transformations, there is still a lot of room for improvement in the algorithm for recursive invariant inference. Namely, likewise the lifted form for products, we are currently researching possible normal forms for such invariants that are more amenable for calculation and optimization.

We also intend to explore mechanisms for a better control of the non-determinism through user-defined quality measures as additional invariants on the domains. In particular, we are studying ways to take advantage of the *shrink* operator proposed in [14], which narrows the output of non-deterministic PF relations, by selecting the "best" values defined by a given order.

Acknowledgements. This work is funded by ERDF - European Regional Development Fund through the COMPETE Programme (operational programme for competitiveness) and by National Funds through the FCT - Fundação para a Ciência e a Tecnologia (Portuguese Foundation for Science and Technology) within project FCOMP-01-0124-FEDER-020532. Nuno Macedo is sponsored by the FCT under grant SFRH/BD/69585/2010. Exchange of ideas with J. N. Oliveira (HASLab) is gratefully acknowledged.

References

1. Abramov, S., Glück, R.: The Universal Resolving Algorithm: Inverse Computation in a Functional Language. In: Backhouse, R., Oliveira, J.N. (eds.) MPC 2000. LNCS, vol. 1837, pp. 187–212. Springer, Heidelberg (2000)
2. Backhouse, R., Hoogendijk, P., Voermans, E., van der Woude, J.: A relational theory of datatypes (December 1992), draft of book in preparation, http://www.cs.nott.ac.uk/~rcb/MPC/papers
3. Bird, R., de Moor, O.: Algebra of Programming. International Series in Computer Science, vol. 100. Prentice-Hall (1996)

4. Cunha, A., Visser, J.: Transformation of structure-shy programs with application to XPath queries and strategic functions. Sci. Comput. Program. 76(6), 512–539 (2011)
5. Foster, N., Greenwald, M., Moore, J., Pierce, B., Schmitt, A.: Combinators for bidirectional tree transformations: A linguistic approach to the view-update problem. TOPLAS 2007 29(3) (2007)
6. Foster, N., Pilkiewicz, A., Pierce, B.: Quotient lenses. In: ICFP 2008, pp. 383–396. ACM (2008)
7. Hanus, M.: Multi-paradigm Declarative Languages. In: Dahl, V., Niemelä, I. (eds.) ICLP 2007. LNCS, vol. 4670, pp. 45–75. Springer, Heidelberg (2007)
8. Hidaka, S., Hu, Z., Inaba, K., Kato, H., Matsuda, K., Nakano, K.: Bidirectionalizing graph transformations. In: ICFP 2010, pp. 205–216. ACM (2010)
9. Hu, Z., Mu, S.C., Takeichi, M.: A programmable editor for developing structured documents based on bidirectional transformations. Higher-Order and Symbolic Computation 21(1-2), 89–118 (2008)
10. Liu, D., Hu, Z., Takeichi, M.: Bidirectional interpretation of XQuery. In: PEPM 2007, pp. 21–30. ACM (2007)
11. Macedo, N., Pacheco, H., Cunha, A.: Relations as executable specifications: Taming partiality and non-determinism using invariants. Technical Report TR-HASLab:03:2012, University of Minho (July 2012), http://www.di.uminho.pt/~nfmmacedo/publications/invariants-tr.pdf
12. Matsuda, K., Hu, Z., Nakano, K., Hamana, M., Takeichi, M.: Bidirectionalization transformation based on automatic derivation of view complement functions. In: ICFP 2007, pp. 47–58. ACM (2007)
13. Mu, S.-C., Hu, Z., Takeichi, M.: An Algebraic Approach to Bi-directional Updating. In: Chin, W.-N. (ed.) APLAS 2004. LNCS, vol. 3302, pp. 2–20. Springer, Heidelberg (2004)
14. Mu, S.-C., Oliveira, J.N.: Programming from Galois Connections. In: de Swart, H. (ed.) RAMICS 2011. LNCS, vol. 6663, pp. 294–313. Springer, Heidelberg (2011)
15. Oliveira, J.N.: Transforming Data by Calculation. In: Lämmel, R., Visser, J., Saraiva, J. (eds.) GTTSE 2007. LNCS, vol. 5235, pp. 134–195. Springer, Heidelberg (2008)
16. Oliveira, J.N.: Extended Static Checking by Calculation Using the Pointfree Transform. In: Bove, A., Barbosa, L.S., Pardo, A., Pinto, J.S. (eds.) LerNet ALFA Summer School 2008. LNCS, vol. 5520, pp. 195–251. Springer, Heidelberg (2009)
17. Pacheco, H., Cunha, A.: Generic Point-free Lenses. In: Bolduc, C., Desharnais, J., Ktari, B. (eds.) MPC 2010. LNCS, vol. 6120, pp. 331–352. Springer, Heidelberg (2010)
18. Pacheco, H., Cunha, A.: Calculating with lenses: Optimising bidirectional transformations. In: PEPM 2011, pp. 91–100. ACM (2011)
19. Schmidt, G.: Relational Mathematics. Encyclopedia of Mathematics and its Applications, vol. 132. Cambridge University Press (2010)
20. Voigtländer, J.: Bidirectionalization for free! (Pearl). In: POPL 2009, pp. 165–176. ACM (2009)
21. Voigtländer, J., Hu, Z., Matsuda, K., Wang, M.: Combining syntactic and semantic bidirectionalization. In: ICFP 2010, pp. 181–192. ACM (2010)
22. Wang, M., Gibbons, J., Matsuda, K., Hu, Z.: Gradual Refinement: Blending Pattern Matching with Data Abstraction. In: Bolduc, C., Desharnais, J., Ktari, B. (eds.) MPC 2010. LNCS, vol. 6120, pp. 397–425. Springer, Heidelberg (2010)

Left-Handed Completeness

Dexter Kozen[1] and Alexandra Silva[2,*]

[1] Computer Science Department, Cornell University, Ithaca, NY 14853-7501, USA
[2] Radboud University, Postbus 9010, 6500 GL Nijmegen, The Netherlands

Abstract. We give a new, significantly shorter proof of the completeness of the left-handed star rule of Kleene algebra. The proof exposes the rich interaction of algebra and coalgebra in the theory of Kleene algebra.

1 Introduction

Axiomatizations of the equational theory of the regular sets over a finite alphabet have received much attention over the years. The topic was introduced in the seminal 1956 paper of Kleene [6], who left axiomatization as an open problem. Salomaa [15] gave two complete axiomatizations, but these depended on rules of inference that were sound under the standard interpretation but not under other natural interpretations. Conway, in his monograph [3], coined the term *Kleene algebra* (KA) and contributed substantially to the understanding of the question of axiomatization. An algebraic solution was presented by Kozen [9], who postulated two equational implications, similar to the inference rules of Salomaa; but unlike Salomaa's rules, they are universal Horn formulas, therefore sound over a variety of nonstandard interpretations. The main goal of this paper is to show that only one of the two implications is enough to guarantee completeness.

This result, which we shall call *left-handed completeness*, is a known result. It was claimed without proof by Conway [3, Theorem 12.5]. The only extant proof, by Boffa [1], relies on a lengthy (137 journal pages!) result of Krob [12], who presented a schematic equational axiomatization representing infinitely many equations. Krob's result was also later reworked and generalized in the framework of iteration theories [4].

Purely equational axiomatizations are undesirable for several reasons. From a practical point of view, they are inadequate for reasoning in the presence of other equational assumptions, which is almost always the case in real-life applications. For example, consider the redundant assignment $x := 1 \,;\, x := 1$ and let a stand for $x := 1$. We have $aa = a$, since the assignment is redundant. We would expect this equation to imply $a^* = 1 + a$ (intuitively, performing the assignment $x := 1$ any number of times is equivalent to performing it zero or one times), but this is not entailed by the equational theory plus the extra equation $aa = a$. To see this, consider the free R-algebra (Conway's terminology for an algebra satisfying all the equations of the regular sets) on the finite monoid $\{1, a\}$, where $aa = a$. This

---

* Also affiliated to Centrum Wiskunde & Informatica (Amsterdam, The Netherlands) and HASLab / INESC TEC, Universidade do Minho (Braga, Portugal).

W. Kahl and T.G. Griffin (Eds.): RAMiCS 2012, LNCS 7560, pp. 162–178, 2012.

algebra contains six elements: $0, 1, a, 1 + a, a^*, aa^*$. The elements a^* and $1 + a$ are distinct, even under the assumption $aa = a$, which is not at all desirable. This is an example of a finite algebra that satisfies all the equations of KA but is not a KA itself, because in a finite KA a star is always equal to a finite sum of powers. This example shows that purely equational axiomatizations would be inadequate for even the simplest verification tasks involving iteration in the presence of other equations.

On the other hand, characterizing $.a^*$ as a least fixpoint is a natural and powerful device, and is satisfied in virtually all models that arise in real life. However, there are interesting and useful models that satisfy only one of the two star rules [7,8,11], so it is useful to know that only one of the rules is needed for equational completeness.

Even though we present a new proof of a known result, there is added value in the exploration of the exquisite interplay between algebra and coalgebra in the theory of regular sets, which is visible throughout the technical development of the paper and notably in the novel definition of a *differential Kleene algebra*, which captures abstractly the relationship between the algebraic and coalgebraic structure of KA. The (syntactic) Brzozowski derivative provides the link from the algebraic to the coalgebraic view of regular expressions, whereas the canonical embedding of a given coalgebra into a matrix algebra plays the converse role. This interplay between algebra and coalgebra, first explored in [5,13], has opened the door to far-reaching extensions of Kleene's theorem and Kleene algebras [16].

Another contribution is a clear characterization of how far one can go in the proof of completeness with just equations. We show that the equational implication is needed only at two places (Lemmas 6 and 16). Furthermore, we show that the existence of least solutions implies uniqueness of solutions in the free algebra, which neatly ties our axiomatization with the original axiomatization of Salomaa.

Proofs omitted from the main text can be found in the extended version of this paper [10].

2 Axiomatization

2.1 Left-Handed Kleene Algebra

A *weak Kleene algebra* (*weak* KA) is an idempotent semiring with star satisfying (1)–(4):

$$a^* = 1 + aa^* \tag{1}$$
$$(ab)^* a - a(bu)^* \tag{2}$$
$$(a + b)^* = a^*(ba^*)^* \tag{3}$$
$$a^{**} = a^* \tag{4}$$

Axioms (2) and (3) are called *sliding* and *denesting*, respectively. These axioms were studied in depth by Conway [3] under the names *productstar* (for the combination of (1) and (2) in the single equation $(ab)^* = 1 + a(ba)^*b$), *sumstar*, and *starstar*, respectively. Although incomplete, these equations are sufficient for many arguments involving the star operator.

Conway studied many other useful families of axioms, including the *powerstar rules*

$$a^* = (a^n)^* \sum_{i=0}^{n-1} a^i, \tag{5}$$

although we will not need them here.

A *left-handed Kleene algebra* (LKA) is a weak KA satisfying a certain universal Horn formula, called the *left-handed star rule*, which may appear in either of the two equivalent forms

$$b + ax \le x \Rightarrow a^* b \le x \qquad\qquad ax \le x \Rightarrow a^* x \le x, \tag{6}$$

where $\le$ is the natural partial order given by $a \le b \Leftrightarrow a + b = b$. One consequence is the *left-handed bisimulation rule*

$$ax \le xb \Rightarrow a^* x \le xb^*. \tag{7}$$

2.2 Matrices

Let $\mathrm{Mat}(S, K)$ be the family of square matrices with rows and columns indexed by a finite set S with entries in a semiring K. Conway [3] shows that under the appropriately defined matrix operations, axioms (1)–(4) imply themselves for matrices. This is also true for (6) [9]. It is known for the powerstar rules (5) too, but only in a weaker form [3].

The *characteristic matrix* P_f of a function $f : S \to S$ has $(P_f)_{st} = 1$ if $f(s) = t$, 0 otherwise. A matrix is a *function matrix* if it is P_f for some f; that is, each row contains exactly one 1 and all other entries are 0.

Let $S_1, \ldots, S_n \subseteq S$ be a partition of S. A matrix $A \in \mathrm{Mat}(S, K)$ is said to be *block diagonal with blocks* $S_1, \ldots, S_n$ if $A_{st} = 0$ whenever s and t are in different blocks.

Lemma 1. *Let* $A, P_f \in \mathrm{Mat}(S, K)$ *with* P_f *the characteristic matrix of a function* $f : S \to S$. *The following are equivalent:*

(i) *A is block diagonal with blocks refining the kernel of f; that is, if $A_{st} \ne 0$, then $f(s) = f(t)$;*
(ii) *$AP_f = DP_f$ for some diagonal matrix D;*
(iii) *$AP_f = DP_f$, where D is the diagonal matrix $D_{ss} = \sum_{f(s)=f(t)} A_{st}$.*

2.3 Differential Kleene Algebra

A *differential Kleene algebra* (DKA) K is a weak KA containing a (finite) set $\Sigma \subseteq K$, called the *actions*, and a subalgebra C, called the *observations*, such that

(i) $ac = ca$ for all $a \in \Sigma$ and $c \in C$, and

(ii) C and Σ generate K,

and supporting a *Brzozowski derivative* consisting of a pair of functions $\varepsilon \colon K \to C$ and $\delta_a \colon K \to K$ for $a \in \Sigma$ satisfying the equations in Fig. 1. Thus $\varepsilon \colon K \to C$ is

$$\delta_a(e_1 + e_2) = \delta_a(e_1) + \delta_a(e_2) \qquad\qquad \varepsilon(e_1 + e_2) = \varepsilon(e_1) + \varepsilon(e_2)$$

$$\delta_a(e_1 e_2) = \delta_a(e_1)e_2 + \varepsilon(e_1)\delta_a(e_2) \qquad \varepsilon(e_1 e_2) = \varepsilon(e_1)\varepsilon(e_2)$$

$$\delta_a(e^*) = \varepsilon(e^*)\delta_a(e)\, e^* \qquad\qquad\qquad \varepsilon(e^*) = \varepsilon(e)^*$$

$$\delta_a(b) = \begin{cases} 1 & \text{if } a = b, \\ 0 & \text{if } a \neq b, \end{cases} \quad b \in \Sigma \qquad\qquad \varepsilon(b) = 0,\ b \in \Sigma$$

$$\delta_a(c) = 0,\ c \in C \qquad\qquad\qquad\qquad\qquad \varepsilon(c) = c,\ c \in C$$

Fig. 1. Brzozowski derivatives

a retract (a KA homomorphism that is the identity on C, which immediately implies $0, 1 \in C$). The functions δ_a and ε impart a coalgebra structure of signature $-^{\Sigma} \times C$ in addition to the Kleene algebra structure.

This definition is a slight generalization of the usual situation in which $C = 2 = \{0, 1\}$ and the function ε and δ_a are the (syntactic) Brzozowski derivatives. We will be primarily interested in matrix KAs in which C is the set of square matrices over 2.

2.4 Examples

One example of a DKA with observations 2 is $\mathsf{Brz} = (2^{\Sigma^*}, \delta, \varepsilon)$, where $\varepsilon(A) = 1$ iff A contains the null string and 0 otherwise, and $\delta_a : 2^{\Sigma^*} \to 2^{\Sigma^*}$ is the classical *Brzozowski derivative*

$$\delta_a(A) = \{x \in \Sigma^* \mid ax \in A\}.$$

This is the final coalgebra of the functor $-^{\Sigma} \times 2$ [13]. It is also an LKA under the usual set-theoretic operations.

Another example is the free LKA K_Σ on generators Σ. It is also a DKA, where δ_a and ε are defined inductively on the syntax of regular expressions according to Fig. 1. The maps δ_a and ε are easily shown to be well defined modulo the axioms of LKA.

These structures possess both an algebra and a coalgebra structure, and in fact are bialgebras [5]. Our main result essentially shows that the latter is isomorphically embedded in the former.

2.5 Properties of DKAs

Silva [16] calls the following result the *fundamental theorem* in analogy to a similar result proved for infinite streams by Rutten [14], closely related to the fundamental theorem of calculus. It is fundamental in the sense that it connects the differential structure, given by δ_a and ε, with the axioms of LKA. We show here that the result holds under weaker assumptions than those assumed in [16]: in fact, we prove this theorem only using equations.

Theorem 1. *Let K be a* DKA. *For all elements $e \in K$,*

$$e = \sum_{a \in \Sigma} a\delta_a(e) + \varepsilon(e). \tag{8}$$

Proof. We proceed by induction on the generation of e from Σ and C using only equations of weak KA and properties of derivatives. For $e \in C$, $\varepsilon(e) = e$ and $\delta_a(e) = 0$, thus (8) holds. For $e = a \in \Sigma$, the right-hand side of (8) reduces to a, thus (8) holds in this case as well.

For the induction step, the case of $+$ is straightforward. For multiplication,

$$e_1 e_2 = (\sum_{a \in \Sigma} a\delta_a(e_1) + \varepsilon(e_1))e_2 = \sum_{a \in \Sigma} a\delta_a(e_1)e_2 + \varepsilon(e_1)(\sum_{a \in \Sigma} a\delta_a(e_2) + \varepsilon(e_2))$$

$$= \sum_{a \in \Sigma} a\delta_a(e_1)e_2 + \sum_{a \in \Sigma} a\varepsilon(e_1)\delta_a(e_2) + \varepsilon(e_1)\varepsilon(e_2)$$

$$= \sum_{a \in \Sigma} a(\delta_a(e_1)e_2 + \varepsilon(e_1)\delta_a(e_2)) + \varepsilon(e_1 e_2) = \sum_{a \in \Sigma} a\delta_a(e_1 e_2) + \varepsilon(e_1 e_2).$$

For e^*, we use the KA identity

$$(x + y)^* = y^* x (x + y)^* + y^*, \tag{9}$$

which follows equationally from (1), (3), and distributivity. Using this identity with $x = \sum_{a \in \Sigma} a\delta_a(e)$ and $y = \varepsilon(e)$,

$$e^* = (\sum_{a \in \Sigma} a\delta_a(e) + \varepsilon(e))^* = \varepsilon(e)^* \sum_{a \in \Sigma} a\delta_a(e)e^* + \varepsilon(e)^* \quad \text{by (9)}$$

$$= \sum_{a \in \Sigma} a\varepsilon(e)^*\delta_a(e)e^* + \varepsilon(e)^* = \sum_{a \in \Sigma} a\delta_a(e^*) + \varepsilon(e^*). \qquad \square$$

Let K be a DKA with actions Σ and observations C. We define the *C-free part* of $e \in K$ to be

$$e' = \sum_{a \in \Sigma} a\delta_a(e). \tag{10}$$

By the fundamental theorem, every element of K can be decomposed into its C-free part e' and $\varepsilon(e) \in C$.

$$e = e' + \varepsilon(e) \qquad\qquad \varepsilon(e') = 0. \tag{11}$$

The map $e \mapsto e'$ is linear and satisfies properties akin to *derivations* in calculus:

$$1' = 0 \qquad (de)' = d'e + de' \qquad e^{*\prime} = \varepsilon(e^*)(e' \cdot \varepsilon(e^*))^+. \tag{12}$$

For the last two,

$$\sum_{a \in \Sigma} a\delta_a(de) = \sum_{a \in \Sigma} a\delta_a(d)e + \sum_{a \in \Sigma} a\varepsilon(d)\delta_a(e)$$

$$= d'e + \varepsilon(d)e' = d'e' + d'\varepsilon(e) + \varepsilon(d)e' = d'e + de',$$

$$\sum_{a \in \Sigma} a\delta_a(e^*) = \sum_{a \in \Sigma} a\varepsilon(e^*)\delta_a(e)e^* = \varepsilon(e^*) \left(\sum_{a \in \Sigma} a\delta_a(e) \right) e^*$$

$$= \varepsilon(e^*)e'e^* = \varepsilon(e^*)e'(e' + \varepsilon(e))^*$$

$$= \varepsilon(e^*)e'\varepsilon(e)^*(e'\varepsilon(e)^*)^* = \varepsilon(e^*)(e' \cdot \varepsilon(e^*))^+.$$

Note that for $C = 2$, the rightmost equation in (12) simplifies to $e^{*'} = e'^+$, using the fact that $\varepsilon(e^*) = 1$.

Moreover, the decomposition is unique: if $e = b + c$ with $\varepsilon(b) = 0$ and $c' = 0$, then

$$b = b' + \varepsilon(b) = b' + c' = e' \qquad c = c' + \varepsilon(c) = \varepsilon(b) + \varepsilon(c) = \varepsilon(e).$$

The following consequences of the above observations will be useful in our application (more precisely in Lemma 13). If $G \subseteq K$, let $\langle G \rangle$ denote the subalgebra of K generated by G.

Lemma 2. *Let K be a* DKA *with derivation* $'$. *Let $G \subseteq K$ and $x \in K$. If $e' = e'x$ and $\varepsilon(e) \in 2$ for all $e \in G$, then $e' = e'x$ and $\varepsilon(e) \in 2$ for all $e \in \langle G \rangle$.*

Proof. We have $1'x = 0'x = 0$ and $e'x = e'$ for $e \in G$, and by induction,

$$(d + e)'x = d'x + e'x = d' + e' = (d + e)',$$

$$(de)'x = d'ex + de'x = d'e'x + d'\varepsilon(e)x + de'x = d'e + de' = (de)',$$

$$(e^*)'x = e'^+x = e'^*e'x = e'^*e' = e'^+ = e^{*'}.$$

Also, $\varepsilon(e) \in 2$ for all $e \in \langle G \rangle$ because ε is a homomorphism.   □

Lemma 3. *Let K be a* DKA *with derivation* $'$. *Suppose $G \subseteq K$ and $x, x^- \in C$ such that $x^-x = 1$ and $e'xx^- = e'$ and $\varepsilon(e) \in 2$ for all $e \in G$. Then the map $e \mapsto x^-ex$ is a* KA *homomorphism on $\langle G \rangle$.*

Proof. It is clearly a homomorphism with respect to 0, 1, and $+$. By Lemma 2, we can assume that $e'xx^- = e'$ and $\varepsilon(e) \in 2$ for all $e \in \langle G \rangle$. Now to show that the map preserves multiplication and star,

$$x^-dex = x^-(d' + \varepsilon(d))ex = x^-d'ex + x^-\varepsilon(d)ex = x^-d'xx^-ex + x^-xx^-\varepsilon(d)ex$$

$$= x^-d'xx^-ex + x^-\varepsilon(d)xx^-ex = x^-dxx^-ex,$$

$$x^-e^*x = x^-(e' + \varepsilon(e))^*x = x^-e'^*x = x^-(e'xx^-)^*x$$

$$= (x^-e'x)^*x^-x = (x^-e'x + x^-x\varepsilon(e))^* = (x^-(e' + \varepsilon(e))x)^* = (x^-ex)^*.$$

Above, we use the fact that $\varepsilon(d)f = f\varepsilon(d)$, for any expression f, since $\varepsilon(d) \in 2$.   □

2.6   Systems of Linear Equations

A *system of (left-)linear equations* over a weak KA K is a coalgebra (S, D, E) of signature $-^{\Sigma} \times K$, where $\Sigma \subseteq K$, $D: S \to S^{\Sigma}$, and $E: S \to K$. A finite system corresponds to a finite coalgebra, that is the set of states S is finite. We curry D so as to write $D_a: S \to S$ for $a \in \Sigma$. The map $D: \Sigma \to S \to S$ extends uniquely to a monoid homomorphism $D: \Sigma^* \to S \to S$, thus we have $D_x: S \to S$ for $x \in \Sigma^*$. A *solution* in K is a map $\phi: S \to K$ such that

$$\phi(s) = \sum_{a \in \Sigma} a\phi(D_a(s)) + E(s). \tag{13}$$

Every finite system of linear equations has a solution. To see this, form an associated matrix $A \in \mathrm{Mat}(S, K)$, where

$$A = \sum_{a \in \Sigma} \Delta(a)P(a) \in \mathrm{Mat}(S, K),$$

where $\Delta(a)$ is the diagonal matrix with diagonal entries a and $P(a)$ is the characteristic matrix of the function D_a. Regarding ϕ and E as column vectors indexed by S, the solution condition (13) takes the form $\phi = A\phi + E$. Since $\mathrm{Mat}(S, K)$ is a weak KA, the vector A^*E is a solution by (1). We call this solution the *canonical solution*. If in addition K is an LKA, then the canonical solution is also the least solution.

   If K is freely generated by Σ, then the map $a \mapsto \Delta(a)P(a)$ extends uniquely to a KA homomorphism $\chi: K \to \mathrm{Mat}(S, K)$, called the *standard embedding*. It will follow from our results that χ is injective.

3   Decompositions

3.1   Simple Strings

Let (S, D, E) be a finite coalgebra of type $-^{\Sigma} \times 2$. Let K_{Σ} be the free LKA on generators Σ. Extend D to a monoid homomorphism $D: \Sigma^* \to S \to S$. The corresponding characteristic matrices P also extend homomorphically by matrix multiplication. Let $\chi: K_{\Sigma} \to \mathrm{Mat}(S, K_{\Sigma})$ with $\chi(a) = \Delta(a)P(a)$ be the standard embedding as defined in Section 2.6.

   Call $x \in \Sigma^*$ *simple* if $P(y) \neq P(z)$ for all distinct suffixes y, z of x. If x is simple, then so are all its suffixes. Define

$M = \{x \mid x \text{ is simple}\}$

$M_x = \{y \mid |y| > 0 \text{ and } P(yx) = P(x), \text{ but all proper suffixes of } yx \text{ are simple}\}.$

Every string can be reduced to a simple string by repeatedly removing certain substrings while preserving $P(\text{-})$. This is the well-known *pumping lemma* from automata theory. If y is not simple, find a suffix vw such that $P(vw) = P(w)$

and $v \neq \varepsilon$, and remove v. The resulting string is shorter and $P(\text{-})$ is preserved. Repeating this step eventually produces a string $x \in M$ such that $P(y) = P(x)$. If we always choose the shortest eligible suffix vw, so that $v \in M_w$—this strategy is called *right-to-left greedy*—we obtain a particular element $\gamma(y) \in M$ related to the construction of V_y (see Lemma 7).

Let $n = |S|$. If $y \in M_x$, then $1 + |x| \leq |yx| \leq n^n$, as each function $S \to S$ is represented at most once as $P(z)$ for a proper suffix z of yx.

We now define a family of elements R_x, $T_{y,x}$, and V_x of K_Σ for $x, y \in \Sigma^*$.

$$R_x = \left(\sum_{y \in M_x} T_{y,x} \right)^* \qquad T_{1,x} = 1 \qquad T_{ay,x} = R_{ayx} a T_{y,x}, \ a \in \Sigma \qquad (14)$$

$$V_x = T_{x,1} R_1 \qquad\qquad V = \sum_{x \in M} V_x. \qquad (15)$$

Intuitively, if x is a simple word labeling a path from s to t, then all words represented by V_x lead from s to t, and V represents all words in Σ^*. The expressions R_x and $T_{y,x}$ allow the encoding of loops.

The definitions of R_x and $T_{y,x}$ in (14) are by mutual induction, but it is not immediately clear that the definition is well-founded: note that R_x depends on $T_{y,x}$ for $y \in M_x$, which depends on R_{yx}. To prove well-foundedness, we define a binary relation $\succ$ on tuples (R, x) and (T, y, x) defined as follows. For $x, y \in \Sigma^*$ and $a \in \Sigma$, let

$$(R, x) \succ (T, y, x), \ y \in M_x \qquad (T, ay, x) \succ (R, ayx) \qquad (T, ay, x) \succ (T, y, x).$$

The relation $\succ$ describes the dependencies in the definition (14).

Lemma 4. *The relation $\succ$ is well-founded; that is, there are no infinite $\succ$-paths.*

Note that $R_x = 1$ for $|x| \geq n^n$, since the sum in the definition of R_x in (14) is vacuous in that case. It follows inductively that $T_{y,x} = y$ for $|x| \geq n^n$.

Lemma 5. *For all $x, y \in \Sigma^*$ and $a \in \Sigma$,*

(i) $V_1 = R_1$ *and* $V_{ax} = R_{ax} a V_x$.
(ii) $V_{yx} = T_{y,x} V_x$.

Proof. For (i), we have $V_1 = T_{1,1} R_1 = R_1$ and $V_{ax} = T_{ax,1} R_1 = R_{ax} a T_{x,1} R_1 = R_{ax} a V_x$. For (ii), we proceed by induction on $|y|$. The basis $V_x = T_{1,x} V_x$ is immediate. For the induction step, using (i), $V_{ayx} = R_{ayx} a V_{yx} = R_{ayx} a T_{y,x} V_x = T_{ay,x} V_x$. $\square$

In the following two lemmas, we will exploit the fact that $R_z V_z = V_z$, which can be proven by case analysis on z using the fact that $R_z R_z = R_z$.

Lemma 6. $\left(\sum_{a \in \Sigma} a \right)^* = V.$

Proof. For the forward inequality, we use the left-handed star rule (6). Let $x \in M$ and $a \in \Sigma$. By Lemma 5(i), $aV_x \le R_{ax}aV_x = V_{ax}$. If $ax \in M$, then $V_{ax} \le V$. If $ax \notin M$, say $x = yz$ with $P(ax) = P(ayz) = P(z)$, then $ay \in M_z$ and $z \in M$. By Lemma 5, $V_{ax} = V_{ayz} = T_{ay,z}V_z \le R_z V_z = V_z \le V$. In either case, $aV_x \le V$. Since $a \in \Sigma$ and $x \in M$ were arbitrary, $(\sum_{a \in \Sigma} a)V \le V$. Also $1 \le V$, since $1 \le R_1 = V_1$. By (6), $(\sum_{a \in \Sigma} a)^* \le V$. The reverse inequality follows from monotonicity. □

Lemma 7. *For all* $y \in \Sigma^*$, $V_y \le V_{\gamma(y)}$.

Proof. If $v \in M_w$, then, by Lemma 5(ii), we have that $V_{vw} = T_{v,w}V_w \le V_w$, since $T_{v,w} \le R_w$ and $R_w V_w \le V_w$. The result follows inductively from the right-to-left construction of $\gamma(y)$. □

3.2   Decompositions

Let (S, D, E) be a finite coalgebra of type $-^\Sigma \times 2$ with standard embedding

$$\chi : K_\Sigma \to \mathrm{Mat}(S, K_\Sigma) \qquad\qquad \chi(a) = \Delta(a)P(a).$$

Let $e \in K_\Sigma$. A *decomposition* of e (with respect to χ) is a family of expressions $e_x \in K_\Sigma$ indexed by $x \in M$ (recall that M is the set of simple strings) such that

(a) $e = \sum_x e_x$, and
(b) $\chi(e_x) = \Delta(e_x)P(x)$ for all $x \in M$.

It follows that

$$\chi(e) = \sum_x \Delta(e_x)P(x). \tag{16}$$

If P, Q are matrices, we say that the decomposition *respects* P, Q if in addition

(c) $P(x)Q = P$ for all x such that $e_x \ne 0$.

We say that e is *decomposable* if it has a decomposition. We will eventually show that all expressions are decomposable.

Lemma 8. *Let* $x \mapsto e_x$ *be a decomposition of* e. *The decomposition respects* P, Q *iff* $\chi(e)Q = \Delta(e)P$.

Proof. If the decomposition respects P, Q, then

$$\chi(e)Q = \sum_x \Delta(e_x)P(x)Q = \sum_x \Delta(e_x)P = \Delta(\sum_x e_x)P = \Delta(e)P.$$

Conversely, if $e_x \ne 0$ and $P(x)Q \ne P$, then $\Delta(e_x)P(x)Q \ne \Delta(e)P$, therefore

$$\chi(e)Q = \sum_x \Delta(e_x)P(x)Q \ne \Delta(e)P. \qquad □$$

We have specified the index set M in the definition of decomposition to emphasize that the $P(x)$ must be generated by the $P(a)$, but in fact any finite index set will do, provided the function matrices are so generated.

Lemma 9. *Let e_α and P_α be finite indexed collections of elements of K_Σ and function matrices, respectively, such that*

$$e = \sum_\alpha e_\alpha \qquad\qquad \chi(e_\alpha) = \Delta(e_\alpha)P_\alpha$$

and such that each P_α is $P(y_\alpha)$ for some $y_\alpha \in \Sigma^*$. Then $e_x = \sum_{x=\gamma(y_\alpha)} e_\alpha$ is a decomposition of e.

Proof. By Lemma 7, if $x = \gamma(y_\alpha)$, then $P(x) = P(y_\alpha)$. Easy calculations then show $e = \sum_x e_x$ and $\chi(e_x) = \Delta(e_x)P(x)$. $\qquad\qquad\square$

Decompositions can be combined additively or multiplicatively. The *sum* and *product* of two decompositions $F : M \to K_\Sigma$ and $G : M \to K_\Sigma$ are, respectively, the decompositions

$$(F+G)(x) = F(x) + G(x) \qquad (F \times G)(x) = \sum_{x=\gamma(yz)} F(y)G(z).$$

Lemma 10

(i) *If F is a decomposition of e and G is a decomposition of d, then $F + G$ is a decomposition of $e + d$. If F and G both respect P, Q, then so does $F + G$.*

(ii) *If F is a decomposition of e and G is a decomposition of d, then $F \times G$ is a decomposition of ed. If F respects P, Q and G respects Q, R, then $F \times G$ respects P, R.*

To handle star, we describe a monad structure on systems built on top of the string monad. The motivation is that we wish to consider the elements of M as single letters of an alphabet. To avoid confusion, we use $\alpha, \beta, \ldots$ to denote words in M^*. In §2.6, we constructed the standard embedding χ with respect to a coalgebra (S, D, E) of type $-^\Sigma \times C$. Now we wish to do the same for the alphabet M. We thus have a coalgebra $(S, \hat{D})$ with $\hat{D}_x : S \to S$ of type $-^M$ with $\hat{D}_x = D_x$. The only difference is that on the left-hand side, x is considered as a single letter, whereas on the right-hand side, D_x is defined inductively from D_a for $a \in \Sigma$. The standard embedding is η, defined in the same way for (S, M) as χ was defined for (S, D):

$$\eta : K_M \to \mathrm{Mat}(S, K_M) \qquad\qquad \eta(x) = \Delta(x)P(x), \ x \in M.$$

Now let $\hat{M}$ be constructed as in §3.1 for the alphabet M as M was constructed for Σ.

Lemma 11. *Suppose that $(\sum_{x \in M} x)^* \in K_M$ has a decomposition d_α, $\alpha \in \hat{M}$ with respect to η and that $e \in K_\Sigma$ has a decomposition $\sigma : x \mapsto e_x$ with respect to χ. Let $\mu(x) = \sum_{x=\gamma(\alpha)} d_\alpha$. Then $\sigma\mu : x \mapsto \sigma(\sum_{x=\gamma(\alpha)} d_\alpha)$ is a decomposition of e^* with respect to χ. Moreover, if the decomposition of e respects Q, Q, then so does the decomposition e^*.*

3.3   Existence of Decompositions

Let (S, D, E) be a finite coalgebra of type $-^\Sigma \times C$ with standard embedding $\chi \colon K_\Sigma \to \mathrm{Mat}(S, K_\Sigma)$. Let $M \subseteq \Sigma^*$ and $M_x \subseteq \Sigma^*$ for $x \in M$ be defined as in §3.1. Let R_x, $T_{y,x}$, and $V_x \in K_\Sigma$ be as defined in §3.1 with respect to M and M_x.

In the following, the term *decomposition* refers to decompositions with respect to χ. A *universal decomposition* is a decomposition for the universal expression $(\sum_{a \in \Sigma} a)^*$.

We remark that Lemmas 12 and 13 are actually co-dependent and require proof by mutual induction on the well-founded relation $\succ$ and on dimension of the associated matrices. Lemma 12 can be proved for permutations without reference to Lemma 13 (this is the basis of the induction), but in the general case requires Lemma 13 for lower dimension; and the proof of Lemma 13 depends on Lemma 12 for permutations.

Lemma 12. *For $x, y \in \Sigma^*$,*

  (i) *$T_{y,x}$ has a decomposition respecting $P(yx), P(x)$.*
  (ii) *R_x has a decomposition respecting $P(x), P(x)$.*
  (iii) *$x \mapsto V_x$ is a universal decomposition.*

Proof. The proof is by induction on the well-founded relation $\succ$, using the fact that χ and Δ are homomorphisms, and on dimension. Let us assume that the lemma is true for all matrices of smaller dimension.

For (i), $T_{1,x} = 1$ has the trivial decomposition $1 \mapsto 1$ and $x \mapsto 0$ for all $x \in M - \{1\}$, and this clearly respects $P(x), P(x)$.

For ay, we have $T_{ay,x} = R_{ayx} a T_{y,x}$. By the induction hypothesis, we have a decomposition for R_{ayx} respecting $P(ayx), P(ayx)$ and a decomposition for $T_{y,x}$ respecting $P(yx), P(x)$. We also have the trivial decomposition $a \mapsto a$ and $x \mapsto 0$ for all $x \in M - \{a\}$, which respects $P(ayx), P(yx)$. By Lemma 10(ii), the product of these three decompositions in the appropriate order is a decomposition for $T_{ay,x}$ respecting $P(ayx), P(x)$.

For (ii), we have $R_x = e^*$, where $e = \sum_{y \in M_x} T_{y,x}$. By the induction hypothesis, we can assume decompositions of $T_{y,x}$ for each $y \in M_x$ respecting $P(yx), P(x)$. Since $P(yx) = P(x)$ for $y \in M_x$, these decompositions also respect $P(x), P(x)$. By Lemma 10(i), the sum of these decompositions gives a decomposition of e respecting $P(x), P(x)$. By Lemma 8, $\chi(e)P(x) = \Delta(e)P(x)$.

If $P(x)$ is invertible, then $\chi(e) = \Delta(e)$, therefore

$$\chi(R_x) = \chi(e)^* = \Delta(e)^* = \Delta(R_x).$$

In this case, we can decompose R_x trivially as $1 \mapsto R_x$ and $y \mapsto 0$ for $y \in M - \{1\}$, which respects $P(x), P(x)$, and we are done.

If $P(x)$ is not invertible, we can use Lemma 13 to reduce the problem to a lower dimension. By that lemma, we have a universal decomposition that we can use with Lemma 11 to obtain a decomposition of e^* respecting $P(x), P(x)$.

For (iii),

$$\chi(V_x) = \chi(T_{x,1})\chi(R_1) = \chi(T_{x,1})P(1)\chi(R_1) = \Delta(T_{x,1})P(x)\Delta(R_1)$$
$$= \Delta(T_{x,1}R_1)P(x) = \Delta(V_x)P(x).$$

Combined with Lemma 6, this makes $x \mapsto V_x$ a universal decomposition.   □

Lemma 13. *There exists a universal decomposition.*

Proof. The proof is by induction on dimension and on the number of letters of Σ. We can assume by Lemma 12 that we already have a universal decomposition for the subalphabet of Σ consisting of all a such that $P(a)$ is invertible. Now we show how to add in the rest of the elements of Σ one by one.

Suppose we have constructed a universal decomposition $x \mapsto e_x$ for a subalphabet $\Gamma \subseteq \Sigma$ including all a such that $P(a)$ is invertible. Let $e = \sum_{a \in \Gamma} a$ and $a \in \Sigma - \Gamma$. We have

$$e^* = \sum_x e_x \qquad\qquad \chi(e^*) = \Delta(e_x)P(x),$$

and we wish now to construct a decomposition for $(a + e)^*$.

Since $P(a)$ is not a permutation, the range of the corresponding function is a proper subset $C \subset S$. Equivalently stated, the $S \times (S - C)$ submatrix of $P(a)$ is the zero matrix. Let X be the $S \times C$ matrix whose $C \times C$ submatrix is the identity matrix and whose other entries are 0, and let X^T be its transpose. The following facts are easy to verify:

$$P(a) = P(a)XX^T \qquad\qquad X^TX = I. \qquad (17)$$

These are square matrices of dimension $S \times S$ and $C \times C$, respectively. Now

$$(a + e)^* = (e^*a)^*e^* = (1 + e^*a(e^*a)^*)e^*.$$

By Lemma 10, we know how to combine decompositions additively and multiplicatively, and we have decompositions of a, e^*, and 1. It thus suffices to construct a decomposition of $a(e^*a)^*$.

We can reduce to a lower dimensional $C \times C$ problem. Let

$$R(x) = XX^TP(xa) \qquad\qquad Q(x) = X^TP(xa)X.$$

The matrix $R(x)$ is the $S \times S$ matrix whose $C \times C$ submatrix is $Q(x)$ and whose other entries are 0. It follows from (17) that

$$R(x) = XQ(x)X^T \qquad\qquad R(\alpha) = XQ(\alpha)X^T \qquad (18)$$

for any $\alpha \in M^*$. Now consider the system

$$\eta : K_M \to \mathrm{Mat}(C, K_M) \qquad\qquad \eta(x) = \Delta(x)Q(x)$$

of dimension $C \times C$. By the induction hypothesis on dimension, we have a universal decomposition with respect to η:

$$\left(\sum_x x\right)^* = \sum_\alpha d_\alpha \qquad\qquad \eta(d_\alpha) = \Delta(d_\alpha)Q(\alpha)$$

where α ranges over $\hat{M}$. Let

$$P_\alpha = P(a)R(\alpha), \ \alpha \in \hat{M} \qquad\qquad \sigma(x) = e_x a.$$

The map σ extends uniquely to a KA homomorphism $\sigma : K_M \to K_\Sigma$. We claim that $a\sigma(d_\alpha)$ and P_α form a decomposition of $a(e^*a)^*$ with respect to χ. We must show that

$$a(e^*a)^* = \sum_\alpha a\sigma(d_\alpha) \qquad\qquad \chi(a\sigma(d_\alpha)) = \Delta(a\sigma(d_\alpha))P_\alpha. \qquad (19)$$

According to Lemma 9, we must also show that the P_α are generated by the $P(a)$, $a \in \Sigma$. The left-hand equation of (19) is a straightforward calculation:

$$a(e^*a)^* = a\left(\sum_x e_x a\right)^* = a\sigma\left(\left(\sum_x x\right)^*\right) = a\sigma\left(\sum_\alpha d_\alpha\right) = \sum_\alpha a\sigma(d_\alpha).$$

That the P_α are generated by the $P(a)$ can be shown inductively using (17):

$$P_1 = P(a)R(1) = P(a)XX^T P(a) = P(a^2)$$
$$P_{x\alpha} = P(a)R(x)R(\alpha) = P(a)XX^T P(xa)R(\alpha) = P(ax)P(a)R(\alpha) = P(ax)P_\alpha.$$

It remains to prove the right-hand equation of (19). Let G be the image of the map $\chi\sigma : K_M \to \mathrm{Mat}(S, K_\Sigma)$ defined by

$$\chi\sigma(x) = \chi(e_x a) = \Delta(e_x a)P(xa).$$

The generators satisfy $\chi\sigma(x)' = \chi\sigma(x)'XX^T$, so by Lemma 2, this also holds true for all elements of G, and $\varepsilon(A) \in \{0, I\}$ for all $A \in G$. Also, by Lemma 3, the map

$$A \mapsto X^T A X : G \to \mathrm{Mat}(C, K_\Sigma)$$

is a homomorphism on G, therefore so is its composition with $\chi\sigma$, the map $X^T(\chi\sigma)X : K_M \to \mathrm{Mat}(C, K_\Sigma)$.

Now $X^T(\chi\sigma)X = \hat{\sigma}\eta$, as they are both homomorphisms $K_M \to \mathrm{Mat}(C, K_\Sigma)$ and agree on the generators $x \in M$:

$$(X^T(\chi\sigma)X)(x) = X^T(\chi\sigma(x))X = X^T(\chi(e_x a))X$$
$$= X^T(\Delta(e_x a)P(xa))X = \Delta(e_x a)X^T P(xa)X = \Delta(e_x a)Q(x)$$
$$\hat{\sigma}\eta(x) = \hat{\sigma}(\Delta(x)Q(x)) = \Delta(\sigma(x))Q(x) = \Delta(e_x a)Q(x).$$

Thus the value they take on $d_\alpha \in K_M$ is the same:

$$X^T \chi(\sigma(d_\alpha))X = \hat{\sigma}\eta(d_\alpha) = \hat{\sigma}(\Delta(d_\alpha)Q(\alpha)) = \Delta(\sigma(d_\alpha))Q(\alpha). \qquad (20)$$

Calculating, we find

$$
\begin{aligned}
\chi(a\sigma(d_\alpha)) &= \chi(a\sigma(d_\alpha))' && \text{since } \varepsilon(\chi(a\sigma(d_\alpha))) = 0 \\
&= \Delta(a)P(a)XX^T\chi(\sigma(d_\alpha))XX^T && \text{by (17) and Lemma 2} \\
&= \Delta(a)P(a)X\Delta(\sigma(d_\alpha))Q(\alpha)X^T && \text{by (20)} \\
&= \Delta(a)\Delta(\sigma(d_\alpha))P(a)XQ(\alpha)X^T && \\
&= \Delta(a\sigma(d_\alpha))P(a)R(\alpha) && \text{by (18)} \\
&= \Delta(a\sigma(d_\alpha))P_\alpha && \text{by definition of } P_\alpha. \qquad \Box
\end{aligned}
$$

Corollary 1. *All expressions are decomposable.*

Proof. We proceed by induction on structure of the expression. Every element $a \in \{0, 1\} \cup \Sigma$ has a trivial decomposition $1 \mapsto a$ and $x \mapsto 0$ for $x \in M - \{1\}$. Closure under sum and product follow from Lemma 10. For star, suppose we have a decomposition e_x, $x \in M$, of e. By Lemma 13, we have a decomposition for the universal expression $(\sum_{x \in M} x)^*$. Lemma 11 then provides a decomposition for e^* via the substitution $x \mapsto e_x$. $\qquad \Box$

4   Completeness

Let (S, D, E) be a coalgebra of signature $-^\Sigma \times 2$. We say that states $s, t \in S$ are *bisimilar*, and write $s \approx t$, if $E(D_x(s)) = E(D_x(t))$ for all $x \in \Sigma^*$. The relation $\approx$ is the maximal bisimulation on S and is the kernel of the unique coalgebra homomorphism $L_S \colon S \to \mathsf{Brz}$, where

$$L_S(s) = \{x \in \Sigma^* \mid E(D_x(s)) = 1\}.$$

Soundness and completeness can be expressed in the following terms. Let $\mathcal{E}$ be a set of equations or equational implications on regular expressions, and let $\mathsf{Con}\,(\mathcal{E})$ be the set of consequences of $\mathcal{E}$ in ordinary equational logic. The axioms $\mathcal{E}$ are *sound* if $\mathsf{Con}\,(\mathcal{E})$ refines bisimilarity; equivalently, if the Brzozowski derivative is well-defined on the free weak KA modulo $\mathcal{E}$. A sound set of axioms is *complete* if $\mathsf{Con}\,(\mathcal{E})$ and bisimilarity coincide; that is, if the unique coalgebra homomorphism to the final coalgebra Brz is injective. We have mentioned above that the LKA axioms are sound; indeed, soundness has been shown in [9] for a larger set of axioms, namely those of KA. To prove that they are complete, our task is to show that the unique coalgebra homomorphism $L_{K_\Sigma} \colon K_\Sigma \to \mathsf{Brz}$ is injective.

This characterization of soundness and completeness was first observed by Jacobs [5] for classical regular expressions and KA and largely explored in the thesis of Silva [16] for generalized regular expressions. See [16] for a comprehensive introduction to this characterization.

In what follows, recall that for a coalgebra (S, D, E) we form an associated matrix $A \in \mathrm{Mat}(S, K)$, where

$$A = \sum_{a \in \Sigma} \Delta(a) P(a) \in \mathrm{Mat}(S, K),$$

where $\Delta(a)$ is the diagonal matrix with diagonal entries a and $P(a)$ is the characteristic matrix of the function D_a.

Lemma 14. *If $s \approx t$ then $(A^* E)_s = (A^* E)_t$.*

Proof. We have

$$A = \sum_{a \in \Sigma} \Delta(a) P(a) = \sum_{a \in \Sigma} \chi(a) = \chi(\sum_{a \in \Sigma} a),$$

$$A^* = \chi(\sum_{a \in \Sigma} a)^* = \chi((\sum_{a \in \Sigma} a)^*) = \chi(\sum_{x \in M} V_x) = \sum_{x \in M} \chi(V_x) = \sum_{x \in M} \Delta(V_x) P(x).$$

Now for any $s \in S$,

$$(A^* E)_s = (\sum_{x \in M} \Delta(V_x) P(x) E)_s = \sum_{x \in M} V_x (P(x) E)_s$$

$$= \sum_{x \in M} V_x \sum_{u \in S} P(x)_{su} E_u = \sum_{x \in M} V_x E(D_x(s)).$$

If $s \approx t$, then $E(D_x(s)) = E(D_x(t))$ for all $x \in \Sigma^*$, therefore

$$(A^* E)_s = \sum_{x \in M} V_x E(D_x(s)) = \sum_{x \in M} V_x E(D_x(t)) = (A^* E)_t. \qquad \square$$

Consider a finite subcoalgebra (S, δ, ε) of K_Σ, where δ and ε comprise the Brzozowski derivative as defined as in Fig. 1. Recall that every $e \in K_\Sigma$ generates a finite subcoalgebra, since it has finitely[1] many Brzozowski derivatives [13]. Let $\chi : K_\Sigma \to \mathrm{Mat}(S, K_\Sigma)$ be the standard embedding as defined in §2.6.

Lemma 15. $e = (\chi(e) E)_e$.

Proof. If $e_x \neq 0$, then there exists $y \in \Sigma^*$ such that $y \leq e_x$. Since χ is monotone,

$$\Delta(y) P(y) = \chi(y) \leq \chi(e_x) = \Delta(e_x) P(x),$$

therefore $P(y) = P(x)$. Moreover, $1 \leq \delta_y(e_x) \leq \delta_y(e)$, therefore $\varepsilon(\delta_y(e)) = 1$. Since $P(y) = P(x)$, $\varepsilon(\delta_x(e)) = 1$.

We have shown that if $e_x \neq 0$, then $\varepsilon(\delta_x(e)) = 1$; in other words, $e_x = e_x \varepsilon(\delta_x(e))$. It follows that

$$(\chi(e) E)_e = (\sum_x \Delta(e_x) P(x) E)_e = \sum_x e_x (P(x) E)_e = \sum_x e_x \varepsilon(\delta_x(e)) = \sum_x e_x = e. \qquad \square$$

---

[1] The finiteness of the subcoalgebra generated by $e \in K_\Sigma$ only requires the axioms for associativity, commutativity, and idempotency of $+$ (hence, only equations).

Lemma 16. $e = (A^*E)_e$.

Proof. By Lemma 15 and the monotonicity of χ,

$$e = (\chi(e)E)_e \leq (\chi((\sum_{a \in \Sigma} a)^*)E)_e = ((\sum_{a \in \Sigma} \chi(a))^*E)_e = (A^*E)_e.$$

For the reverse inequality, Theorem 1 says that the identity map $e \mapsto e$ is a solution to (13), and as noted in §2.6, A^*E is the least solution of a LKA.   □

Theorem 2 (Completeness). *If* $d \approx e$ *then* $d = e$.

Proof. Immediate from Lemmas 14 and 16.   □

An interesting consequence of Lemma 16 is that the canonical solution in K_Σ is not only the least, but actually the *unique* solution. For further details, we refer the reader to [10].

5   Discussion

In this paper, we have given a new, significantly shorter proof of the completeness of the left-handed star rule of Kleene algebra.

We have shown that the left-handed star rule is needed only to guarantee the existence of least solutions. It would be interesting to explore how one could prove the existence of least solutions just using the equations assumed by Krob [12], which are of the form $M^* = \sum_{m \in M} \varepsilon_M^{-1}(m)$, for M a finite monoid.

A well-known algorithm to obtain the minimal deterministic automaton is the *Brzozowski algorithm* [2]. Starting from a possibly nondeterministic automaton, (i) reverse the transitions, exchanging final and initial states, then (ii) perform the subset construction, removing inaccessible states; then repeat (i) and (ii). The resulting automaton is a minimal automaton for the original language.

Starting from a finite automaton (S, D, E) with a start state s, we can build an automaton $(2^S, \hat{D}, \hat{E})$ with start state E, and $\hat{D}(f) = D \circ f$, $\hat{E} = \xi(s)$, where $\xi(s)$ denotes the characteristic function of the singleton set containing s. This new automaton recognizes the reverse of the original language. Interestingly, this is also reflected in the construction of the expressions V_f for the new automaton. There is apparently a relationship to the Brzozowski construction, but the exact relationship remains to be explored.

Acknowledgments. We are grateful to the RAMiCS reviewers for the many constructive comments, which greatly helped us to improve the paper. The work of second author was partially funded by ERDF – European Regional Development Fund through the COMPETE Programme and by Fundacão para a Ciência e a Tecnologia, Portugal within projects with ref. FCOMP-01-0124-FEDER-020537 and SFRH/BPD/71956/2010.

References

1. Boffa, M.: Une condition impliquant toutes les identités rationnelles. Informatique Théorique et Applications/Theoretical Informatics and Applications 29(6), 515–518 (1995)
2. Brzozowski, J.A.: Canonical regular expressions and minimal state graphs for definite events. In: Mathematical Theory of Automata. MRI Symposia Series, vol. 12, pp. 529–561. Polytechnic Press, Polytechnic Institute of Brooklyn, N.Y (1962)
3. Conway, J.H.: Regular Algebra and Finite Machines. Chapman and Hall, London (1971)
4. Ésik, Z.: Group axioms for iteration. Inf. Comput. 148(2), 131–180 (1999)
5. Jacobs, B.: A Bialgebraic Review of Deterministic Automata, Regular Expressions and Languages. In: Futatsugi, K., Jouannaud, J.-P., Meseguer, J. (eds.) Goguen Festschrift. LNCS, vol. 4060, pp. 375–404. Springer, Heidelberg (2006)
6. Kleene, S.C.: Representation of events in nerve nets and finite automata. In: Shannon, C.E., McCarthy, J. (eds.) Automata Studies, pp. 3–41. Princeton University Press, Princeton (1956)
7. Kot, L., Kozen, D.: Second-order abstract interpretation via Kleene algebra. Technical Report TR2004-1971, Computer Science Department, Cornell University (December 2004)
8. Kot, L., Kozen, D.: Kleene algebra and bytecode verification. In: Spoto, F. (ed.) Proc. 1st Workshop Bytecode Semantics, Verification, Analysis, and Transformation (Bytecode 2005), pp. 201–215 (April 2005)
9. Kozen, D.: A completeness theorem for Kleene algebras and the algebra of regular events. Infor. and Comput. 110(2), 366–390 (1994)
10. Kozen, D., Silva, A.: Left-handed completeness. Technical Report, Computing and Information Science, Cornell University (August 2011), http://hdl.handle.net/1813/23556
11. Kozen, D., Tiuryn, J.: Substructural logic and partial correctness. Trans. Computational Logic 4(3), 355–378 (2003)
12. Krob, D.: A complete system of B-rational identities. Theoretical Computer Science 89(2), 207–343 (1991)
13. Rutten, J.J.M.M.: Automata and Coinduction (An Exercise in Coalgebra). In: Sangiorgi, D., de Simone, R. (eds.) CONCUR 1998. LNCS, vol. 1466, pp. 194–218. Springer, Heidelberg (1998)
14. Rutten, J.J.M.M.: A coinductive calculus of streams. Mathematical Structures in Computer Science 15(1), 93–147 (2005)
15. Salomaa, A.: Two complete axiom systems for the algebra of regular events. J. Assoc. Comput. Mach. 13(1), 158–169 (1966)
16. Silva, A.: Kleene Coalgebra. PhD thesis, University of Nijmegen (2010)

On Completeness of Omega-Regular Algebras

Michael R. Laurence and Georg Struth

Department of Computer Science, University of Sheffield, UK
{mike,g.struth}@dcs.shef.ac.uk

Abstract. Omega-regular algebras axiomatise the equational theory of omega-regular expressions as induced by omega-regular language identity. Wagner presented an omega-regular algebra which requires recursively defined side conditions in some of its axioms. We introduce a first-order Horn axiomatisation for which such conditions can be avoided because additive and multiplicative units are absent. We prove its completeness relative to Wagner's result using categorical constructions for adjoining additive and multiplicative units.

1 Introduction

Regular algebras—algebras that axiomatise the equational theory of regular expressions as induced by regular language identity—have been well understood for several decades. Completeness proofs for various axiom systems have been published by Salomaa [14], Krob [11], Kozen [8], Boffa [1,2] and others.

Similar algebras for ω-regular expressions and languages have received much less attention. To our knowledge, the only published completeness result for an ω-regular algebra is due to Wagner [16]. It uses an expansion of Salomaa's regular algebra. A simpler variant—called ω-algebra—which is very similar to Kozen's *Kleene algebra* [8], has been proposed by Cohen [4], but a proper completeness proof for this system has so far not been given. Wilke [17] has proposed a two-sorted semigroup-based approach to ω-regular languages that adapts the syntactic-monoid approach for regular languages [13]. This approach, however, is not directly related to the question of axiomatising the variety of ω-regular languages, and the axioms used are schematic.

Wagner's approach could be criticised for the same reason Kozen criticised Salomaa's. Both axiomatisations are not given in terms of a finite number of identities or quasi-identities with first-order variables. Instead they use infinite axiom schemas parametrised by regular and ω-regular expressions or ground terms in the language of regular and ω-regular algebras. The reason for this is that an algebraic characterisation of the negation of the empty word property from language theory is used as a side condition in some axioms, and this condition is inductively defined over terms. This makes the approaches sound for (ω-)regular languages, but unsound for other computationally important models. Kozen [8] points out that, for Salomaa's axioms, 'the proviso "α does not have the [empty word property]" is not algebraic in the sense that it is not preserved under substitution. Consequently [one of his axioms] is not valid under

W. Kahl and T.G. Griffin (Eds.): RAMiCS 2012, LNCS 7560, pp. 179–194, 2012.
© Springer-Verlag Berlin Heidelberg 2012

nonstandard interpretations [...] [and] must not be interpreted as a universal Horn formula.' The same can be said about Wagner's axioms. Therefore the question of providing a simple algebraic axiomatisation of the equational theory of ω-regular expressions and languages that is sound and complete for this setting seems to be open.

Such a result is, however, very important for applications in program construction and verification. Cohen [4], von Wright [15], Guttmann [6] and others have pointed out the importance of the ω-operation in areas such as concurrency control, program refinement or total and general program correctness. Theorem proving environments and repositories in which the corresponding algebras are implemented have recently been developed [5,7], for instance in the theorem prover Isabelle [12]. A decision procedure for regular expressions has already been implemented in Isabelle by Krauss and Nipkow [10] and in Coq by Braibant and Pous [3]. It uses the classical regular languages and finite automata correspondence. By completeness of regular algebras this procedure can be used for deciding equations in that class.

Similar procedures for ω-regular algebras based on completeness results with respect to ω-regular expressions and infinite word automata seem equally valuable in the contexts mentioned above.

This paper provides a basis for simple complete axiomatisations for different models of ω-regular expression and languages. In order to avoid the complications in Wagner's (and Salomaa's) axioms that arise due to the empty word property, we consider algebras without a multiplicative unit 1. Our approach is therefore based on regular algebras that axiomatise the operation $^+$ rather than the usual Kleene star *. In Kleene algebra the two operations are related by $x^* = 1 + x^+$ and $x^+ = x \cdot x^*$. To avoid design decisions as to whether the axiom $x \cdot 0 = 0$ should be present in the algebra—this depends on the model of ω-language chosen—we allow the removal of this additive unit from our signature, too.

We present new axiom systems for both classes of algebras and establish their completeness relative to Wagner's rather complex and intricate result. Soundness of our axioms is rather straightforward. Relative completeness means essentially that we can derive Wagner's axioms when units are added. In contrast to Wagner's schematic axiomatisation over regular expressions ours are finite, algebraic and based on universal Horn formulas. They are presented in terms of module-like structures in which finite and infinite expressions have different sorts. This directly reflects a standard model of ω-regular languages in which finite words and infinite words are kept separate.

The main techniques used in the construction are adjunctions of additive and multiplicative units in categories of ω-regular algebras which are based on similar constructions from ring and semigroup theory. In the case of regular algebras, the adjunction of an additive unit 1 has already been studied by Kozen [9]. The derivation of Wagner's axioms from ours is the essential step in this construction.

We also sketch how these results can be adapted to a completeness result for a variant of Cohen's ω-algebra in which the unit 1 and the $x \cdot 0 = 0$ axiom are absent and in which the shape of admissible ω-terms is restricted by rewrite rules. This algebra is sound with respect to an alternative model of ω-languages in which finite and infinite words can be mixed.

2   Regular Algebras

A (non-unital) *dioid* (or *idempotent semiring*) is a structure $(D, +, \cdot, 0)$ such that $(D, +)$ is a semilattice with 0 as its least element, $(D, \cdot)$ is a semigroup, 0 is a left and right annihilator $(0 \cdot x = 0 = x \cdot 0)$, and the distributive axioms $x \cdot (y + z) = x \cdot y + x \cdot z$ and $(y + z) \cdot x = y \cdot x + z \cdot x$ hold. A *unital* dioid is a structure $(D, +, \cdot, 0, 1)$, where $(D, +, \cdot, 0)$ is a dioid and $(D, \cdot, 1)$ is a monoid.

A *regular algebra* is a dioid expanded by the operation $^+$ which satisfies the unfold and induction axiom

$$x + x \cdot x^+ = x^+, \qquad\qquad z + x \cdot y \leq y \Rightarrow x^+ \cdot z \leq y,$$

where $\leq$ is the semilattice order defined by $x \leq y \Leftrightarrow x + y = y$. A regular algebra is *unital* if the underlying dioid is. We write R and R^1 for the class of regular and unital regular algebras as well as for the associated categories.

It is more common to define regular algebras as unital dioids with an operation * that satisfies the unfold and the induction axiom $1 + x \cdot x^* = x^*$ and $z + x \cdot y \leq y \Rightarrow x^* \cdot z \leq y$. These algebras are also known as *left Kleene algebras*. By defining $x^* = 1 + x^+$, the star axioms can be derived from the plus axioms in unital dioids.

We denote the sets of terms in the language of regular algebras and unital regular algebras over a set X of variables by $T_R(X)$ and $T_{R^1}(X)$. For a finite set Σ of constants, elements of the set of ground terms $T_{R^1}(\Sigma)$ are called *regular expressions* over Σ.

Languages form models of unital regular algebras. Let Σ^* be the free monoid over Σ and let ϵ denote the empty word. As usual, a *language* L is a subset of Σ^*. It is well known that the structure $(2^{\Sigma^*}, \cup, \cdot, {}^*, \emptyset, \{\epsilon\})$ forms a unital regular algebra and so do its subalgebras. For languages L and L',

$$L \cdot L' = \{vw \mid v \in L, w \in L'\}, \qquad L^* = \bigcup_{i \geq 0} L^i, \qquad L^+ = \bigcup_{i \geq 1} L^i,$$

where powers L^i are recursively defined.

We write $[\![.]\!]$ for the homomorphism that maps regular expressions over Σ to languages over Σ such that $[\![\sigma]\!] = \{\sigma\}$ for all $\sigma \in \Sigma$. The *regular languages* over Σ are the homomorphic images of the regular expressions over Σ under $[\![.]\!]$. Analogously, languages generated by Σ^+, that is, languages that do not contain the empty word ϵ, form non-unital regular algebras. Non-unital regular languages arise as homomorphic images of $T_R(\Sigma)$ under $[\![.]\!]$.

Equality of regular languages induces a congruence on regular expressions: $s = t \Leftrightarrow [\![s]\!] = [\![t]\!]$ (with = overloaded). Boffa [1,2] has shown that left Kleene algebras axiomatise this congruence on regular expressions, hence these algebras are *complete* with respect to the variety generated by the left Kleene algebra axioms. They are also *sound* because regular languages are among their models. The regular languages therefore form the free algebras in that variety.

The relationship between unital and non-unital regular languages is expressed by the empty word property. A language has the *empty word property* if it contains the empty word ϵ. This can be modelled, for instance, by the endomorphism $o(L) =$ if $\epsilon \in L$ then $\{\epsilon\}$ else $\emptyset$. We inductively define a corresponding endomorphism on regular expressions.

$$o(\sigma) = o(0) = 0 \text{ for all } \sigma \in \Sigma, \qquad o(1) = 1,$$
$$o(s + t) = o(s) + o(t), \qquad o(st) = o(s)o(t), \qquad o(s^+) = o(s)^+$$

for all $s, t \in T_{R^1}(\Sigma)$. Accordingly, $[\![o(s)]\!] = o([\![s]\!])$. The following fact is folklore and can easily be proved by structural induction.

Lemma 1. *For every $s \in T_{R^1}(\Sigma)$ there exists a $t \in T_R(\Sigma)$ with $o(t) = 0$ such that $\mathsf{R}^1 \models s = o(s) + t$.*

This lemma allows one to isolate the empty word in a language whenever it occurs as an element. Lemma 1 reflects this fact at the level of regular expressions.

3   Omega Regular Algebras

Wagner [16] gave a soundness and completeness result for ω-regular languages. We now revisit the definitions of ω-regular expressions, languages and algebras used in this result.

One of the standard definitions of of ω-regular expressions is based on a two-sorted signature. Suppose that R and O are sorts for regular and ω-regular terms. The regular operations $+$, $\cdot$, $^+$, 0 and 1 are typed within R. Define a signature $\omega = \{+, \cdot, ^+, 0\} \cup \{+, :, {}^\omega, 0_\omega\}$, where the operation $^\omega$ has type $R \to O$, the operation : has type $R \times O \to O$, the operation $+$, which is overloaded, has type $O \times O \to O$ and 0_ω has type O. For a finite set Σ and terms $t \in T_R^1(\Sigma)$ with $o(t) = 0$, the set $T_\omega(\Sigma)$ of ω-*regular expressions* over Σ is defined by

$$T ::= 0_\omega \mid t^\omega \mid t : T \mid T + T.$$

We henceforth use letters $r, s, t, \ldots$ and variables $x, y, z, \ldots$ of sort R for regular terms, and letters S, T and variables $X, Y, Z, \ldots$ for ω-regular terms. Strictly, speaking, ω-regular expressions are usually defined with respect to $T_{R^1}(\Sigma)$, but then side conditions are used to prevent regular expressions with the empty word property appearing under an omega.

We now define Wagner's algebra for ω-regular expressions over the two-sorted term structure, which resembles a module.

An R-*module* is a structure $(R, L, :)$ where $R \in \mathsf{R}$ and L is a semilattice (with join $+$ overloaded) and least element 0_ω. The operation : has type $R \times L \to L$, and the following axioms hold.

$$x : (X + Y) = x : X + x : Y, \qquad (x + y) : X = x : X + y : X,$$
$$(x \cdot y) : X = x : (y : X), \qquad 0 : X = 0_\omega, \qquad x : 0_\omega = 0_\omega,$$

An R-module is *unital* if the underlying regular algebra is unital and $1 : X = X$.

A *unital Wagner algebra* [16] is a unital R-module $(R, L, :)$ expanded by an *omega operation* $^\omega : R \to O$ such that, for all terms s, t and S,

$$o(s) = 0 \Rightarrow s^\omega = (s \cdot s^*)^\omega, \qquad o(st) = 0 \Rightarrow (s \cdot t)^\omega = s : (t \cdot s)^\omega,$$
$$o(s + t) = 0 \wedge (s + t)^\omega = t : (s + t)^\omega + S \Rightarrow (s + t)^\omega = t^\omega + t^* : S.$$

In fact, Wagner's original axiomatisation is based on Salomaa's complete axiomatisation of regular algebra [14], which is also not defined in terms of first-order variables, but by infinite axiom schemas that are based on regular expressions. The reason is that he uses an algebraic variant of Arden's rule as an axiom and this requires the absence of the empty word property as a side condition. Stricly speaking, therefore, also the R-module axioms above should have been written as axiom schemas to yield precisely Wagner's axioms, but any complete axiomatisation for regular expressions could be used. In addition, axiom schemas for deduction—for transitivity of equality and replacement—are needed in Salomaa's and Wagner's axiomatisations to induce the desired notion of congruence. Our own axiomatisation of ω-algebras in Section 5 and 6 uses first-order variables and the standard first-order axioms for equality instead. We usually leave this technical difference implicit to keep the presentation simple.

Finally, only the condition $o(t) = 0$ is used in Wagner's original third axiom, which seems a mistake.

We denote the class—and category—of unital Wagner algebras by W^1; the non-unital class, which is axiomatised in Section 5, is denoted W.

We now define the class of (omega) regular languages following Wagner. Let Σ^ω be the set of all countably infinite sequences over the finite set Σ. Then, for a unital regular algebra R and $L = (2^{\Sigma^\omega}, \cup, \emptyset)$, the structure $(R, L, :, ^\omega)$ is a unital Wagner algebra, where $L^\omega = \{\sigma_0\sigma_1 \ldots \mid \sigma_i \in L\}$ for all $L \in 2^{\Sigma^*}$ with $o(L) = \emptyset$ and $L : L' = \{\sigma\sigma' \mid \sigma \in L, \sigma' \in L'\}$ for all $L \in 2^{\Sigma^*}$ and $L' \in 2^{\Sigma^\omega}$. Again, all subalgebras of this algebra are models of unital Wagner algebras, too. The homomorphism $[\![.]\!]$ and the function o can be extended to ω-regular expressions as expected. The ω-*regular languages* over Σ are the homomorphic images of $[\![.]\!]$. It is obvious from these definitions that no ω-regular language has the empty word property. If $S \in T_\omega(\Sigma)$, then $o(S) = 0_\omega$.

Wagner has shown that his axiomatisation is sound and complete with respect to ω-regular languages.

Theorem 1 ([16]). *Let S and T be ω-regular expressions. Then $[\![S]\!] = [\![T]\!]$ if and only if $S = T$ is derivable from the axioms of unital Wagner algebras.*

Again we overload the equality symbol. In other words, unital Wagner algebras axiomatise the congruence $S = T$ on ω-regular expressions induced by ω-regular language equality. The ω-regular expressions are therefore the free algebras in the variety generated by the unital Wagner algebras. We obtain our completeness result for an axiomatisation of non-unital Wagner algebras relative to this result. Soundness of our axiomatisation is evident from Wagner's result and the discussion of the non-unital axioms in Section 5.

4   Forgetting 1 in Regular Algebras

Our main goal is a completeness result for non-unital Wagner algebras, in which the non-algebraic side conditions in Wagner's axioms can be dropped. This requires a completeness result for non-unital regular algebras, which we develop in this section. Part of this result is due to Kozen [9]; it is obtained from a categorical construction that adjoins a unit to a non-unital regular algebra.

In ring theory, the unit 1 is adjoined to a (non-unital) ring R by taking the cartesian product $\mathbb{Z} \times R$ and defining

$$(m, x) + (n, y) = (m + n, x + y), \qquad (m, x)(n, y) = (mn, my + nx + xy).$$

It is well known that this yields a ring with identity $(1, 0)$. In the case of a (non-unital) dioid D, the boolean semiring $\mathbb{B}$ can be used instead of $\mathbb{Z}$. To extend this construction to regular algebras, we define

$$(m, x)^+ = (m, x^+).$$

We write A for the functor from the category of non-unital regular algebras to the category of unital regular algebras that embodies this construction. We write F for the corresponding functor that forgets the unital structure in a unital regular algebra. The next lemma shows that F is well defined.

Lemma 2. *If $K \in \mathsf{R}$ then $A(K) \in \mathsf{R}^1$ with unit $(1, 0)$.*

Assuming the (non-unital) regular algebra axioms hold in K it must be verified that the unital regular algebra axioms hold in $A(K)$. A proof for a similar axiomatisation of regular algebra and a similar encoding has been given by Kozen [9]. Therefore we do not explicitly prove this result.

Next we show universality of the construction.

Proposition 3. *The functors A and F are adjoints.*

Proof. We must check that for each $K \in \mathsf{R}$ and $K' \in \mathsf{R}^1$ there exist maps $\eta_K : 1_K \to FA$ and $\epsilon_{K'} : AF \to 1_{K'}$ such that the following diagrams commute:

Here, η_K is the canonical embedding $\eta_K(a) = (0, a)$ and $\epsilon_{K'}$ is defined by the conditions $\epsilon_{K'} : (0, a) \mapsto a$ and $\epsilon_{K'} : (1, a) \mapsto 1 + a$.

According to the first diagram, $F(\hat{h}) \circ \eta_K = h$ must hold, that is, every non-unital regular algebra homomorphism h can uniquely be extended to a unital regular algebra homomorphism $\hat{h}$. According to the second one, $\epsilon_{K'} \circ A(h) = \hat{h}$ must hold, that is, every unital regular algebra homomorphism $\hat{h}$ can uniquely be restricted to a non-unital Kleene algebra homomorphism h. This is satisfied by $\hat{h}((0, a)) = h(a)$ and $\hat{h}((1, a)) = 1 + h(a)$. Uniqueness is then straightforward in both cases. $\qquad \square$

Theorem 4. *Let $s, t \in T_R(X)$ be (1-free) terms. Then*

$$\mathsf{R}^1 \models s = t \Leftrightarrow \mathsf{R} \models s = t.$$

Proof. $\mathsf{R} \models s = t \Rightarrow \mathsf{R}^1 \models s = t$ holds since very regular algebra is also a non-unital regular algebra, that is, R^1 is a subclass of R.

For the converse direction, assume that $\mathsf{R} \not\models s = t$. Then there exists $K \in \mathsf{R}$ and an interpretation by which s and t are mapped to different elements of K. Now, by the embedding η_K, every algebra K is a subalgebra of $A(K) \in \mathsf{R}^1$. Hence s and t are mapped to different elements by some interpretation in $A(K)$, because validity of universal sentences is closed under subalgebras. Therefore $A(K) \not\models s = t$ and $\mathsf{R}^1 \not\models s = t$ since $A(K) \in \mathsf{R}^1$. $\qquad\square$

In fact, this theorem holds for arbitrary universal formulas, but we are only interested in equations. An immediate consequence of this result is that the decision problem for the variety of non-unital regular algebras is PSPACE-complete, simply because that of unital regular algebras is [8]. Moreover, the decision problem can be reduced to emptiness tests on regular languages and decided by using finite automata.

Corollary 5

(1) *The non-unital regular algebras axiomatise the set of universal equalities $s = t$ (for non-unital regular expressions s, t) that are theorems of regular algebra.*
(2) *The regular languages L that satisfy $o(L) = 0$ are the free algebras in the variety generated by the non-unital regular algebras.*

Obviously, in the subclass $A(\mathsf{R})$ of R^1, the empty word property is now explicitly expressed in the first coordinates of pairs: $o((m, x)) = m$. In addition, it is obvious that $(1, t) = (1, 0) + (0, t)$ holds for every $t \in T_R(X)$; that is, Lemma 1 trivially holds on $A(\mathsf{R})$. In $A(\mathsf{R})$, therefore, the empty word property and its negation can be explicitly defined and the isolation of the unit from a term, as expressed in Lemma 1, is trivial.

However, this isolation property does not generally hold in R. This property makes $A(\mathsf{R})$ quite similar to regular languages. In fact, each (unital) language regular algebra is isomorphic to the algebra obtained by first removing the empty word from all its languages and then adjoining a unit. Conversely, each language unital regular algebra obtained by adjoining a unit to a non-unital language algebra is isomorphic to the algebra obtained by making a copy of each of its languages and adding the empty word to it.

Finally, the following property follows from Theorem 4.

Corollary 6. *For every $s \in T_{R^1}(\Sigma)$ with $o(s) = 0$ there exists a (1-free) term $t \in T_R(\Sigma)$ such that $\mathsf{R}^1 \models s = t$.*

Proof. By completeness of unitary regular algebras with respect to regular languages it suffices to interpret s in the regular languages with ϵ, hence in the

ϵ-free languages with a unit adjoined. Since $o(s) = 0$, it must be interpreted in the subalgebra of elements with first coordinate zero. Hence s is equivalent (with respect to the axioms of unital regular algebras) to all 1-free terms $t \in T_R(\Sigma)$ that are interpreted via η by the same element. This equivalence class must be non-empty because s is mapped to it.   □

Operationally, t in the proof of Corollary 6 can of course be obtained from s by using $1 \cdot x = x$ and $x \cdot 1 = x$ as rewrite rules.

5   Forgetting 1 in Wagner Algebras

We now axiomatise the (non-unital) Wagner algebras and prove their completeness with respect to ω-regular languages relative to Wagner's result, the main result in this paper.

A (non-unital) *Wagner algebra* is a (non-unital) R-module expanded by the operation $^\omega : R \to O$ that satisfies

$$x^\omega = x^{+\omega}, \tag{1}$$
$$x^\omega = x : x^\omega, \tag{2}$$
$$(x \cdot y)^\omega = x : (y \cdot x)^\omega, \tag{3}$$
$$(x \cdot y + y)^\omega = x : (y \cdot x + y)^\omega + (y \cdot x + y)^\omega, \tag{4}$$
$$(x \cdot y + x)^\omega = x : (y \cdot x + x)^\omega, \tag{5}$$
$$(x + y)^\omega = y : (x + y)^\omega + X \Rightarrow (x + y)^\omega = y^\omega + x^+ : X + X. \tag{6}$$

It is an open problem as to whether these axioms, together with the R-module axioms, are redundant. Note that $0^\omega = 0_\omega$ follows from axiom (2) and the module laws.

The axioms can be understood as refinements of Wagner's original axioms for the unital case in the absence of 1. Axiom (1) is Wagner's first axiom. Axiom (2) is obtained from Wagner's second axiom by setting $y = 1$ (setting $x = 1$ yields the trivial identity $y^\omega = y^\omega$). Axiom (3) is Wagner's second axiom. Axiom (4) is obtained from Wagner's second axiom by setting $x = x + 1$ and using distributivity. Axiom (5) is obtained from Wagner's second axiom by setting $y = y + 1$ and using distributivity. Finally, axiom (6) is Wagner's third axiom, replacing x^* by $x^+ + 1$ and using distributivity.

We now extend our adjunction construction to Wagner algebras. We define

$$(0, x) : (m, X) = (m, x : X),$$
$$(1, x) : (m, X) = (m, X) + (m, x : X),$$
$$(0, x)^\omega = (0, x^\omega),$$

whereas $(1, x)^\omega$ is undefined. This is consistent with the standard construction of ω-regular languages, where no regular language with the empty word property is admitted below an omega. It is immediate from this construction that (m, T) always has $m = 0$, as required.

As in Section 4, we need to check that the functor A that embodies adjunction is well defined. We proceed in two steps. First, we verify the unital R-module axioms, and then the ω-axioms of unital Wagner algebra.

Lemma 7. *If W is an R-module, then $A(W)$ is a unital R-module.*

Proof. Assuming that the axioms of (non-unital) R-modules hold in W, we must show that the axioms of unital R-modules hold in $A(W)$. For the regular part this has been proved in Lemma 2. So it remains to check the unital R-module axioms.

- We first consider $1 : X = X$ for $X = (0, X)$. We calculate

$$(1, 0) : (0, X) = (0, X) + (0, 0 : X) = (0, X) + (0, 0_\omega) = (0, X).$$

- Next we check $0 : X = 0$ for $X = (0, X)$:

$$(0, 0) : (0, X) = (0, 0 : X) = (0, 0).$$

- We now check $(m, x) : (0, 0) = (0, x : 0) = (0, 0)$.
- Now we prove $x : (Y + Z) = x : Y + x : Z$. We can assume that $Y = (0, Y)$ and $Z = (0, Z)$.
  - If $m = 0$, then

$$\begin{aligned}
(m, x) : ((0, Y) + (0, Z)) &= (m, x) : (0, Y + Z) \\
&= (0, x : (Y + Z)) \\
&= (0, x : Y + x : Z) \\
&= (0, x : Y) + (0, x : Z) \\
&= (m, x) : (0, Y) + (m, x) : (0, Z).
\end{aligned}$$

  - If $m = 1$, then

$$\begin{aligned}
(m, x) : ((0, Y) + (0, Z)) &= (1, x) : (0, Y + Z) \\
&= (0, x : (Y + Z)) + (0, Y + Z) \\
&= (0, x : Y + x : Z) + (0, Y) + (0, Z) \\
&= (0, x : Y) + (0, x : Z) + (0, Y) + (0, Z) \\
&= (1, x) : (0, Y) + (m, x) : (0, Z).
\end{aligned}$$

- Next we check $(x + y) : Z = x : Z + y : Z$, again for $Z = (0, Z)$. So

$$\begin{aligned}
((m, x) + (n, y)) : (0, Z) &= (m + n, x + y) : (0, Z) \\
&= (0, x + y) : (0, Z) + (m + n, 0)(0, Z) \\
&= (0, (x + y) : Z) + (m + n, 0)(0, Z) \\
&= (0, x : Z + y : Z) + (m + n, 0)(0, Z) \\
&= (0, x : Z) + (0, y : Z) + (m, 0)(0, Z) + (n, 0)(0, Z) \\
&= (m, x) : (0, Z) + (n, y) : (0, Z).
\end{aligned}$$

– Lastly we check $(x \cdot y) : Z = x : (y : Z)$ for $Z = (0, Z)$.
  • Suppose first that $m = n = 0$. Then

$$
\begin{aligned}
(m, x) \cdot (n, y) : (0, Z) &= (0, (x \cdot y) : Z) \\
&= (0, x : (y : Z)) \\
&= (m, x) : (0, y : Z) \\
&= (m, x) : (n, y) : (0, Z).
\end{aligned}
$$

  • If instead $m = 1$ and $n = 0$ then

$$
\begin{aligned}
(m, x) \cdot (n, y) : (0, Z) &= (0, x \cdot y) : (0, Z) + (0, y) : (0, Z) \\
&= (0, (x \cdot y) : Z) + (0, y) : (0, Z) \\
&= (0, x : (y : Z)) + (0, y : Z) \\
&= (0, x) : (0, y : Z) + (0, y : Z) \\
&= (m, x) : (0, y : Z) \\
&= (m, x) : (n, y) : (0, Z).
\end{aligned}
$$

  • If instead $m = 0$ and $n = 1$ then

$$
\begin{aligned}
(m, x) \cdot (n, y) : (0, Z) &= (0, x \cdot y + x) : (0, Z) \\
&= (0, x \cdot y) : (0, Z) + (0, x) : (0, Z) \\
&= (0, (x \cdot y) : Z) + (0, x) : (0, Z) \\
&= (0, x) : (0, y : Z) + (0, x) : (0, Z) \\
&= (0, x) : ((0, y : Z) + (0, Z)) \\
&= (m, x) : (n, y) : (0, Z)
\end{aligned}
$$

  • Finally, if $m = n = 1$ then

$$
\begin{aligned}
(m, x) \cdot (n, y) : (0, Z) &= (1, x \cdot y + y + x) : (0, Z) \\
&= (1, 0) : (0, Z) + (0, x \cdot y) : (0, Z) + (0, y) : (0, Z) \\
&\quad + (0, x) : (0, Z) \\
&= (1, y) : (0, Z) + (0, x \cdot y) : (0, Z) + (0, x) : (0, Z) \\
&= (1, y) : (0, Z) + (0, x) \cdot (0, y) : (0, Z) + (0, x) : (0, Z) \\
&= (1, y) : (0, Z) + (0, x) : ((0, y) : (0, Z)) + (0, x) : (0, Z) \\
&= (1, y) : (0, Z) + (0, x) : ((1, y) : (0, Z)) \\
&= (1, x) : ((1, y) : (0, Z)).
\end{aligned}
$$

Now all R-module axioms have been inspected and $A(W)$ has been shown to be unital.   □

Next we verify Wagner's axioms from our axioms for the non-unital case when 1 has been adjoined.

Proposition 8. *If $W \in \mathsf{W}$, then $A(W) \in \mathsf{W}^1$.*

Proof. Since the R-module part has been verified in Lemma 7, it remains to consider the ω-axioms of unital Wagner algebra.

- We start with Wagner's first axiom. Let $o(s) = 0$, that is $s = (0, s)$. Then, by axiom (1),

$$(0, s)^\omega = (0, s^\omega) = (0, s^{+\omega}) = (0, s)^{+\omega}$$

- Next, we verify Wagner's second axiom. Let $o(s \cdot t) = 0$, that is, either $s = (0, s)$ or $t = (0, t)$. We must show that $(s \cdot t)^\omega = s : (t \cdot s)^\omega$.

  • Let $o(s) = o(t) = 0$. Then, by axiom (3),

$$\begin{aligned}
((0, s) \cdot (0, t))^\omega &= (0, (s \cdot t))^\omega \\
&= (0, (s \cdot t)^\omega) \\
&= (0, s : (t \cdot s)^\omega) \\
&= (0, s) : (0, (t \cdot s)^\omega) \\
&= (0, s) : (0, (t \cdot s))^\omega \\
&= (0, s) : ((0, t) \cdot (0, s))^\omega
\end{aligned}$$

  • Let $o(s) = 1$ and $o(t) = 0$. Then, using axiom (4),

$$\begin{aligned}
((1, s) \cdot (0, t))^\omega &= (((1, 0) + (0, s)) \cdot (0, t))^\omega \\
&= ((1, 0) \cdot (0, t) + (0, s) \cdot (0, t))^\omega \\
&= ((0, t) + (0, s) \cdot (0, t))^\omega \\
&= (0, (s \cdot t + t)^\omega) \\
&= (0, s : (t \cdot s + t)^\omega + (t \cdot s + t)^\omega) \\
&= (0, s) : (0, (t \cdot s + t)^\omega) + (0, t \cdot s + t)^\omega) \\
&= (1, s) : (0, (t \cdot s + t))^\omega \\
&= (1, s) : ((0, t \cdot s) + (0, t))^\omega \\
&= (1, s) : ((0, t) \cdot (1, s) + (0, t) \cdot (1, 0))^\omega \\
&= (1, s) . ((0, t) (1, o))^\omega
\end{aligned}$$

  • Let $o(s) = 0$ and $o(t) = 1$. Then

$$((0, s) \cdot (1, t))^\omega = ((0, s) \cdot ((1, 0) + (0, t)))^\omega = (0, (s + s \cdot t)^\omega).$$

If $t = (1, 0)$, the right-hand term reduces to $(0, s)^\omega = (0, s : s^\omega)$ by axiom (2). This is easily shown to be equivalent to $(0, s) : ((1, t) \cdot (0, s))^\omega$.

Otherwise, we continue by axiom (5)

$$\begin{aligned}
(0, (s + s \cdot t)^\omega) &= (0, s : (t \cdot s + s)^\omega) \\
&= (0, s) : (0, (t \cdot s + s)^\omega) \\
&= (0, s) : (0, (t \cdot s + s))^\omega \\
&= (0, s) : ((0, t \cdot s) + (0, s))^\omega \\
&= (0, s) : ((1, t) \cdot (0, s) + (1, 0) \cdot (0, s))^\omega \\
&= (0, s) : ((1, t) \cdot (0, s))^\omega.
\end{aligned}$$

- Finally we consider Wagner's third axiom. We may assume that $o(s) = o(t) = 0$ and that

$$((0, s) + (0, t))^\omega = (0, t) : ((0, s) + (0, t))^\omega + (0, S),$$

hence, using axiom (6),

$$(0, (s + t)^\omega) = (0, t : (s + t)^\omega + S) = (0, t^\omega + s^+ : S + S).$$

Finally, this yields $((0, s) + (0, t))^\omega = (0, t)^\omega + (0, s)^+ : (0, S) + (0, S)$.   □

As in the case of regular algebras we now need to check universality of the construction.

Lemma 9. *The functors A and F are adjoints.*

The proof is essentially the same as for regular algebras, but now for a two-sorted algebra. We do not show it in detail. The main theorem in this section can then be obtained along the lines of the proof of Theorem 4 from Proposition 3 and Lemma 2.

Theorem 10. *Let $S, T \in T_\omega(X)$ be 1-free. Then*

$$\mathsf{W}^1 \models S = T \Leftrightarrow \mathsf{W} \models S = T.$$

It follows that the decision problem for the variety generated by the (non-unital) Wagner algebras is of doubly exponential complexity; equality between terms can be decided by ω-automata.

Corollary 11

(1) *The non-unital Wagner algebras axiomatise the set of universal equalities $S = T$ (for non-unital ω-regular expressions S, T) that are theorems of Wagner algebra.*

(2) *The ω-regular languages based on non-unital regular languages are the free algebras in the variety generated by the non-unital Wagner algebras.*

The results in this corollary can be strengthened. In contrast to the regular languages, units play no significant role in Wagner algebras and ω-regular languages.

Lemma 12. *Every ω-regular expression is equivalent to a 1-free term in unital Wagner algebra.*

Proof. By Lemma 1, all terms $s \in T_{R^1}(\Sigma)$ can be rewritten as $o(s) + s'$ with $o(s') = 0$. By Corollary 6, s' is therefore equivalent to a 1-free term t. That is, $R^1 \models s = o(s) + t$ for some $t \in T_R(\Sigma)$. By definition of ω-regular terms, s^ω is only defined for $o(s) = 0$. Hence every term s^ω is equivalent to some term t^ω, where $s = t$ and t is 1-free. For $s = 1 + t$, the term $s : S$ is equivalent to $S + t : S$, where t is 1-free. Moreover, this property is preserved when building terms $S + T$.   □

Accordingly, the conditions on the empty word property in Wagner's axioms are unnecessary.

Theorem 13. *(Non-unital) Wagner algebras are sound and complete for the equational theory of ω-regular expressions.*

6   Forgetting 0_ω in Non-unital Wagner Algebras

The axiom $x \cdot 0 = 0$ plays an interesting role in the context of ω-regular languages and, more generally, models of computation in which an element x may contain finite and infinite computations. In our current two-sorted model of ω-regular expressions and languages, finite and infinite computations are strictly separated. It is however easy to define alternative models in which finite and infinite computations are mixed in a one-sorted setting. In this case, the complex product of two languages X and Y consisting of both finite and infinite words would collect the infinite words in X together with the finite words in X concatenated with all words in Y. In this alternative model, the equation $X \cdot \emptyset = \emptyset$ does not hold. In fact, it holds precisely if the set X is regular, thus expressing finiteness of the members of X. Similarly, $X \cdot \emptyset = X$ could be used to express the assertion that X contains only infinite words.

   In this section we build a basis for extending our completeness result to such more general situations. At the level of (non-unital) Wagner algebras this means that also the constant 0_ω and, accordingly, the module axiom $x : 0_\omega = 0_\omega$ are dropped.

   We now study the addition of 0_ω to a zero-free (non-unital) Wagner algebra (often writing 0 instead of 0_ω). In order to define the corresponding class W^{-0} of zero-free (non-unital) Wagner algebras, we need to add the following axiom:

$$(x + y)^\omega = y : (x + y)^\omega \Rightarrow y^\omega = (x + y)^\omega. \tag{7}$$

This axiom is obtained from Wagner's third axiom by setting $X = 0_\omega$. The adjunctions are based on that of a (multiplicative) unit to a semigroup. Hence we expand the signature of the semilattice L of an R-module by a new element 0 and stipulate $X + 0 = X$ and $x : 0 = 0$ for all elements $x \in R$ and $X \in L$ in the zero-free R-module $(R, L, :)$. Let A be the functor that embodies this construction and F the associated forgetful functor mapping between the respective categories with and without zero.

As usual, we first verify that F is well defined.

Lemma 14. *Let W be a zero-free (non-unital) Wagner algebra. Then $A(W)$ is a (non-unital) Wagner algebra.*

Proof. The R-module axioms, in particular $x : 0 = 0$ and $X + 0 = X$, hold by definition. It remains to verify Wagner's last axiom,

$$o(s + t) = 0 \wedge (s + t)^\omega = t : (s + t)^\omega + S \Rightarrow (s + t)^\omega = t^\omega + t^* : S,$$

using (7) and (6). Since we are in the non-unital case, the assumption that $o(s + t) = 0$ trivially holds and can be dropped. At the same time, regular expressions can be replaced by first-order variables. This yields axiom (6), that is,

$$(x + y)^\omega = y : (x + y)^\omega + X \Rightarrow (x + y)^\omega = y^\omega + x^+ : X + X.$$

For $X = 0$, this simplifies to axiom (7). In the absence of 0, therefore, axiom (7) simulates the effect of $X = 0$ in axiom (6) and in both cases, $X = 0$ and $X \neq 0$, Wagner's third axiom can be derived.   □

Next we check universality of the construction.

Lemma 15. *The functor A is the left adjoint of the forgetful functor F.*

Proof. Again we must show that the diagrams, which are the same as those in the proof of Proposition 3, commute. Let now $K \in W^{-0}$ and $K' \in W$. Here, η_K and $\epsilon_{K'}$ are simply the identity maps (up to coercion), since the carrier sets of K, K', $FA(K)$ and $AF(K')$ are all the same. Extension and restriction of the homomorphisms therefore amounts to adding and forgetting the case $h(0) = 0$.   □

Proposition 16. *Let $S, T \in T_\omega(X)$ be 01-free ω-regular terms. Then*

$$\mathsf{W}^{-0} \models S = T \Leftrightarrow \mathsf{W} \models S = T.$$

Proof. First, if $\mathsf{W}^{-0} \models S = T$, then $\mathsf{W} \models S = T$ because W is a subclass of W^{-0}. Second, if $\mathsf{W}^{-0} \not\models S = T$, then, for each algebra $K \in \mathsf{W}^{-0}$, the interpretation function cannot map S and T to the same element, hence this cannot happen in W, where all interpretation functions remain the same.   □

These results show that the multiplicative unit 1 and the additive unit 0_ω for infinite languages can be discarded in a modular way and separate completeness results can be obtained for both variants. Obviously, 0 could be adjoined and forgotten to regular algebras in a similar way, but this shall not further concern us in this paper.

7   Conclusion

Based on Wagner's axioms for the equational theory of ω-regular expressions we have presented two alternative axiomatisations that are complete with respect to a model of ω-regular expressions and languages without multiplicative, and without additive and multiplicative units. This circumvents the non-algebraic restrictions in some of Wagner's axioms. It leads to a finitary axiomatisation using quasi-identities that admits models beyond ω-regular languages and is therefore applicable in additional program construction and verification contexts.

There is an interesting extension of our results that we can only sketch in this paper. We have defined a variant of Cohen's one-sorted ω-algebra [4] that uses $^+$ instead of * in its signature and that discards the $x \cdot 0 = 0$ axiom of ω-algebra. We can provide an alternative model of ω-languages for this algebra in which finite and infinite words can occur in the same language, defining, for instance, the complex product $X \cdot Y = \{xy \mid x \in \mathsf{fin}(X) \wedge y \in Y\} \cup \mathsf{inf}(X)$ where $\mathsf{fin}(X)$ and $\mathsf{inf}(X)$ stand for the sets of finite and infinite words in X. We can add further axioms to these weak ω-algebras in order to ensure that terms have the appropriate form for ω-languages; for instance, terms of the form $x^\omega \cdot y$, which are well formed in ω-algebra, but meaningless in our model, are reduced to x^ω. Finally, we have obtained a completeness result for this one-sorted algebra relative to that for non-unital Wagner algebras without 0_ω.

References

1. Boffa, M.: Une remarque sur les systèmes complets d'identités rationnelles. Informatique Théorique et Applications 24(4), 419–423 (1990)
2. Boffa, M.: Une condition impliquant toutes les identités rationnelles. Informatique Théorique et Applications 29(6), 515–518 (1995)
3. Braibant, T., Pous, D.: An Efficient Coq Tactic for Deciding Kleene Algebras. In: Kaufmann, M., Paulson, L. (eds.) ITP 2010. LNCS, vol. 6172, pp. 163–178. Springer, Heidelberg (2010)
4. Cohen, E.: Separation and Reduction. In: Backhouse, R., Oliveira, J.N. (eds.) MPC 2000. LNCS, vol. 1837, pp. 45–59. Springer, Heidelberg (2000)
5. Foster, S., Struth, G., Weber, T.: Automated Engineering of Relational and Algebraic Methods in Isabelle/HOL. In: de Swart, H. (ed.) RAMICS 2011. LNCS, vol. 6663, pp. 52–67. Springer, Heidelberg (2011)
6. Guttmann, W.: Partial, Total and General Correctness. In: Bolduc, C., Desharnais, J., Ktari, B. (eds.) MPC 2010. LNCS, vol. 6120, pp. 157–177. Springer, Heidelberg (2010)
7. Guttmann, W., Struth, G., Weber, T.: Automating Algebraic Methods in Isabelle. In: Qin, S., Qiu, Z. (eds.) ICFEM 2011. LNCS, vol. 6991, pp. 617–632. Springer, Heidelberg (2011)
8. Kozen, D.: A completeness theorem for Kleene algebras and the algebra of regular events. Information and Computation 110(2), 366–390 (1994)
9. Kozen, D.: Typed Kleene algebra. Technical Report TR98-1669, Computer Science Department, Cornell University, USA (1998)
10. Krauss, A., Nipkow, T.: Proof pearl: Regular expression equivalence and relation algebra. J. Automated Reasoning 49(1), 95–106 (2012)

11. Krob, D.: Complete systems of $\mathcal{B}$-rational identities. Theoretical Computer Science 89, 207–343 (1991)
12. Nipkow, T., Paulson, L.C., Wenzel, M.: Isabelle/HOL. LNCS, vol. 2283. Springer, Heidelberg (2002)
13. Pin, J.-E.: Syntactic semigroups. In: Rozenberg, G., Salomaa, A. (eds.) Handbook of Formal Language Theory, vol. 1, pp. 679–746. Springer (1997)
14. Salomaa, A.: Two complete axiom systems for the algebra of regular events. J. ACM 13(1), 158–169 (1966)
15. von Wright, J.: Towards a refinement algebra. Science of Computer Programming 51(1-2), 23–45 (2004)
16. Wagner, K.W.: Eine Axiomatisierung der Theorie der regulären Folgenmengen. Elektronische Informationsverarbeitung und Kybernetik 12(7), 337–354 (1976)
17. Wilke, T.: An Eilenberg Theorem for Infinity-languages. In: Leach Albert, J., Monien, B., Rodríguez-Artalejo, M. (eds.) ICALP 1991. LNCS, vol. 510, pp. 588–599. Springer, Heidelberg (1991)

Categories of Algebraic Contexts Equivalent to Idempotent Semirings and Domain Semirings

Peter Jipsen

Chapman University

Abstract. A categorical equivalence between algebraic contexts with relational morphisms and join-semilattices with homomorphisms is presented and extended to idempotent semirings and domain semirings. These contexts are the Kripke structures for idempotent semirings and allow more efficient computations on finite models because they can be logarithmically smaller than the original semiring. Some examples and constructions such as matrix semirings are also considered.

Keywords: Algebraic contexts, idempotent semirings, domain semirings.

1 Introduction

The characterization of complete and atomic Boolean algebras as powerset algebras, essentially due to Tarski, is the basis of the categorical duality between the category of complete and atomic Boolean algebras with complete homomorphisms and the category of sets with arbitrary functions. This duality has been extended in many ways to other dualities such as between modal algebras and Kripke frames, between Boolean algebras with operators (which include relation algebras) and atom structures, between distributive lattices with operators and partially ordered frames, and between residuated lattices and residuated frames, to name some of the main examples. Here we present a categorical equivalence that is suitable for idempotent semirings with additional operations and which is based on notions from formal concept analysis. Recall that two categories $\mathbf{C}$, $\mathbf{D}$ are *equivalent* if they are "essentially the same", i.e., there are covariant functors $F : \mathbf{C} \to \mathbf{D}$, $G : \mathbf{D} \to \mathbf{C}$ such that GF and FG are naturally isomorphic to the identity functors $1_{\mathbf{C}}$ and $1_{\mathbf{D}}$ respectively. If the same condition holds for contravariant functors, then the categories are *dually equivalent*.

In formal concept analysis a complete lattice is represented by a *context*, i.e., a triple $\mathbb{X} = (X_-, X_+, X)$ where $X \subseteq X_- \times X_+$ is the *incidence relation*. Since we are interested in idempotent semirings, we consider (completely) join-preserving maps as morphisms between complete lattices, which places us in the category called **SUP** (since joins are also called suprema). In a recent development M. A. Moshier [12] defined morphisms for contexts to obtain a *relational* category **Cxt** that is dual to the category **INF** of complete meet semilattices with completely meet-preserving homomorphisms. Specifically, a morphism R from $\mathbb{X}$ to $\mathbb{Y} = (Y_-, Y_+, Y)$ is a binary relation $R \subseteq X_- \times Y_+$ that satisfies

W. Kahl and T.G. Griffin (Eds.): RAMiCS 2012, LNCS 7560, pp. 195–206, 2012.

a natural compatibility condition, and composition of morphisms is defined in such a way that the incidence relation of a context is the identity morphism for that object. Since **INF** is both dual and equivalent to **SUP** (a symmetry that is made explicit in the category **Cxt**) we find this setting well suited for our purposes. For the categories of lattices and complete perfect lattices, similar dualities are contained in [9,3,7]

We extend this duality to an equivalence between semilattices and algebraic contexts (i.e., contexts where the incidence relation induces an algebraic closure operator) with morphisms that are directed-join-preserving relations between contexts. An alternative presentation of algebraic contexts with approximable maps as morphisms, and their relation to various other categories in domain theory, is given in [10]. For applications to domain semirings, we add a multiplication on the context side via a ternary relation; the identity element corresponds to a unary relation; and a domain operation is given by a binary relation. One of the advantages of the equivalent category of contexts with relations is that the objects can be "logarithmic" in size relative to their algebraic counterparts. In fact objects on the algebraic side correspond to contexts of many different sizes which are all isomorphic in the category **Cxt** (this is possible since the notion of isomorphism in **Cxt** is not based on bijections). This makes the construction of examples on the context side less restricted, so for example idempotent semirings can be obtained from Gentzen-style proof systems. Another interesting aspect is that the equivalence maps products of domain semirings to certain disjoint unions of contexts, and other constructions like ordinal products and poset products can also be obtained by combinatorial means on the context side.

Furthermore the relational morphisms in the category of contexts give this setting a flavor of the category **Rel** (where the objects are sets and morphisms are binary relations), and it is indeed the case that **Rel** is isomorphic to a full subcategory of **Cxt**. We also observe that the ideal completions of Kleene algebras are related to the equivalence with algebraic contexts. Since the notion of relational context morphism is relatively recent, this area is currently still developing and is likely to yield further insight into proof-theoretic and algebraic properties of idempotent semirings with additional operations.

2   Background

We first recall some standard definitions and fix the notation that is convenient for our approach. A *context* is a structure $\mathbb{X} = (X_-, X_+, X)$ such that X_-, X_+ are sets and $X \subseteq X_- \times X_+$. Thus a context is simply a typed relation, called the *incidence relation*, and we will usually not distinguish between the relation X and the context $\mathbb{X}$ that it defines. The relation X determines two functions $X^\uparrow : \mathcal{P}(X_-) \to \mathcal{P}(X_+)$ and $X^\downarrow : \mathcal{P}(X_+) \to \mathcal{P}(X_-)$ by $X^\uparrow A = \{b : \forall a \in A\ aXb\}$ and $X^\downarrow B = \{a : \forall b \in B\ aXb\}$. As usual, these maps form a Galois connection from $\mathcal{P}(X_-)$ to $\mathcal{P}(X_+)$, which means that $A \subseteq X^\downarrow B \Leftrightarrow B \subseteq X^\uparrow A$ for all $A \subseteq X_-$ and $B \subseteq X_+$. Moreover, $X^\downarrow X^\uparrow$ and $X^\uparrow X^\downarrow$ are *closure operators* on X_- and X_+ respectively. The sets $\mathrm{Cl}_-(X) = \{X^\downarrow X^\uparrow A : A \subseteq X_-\}$ and $\mathrm{Cl}_+(X) = \{X^\uparrow X^\downarrow B :$

$B \subseteq X_+\}$ of *Galois-closed sets* are dually isomorphic complete lattices with intersection as meet and Galois-closure of union as join. Note that $\mathrm{Cl}_-(X)$ can also be defined as $\{X^{\downarrow}B : B \subseteq X_+\}$, so a set is Galois-closed if and only if it is in the image of $X^{\downarrow}$ (or similarly $X^{\uparrow}$ for $\mathrm{Cl}_+(X)$). The operations $X^{\uparrow}$ and $X^{\downarrow}$ both map unions to intersections, and sets of the form $X^{\downarrow}\{x\}$ are a *basis* from which all Galois-closed sets can be obtained by intersections.

For contexts X, Y a *context morphism* $R : X \to Y$ is a relation $R \subseteq X_- \times Y_+$ such that $X^{\downarrow}X^{\uparrow}R^{\downarrow} = R^{\downarrow} = R^{\downarrow}Y^{\uparrow}Y^{\downarrow}$ (or equivalently $X^{\downarrow}X^{\uparrow}A \subseteq R^{\downarrow}R^{\uparrow}A$ and $Y^{\uparrow}Y^{\downarrow}B \subseteq R^{\uparrow}R^{\downarrow}B$ for all $A \subseteq X_-, B \subseteq Y_+$), in which case the relation is said to be *compatible* with X, Y (see Fig. 1). Here the operations $R^{\uparrow} : \mathcal{P}(X_-) \to \mathcal{P}(Y_+)$ and $R^{\downarrow} : \mathcal{P}(Y_+) \to \mathcal{P}(X_-)$ are defined in the same way as for the binary relation X above, and we use juxtaposition for both function composition and function application (associating to the right). Note that the incidence relation X is itself

$$X_+ \quad Y_+$$
$$X \uparrow \quad \overset{R}{\nearrow} \quad \uparrow Y$$
$$X_- \quad Y_-$$

$$R \subseteq X_- \times X_+ \text{ and } X^{\downarrow}X^{\uparrow}R^{\downarrow} = R^{\downarrow} = R^{\downarrow}Y^{\uparrow}Y^{\downarrow}$$

Fig. 1. Context morphism

a morphism from $\mathbb{X}$ to $\mathbb{X}$ (since $X^{\downarrow}X^{\uparrow}X^{\downarrow} = X^{\downarrow}$), and with the composition defined below it is, in fact, the identity morphism on $\mathbb{X}$. Furthermore, compatibility implies that the map $R^{\downarrow}Y^{\uparrow}$ maps closed sets to closed sets, and it is easy to see that if B is a closed set then $A \subseteq R^{\downarrow}Y^{\uparrow}B$ if and only if $Y^{\uparrow}B \subseteq R^{\uparrow}A$ or equivalently $Y^{\downarrow}R^{\uparrow}A \subseteq B$. Hence the maps $Y^{\downarrow}R^{\uparrow}$ and $R^{\downarrow}Y^{\uparrow}$ are *residuals*, which implies that $Y^{\downarrow}R^{\uparrow} : \mathrm{Cl}_-(X) \to \mathrm{Cl}_-(Y)$ preserves arbitrary joins and $R^{\downarrow}Y^{\uparrow} : \mathrm{Cl}_-(Y) \to \mathrm{Cl}_-(X)$ preserves arbitrary intersections. Given a context Z and morphism $S : Y \to Z$, the composite morphism $R \,\overset{\circ}{,}\, S : X \to Z$ is defined by $x\, R \,\overset{\circ}{,}\, S\, y \Leftrightarrow x \in R^{\downarrow}Y^{\uparrow}S^{\downarrow}\{y\}$ (see Fig. 2).

Finally, the *Dedekind-MacNeille context* of a poset L is $\mathrm{DM}(L) = (L, L, \leq)$.

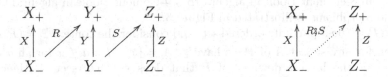

Fig. 2. Composition of context morphisms

Theorem 1. *[12]*

1. *The collection* **Cxt** *of all contexts with compatible relations as morphism is a category with the incidence relation of each context as the identity morphism.*
2. *The category* **Cxt** *is dually equivalent to the category* **INF** *of complete semilattices with completely meet-preserving homomorphisms. The adjoint functors are* $\mathrm{Cl}_- : \textbf{Cxt} \to \textbf{INF}$ *and DM:* $\textbf{INF} \to \textbf{Cxt}$. *On morphisms,* $\mathrm{Cl}_-(R) = R^{\downarrow}Y^{\uparrow} : \mathrm{Cl}_-(Y) \to \mathrm{Cl}_-(X)$ *and for an* **INF** *morphism* $h : L \to M$, $DM(h) = \{(x,y) \in M \times L : x \leq h(y)\}$.

Lemma 2. *(i)* $R : X \to Y$ *is a monomorphism in* **Cxt** *if and only if* $R^{\downarrow}R^{\uparrow} = X^{\downarrow}X^{\uparrow}$.

(ii) $R : X \to Y$ *is an epimorphism in* **Cxt** *if and only if* $R^{\uparrow}R^{\downarrow} = Y^{\uparrow}Y^{\downarrow}$.

(iii) $R : X \to Y$ *is an isomorphism if and only if it is both mono and epi, or equivalently if* $R^{\downarrow}R^{\uparrow}X^{\downarrow} = X^{\downarrow}$ *and* $R^{\uparrow}R^{\downarrow}Y^{\uparrow} = Y^{\uparrow}$.

To illustrate the duality in Theorem 1(2) we discuss a few examples. On objects, the duality is simply Birkhoff's [1] notion of polarity that represents any complete lattice as the lattice of Galois-closed sets of some binary relation. This has been studied extensively in formal concept analysis [6] and many tools have been developed to compute with finite contexts. However the notion of morphism for contexts has not gotten as much attention, and several competing definitions have appeared [4,7,10]. The category **Cxt** of [12], defined above, has one of the most natural notions of morphism and fits best with the applications we are interested in.

1. Let S be any set, and consider the context $\mathbb{S} = (S, S, \neq)$. Then for any subset A of S we have $\neq^{\uparrow}A = S \backslash A$, hence the Galois closure of A is $S \backslash (S \backslash A) = A$. It follows that $\mathrm{Cl}_-(\mathbb{S}) = (\mathcal{P}(S), \cap)$, so the complete semilattice corresponding to the context $\mathbb{S}$ is the complete and atomic Boolean algebra of all subsets of S. Of course the duality between the category of complete and atomic Boolean algebras and the category of sets and functions is well known, but here the duality between **INF** and **Cxt** restricts to a duality of complete and atomic Boolean algebras with $\bigwedge$-preserving functions and the category **Rel** of sets and binary relations (since all relations are compatible in this case, hence **Cxt** morphisms. For example if $S = \{0, 1\}$ and $T = \{0, 1, 2\}$ then there are $2^{2 \cdot 3} = 64$ binary relations from S to T, hence there are 64 morphisms from context $\mathbb{S}$ to $\mathbb{T} = (T, T, \neq)$, corresponding to 64 $\bigwedge$-preserving maps from an 8-element Boolean algebra to a 4-element Boolean algebra. One of these morphisms is illustrated in Figure 3.
2. Let $(P, \leq)$ be a partially ordered set, and consider the context $\mathbb{P} = (P, P, \nleq)$. Then for any subset A of P we have $\nleq^{\uparrow}A = \{x \in P : a \nleq x \text{ for all } a \in A\}$, which is the largest downset of P that does not intersect A. Hence the Galois closure of A is $P \setminus (\nleq^{\uparrow}A)$ = the smallest upset containing A, and therefore $\mathrm{Cl}_-(\mathbb{P})$ is the complete $\cap$-semilattice of upsets of the poset P. This is in fact a complete and perfect distributive lattice (a lattice is *perfect* if every element is a join of completely join irreducible elements and a meet

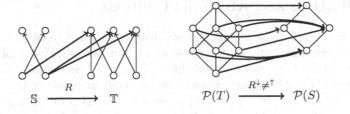

Fig. 3. One of 64 morphisms from $\mathbb{S}$ to $\mathbb{T}$

of completely meet irreducible elements). The duality between finite posets and finite distributive lattices is due to Birkhoff, but as in 1., the morphisms are compatible relations for the contexts and $\bigwedge$-preserving functions for the complete $\bigwedge$-semilattices.

3. Both 1. and 2. have dualities that can be described without the use of contexts (sets for 1. and posets for 2.). However for semilattices in general, contexts are required. Note that complete $\bigwedge$-semilattices are complete lattices since the $\bigvee A = \bigwedge \{b : a \leq b \text{ for all } a \in A\}$. If a lattice L is also perfect, then we can obtain a smallest context by taking $(J_\infty(L), M_\infty(L), \leq)$, where $J_\infty(L)$ is the set of completely join irreducible elements and $M_\infty(L)$ is the set of completely meet irreducible elements of L. Figure 4 shows some 6-element (semi)lattices with their contexts.

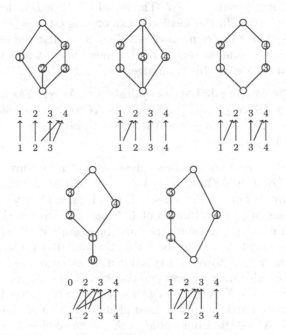

Fig. 4. Examples of 6-element (semi)lattices and corresponding contexts

3   Semilattices and Algebraic Contexts

Since we are interested in idempotent semirings, we want to obtain a similar categorical connection between the category of (not necessarily complete) join-semilattices with bottom element and (finite) join-preserving maps, and the category of so-called *algebraic* contexts and certain context morphisms.

Recall [2] that a family $\{A_i : i \in I\}$ of sets is *directed* if for all $i, j \in I$ there exists $k \in I$ such that $A_i \cup A_j \subseteq A_k$. For a context X, the closure operator $X^{\downarrow}X^{\uparrow}$ is *algebraic* if the closure of any subset is the union of the closures of its finite subsets, or equivalently ([2], Theorem 7.14), if for any directed family of sets $\{A_i \subseteq X_- : i \in I\}$ we have $X^{\downarrow}X^{\uparrow}\bigcup_i A_i = \bigcup_i X^{\downarrow}X^{\uparrow}A_i$. The *compact sets* of a context X are given by $K(X) = \{X^{\downarrow}X^{\uparrow}A : A \text{ is a finite subset of } X_-\}$. For algebraic contexts X, Y, an *algebraic* morphism $R : X \to Y$ is a context morphism such that $R^{\downarrow}Y^{\uparrow}$ preserves directed unions, i.e., for any directed family $\{A_i \subseteq X_- : i \in I\}$ we have $R^{\downarrow}Y^{\uparrow}\bigcup_i A_i = \bigcup_i R^{\downarrow}Y^{\uparrow}A_i$. This property holds for identity morphisms and is preserved by composition, hence algebraic contexts form a subcategory of **Cxt** denoted by **ACxt**. Observe that all finite contexts are algebraic, and in this case the compact sets $K(X)$ coincide with the closed sets $\mathrm{Cl}_-(X)$.

The category of join-semilattices with bottom element 0 and join-preserving homomorphisms that preserve 0 is denoted by **JSLat$_0$**. We will use $+$ to denote the join operation, since this is in agreement with idempotent semirings. For a join-semilattice L, an *ideal* is a subset J of L such that for all $x, y \in J$, $x + y \in J$ and for all $z \in L$ if $z \leq x$ then $z \in J$. The set of ideals of L is denoted by $I(L)$. Given a join-semilattice L, the ideal context of L is $C(L) = (L, I(L), \in)$, i.e., the incidence relation is the element-of relation. Note that the closure operator $\in^{\downarrow}\in^{\uparrow}$ generates ideals from subsets of L, and since every ideal is the union of its finitely generated subideals, this is an algebraic context.

Theorem 3. *The category* **JSLat$_0$** *is equivalent to* **ACxt**. *The adjoint functors are* $K :$ **ACxt** $\to$ **JSLat$_0$** *and* $C :$ **JSLat$_0$** $\to$ **ACxt**. *On morphisms,* $K(R) = Y^{\downarrow}R^{\uparrow} : K(X) \to K(Y)$ *and for a* **JSLat$_0$** *morphism* $h : L \to M$, $C(h) = \{(a, D) \in L \times I(M) : h(a) \in D\}$.

Proof. We first show that up to isomorphism C and K are inverses on objects of the category. For L in **JSLat$_0$** and $A \subseteq L$, the closure $\in^{\downarrow}\in^{\uparrow} A = \langle A \rangle$ is the ideal generated by A. For a finite subset A, this ideal is always principal, hence $K(C(L))$ is the set of principal ideals of L. Since any ideal is the union of the principal ideals it contains, and since they are the closure of a singleton, it follows that $C(L)$ is algebraic. It is easy to check that the map $a \mapsto \downarrow a = \{b \in L : b \leq a\}$ is an isomorphism from L to $K(C(L))$ ordered by inclusion.

Now let X be an algebraic context and consider $A, B \in K(X)$. Then $A + B = X^{\downarrow}X^{\uparrow}A_0 + X^{\downarrow}X^{\uparrow}B_0 = X^{\downarrow}X^{\uparrow}(A_0 \cup B_0)$ for some finite $A_0 \subseteq A$ and $B_0 \subseteq B$. Hence $K(X)$ is a semilattice, and the least element is $X^{\downarrow}X^{\uparrow}\emptyset$. We need to prove that $(K(X), I(K(X)), \in)$ is isomorphic to X, so we define a relation R from X_- to $I(K(X))$ by xRD if and only if $X^{\downarrow}X^{\uparrow}\{x\} \in D$. To see that R is an isomorphism, it suffices by Lemma 2(iii) to check that R is compatible and that

$R^{\downarrow}R^{\uparrow}X^{\downarrow} = X^{\downarrow}$ and $R^{\uparrow}R^{\downarrow}\in^{\uparrow} = \in^{\uparrow}$. Note also that each of these equations holds for all subsets if it is valid for singleton subsets.

$X^{\downarrow}X^{\uparrow}R^{\downarrow} = R^{\downarrow}$: Let D be an ideal of $K(X)$. Then $R^{\downarrow}\{D\} = \{x \in X_- : X^{\downarrow}X^{\uparrow}\{x\} \in D\} = \{x \in X_- : x \in A$ for some $A \in D\} = \bigcup D$. Since D is a directed set and X is an algebraic context, $X^{\downarrow}X^{\uparrow}\bigcup D = \bigcup_{A \in D} X^{\downarrow}X^{\uparrow}A = \bigcup D$.

$\in^{\uparrow}\in^{\downarrow}R^{\uparrow} = R^{\uparrow}$: For $x \in X_-$ we have

$$R^{\uparrow}\{x\} = \{D \in I(K(X)) : X^{\downarrow}X^{\uparrow}\{x\} \in D\}$$

which is the collection of all ideals that include the principal ideal, say J, generated by $X^{\downarrow}X^{\uparrow}\{x\}$ in $K(X)$. Now $\in^{\downarrow}$ of this collection is the intersection of all these ideals, hence is equal to the ideal J. Since J is principal, it follows that $\in^{\uparrow}J = \{D : X^{\downarrow}X^{\uparrow}\{x\} \in D\} = R^{\uparrow}\{x\}$.

$R^{\downarrow}R^{\uparrow}X^{\downarrow} = X^{\downarrow}$: Note that for $A \subseteq X_-$ we have $R^{\uparrow}A = \{D : X^{\downarrow}X^{\uparrow}\{a\} \in D$ for all $a \in A\} = \{D : \{X^{\downarrow}X^{\uparrow}\{a\} : a \in A\} \subseteq D\}$. Hence $x \in R^{\downarrow}R^{\uparrow}A$ implies xRD_A, where D_A is the ideal generated by the set $\{X^{\downarrow}X^{\uparrow}\{a\} : a \in A\}$. Therefore $X^{\downarrow}X^{\uparrow}\{x\} \in D_A$, so $X^{\downarrow}X^{\uparrow}\{x\} \subseteq X^{\downarrow}X^{\uparrow}\{a_1, \ldots, a_n\}$ for some finite subset of A. It follows that $x \in X^{\downarrow}X^{\uparrow}A$, and replacing A with $X^{\downarrow}B$ proves the result.

$R^{\uparrow}R^{\downarrow}\in^{\uparrow} = \in^{\uparrow}$: We first observe that for $C \in K(X)$ we have

$$R^{\downarrow}\in^{\uparrow}\{C\} = R^{\downarrow}\{D : C \in D\} = \bigcap\{\bigcup D : C \in D\}$$

since $R^{\downarrow}\{D\} = \bigcup D$. But one of the ideals D is $\downarrow C$, and $\bigcup \downarrow C = C$, hence $R^{\downarrow}\in^{\uparrow}\{C\} = C$. It follows that $R^{\uparrow}R^{\downarrow}\in^{\uparrow}\{C\} = R^{\uparrow}C = \{D : X^{\downarrow}X^{\uparrow}\{x\} \in D$ for all $x \in C\}$. Therefore $R^{\uparrow}C \subseteq \{D : C \in D\} = \in^{\uparrow}\{C\}$ as required.

This isomorphism also shows that R is an algebraic morphism. To check that K is a functor, recall that the identity morphism id_X of a context is the incidence relation X. Hence $K(id_X) = X^{\downarrow}X^{\uparrow}$ which is the identity map on the semilattice $K(X)$ since the elements of $K(X)$ are closed. Composition is preserved since if $R : X \to Y$ and $S : Y \to Z$ then

$$K(R; S) = Z^{\downarrow}(R; S)^{\uparrow} = Z^{\downarrow}S^{\uparrow}Y^{\downarrow}R^{\uparrow} = K(S)K(R).$$

For the map C we have $C(id_L) = \{(a, D) \in L \times I(L) : id_L(a) \in D\} = \{(a, D) : a \in D\} = \in$, and this is the incidence relation of the context $C(L)$. Let $h : L \to M$ and $g : M \to N$ be **JSLat$_0$** homomorphisms, then $a\, C(h); C(g)\, D$ if and only if

$$a \in C(h)^{\downarrow}\in^{\uparrow}C(g)^{\downarrow}\{D\} = C(h)^{\downarrow}\in^{\uparrow}g^{-1}[D] = C(h)^{\downarrow}\{g^{-1}[D]\}$$
$$= h^{-1}[g^{-1}[D]] = (gh)^{-1}[D]$$

and this is equivalent to $a\, C(gh)\, D$. Moreover, it is not difficult to check that $C(h)$ is an algebraic morphism, that $K(R)$ is a **JSLat$_0$** homomorphism, and that $K(C(h))$ and $C(K(R))$ are naturally isomorphic to h and R respectively. □

The above equivalence is of course closely related to the Hofmann-Mislove-Stralka duality [11] between join-semilattices and algebraic lattices with maps that preserve all meets and directed joins (see also [8], p. 274). However, the category of algebraic contexts is much "bigger" than the category of algebraic lattices, since there are many contexts of different sizes that correspond to the same algebraic lattice. Consequently one has much more freedom constructing contexts, and for many semilattices one can obtain contexts that are logarithmically smaller. For example, if a semilattice is given by the compact elements of a complete and atomic Boolean algebra B, then one may take the algebraic context $(At(B), At(B), \neq)$ to represent the semilattice. As mentioned at the end of the previous section, in formal concept analysis many algorithms and visualization tools have been developed for contexts, and with the above equivalence they can be readily applied to arbitrary semilattices.

We have also noted that the category of algebraic contexts contains a subcategory that is isomorphic to **Rel**, the category of sets and binary relations. The objects of this subcategory are the contexts $(A, A, \neq)$ where A is any set, and the morphisms are any binary relation, since the compatibility conditions are automatically satisfied. Another interesting subcategory is obtained by considering posets $(P, \leq)$ and defining the contexts $(P, P, \not\geq)$. The algebraic lattice in this case is the lattice of all downsets of P, which is a complete perfect distributive lattice, and the semilattice of compact elements consists of the downsets of finite subsets of P.

4 Contexts for Idempotent Semirings and Domain Semirings

We now show how additional join-preserving operations on the semilattice are represented on the context side. We use the example of domain semirings, but it will be clear that the framework can handle semilattices with join-preserving operations of any arity. Thus the categorical equivalence with algebraic contexts is extended to a proper generalization of the duality for complete and atomic Boolean algebras with operators and relational structures with bounded morphisms.

Recall that an *idempotent semiring* is an algebra $(L, +, 0, \cdot, 1)$ such that $(L, +, 0)$ is in $\mathbf{JSLat}_0$, $(L, \cdot, 1)$ is a monoid, $\cdot$ is join-preserving in both arguments, and $0x = 0 = x0$. A *domain semiring* is of the form $\mathbf{L} = (L, +, 0, \cdot, 1, d)$ such that $(L, +, 0, \cdot, 1)$ is an idempotent semiring, d is join-preserving, $d(0) = 0$,

$$d(x) + 1 = 1$$

$$d(x)x = x \text{ and}$$

$$d(xd(y)) = d(xy).$$

When confusion is unlikely, we usually refer to a domain semiring $\mathbf{L}$ simply by the name of its underlying set L.

Let X be an algebraic context. To capture the operations of the domain semiring on the semilattice $K(X)$, we need a ternary relation $\circ \subseteq X_-^3$, a unary relation $E \subseteq X_-$ and a binary relation $D \subseteq X_-^2$. For $A, B \subseteq X_-$, we define the notation

$$A \circ B = \{c \in X_- : (a, b, c) \in \circ \text{ for some } a \in A, b \in B\} \text{ and}$$

$$D[A] = \{b \in X_- : aDb \text{ for some } a \in A\}.$$

For $x, y \in X_-$ we further abbreviate $x \circ y = \{x\} \circ \{y\}$ and $D(x) = D[\{x\}]$. The closure operation $X^\downarrow X^\uparrow$ is called a *nucleus with respect to* $\circ$ if for all $A, B \subseteq X_-$ we have

$$(X^\downarrow X^\uparrow A) \circ (X^\downarrow X^\uparrow B) \subseteq X^\downarrow X^\uparrow (A \circ B)$$

and a *nucleus with respect to* D if for all $A \subseteq X_-$ we have

$$D[X^\downarrow X^\uparrow A] \subseteq X^\downarrow X^\uparrow D[A].$$

The nucleus property ensures that the operations $X^\downarrow X^\uparrow (A \circ B)$ and $X^\downarrow X^\uparrow D[A]$ are join-preserving in each argument. For example, the following calculation shows that the first operation is join-preserving in the second argument (recall that join $\sum$ is the closure of union):

$$X^\downarrow X^\uparrow (A \circ \sum_i B_i) = X^\downarrow X^\uparrow (A \circ X^\downarrow X^\uparrow \bigcup_i B_i) \subseteq X^\downarrow X^\uparrow (A \circ \bigcup_i B_i)$$

$$= X^\downarrow X^\uparrow \bigcup_i (A \circ B_i) = \sum_i (A \circ B_i) \subseteq \sum_i X^\downarrow X^\uparrow (A \circ B_i)$$

where the first $\subseteq$ follows from the nucleus property, and the reverse inclusion always holds.

The relations $\circ$ and D are called *algebraic* if for all $A, B \in K(X)$ the operations $X^\downarrow X^\uparrow (A \circ B)$ and $X^\downarrow X^\uparrow D[A]$ are also in $K(X)$.

An *idempotent semiring context* is of the form $(X_-, X_+, X, \circ, E)$ such that (X_-, X_+, X) is an algebraic context, $\circ, E$ are an algebraic ternary and unary relation on X_-, the closure operator is a nucleus with respect to $\circ$, and for all $x, y, z \in X_-$ we have

$$X^\uparrow ((x \circ y) \circ z) = X^\uparrow (x \circ (y \circ z)) \text{ and}$$

$$X^\uparrow (x \circ E) = X^\uparrow \{x\} = X^\uparrow (E \circ x).$$

A *domain context* is a structure $\mathbb{X} = (X_-, X_+, X, \circ, E, D)$ such that $(X_-, X_+, X, \circ, E)$ is an idempotent semiring context, the closure operator is also a nucleus with respect to D, and for all $x, y \in X_-$ we have

$$D(x) \subseteq X^\downarrow X^\uparrow E,$$

$$X^\uparrow (D(x) \circ x) = X^\uparrow \{x\} \text{ and}$$

$$X^\uparrow D[x \circ D(y)] = X^\uparrow D[x \circ y]$$

corresponding to the axioms for the domain operation d. Note that the last 3 conditions need only hold for all elements of X_-, whereas the domain axioms would have to be checked for all elements of the potentially much bigger semilattice of compact sets.

Let $\mathbb{X}, \mathbb{Y}$ be two domain contexts. A relation $R \subseteq X_- \times Y_+$ is a *domain context morphism* if it is compatible, algebraic, $R^\uparrow(E^\mathbb{X}) = Y^\uparrow(E^\mathbb{Y})$, and for all $A, B \in \text{Cl}_-(\mathbb{X})$ we have

$$R^\uparrow(x \circ y) = Y^\uparrow(Y^\downarrow R^\uparrow\{x\} \circ Y^\downarrow R^\uparrow\{y\}) \text{ and}$$

$$R^\uparrow D(x) = Y^\uparrow D[Y R^\uparrow\{x\}].$$

An *idempotent semiring context morphism* is defined likewise, but without the last equation.

As with bounded morphisms (also called p-morphisms) in modal logic the notion of domain context morphism can be written as a first-order formula with variables ranging only over elements of the context. We have not done this here since it is less compact and is no more efficient in implementations than the given formulation.

The functor K from contexts to join-semilattices is extended to domain contexts by defining $K(\mathbb{X}) = (K(X), +, 0, \cdot, 1, d)$ where $A + B = X^\downarrow X^\uparrow(A \cup B)$, $0 = X^\downarrow X^\uparrow \emptyset$, $A \cdot B = X^\downarrow X^\uparrow(A \circ B)$, $1 = X^\downarrow X^\uparrow E$ and $d(A) = X^\downarrow X^\uparrow D[A]$. Likewise the functor C is extended to domain semirings by $C(\mathbf{L}) = (L, I(L), \in, \circ, \{1\}, D)$ where $\circ = \{(x, y, z) \in L^3 : x \cdot y = z\}$ and $D = \{(x, y) \in L^2 : d(x) = y\}$. With these definitions one can check that $K(\mathbb{X})$ is a domain semiring and $C(\mathbf{L})$ is a domain context. For example to check that $X^\downarrow X^\uparrow$ is a nucleus with respect to the relation D, recall that the closure operator generates an ideal from a subset of L. So $D[X^\downarrow X^\uparrow A] = D[\langle A \rangle] = \{y : d(x) = y \text{ for some } x \in \langle A \rangle\} = \{d(a_1 + \cdots + a_n) : a_i \in A, n \in \mathbb{N}\} = \{d(a_1) + \cdots + d(a_n) : a_i \in A, n \in \mathbb{N}\} \subseteq \langle D[A] \rangle = X^\downarrow X^\uparrow D[A]$. We are now ready to state the extended versions of the previous result.

Theorem 4. *The category* **IS** *of idempotent semirings is equivalent to the category* **ISCxt** *of idempotent semiring contexts. The adjoint functors are* $K :$ **ISCxt** $\to$ **IS** *and* $C :$ **IS** $\to$ **ISCxt**. *On morphisms,* $K(R) = Y^\downarrow R^\uparrow : K(X) \to K(Y)$ *and* $C(h) = \{(a, D) \in L \times I(M) : h(a) \in D\}$.

Similarly the category **DS** *of domain semirings is equivalent to the category* **DSCxt** *of domain semiring contexts. The adjoint functors are* $K :$ **DSCxt** $\to$ **DS** *and* $C :$ **DS** $\to$ **DSCxt** *with the operation on morphisms as for idempotent semirings.*

With this result one can specify any domain semiring by a context, a subset, a binary relation and a ternary relation on the first component of the context. For example the first (semi)lattice in Figure 4 can be expanded into 5 nonisomorphic domain semirings where 1 is the identity element. Using the context from the figure with $X_- = \{1, 2, 3\}$, in each case $E = \{1\}$, the binary relation $D = \{(1, 1), (2, 1), (3, 1)\}$, and the 5 ternary relations are

$\circ_1$	2	3
2	{2}	{2}
3	{2}	{2}

$\circ_2$	2	3
2	{2}	{2}
3	{2}	{2,3}

$\circ_3$	2	3
2	{2,3}	{2,3}
3	{2,3}	{2}

$\circ_4$	2	3
2	{2,3}	{2,3}
3	{2,3}	{2,3}

$\circ_5$	2	3
2	X_-	X_-
3	X_-	X_-

where $1 \circ x = x = x \circ 1$ for all $x \in X_-$. Clearly this is more economical than giving the multiplication tables for five 6-element monoids.

Given a semiring L, one can construct the semiring $M_n(L)$ of all $n \times n$ matrices with entries from L in the usual way. This object has $|L|^{n^2}$ many elements, but for idempotent semirings the context $\mathbb{Y}$ of $M_n(L)$ is much smaller since it can be constructed from n^2 disjoint copies of the idempotent semiring context $\mathbb{X} = C(L)$ as follows. Let $Y_- = \{(i,j,a) : a \in X_-, i,j = 1, \ldots, n\}$ and define $(i,j,a)Y(i',j',a')$ iff $i \neq i'$ or $j \neq j'$ or aXa', $E = \{(i,i,a) : a \in E, i = 1, \ldots, n\}$, and $(i,j,a) \circ (k,l,b) = \{(i,l,c) : j = k \text{ and } c \in a \circ b\}$.

Kripke-style semantics, as provided for semilattice-expansions by algebraic contexts, are also nicely related to completions. Instead of the functor K to the category of semilattices, one can use the functor Cl_- to the category of complete semilattices. Indeed, the functor C followed by Cl_- is simply the ideal completion of semilattices, which also applies to idempotent semirings, domain semirings and (domain) Kleene algebras. For example there is a Kleene $*$ induced on the completion of a domain semiring by defining $x^* = \sum_{i=0}^{\omega} x^i$, where $x^0 = 1$ and $x^{i+1} = x^i x$ for $i \geq 1$. It is currently ongoing research to adapt the equivalence with algebraic contexts so that it can represent a given Kleene $*$ directly on contexts.

The equivalences with contexts (plus relations) can also be used to get insight into constructions on the algebraic side. In particular, products and coproducts in the algebraic categories are mapped to significantly different types of constructions on the context side. It remains to be seen whether this produces new results about, for example, the structure of free objects on the algebraic side. In the related area of residuated lattices the notion of residuated frame (= residuated context) has already produced significant results about decidability and finite embeddability [5]. Exploring connections with proof theory and coalgebras are other promising directions.

5   Conclusion

Based on a duality by Moshier [12] between complete semilattices and contexts, we have defined algebraic contexts and algebraic morphisms and proved an equivalence between the category of semilattices and the category of algebraic contexts. We then extended this equivalence to idempotent semirings and domain semirings, thereby obtaining Kripke-style semantics for these two categories. The same approach can be used to define categories of algebraic contexts with additional relations that are equivalent to categories of idempotent semilattices with operations that are join-preserving in each argument, thus generalizing the duality between complete and atomic Boolean algebras with operators and relational Kripke structures.

References

1. Birkhoff, G.: Lattice Theory, 3rd edn. AMS Colloquium Publications, vol. XXV. American Mathematical Society, Providence (1967)
2. Davey, B., Priestley, H.A.: Introduction to Lattices and Order, 2nd edn. Cambridge University Press (2002)
3. Dunn, J.M., Gehrke, M., Palmigiano, A.: Canonical extensions and relational completeness of some substructural logics. Journal of Symbolic Logic 70(3), 713–740 (2005)
4. Erné, M.: Categories of contexts (preprint), http://www.iazd.uni-hannover.de/~erne/preprints/CatConts.pdf
5. Galatos, N., Jipsen, P.: Residuated frames with applications to decidability. To appear in Transactions of the American Math. Soc.
6. Ganter, B., Wille, R.: Formal concept analysis. Mathematical foundations. Springer, Berlin (1999)
7. Gehrke, M.: Generalized Kripke frames. Studia Logica 84, 241–275 (2006)
8. Gierz, G., Hofmann, K.H., Keimel, K., Lawson, J.D., Mislove, M., Scott, D.S.: Continuous Lattices and Domains. Encyclopedia of Mathematics and its Applications, vol. 93. Cambridge University Press (2003)
9. Hartung, G.: A topological representation of lattices. Algebra Universalis 29(2), 273–299 (1992)
10. Hitzler, P., Krötzsch, M., Zhang, G.-Q.: A categorical view on algebraic lattices in formal concept analysis. Fundamenta Informaticae 74, 301–328 (2006)
11. Hofmann, K.H., Mislove, M.W., Stralka, A.R.: The Pontryagin Duality of Compact 0-Dimensional Semilattices and Its Applications. Lecture Notes in Mathematics, vol. 396. Springer (1974)
12. Moshier, M.A.: A relational category of formal contexts (preprint)

Relational Representation Theorem
for Powerset Quantales

Koki Nishizawa[1] and Hitoshi Furusawa[2]

[1] Department of Information Systems, Faculty of Environmental and Information
Studies, Tottori University of Environmental Studies
koki@kankyo-u.ac.jp
[2] Department of Mathematics and Computer Science, Kagoshima University
furusawa@sci.kagoshima-u.ac.jp

Abstract. The paper gives a sufficient condition for a quantale to be
isomorphic to a sub-quantale of the quantale whose elements are binary
relations on a set and whose order and monoid structure are respectively
given by inclusion and relational composition and the identity relation.
A quantale has such a relational representation, if its underlying lattice
is a powerset of some set. We also show some other equivalent conditions
of the sufficient condition.

1 Introduction

This paper shows a relational representation theorem for quantales. Quantales
were introduced by Mulvey [1] in order to provide a constructive formulation of
foundations of quantum mechanics. They are complete join semilattices together
with a monoid structure satisfying the distributive laws. In the literature, they
are also known as complete idempotent semirings or standard Kleene algebras [2].

There is a relational quantale whose elements are binary relations on a set,
whose order is given by inclusion, and whose monoid structure is given by relational composition and the identity relation. Relational quantales play an important role in computer science. For example, they are models for the semantics of
non-deterministic while-programs [3,4], they also provides a sound and complete
class of models for linear intuitionistic logic [5], and so on.

In Stone's representation theorem for Boolean algebra or Priestley's representation theorem for bounded distributive lattices, a powerset is regarded as a
standard Boolean algebra or a standard bounded distributive lattice [6,7]. On
the other hand, since a relational quantale has been regarded as a 'standard
quantale', some results called 'relational representation theorem for quantales'
have been shown in the literature.

However, relational quantales in their results are not equal to the standard
relational quantales. For example, Valentini [8] shows that a quantale Q is isomorphic to a sub-quantale of the quantale whose elements are binary relations on
Q. However, the order of the quantale is not given by inclusion, but the opposite
order of inclusion.

W. Kahl and T.G. Griffin (Eds.): RAMiCS 2012, LNCS 7560, pp. 207–218, 2012.
© Springer-Verlag Berlin Heidelberg 2012

Brown and Gurr [9] show that a completely coprime algebraic quantale Q is isomorphic to a sub-quantale of the quantale whose elements are binary relations on Q and whose order is given by inclusion. However, the unit of the monoid structure of the quantale is not equal to the identity relation.

Palmigiano and Re [10] give a sufficient condition for a quantale to be isomorphic to a sub-quantale of the quantale whose elements are binary relations on a set and whose order is given by inclusion and whose monoid structure is given by relational composition and the identity relation. Indeed, this result means 'relational representation theorem'. It is important point to embed a quantale Q in the relations not on Q but on the set of all atoms of Q. However, the result given in [10] is a relational representation theorem for 'unital involutive quantales'. A quantale Q is involutive if it is endowed with a unary operation $*$ such that, for all $a, b \in Q$ and every $S \subseteq Q$,

1. $a^{**} = a$;
2. $(a \cdot b)^* = b^* \cdot a^*$;
3. $(\bigvee S)^* = \bigvee \{q^* | q \in S\}$.

This paper gives a relational representation theorem for quantales which are not in general involutive. Similarly to the papers [11,10], our representation theorem shows that a quantale Q satisfying some condition is isomorphic to a sub-quantale of the quantale whose elements are binary relations on the set of all atoms of Q.

The main theorem Theorem 4 of this paper says that for a quantale Q, the following are equivalent.

1. Q has a relational representation in our way and it is CCP-invertible.
2. Q is isomorphic to $\wp(X)$ as complete join semilattice for some set X.
3. Q is atom-algebraic and it is a frame.
4. Q is atom-algebraic and its atoms are completely coprime.
5. Q is completely coprime algebraic and its completely coprime elements are atoms.
6. Q is completely coprime algebraic and the order of its completely coprime elements is discrete.

This theorem asserts that a quantale has a relational representation, if it is isomorphic to a powerset as complete join semilattice. A powerset quantale has four other equivalent conditions. The notion of a CCP-invertible quantale is defined in this paper. When a quantale is CCP-invertible, it has a relational representation in our way if and only if it is isomorphic to a powerset as complete join semilattice.

Powerset quantales are examples of 'completely coprime algebraic quantales' which is the sufficient condition given in the paper [9]. However, our result is not an application of the result in [9] to powerset quantales, since our representation theorem embeds a quantale Q in the relations not on Q but on the set of all atoms of Q.

This paper is organized as follows. Section 2 defines completely coprime algebraic quantales and Section 3 defines powerset quantales. Section 4 shows that a quantale has a relational representation, if it is isomorphic to a powerset as complete join semilattice. Section 5 shows that a quantale satisfies the condition, if it has a relational representation in our way and it is CCP-invertible. Section 6 summarizes this work and discusses future work.

2   Completely Coprime Algebraic Quantales

In this section, we recall the notion of completely coprime algebraic quantales and show some examples. A quantale is called completely coprime algebraic depending only on its underlying complete join semilattice structure. The terminology in this section relies on the paper [9].

Definition 1 (complete join semilattice). *A complete join semilattice is a tuple* $(K, \leq, \bigvee)$ *with the following properties:*

1. $(K, \leq)$ *is a partially ordered set.*
2. $\bigvee S$ *is the join (i.e., the least upper bound) for each subset S of K.*

A complete join semilattice must have the least element, which is the join of the empty subset. We write $\perp$ for it.

A complete join semilattice must be a complete lattice, since the meet of a subset S is the join of all lower bounds of S. We write $a \wedge b$ for the meet of $\{a, b\}$.

Definition 2. *An element x of a complete join semilattice Q is called* completely coprime (or completely join-prime), *if*

$$x \leq \bigvee S \iff \exists a \in S. x \leq a$$

for each subset S of Q.

We write $\mathbf{CCP}(Q)$ for the set of all completely coprime elements of Q.

Remark 1. $\perp$ is not completely coprime, since $\perp = \bigvee \emptyset$.

Definition 3 (completely coprime algebraic). *A complete join semilattice Q is called* completely coprime algebraic (or CCPA) *if for each $a \in Q$,*

$$a = \bigvee \{x \in \mathbf{CCP}(Q) \mid x \leq a\}.$$

Definition 4 (quantale (or complete idempotent semiring)). *A quantale is a tuple* $(K, \leq, \bigvee, \cdot, 1)$ *with the following properties:*

1. $(K, \cdot, 1)$ *is a monoid.*
2. $(K, \leq, \bigvee)$ *is a complete join semilattice.*
3. $(\bigvee S) \cdot a = \bigvee \{b \cdot a \mid b \in S\}$ *for each element a and each subset S of K.*
4. $a \cdot (\bigvee S) = \bigvee \{a \cdot b \mid b \in S\}$ *for each element a and each subset S of K.*

A quantale $(K, \leq, \bigvee, \cdot, 1)$ is called completely coprime algebraic if $(K, \leq, \bigvee)$ is completely coprime algebraic.

Example 1. For a set A, the tuple $\mathbf{Rel}(A) = (K, \leq, \bigvee, \cdot, 1)$ forms a quantale where

- K is the set of all binary relations on A,
- $\leq$ is the inclusion $\subseteq$,
- $\bigvee$ is the union operator $\bigcup$,
- $R \cdot Q$ is the composition of R and Q, (i.e., $(a, b) \in R \cdot Q \Leftrightarrow \exists c \in A.(a, c) \in R, (c, b) \in Q$), and
- 1 is the identity (diagonal) relation on A.

A binary relation on A is completely coprime in $\mathbf{Rel}(A)$ if and only if it is a singleton subset of $A \times A$. $\mathbf{Rel}(A)$ is completely coprime algebraic, since for $R \in \mathbf{Rel}(A)$,

$$\begin{aligned}
&\bigcup \{Q \in \mathbf{CCP}(\mathbf{Rel}(A)) \mid Q \subseteq R\} \\
&= \bigcup \{\{(a, a')\} \mid \{(a, a')\} \subseteq R\} \\
&= \{(a, a') \mid (a, a') \in R\} \\
&= R.
\end{aligned}$$

$\square$

The powerset of a monoid forms a quantale. This paper gives only two examples for monoids.

Example 2. For a set Σ, the tuple $\wp(\Sigma^*) = (K, \leq, \bigvee, \cdot, 1)$ forms a quantale where

- K is the powerset of Σ^* where Σ^* is the set of all finite sequences of elements of Σ,
- $\leq$ is the inclusion $\subseteq$,
- $\bigvee$ is the union operator $\bigcup$,
- $R \cdot Q$ is $\{\sigma\pi \mid \sigma \in R, \pi \in Q\}$, and
- 1 is the singleton set of the empty sequence on A.

A subset of Σ^* is completely coprime if and only if it is a singleton subset. $\wp(\Sigma^*)$ is completely coprime algebraic.

Example 3. The tuple $\wp(\mathbf{Z}_2) = (K, \leq, \bigvee, \cdot, 1)$ forms a quantale where

- K is the powerset of the group $\mathbf{Z}_2 = (\{0, 1\}, +, 0, -)$,
- $\leq$ is the inclusion $\subseteq$,
- $\bigvee$ is the union operator $\bigcup$,
- $A \cdot B$ is $\{a + b \mid a \in A, b \in B\}$, and
- 1 is the set $\{0\}$.

A subset of $\mathbf{Z}_2$ is completely coprime if and only if it is a singleton subset. This quantale is completely coprime algebraic.

Example 4. For a set A, the tuple $\mathbf{N} \cup \{\omega\} = (K, \leq, \bigvee, \cdot, 1)$ forms a quantale where

- K is the set of all natural numbers and the additional element ω,
- $a \leq b$ if and only if $a = \omega$, $a = b$, or a is a natural number greater than b,
- $\bigvee S$ is the minimum number of S except for $\bigvee \{\omega\} = \bigvee \emptyset = \omega$,
- $a \cdot b$ is $a + b$ except for $\omega \cdot a = a \cdot \omega = \omega$, and
- 1 is the zero number.

This quantale is completely coprime algebraic and $\mathbf{CCP}(\mathbf{N} \cup \{\omega\}) = \mathbf{N}$.

3   Powerset Quantales

In this section, we recall the notion of atom and define the notion of atom-algebraic. We compare a powerset semilattice with the four conditions based on completely coprime elements or atoms. Finally, we define the notion of powerset quantale.

Definition 5 (atom). *An* atom *of a complete join semilattice Q is an element x with the following properties:*

1. *$x \neq \bot$.*
2. *$a < x$ implies $a = \bot$.*

We write **Atom**(Q) for the set of all atoms of Q.

Definition 6 (atom-algebraic). *A complete join semilattice Q is called atom-algebraic if for each $a \in Q$,*

$$a = \bigvee \{x \in \mathbf{Atom}(Q) \mid x \leq a\}.$$

We also recall the notion of frame.

Definition 7 (frame). *A complete join semilattice Q is called a frame if*

$$a \wedge \bigvee S = \bigvee \{a \wedge s \mid s \in S\}$$

for each element a and each subset S of K.

Example 5. In **Rel**(A) of Example 1, $\wp(\Sigma^*)$ of Example 2, and $\wp(\mathbf{Z}_2)$ of Example 3, an element is an atom if and only if it is a singleton subset. They are atom-algebraic and they are frames.

Theorem 1. *For a complete join semilattice Q, the following are equivalent.*

1. *Q is isomorphic to $\wp(X)$ for some set X.*
2. *Q is atom-algebraic and it is a frame.*
3. *Q is atom-algebraic and its atoms are completely coprime.*
4. *Q is completely coprime algebraic and its completely coprime elements are atoms.*
5. *Q is completely coprime algebraic and the order of its completely coprime elements is discrete.*

Proof. (1$\Longrightarrow$2) A subset Y of X is an atom in $\wp(X)$ if and only if Y is a singleton subset. $\wp(X)$ is an atom-algebraic frame.

(2$\Longrightarrow$3) Let us show that an atom x of a frame Q is completely coprime, that is,

$$x \leq \bigvee S \iff \exists s \in S. x \leq s$$

for each subset S of Q. RHS implies LHS, since $x \leq s \leq \bigvee S$. To show that LHS implies RHS, assume that $x \leq \bigvee S$. Since Q is a frame and $x \leq \bigvee S$, we have

$x = x \wedge \bigvee S = \bigvee \{x \wedge s \mid s \in S\}$. Since x is an atom, we have $\bigvee \{x \wedge s \mid s \in S\} \neq \bot$. Therefore, there exists $s \in S$ satisfying $x \wedge s \neq \bot$. Since x is an atom and $\bot \neq x \wedge s \leq x$, we have $x \wedge s = x$. Therefore, we have $x \leq s$.

(3$\Longrightarrow$4) Let Q be atom-algebraic and assume that its atoms are completely coprime. Q is completely coprime algebraic, since

$$a = \bigvee \{x \in \mathbf{Atom}(Q) \mid x \leq a\} \leq \bigvee \{x \in \mathbf{CCP}(Q) \mid x \leq a\} \leq a.$$

Let a be completely coprime.

$$a \leq a$$
$$\Longleftrightarrow a \leq \bigvee \{x \in \mathbf{Atom}(Q) \mid x \leq a\}$$
$$\Longleftrightarrow \exists x \in \mathbf{Atom}(Q).a \leq x \leq a$$
$$\Longleftrightarrow a \in \mathbf{Atom}(Q)$$

Therefore, a is an atom.

(4$\Longrightarrow$5) Let Q be completely coprime algebraic and assume that its completely coprime elements are atoms. Assume that there are completely coprime elements a, b satisfying $a \leq b$ and $a \neq b$. Since b is also an atom, $a = \bot$. But since a is also an atom, $a \neq \bot$. It is a contradiction. Therefore, the order of completely coprime elements is discrete.

(5$\Longrightarrow$1) Let Q be completely coprime algebraic and assume that the order of its completely coprime elements is discrete. Let f be a function $f \colon Q \to \wp(\mathbf{CCP}(Q))$ such that $f(a) = \{x \in \mathbf{CCP}(Q) \mid x \leq a\}$. Let g be a function $g \colon \wp(\mathbf{CCP}(Q)) \to Q$ such that $g(S) = \bigvee S$. We have $a = g(f(a))$ for all $a \in Q$, since $g(f(a)) = \bigvee \{x \in \mathbf{CCP}(Q) \mid x \leq a\}$ and Q is completely coprime algebraic. Since the order of $\mathbf{CCP}(Q)$ is discrete, an arbitrary subset of $\mathbf{CCP}(Q)$ is down-closed. Therefore, $Y \subseteq \mathbf{CCP}(Q)$ satisfies

$$f(g(Y))$$
$$= \{x \in \mathbf{CCP}(Q) \mid x \leq \bigvee Y\}$$
$$= \{x \in \mathbf{CCP}(Q) \mid \exists y \in Y.x \leq y\}$$
$$= Y. \hspace{3cm} \square$$

When these conditions are satisfied by a quantale, we call it a *powerset quantale*. Every powerset quantale Q satisfies $\mathbf{CCP}(Q) = \mathbf{Atom}(Q)$.

Example 6. $\mathbf{Rel}(A)$ in Example 1, $\wp(\Sigma^*)$ in Example 2, and $\wp(\mathbf{Z}_2)$ in Example 3 are powerset quantales.

We also give an example of completely coprime algebraic quantale which is not a powerset quantale.

Example 7. $\mathbf{N} \cup \{\omega\}$ in Example 4 is completely coprime algebraic and it is a frame. However, it has no atoms. Therefore, it is not a powerset quantale.

4   Representation Theorem for Powerset Quantales

This section shows that a powerset quantale has a relational representation.

Theorem 2. *Let $(Q, \leq, \bigvee, \cdot, 1)$ be a quantale. If Q is a powerset quantale, then the following function $\eta \colon Q \to \mathbf{Rel}(\mathbf{CCP}(Q))$ is an injective homomorphism of quantales.*

$$\eta(a) = \{(x, y) \mid x \in \mathbf{CCP}(Q), y \in \mathbf{CCP}(Q), x \leq a \cdot y\}$$

Proof. By Theorem 1, Q is completely coprime algebraic and the order of its completely coprime elements is discrete.

(η preserves joins)

$$
\begin{aligned}
&(x, y) \in \eta(\textstyle\bigvee S) \\
\Longleftrightarrow\; & x \leq (\textstyle\bigvee S) \cdot y \\
\Longleftrightarrow\; & x \leq \textstyle\bigvee \{a \cdot y \mid a \in S\} \\
\Longleftrightarrow\; & \exists a \in S. x \leq a \cdot y && \text{(by } x \in \mathbf{CCP}(Q)) \\
\Longleftrightarrow\; & \exists a \in S. (x, y) \in \eta(a) \\
\Longleftrightarrow\; & (x, y) \in \textstyle\bigcup \{\eta(a) \mid a \in S\}
\end{aligned}
$$

(η preserves $\cdot$)

$$
\begin{aligned}
&(x, y) \in \eta(a \cdot a') \\
\Longleftrightarrow\; & x \leq a \cdot a' \cdot y \\
\Longleftrightarrow\; & x \leq a \cdot \textstyle\bigvee \{z \in \mathbf{CCP}(Q) \mid z \leq a' \cdot y\} && \text{(since } Q \text{ is CCPA)} \\
\Longleftrightarrow\; & x \leq \textstyle\bigvee \{a \cdot z \mid z \in \mathbf{CCP}(Q), z \leq a' \cdot y\} \\
\Longleftrightarrow\; & \exists z \in \mathbf{CCP}(Q). x \leq a \cdot z, z \leq a' \cdot y && \text{(by } x \in \mathbf{CCP}(Q)) \\
\Longleftrightarrow\; & \exists z \in \mathbf{CCP}(Q). (x, z) \in \eta(a), (z, y) \in \eta(a')
\end{aligned}
$$

(η preserves 1)

$$
\begin{aligned}
&(x, y) \in \eta(1) \\
\Longleftrightarrow\; & x \leq 1 \cdot y \\
\Longleftrightarrow\; & x \leq y \\
\Longleftrightarrow\; & x = y && \text{(since } \mathbf{CCP}(Q) \text{ is discrete)}
\end{aligned}
$$

(η is injective) Assume $\eta(a) \subseteq \eta(a')$.

$$
\begin{aligned}
& a \\
=\; & \textstyle\bigvee \{x \in \mathbf{CCP}(Q) \mid x \leq a\} && \text{(since } Q \text{ is CCPA)} \\
=\; & \textstyle\bigvee \{x \in \mathbf{CCP}(Q) \mid x \leq a \cdot 1\} \\
=\; & \textstyle\bigvee \{x \in \mathbf{CCP}(Q) \mid x \leq a \cdot \textstyle\bigvee \{y \in \mathbf{CCP}(Q) \mid y \leq 1\}\} && \text{(since } Q \text{ is CCPA)} \\
=\; & \textstyle\bigvee \{x \in \mathbf{CCP}(Q) \mid x \leq \textstyle\bigvee \{a \cdot y \mid y \in \mathbf{CCP}(Q), y \leq 1\}\} \\
=\; & \textstyle\bigvee \{x \in \mathbf{CCP}(Q) \mid \exists y \in \mathbf{CCP}(Q), x \leq a \cdot y, y \leq 1\} && \text{(by } x \in \mathbf{CCP}(Q)) \\
=\; & \textstyle\bigvee \{x \in \mathbf{CCP}(Q) \mid \exists y \in \mathbf{CCP}(Q), (x, y) \in \eta(a), y \leq 1\} \\
\leq\; & \textstyle\bigvee \{x \in \mathbf{CCP}(Q) \mid \exists y \in \mathbf{CCP}(Q), (x, y) \in \eta(a'), y \leq 1\} \\
=\; & \textstyle\bigvee \{x \in \mathbf{CCP}(Q) \mid \exists y \in \mathbf{CCP}(Q), x \leq a' \cdot y, y \leq 1\} \\
=\; & \textstyle\bigvee \{x \in \mathbf{CCP}(Q) \mid x \leq \textstyle\bigvee \{a' \cdot y \mid y \in \mathbf{CCP}(Q), y \leq 1\}\} && \text{(by } x \in \mathbf{CCP}(Q)) \\
=\; & \textstyle\bigvee \{x \in \mathbf{CCP}(Q) \mid x \leq a' \cdot \textstyle\bigvee \{y \in \mathbf{CCP}(Q) \mid y \leq 1\}\} \\
=\; & \textstyle\bigvee \{x \in \mathbf{CCP}(Q) \mid x \leq a' \cdot 1\} && \text{(since } Q \text{ is CCPA)}
\end{aligned}
$$

$$= \bigvee \{x \in \mathbf{CCP}(Q) \mid x \le a'\}$$
$$= a' \qquad \qquad \text{(since } Q \text{ is CCPA)} \qquad \qquad \square$$

Since the above function η is an injective homomorphism, the image of Q by η is isomorphic to Q and it is a sub-quantale of $\mathbf{Rel}(\mathbf{CCP}(Q))$.

We give some examples.

Example 8. $\mathbf{Rel}(A)$ in Example 1 is a powerset quantale. Therefore, by Theorem 2, $\mathbf{Rel}(A)$ has a relational representation. The injective map η from $\mathbf{Rel}(A)$ to $\mathbf{Rel}(\mathbf{CCP}(\mathbf{Rel}(A)))$ is given as follows.

$$\eta(R) = \{(P, Q) \mid P, Q \in \mathbf{CCP}(\mathbf{Rel}(A)), P \le R \cdot Q\}$$

Since $\mathbf{CCP}(\mathbf{Rel}(A))$ is the set of singleton subsets of $A \times A$, η can be also given as follows.

$$\begin{aligned} \eta(R) &= \{(\{(s,t)\}, \{(u,v)\}) \mid s,t,u,v \in A, \{(s,t)\} \subseteq R \cdot \{(u,v)\}\} \\ &= \{(\{(s,t)\}, \{(u,v)\}) \mid s,t,u,v \in A, (s,t) \in R \cdot \{(u,v)\}\} \\ &= \{(\{(s,t)\}, \{(u,t)\}) \mid s,t,u \in A, (s,t) \in R \cdot \{(u,t)\}\} \\ &= \{(\{(s,t)\}, \{(u,t)\}) \mid t \in A, (s,u) \in R\} \end{aligned}$$

Example 9. Similarly, $\wp(\Sigma^*)$ in Example 2 has a relational representation. The map η is injective from $\wp(\Sigma^*)$ to $\mathbf{Rel}(\mathbf{CCP}(\wp(\Sigma^*))) \cong \mathbf{Rel}(\Sigma^*)$. The injective map η from $\wp(\Sigma^*)$ to $\mathbf{Rel}(\mathbf{CCP}(\wp(\Sigma^*)))$ is given as follows.

$$\eta(R) = \{(P, Q) \mid P, Q \in \mathbf{CCP}(\wp(\Sigma^*)), P \le R \cdot Q\}$$

Since $\mathbf{CCP}(\wp(\Sigma^*))$ is the set of singleton subsets of Σ^*, η can be also given as follows.

$$\begin{aligned} \eta(R) &= \{(\{\sigma\}, \{\pi\}) \mid \sigma, \pi \in \Sigma^*, \{\sigma\} \subseteq R \cdot \{\pi\}\} \\ &= \{(\{\sigma\}, \{\pi\}) \mid \sigma, \pi \in \Sigma^*, \sigma \in R \cdot \{\pi\}\} \\ &= \{(\{\tau\pi\}, \{\pi\}) \mid \pi \in \Sigma^*, \tau \in R\} \end{aligned}$$

Example 10. Similarly, $\wp(\mathbf{Z}_2)$ in Example 3 has a relational representation. The injective map η from $\wp(\mathbf{Z}_2)$ to $\mathbf{Rel}(\mathbf{CCP}(\wp(\mathbf{Z}_2)))$ is given as follows.

$$\eta(A) = \{(B, C) \mid B, C \in \mathbf{CCP}(\wp(\mathbf{Z}_2)), B \le A \cdot C\}$$

Since $\mathbf{CCP}(\wp(\mathbf{Z}_2))$ is the set of singleton subsets of $\mathbf{Z}_2$, η can be also given as follows.

$$\begin{aligned} \eta(A) &= \{(\{b\}, \{c\}) \mid b, c \in \mathbf{Z}_2, \{b\} \subseteq A \cdot \{c\}\} \\ &= \{(\{b\}, \{c\}) \mid b, c \in \mathbf{Z}_2, b \in A \cdot \{c\}\} \\ &= \{(\{a+c\}, \{c\}) \mid c \in \mathbf{Z}_2, a \in A\} \end{aligned}$$

Example 11. For a frame $(Q, \le, \bigvee)$, the tuple $(Q, \le, \bigvee, \wedge, \top)$ is a quantale. If it is also atom-algebraic, then it has a relational representation by Theorem 2 and Theorem 1.

5   CCP-Invertible Quantale

Section 4 shows that a powerset quantale has a relational representation. Conversely, if a quantale has a relational representation in the same way as Section 4, is it then a powerset quantale? The answer is 'Yes', if it is CCP-invertible.

Definition 8 (CCP-invertible). *A quantale Q is called* CCP-invertible, *if for all $x, y \in \mathbf{CCP}(Q)$, for all $a \in Q$, it holds that*

$$x \leq a \cdot y \Longleftrightarrow \exists z \in \mathbf{CCP}(Q). \, x \leq z \cdot y \text{ and } z \leq a.$$

Theorem 3. *Let $(Q, \leq, \bigvee, \cdot, 1)$ be a quantale. The following are equivalent.*

1. *The following function $\eta \colon Q \to \mathbf{Rel}(\mathbf{CCP}(Q))$*

$$\eta(a) = \{(x, y) \mid x \in \mathbf{CCP}(Q), y \in \mathbf{CCP}(Q), x \leq a \cdot y\}$$

   is an injective homomorphism of quantales and Q is CCP-invertible.
2. *Q is a powerset quantale.*

Proof. $(1 \Longrightarrow 2)$ Let a be an element of Q. Since η is a homomorphism of quantales and Q is CCP-invertible, we have

$$\begin{aligned}
&\eta(\bigvee\{z \in \mathbf{CCP}(Q) \mid z \leq a\}) \\
&= \bigcup\{\eta(z) \mid z \in \mathbf{CCP}(Q), \, z \leq a\} \\
&= \{(x, y) \mid \exists z \in \mathbf{CCP}(Q). \, (x, y) \in \eta(z), \, z \leq a\} \\
&= \{(x, y) \mid x \in \mathbf{CCP}(Q), \, y \in \mathbf{CCP}(Q), \, \exists z \in \mathbf{CCP}(Q). \, x \leq z \cdot y, \, z \leq a\} \\
&= \{(x, y) \mid x \in \mathbf{CCP}(Q), \, y \in \mathbf{CCP}(Q), \, x \leq a \cdot y\} \\
&= \eta(a).
\end{aligned}$$

Moreover, since η is injective, we have

$$a = \bigvee\{z \in \mathbf{CCP}(Q) \mid z \leq a\}.$$

Therefore, Q is completely coprime algebraic.

Since η preserves 1, for $x, y \in \mathbf{CCP}(Q)$, $x \leq y$ if and only if $x = y$. Therefore, the order of completely coprime elements of Q is discrete.

$(2 \Longrightarrow 1)$ By Theorem 2, η is an injective homomorphism of quantales. Let a be an element of Q and x, y elements of $\mathbf{CCP}(Q)$. Since Q is completely coprime algebraic, we have

$$\begin{aligned}
&x \leq a \cdot y \\
&\Longleftrightarrow x \leq \bigvee\{z \in \mathbf{CCP}(Q) \mid z \leq a\} \cdot y \\
&\Longleftrightarrow x \leq \bigvee\{z \cdot y \mid z \in \mathbf{CCP}(Q), \, z \leq a\} \\
&\Longleftrightarrow \exists z \in \mathbf{CCP}(Q). \, x \leq z \cdot y, \, z \leq a.
\end{aligned}$$

Therefore, Q is CCP-invertible. $\square$

Example 12. $\mathbf{Rel}(A)$ in Example 1 is CCP-invertible, since z in Definition 8 is given by $x \cdot y^\circ$ where y° is the opposite relation of y.

Example 13. $\wp(\Sigma^*)$ in Example 2 is CCP-invertible, since z in Definition 8 is given by $\{\sigma\}$ where $x = \{\sigma\pi\}$ and $y = \{\pi\}$.

Example 14. $\wp(\mathbf{Z}_2)$ in Example 3 is CCP-invertible, since z in Definition 8 is given by $z = \{s+t\} = \{s-t\}$ where $x = \{s\}$ and $y = \{t\}$.

Example 15. $\mathbf{N} \cup \{\omega\}$ in Example 4 is CCP-invertible, since z in Definition 8 is given by $x - y$. However, the order of its completely coprime elements is not discrete. Therefore, by Theorem 3, η is not an injective homomorphism of quantales.

Example 16. The ordered set of Fig 1 forms a quantale where $a \cdot b = a \wedge b$ except for $1 \cdot a = a \cdot 1 = a$. It is not CCP-invertible, since the set of its completely coprime elements is $\{s, t\}$ and $t \leq 1 \cdot t$ but $t \not\leq s \cdot t$.

Fig. 1. A quantale which is not CCP-invertible

Example 17. The ordered set of Fig 2 forms a quantale where $a \cdot b = a \wedge b$ except for $1 \cdot a = a \cdot 1 = a$. This quantale is CCP-invertible. However, this quantale is not completely coprime algebraic, since the set of its completely coprime elements is $\{s, 1\}$, but $t \neq \bigvee\{s\}$. Therefore, by Theorem 3, η is not an injective homomorphism of quantales.

Remark that there exist other relational representations of Example 17, for example, the following $\eta'\colon Q \to \mathbf{Rel}(\{\alpha, \beta, \gamma\})$ [9].

$$
\begin{aligned}
\eta'(\top) &= \{(\alpha, \beta), (\alpha, \alpha), (\beta, \beta), (\gamma, \gamma)\} \\
\eta'(t) &= \{(\alpha, \beta), (\alpha, \alpha), (\beta, \beta)\} \\
\eta'(s) &= \{(\alpha, \beta), (\alpha, \alpha)\} \\
\eta'(1) &= \{(\alpha, \alpha), (\beta, \beta), (\gamma, \gamma)\} \\
\eta'(\bot) &= \emptyset
\end{aligned}
$$

Example 18. $(Q, \leq, \bigvee, \wedge, \top)$ in Example 11 is CCP-invertible, since z in Definition 8 is given by $z = x$.

Fig. 2. A quantale which is CCP-invertible

6 Conclusion

We can summarize the theorems of this paper as follows. Palmigiano and Re [10] show 'relational representation theorem' for 'unital involutive quantales'. On the other hand, this paper gives a relational representation theorem for quantales which are not in general involutive.

Theorem 4. *For a quantale Q, the following are equivalent.*

1. *The following function $\eta\colon Q \to \mathbf{Rel}(\mathbf{CCP}(Q))$*

$$\eta(a) = \{(x, y) \mid x \in \mathbf{CCP}(Q), y \in \mathbf{CCP}(Q), x \leq a \cdot y\}$$

   is an injective homomorphism of quantales and Q is CCP-invertible.
2. *Q is isomorphic to $\wp(X)$ as complete join semilattice for some set X.*
3. *Q is atom-algebraic and it is a frame.*
4. *Q is atom-algebraic and its atoms are completely coprime.*
5. *Q is completely coprime algebraic and its completely coprime elements are atoms.*
6. *Q is completely coprime algebraic and the order of its completely coprime elements is discrete.*

Proof. This theorem is implied by Theorem 1 and Theorem 3.   □

This theorem contains the relational representation theorem for a powerset quantale. Conversely, if a quantale has a relational representation in the same way and it is CCP-invertible, then it is a powerset quantale.

It is future work to extend our representation theorem to a Stone-type duality [7].

As shown in Example 15, η for $\mathbf{N} \cup \{\omega\}$ is not an injective homomorphism. However, we do not know whether there exist other relational representations of $\mathbf{N} \cup \{\omega\}$ than η. It is also future work.

Acknowledgments. The authors thank Norihiro Tsumagari for valuable comments about Example 16. This work was supported in part by Grants-in-Aid for Scientific Research (C) 22500016 and by Grants-in-Aid for Young Scientists (B) 24700017 from Japan Society for the Promotion of Science (JSPS). The authors also thank all anonymous reviewers for many helpful comments and discussions.

References

1. Mulvey, C.J.: Second Topology Conference. Rendiconti del Circolo Matematico di Palermo, Series 2, Supplement number 12, pp. 99–104 (1986)
2. Conway, J.H.: Regular Algebra and Finite Machines. Chapman and Hall, London (1971)
3. Jifeng, H., Hoare, C.A.R.: Weakest prespecification. Information Processing Letters 24 (1987)
4. Vickers, S.: Topology via Logic. Cambridge University Press (1989)
5. Yetter, D.N.: Quantales and (noncommutative) linear logic. Journal of Symbolic Logic 55, 41–64 (1990)
6. Davey, B.A., Priestley, H.A.: Introduction to Lattices and Order, 2nd edn. Cambridge University Press (2002)
7. Johnstone, P.T.: Stone Spaces. Cambridge Univ. Press, Cambridge (1982)
8. Valentini, S.: Representation theorems for quantales. Math. Log. Q. 40, 182–190 (1994)
9. Brown, C., Gurr, D.: A representation theorem for quantales. Journal of Pure and Applied Algebra 85, 27–42 (1993)
10. Palmigiano, A., Re, R.: Relational representation of groupoid quantales. Order, 1–19, doi:10.1007/s11083-011-9227-z
11. Jónsson, B., Tarski, A.: Boolean algebras with operators. Part II. Amer. J. Math. 74, 127–162 (1952)

Point Axioms in Dedekind Categories[*]

Hitoshi Furusawa[1] and Yasuo Kawahara[2]

[1] Department of Mathematics and Computer Science, Kagoshima University,
Kagoshima, Japan
[2] Professor Emeritus, Kyushu University, Fukuoka, Japan

Abstract. A Dedekind category is a convenient algebraic framework to
treat relations. Concepts of points and some axioms such as the point
axiom, the axiom of totality, the axiom of subobject, the axiom of complement, and the relational axiom of choice are introduced in Dedekind
categories to connect functional ideas to set-theoretical intuition. This
paper summarises interrelations of these axioms.

1 Introduction

The notion of (binary) relations has been extensively generalised to fuzzy relations and L-relations (or L-fuzzy relations) by Zadeh [11] and Goguen [2],
respectively. Dedekind categories [6] and allegories [1] work as convenient algebraic frameworks to treat relations. The category Rel of sets and (binary)
relations, and the category $FRel$ of sets and fuzzy relations are typical examples of Dedekind categories. The concept of points is explicitly or unintentionally
used in mathematics and computer science. In the relational frameworks Schmidt
and Ströhlein [7] and Kawahara and Furusawa [4] studied concepts of points to
connect functional ideas to set-theoretical intuition.

The paper aims to summarise interrelations of point axioms and some related
axioms in Dedekind categories. The point axioms would be set up based on
the existence of a unit [1] which is a substitute for a singleton set. Of course,
a singleton set serves as a unit in both Rel and $FRel$. Let X be an object
in a Dedekind category $\mathcal{D}$ with a unit I. An I-*point* of X is defined to be a
univalent and total relation $x : I \rightarrow X$. In what follows, the word *relation*
is a synonym for 'morphism' or 'arrow' of a Dedekind category. A univalent
and total relation will be called a *total function* (tfn, for short). Although I-
points in Rel and $FRel$ correspond to ordinary points of sets, they behave a
little bit complicated in general. In the Dedekind category $Rel(L)$ of sets and
L-relations there exist I-points which are not crisp, when L contains an element
with a complement besides the least one and the greatest one. Intuitively, this
means that I-points in $Rel(L)$ may correspond to not only ordinary points but
also "fuzzy" points of sets. Winter [9] pointed out that there is no formula
in the language of Dedekind categories which can determine the crispness of L-
relations, and proposed a theory of Goguen categories possessing a new operation
to characterise the crispness.

---

[*] This work was supported in part by Grants-in-Aid for Scientific Research (C)
22500016 from Japan Society for the Promotion of Science (JSPS).

W. Kahl and T.G. Griffin (Eds.): RAMiCS 2012, LNCS 7560, pp. 219–234, 2012.

It is helpful for better understanding the concept of I-points to introduce the product category of Dedekind categories. For a Dedekind category $\mathcal{D}$ with a unit I, the *product category* $\mathcal{D}^2$ is a category whose objects and relations are pairs of objects in $\mathcal{D}$ and pairs of relations in $\mathcal{D}$, respectively, and relational operations are defined component-wise. $\mathcal{D}^2$ is obviously a Dedekind category with a unit $\langle I, I \rangle$. Also it is trivial that an object $\langle X, Y \rangle$ in $\mathcal{D}^2$ has no $\langle I, I \rangle$-point if and only if at least one of X and Y has no I-point.

Consider the following additional axioms for Dedekind categories (which are a part of axioms discussed later in the paper):

(Tot)   All nonzero universal relations $\nabla_{IX} : I \rightharpoonup X$ are total.
(Tot$_*$)   All nonzero relations $\rho : I \rightharpoonup X$ are total.
(PA)   All universal relations ∇_{IX} are the supremum of all I-points of X.
(PA$_*$)   All relations $\rho : I \rightharpoonup X$ are the supremum of I-points in ρ.
(Sub)   All relations $\rho : I \rightharpoonup X$ are supported by a subobject.
(Ba)   All relations $\alpha : X \rightharpoonup Y$ have a complement.
(AC)   All total relations $\alpha : X \rightharpoonup Y$ contain at least one tfn.

The above axioms (Tot), (PA), (Sub), (Ba) and (AC) will be called the axiom of totality, the point axiom, the axiom of subobject, the axiom of complement, and the (relational) axiom of choice, respectively. The following table shows which of these axioms are satisfied in the four Dedekind categories Rel, $FRel$, Rel^2 and $FRel^2$.

	(Unit)	(PA)	(PA$_*$)	(Tot)	(Tot$_*$)	(Sub)	(Ba)	(AC)
Rel	I	✓	✓	✓	✓	✓	✓	✓
$FRel$	I	✓		✓				
Rel^2	$\langle I, I \rangle$					✓	✓	✓
$FRel^2$	$\langle I, I \rangle$							

For example, the universal relation $\nabla_{\langle I,I \rangle \langle I,\emptyset \rangle} : \langle I, I \rangle \rightharpoonup \langle I, \emptyset \rangle$ in Rel^2 is not total but nonzero, where I is a singleton set. Thus the product construction for Dedekind categories confounds the condition for a relation to be nonzero, and breaks the point axiom (PA). Hence it may be meaningful to study the logical interrelations between the these axioms. The gist of the paper are the following results:

Let $\mathcal{D}$ be a Dedekind category with a unit I.

- $\bullet_1$ (Tot$_*$) holds in $\mathcal{D}$ iff all I-points in $\mathcal{D}$ are atomic iff $\mathcal{D}(I, I) = \{0_{II}, \mathrm{id}_I\}$ iff the (extended) Tarski rule holds in $\mathcal{D}$.
- $\bullet_2$ If $\mathcal{D}$ satisfies (PA) $\wedge$ (Tot$_*$), then each relation of $\mathcal{D}$ has a complement.

We should mention that apart from the point axiom, Winter [10] proved an interesting result that if a distributive allegory $\mathcal{A}$ with relational sums, relational products and total splittings satisfies the axiom (AC), then all relations of $\mathcal{A}$ have a complement. However, in the allegory satisfying the sufficient condition of Winter's theorem the axiom of totality (Tot$_*$) is equivalent to the point axiom (PA$_*$).

The paper is organised as follows: In section 2 we recall the definition of Dedekind categories and relational notations. Also, the basic properties of units are given. In section 3 the interrelation between several axioms related to point axioms are discussed. In section 4 the statement $\bullet_1$ is proved. Also the equivalence between the axiom of totality (Tot_*) and the point axiom (PA_*) is shown under the sufficient condition of Winter's theorem. Section 5 provides two kinds of proofs of the statement $\bullet_2$. One of these two is a consequence from the representation theorem. In section 6 we remark some fundamental facts on L-relations.

2   Dedekind Categories

In this section we recall the definition of Dedekind categories [6]. Dedekind categories are equivalent to locally complete division allegories introduced in [1].

A morphism α from an object X into an object Y in a Dedekind category (which will be defined below) will be denoted by a half arrow $\alpha : X \rightharpoonup Y$, and the composite of a morphism $\alpha : X \rightharpoonup Y$ followed by a morphism $\beta : Y \rightharpoonup Z$ will be written as $\alpha\beta : X \rightharpoonup Z$. Also we will denote the identity morphism on X as id_X.

Definition 1. A Dedekind category $\mathcal{D}$ is a category satisfying the following:
DC1. [Complete Heyting Algebra] For all pairs of objects X and Y the hom-set $\mathcal{D}(X, Y)$ consisting of all morphisms of X into Y is a complete Heyting algebra (namely, a complete distributive lattice) with the least morphism 0_{XY} and the greatest morphism ∇_{XY}. Its algebraic structure will be denoted by

$$(\mathcal{D}(X,Y), \sqsubseteq, \sqcap, \sqcup, \Rightarrow, 0_{XY}, \nabla_{XY}).$$

DC2. [Converse] There is given a converse operation $^\sharp : \mathcal{D}(X,Y) \to \mathcal{D}(Y,X)$. That is, for all morphisms $\alpha, \alpha' : X \rightharpoonup Y$ and $\beta : Y \rightharpoonup Z$ the converse laws hold:
(a) $(\alpha\beta)^\sharp = \beta^\sharp\alpha^\sharp$, (b) $(\alpha^\sharp)^\sharp = \alpha$, (c) If $\alpha \sqsubseteq \alpha'$, then $\alpha^\sharp \sqsubseteq \alpha'^\sharp$.
DC3. [Dedekind Formula] For all morphisms $\alpha : X \rightharpoonup Y$, $\beta : Y \rightharpoonup Z$ and $\gamma : X \rightharpoonup Z$ the Dedekind formula $\alpha\beta \sqcap \gamma \sqsubseteq \alpha(\beta \sqcap \alpha^\sharp\gamma)$ holds.
DC4. [Residual Composition] For all morphisms $\alpha : X \rightharpoonup Y$ and $\beta : Y \rightharpoonup Z$ the residual composite $\alpha \rhd \beta : X \rightharpoonup Z$ is a morphism such that $\delta \sqsubseteq \alpha \rhd \beta$ if and only if $\alpha^\sharp\delta \sqsubseteq \beta$ for all morphisms $\delta : X \rightharpoonup Z$. $\qquad\qquad\square$

In what follows, the word *relation* is a synonym for 'arrow' of a Dedekind category. A relation $\alpha : X \rightharpoonup Y$ is called *univalent* if $\alpha^\sharp\alpha \sqsubseteq \text{id}_Y$, and it is called *total* if $\text{id}_X \sqsubseteq \alpha\alpha^\sharp$. A univalent and total relation is called a *total function* (tfn, for short) and will be written as $f : X \to Y$.

Letting $\alpha : X \rightharpoonup Y$, $\beta : Y \rightharpoonup Z$, $\gamma : Z \rightharpoonup W$ and $\delta : X \rightharpoonup Z$ be relations in a Dedekind category $\mathcal{D}$, Table 1 formally summarises the axioms of Dedekind categories given above. Also, Table 2 summarises fundamental properties of Dedekind categories. For the sake of simplicity, in these tables, some universal quantifiers such as '$\forall X.$' and '$\forall \alpha : X \rightharpoonup Y.$' will be often omitted.

Table 1. Axioms for Dedekind categories

$$(\cdot) \quad (\alpha\beta)\gamma = \alpha(\beta\gamma)$$
$$(\text{id}) \quad (\text{id}_X\alpha = \alpha) \wedge (\alpha\,\text{id}_Y = \alpha)$$
$$(\sqsubseteq_0) \quad \alpha \sqsubseteq \alpha$$
$$(\sqsubseteq_1) \quad (\alpha \sqsubseteq \alpha') \wedge (\alpha' \sqsubseteq \alpha'') \rightarrow (\alpha \sqsubseteq \alpha'')$$
$$(\sqsubseteq_2) \quad (\alpha \sqsubseteq \alpha') \wedge (\alpha' \sqsubseteq \alpha) \rightarrow (\alpha = \alpha')$$
$$(\emptyset) \quad 0_{XY} \sqsubseteq \alpha$$
$$(\nabla) \quad \alpha \sqsubseteq \nabla_{XY}$$
$$(\sqcap) \quad (\alpha \sqsubseteq \alpha' \sqcap \alpha'') \leftrightarrow (\alpha \sqsubseteq \alpha') \wedge (\alpha \sqsubseteq \alpha'')$$
$$(\sqcup) \quad (\alpha' \sqcup \alpha'' \sqsubseteq \alpha) \leftrightarrow (\alpha' \sqsubseteq \alpha) \wedge (\alpha'' \sqsubseteq \alpha)$$
$$(\sqcap_*) \quad (\alpha \sqsubseteq \sqcap_k\alpha_k) \leftrightarrow \forall k.\ (\alpha \sqsubseteq \alpha_k)$$
$$(\sqcup_*) \quad (\sqcup_k\alpha_k \sqsubseteq \alpha) \leftrightarrow \forall k.\ (\alpha_k \sqsubseteq \alpha)$$
$$(\Rightarrow) \quad (\alpha \sqsubseteq \alpha' {\Rightarrow} \alpha'') \leftrightarrow (\alpha \sqcap \alpha' \sqsubseteq \alpha'')$$
$$(\sharp_0) \quad (\alpha^\sharp)^\sharp = \alpha$$
$$(\sharp_1) \quad (\alpha\beta)^\sharp = \beta^\sharp\alpha^\sharp$$
$$(\sharp_2) \quad (\alpha \sqsubseteq \alpha') \rightarrow (\alpha^\sharp \sqsubseteq \alpha'^\sharp)$$
$$(\text{DF}) \quad \alpha\beta \sqcap \delta \sqsubseteq \alpha(\beta \sqcap \alpha^\sharp\delta)$$
$$(\rhd) \quad (\delta \sqsubseteq \alpha \rhd \beta) \leftrightarrow (\alpha^\sharp\delta \sqsubseteq \beta)$$

Table 2. Basic properties of Dedekind categories

$$(\cdot_\emptyset) \quad (\alpha 0_{YZ} = 0_{XZ}) \wedge (0_{VX}\alpha = 0_{VY})$$
$$(\cdot_\sqsubseteq) \quad (\alpha \sqsubseteq \alpha') \wedge (\beta \sqsubseteq \beta') \rightarrow (\alpha\beta \sqsubseteq \alpha'\beta')$$
$$(\rhd_\sqsubseteq) \quad (\alpha \sqsubseteq \alpha') \wedge (\beta \sqsubseteq \beta') \rightarrow (\alpha' \rhd \beta \sqsubseteq \alpha \rhd \beta')$$
$$(\sqcap_\sqcup) \quad \alpha \sqcap (\sqcup_k\alpha_k) = \sqcup_k(\alpha \sqcap \alpha_k)$$
$$(\cdot_\sqcap) \quad \alpha(\sqcap_j\beta_j)\gamma \sqsubseteq \sqcap_j\alpha\beta_j\gamma$$
$$(\cdot_\sqcup) \quad \alpha(\sqcup_j\beta_j)\gamma = \sqcup_j\alpha\beta_j\gamma$$
$$(\text{id}^\sharp) \quad \text{id}_X^\sharp = \text{id}_X$$
$$(\emptyset^\sharp) \quad 0_{XY}^\sharp = 0_{YX}$$
$$(\nabla^\sharp) \quad \nabla_{XY}^\sharp = \nabla_{YX}$$
$$(\sqcap^\sharp) \quad (\sqcap_k\alpha_k)^\sharp = \sqcap_k\alpha_k^\sharp$$
$$(\sqcup^\sharp) \quad (\sqcup_k\alpha_k)^\sharp = \sqcup_k\alpha_k^\sharp$$
$$(\rhd_\sqcup) \quad (\sqcup_k\alpha_k) \rhd \beta = \sqcap_k(\alpha_k \rhd \beta)$$
$$(\text{DF}_0) \quad \alpha \sqsubseteq \alpha\alpha^\sharp\alpha$$
$$(\text{DF}_*) \quad \alpha\beta \sqcap \delta \sqsubseteq (\alpha \sqcap \delta\beta^\sharp)(\beta \sqcap \alpha^\sharp\delta)$$
$$(\text{tfn}_0) \quad \forall f : X \to Y\, \forall g : W \to Z.\ [f(\beta \sqcap \beta')g^\sharp = f\beta g^\sharp \sqcap f\beta'g^\sharp]$$
$$(\text{tfn}_1) \quad \forall f : X \to Y\, \forall g : W \to Z.\ [f(\beta \Rightarrow \beta')g^\sharp = f\beta g^\sharp \Rightarrow f\beta'g^\sharp]$$
$$(\text{tfn}_2) \quad \forall f, g : X \to Y.\ [(f \sqsubseteq g) \rightarrow (f = g)]$$
$$(\text{id}_0) \quad (u \sqsubseteq \text{id}_X) \rightarrow (u^\sharp = u)$$
$$(\text{id}_1) \quad (u \sqsubseteq \text{id}_X) \wedge (v \sqsubseteq \text{id}_X) \rightarrow (uv = u \sqcap v)$$
$$(\text{id}_2) \quad (u \sqsubseteq \text{id}_X) \wedge (v \sqsubseteq \text{id}_X) \rightarrow [(u \rhd v) \sqcap \text{id}_X = (u \Rightarrow v) \sqcap \text{id}_X]$$

The object I in a Dedekind category $\mathcal{D}$, which exists by the axiom of unit

$$(\text{Unit}) \quad \exists I.\ (0_{II} \neq \text{id}_I = \nabla_{II}) \wedge \forall X.\ (\nabla_{XI}\nabla_{IX} = \nabla_{XX}),$$

is called a (strict) *unit*. For units I and I', $\nabla_{II'}\nabla_{II'}^{\sharp} = \mathrm{id}_I$ and $\nabla_{II'}^{\sharp}\nabla_{II'} = \mathrm{id}_{I'}$ hold. So, if it exists, the unit is unique up to isomorphism. The unit I plays a rôle as a substitute for a singleton set. In the rest of the paper, we assume that Dedekind categories satisfy (Unit).

An I-point x of X is a tfn $x : I \to X$ and denoted by $x \in X$. In formulae

$$(x \mathbin{\dot{\in}} X) \leftrightarrow (x : I \to X)$$
$$\leftrightarrow (x : I \to X) \wedge (x^{\sharp}x \sqsubseteq \mathrm{id}_X) \wedge (\mathrm{id}_I = xx^{\sharp}) .$$

Equivalent conditions for all I-points to be atomic will be provided in Proposition 5. It is easy to see that $xx^{\sharp} = x\nabla_{XI} = \mathrm{id}_I$.

Proposition 1. (a) $\forall X, Y. (\nabla_{XY} = \nabla_{XI}\nabla_{IY})$,
(b) $\forall X. [(\nabla_{IX} = 0_{IX}) \leftrightarrow (\nabla_{XX} = 0_{XX}) \leftrightarrow (\mathrm{id}_X = 0_{XX})]$,
(c) $\forall x, y \mathbin{\dot{\in}} X. [(x = y) \leftrightarrow (xy^{\sharp} = \mathrm{id}_I)]$.

Proof. (a) $\nabla_{XI}\nabla_{IY} \sqsubseteq \nabla_{XY}$ is trivial by (∇). Conversely,

$$\begin{aligned}
\nabla_{XY} &\sqsubseteq \nabla_{XX}\nabla_{XY} && \{\ \mathrm{id}_X \sqsubseteq \nabla_{XX},\ (\nabla)\ \} \\
&= \nabla_{XI}\nabla_{IX}\nabla_{XY} && \{\ (\text{Unit})\ \nabla_{XX} = \nabla_{XI}\nabla_{IX}\ \} \\
&\sqsubseteq \nabla_{XI}\nabla_{IY}. && \{\ \nabla_{IX}\nabla_{XY} \sqsubseteq \nabla_{IY},\ (\nabla)\ \}
\end{aligned}$$

(b) $(\nabla_{IX} = 0_{IX}) \to (\nabla_{XX} = 0_{XX})$:

$$\begin{aligned}
\nabla_{XX} &= \nabla_{XI}\nabla_{IX} && \{\ (\text{Unit})\ \nabla_{XX} = \nabla_{XI}\nabla_{IX}\ \} \\
&= \nabla_{XI}0_{IX} && \{\ \nabla_{IX} = 0_{IX}\ \} \\
&= 0_{XX}. && \{\ (\cdot_{\emptyset})\ \}
\end{aligned}$$

$(\nabla_{XX} = 0_{XX}) \to (\mathrm{id}_X = 0_{XX})$: $0_{XX} \sqsubseteq \mathrm{id}_X$ is trivial by ($\emptyset$). Also, $\mathrm{id}_X \sqsubseteq \nabla_{XX}$ by (∇). Since $\nabla_{XX} = 0_{XX}$, we have $0_{XX} \sqsubseteq \mathrm{id}_{XX} \sqsubseteq 0_{XX}$. Hence $\mathrm{id}_X = 0_{XX}$.
$(\mathrm{id}_X = 0_{XX}) \to (\nabla_{IX} = 0_{IX})$:

$$\begin{aligned}
\nabla_{IX} &= \nabla_{IX}\mathrm{id}_X && \{\ (\text{id})\ \} \\
&= \nabla_{IX}0_{XX} && \{\ \mathrm{id}_X = 0_{XX}\ \} \\
&= 0_{IX}. && \{\ (\cdot_{\emptyset})\ \}
\end{aligned}$$

(c) ($\to$) It follows at once from $\nabla_{II} = \mathrm{id}_I \sqsubseteq xx^{\sharp}$.
($\leftarrow$) Assume $xy^{\sharp} = \mathrm{id}_I$. Then

$$\begin{aligned}
y &= xy^{\sharp}y && \{\ xy^{\sharp} = \mathrm{id}_I\ \} \\
&\sqsubseteq x, && \{\ y^{\sharp}y \sqsubseteq \mathrm{id}_X \leftarrow y \mathbin{\dot{\in}} X\ \}
\end{aligned}$$

which shows $x = y$, since x and y are tfns (ttn$_2$). $\qquad\square$

For a relation $\rho : I \to X$ and an I-point $x : I \to X$ the notation $x \mathbin{\dot{\in}} \rho$ denotes $x \sqsubseteq \rho$. In formulae

$$(x \mathbin{\dot{\in}} \rho) \leftrightarrow (x \mathbin{\dot{\in}} X) \wedge (x \sqsubseteq \rho) .$$

A relation $\alpha : X \to Y$ is called *nonzero* if $\alpha \neq 0_{XY}$. An I-point $x : I \to X$ is nonzero by $xx^{\sharp} = \mathrm{id}_I \neq 0_{II}$. We call a relation $\rho : I \to X$ *nonempty* if ρ contains an I-point. Note that all nonempty relations $\rho : I \to X$ are total.

Proposition 2. *Let $\rho : I \rightharpoonup X$ be a relation.*

(a) $(\mathrm{id}_I \sqsubseteq \rho\rho^\sharp) \rightarrow (\rho\rho^\sharp = \rho\nabla_{XI} = \mathrm{id}_I)$,
(b) $(\exists x \,\dot{\in}\, \rho) \rightarrow (\mathrm{id}_I \sqsubseteq \rho\rho^\sharp) \rightarrow (\rho \neq 0_{IX})$,
(c) $(\exists x \,\dot{\in}\, X) \rightarrow (\mathrm{id}_I \sqsubseteq \nabla_{IX}\nabla_{XI}) \rightarrow (\nabla_{IX} \neq 0_{IX})$.

Proof. (a) For arbitrary $\rho : I \rightharpoonup X$,

$$\begin{aligned}
\rho\rho^\sharp &\sqsubseteq \rho\nabla_{XI} \quad \{ \ \rho^\sharp \sqsubseteq \nabla_{XI}, \ (\nabla) \ \} \\
&\sqsubseteq \nabla_{II} \quad \{ \ (\nabla) \ \} \\
&= \mathrm{id}_I. \quad \{ \ (\text{Unit}) \ \mathrm{id}_I = \nabla_{II} \ \}
\end{aligned}$$

Together with the assumption $\mathrm{id}_I \sqsubseteq \rho\rho^\sharp$ this implies the claim.
(b) $(\exists x \,\dot{\in}\, \rho) \rightarrow (\mathrm{id}_I \sqsubseteq \rho\rho^\sharp)$:

$$\begin{aligned}
\mathrm{id}_I &= xx^\sharp \quad \{ \ x : \text{total}, \ (\text{a}) \ \} \\
&\sqsubseteq \rho\rho^\sharp. \quad \{ \ x \sqsubseteq \rho \leftarrow x \,\dot{\in}\, \rho \ \}
\end{aligned}$$

$(\mathrm{id}_I \sqsubseteq \rho\rho^\sharp) \rightarrow (\rho \neq 0_{IX})$: If $\rho : I \rightharpoonup X$ is total, then $\mathrm{id}_I = \rho\rho^\sharp$ by (a), and so $\rho \neq 0_{IX}$ because $\mathrm{id}_I \neq 0_{II}$.
(c) It is a particular case of (b). $\square$

It is easy to see that the existence of units in Dedekind category follows from the existence of objects E and X such that $\mathrm{id}_E = 0_{EE}$ and $\mathrm{id}_X \neq 0_{XX}$ and the existence of power objects [3].

3 Point Axioms

Now we focus on the so-called point axiom

$$(\text{PA}) \quad \forall X. \ (\nabla_{IX} = \textstyle\bigsqcup_{x \,\dot{\in}\, X} x)$$

in Dedekind categories. The point axiom (PA) states that each universal relation ∇_{IX} is the supremum of all I-points of X. The next proposition shows some variants of the point axiom (PA), which make us realize that (PA) connect functional ideas in Dedekind categories to set-theoretical intuition.

Proposition 3. *Let X be an object of a Dedekind category $\mathcal{D}$. Then the following four statements are equivalent.*

(a) $\nabla_{IX} = \bigsqcup_{x \,\dot{\in}\, X} x$,
(b) $\mathrm{id}_X = \bigsqcup_{x \,\dot{\in}\, X} x^\sharp x$,
(c) $\forall Y \,\forall \alpha, \alpha' : X \rightharpoonup Y. \ [\forall x \,\dot{\in}\, X. \ (x\alpha = x\alpha') \rightarrow (\alpha = \alpha')]$,
(d) $\forall \mu, \mu' : X \rightharpoonup I. \ [\forall x \,\dot{\in}\, X. \ (x\mu = x\mu') \rightarrow (\mu = \mu')]$.   (extensionality)

Proof. (a)→(b) Note that $\forall x \doteq X.\ \mathrm{id}_X \sqcap \nabla_{XI} x = x^{\sharp} x$. If $f : X \to Y$, then

$$
\begin{aligned}
f^{\sharp} f &\sqsubseteq \mathrm{id}_Y \sqcap \nabla_{YX} f && \{\ f^{\sharp} f \sqsubseteq \mathrm{id}_Y,\ f^{\sharp} \sqsubseteq \nabla_{YX}\ \} \\
&\sqsubseteq (\mathrm{id}_Y f^{\sharp} \sqcap \nabla_{YX}) f && \{\ (\mathrm{DF}_*),\ \nabla^{\sharp}_{YX} \mathrm{id}_Y \sqcap f \sqsubseteq f\ \} \\
&= (f^{\sharp} \sqcap \nabla_{YX}) f && \{\ (\mathrm{id})\ \} \\
&\sqsubseteq f^{\sharp} f. && \{\ (\cdot\sqsubseteq)\ \}
\end{aligned}
$$

Hence

$$
\begin{aligned}
\mathrm{id}_X &= \mathrm{id}_X \sqcap \nabla_{XX} && \{\ \mathrm{id}_X \sqsubseteq \nabla_{XX},\ (\nabla)\ \} \\
&= \mathrm{id}_X \sqcap \nabla_{XI} \nabla_{IX} && \{\ (\mathrm{Unit})\ \nabla_{XX} = \nabla_{XI} \nabla_{IX}\ \} \\
&= \mathrm{id}_X \sqcap \nabla_{XI} (\sqcup_{x \doteq X} x) && \{\ (\mathrm{a})\ \} \\
&= \mathrm{id}_X \sqcap (\sqcup_{x \doteq X} \nabla_{XI} x) && \{\ (\cdot\sqcup)\ \} \\
&= \sqcup_{x \doteq X} (\mathrm{id}_X \sqcap \nabla_{XI} x) && \{\ (\sqcap\sqcup)\ \} \\
&= \sqcup_{x \doteq X} x^{\sharp} x. && \{\ \mathrm{id}_X \sqcap \nabla_{XI} x = x^{\sharp} x\ \}
\end{aligned}
$$

(b)→(c) Assume $\forall x \doteq X.\ (x\alpha = x\alpha')$. Then

$$
\begin{aligned}
\alpha &= (\sqcup_{x \doteq X} x^{\sharp} x) \alpha && \{\ (\mathrm{b})\ \} \\
&= \sqcup_{x \doteq X} x^{\sharp} x \alpha && \{\ (\cdot\sqcup)\ \} \\
&= \sqcup_{x \doteq X} x^{\sharp} x \alpha' && \{\ x\alpha = x\alpha'\ \} \\
&= (\sqcup_{x \doteq X} x^{\sharp} x) \alpha' && \{\ (\cdot\sqcup)\ \} \\
&= \alpha'. && \{\ (\mathrm{b})\ \}
\end{aligned}
$$

(c)→(d) Trivial.

(d)→(a) First note $\forall y \doteq X.\ y(\sqcup_{x \doteq X} x)^{\sharp} = y \nabla_{XI}$:

$$
\begin{aligned}
\mathrm{id}_I &\sqsubseteq y y^{\sharp} && \{\ y : \mathrm{tfn}\ \} \\
&\sqsubseteq y(\sqcup_{x \doteq X} x)^{\sharp} && \{\ y \sqsubseteq \sqcup_{x \doteq X} x\ \} \\
&\sqsubseteq y \nabla_{XI}. && \{\ (\sqcup_{x \doteq X} x)^{\sharp} \sqsubseteq \nabla_{XI},\ (\nabla)\ \}
\end{aligned}
$$

As $\mathrm{id}_I = \nabla_{II}$, this implies $y(\sqcup_{x \doteq X} x)^{\sharp} = y \nabla_{XI}\ (= \mathrm{id}_I = \nabla_{II})$. Hence by (d) we have $(\sqcup_{x \doteq X} x)^{\sharp} = \nabla_{XI}$ and so $\sqcup_{x \doteq X} x = \nabla_{IX}$.   □

Next, we introduce a few axioms related to the point axiom (PA) and study the interrelationship among them. Another known point axiom in Dedekind categories is as follows:

$$(\mathrm{PA}_*) \quad \forall \rho : I \to X.\ (\rho = \sqcup_{x \doteq \rho} x)$$

Clearly, (PA_*) requires that each relation $\rho : I \to X$ is the supremum of I-points in ρ.

The point axioms (PA) and (PA_*) obviously imply the nonemptiness (NE) and (NE_*), respectively:

$$
\begin{aligned}
&(\mathrm{NE})\quad \forall X.\ [(\nabla_{IX} \neq 0_{IX}) \to (\exists x \doteq X)] \\
&(\mathrm{NE}_*)\ \forall \rho : I \to X.\ [(\rho \neq 0_{IX}) \to (\exists x \doteq \rho)]
\end{aligned}
$$

(NE) and (NE_*) respectively states that all nonzero universal relations ∇_{IX} are nonempty and that all nonzero relations $\rho : I \to X$ are nonempty.

Proposition 4. (a) (PA) $\to$ (NE),
(b) (PA$_*$) $\to$ (NE$_*$).

Proof. Note that all I-points $x \dot{\in} X$ are nonzero, that is, $\neg(x \sqsubseteq 0_{IX})$.
(a) follows from

$$
\begin{aligned}
&\nabla_{IX} = \sqcup_{x \dot{\in} X} x && \{ \text{ (PA) } \} \\
&\leftrightarrow \forall \mu : I \to X. \, [\forall x \dot{\in} X. \, (x \sqsubseteq \mu) \to (\mu = \nabla_{IX})] && \{ \text{ ($\sqcup_*$) } \} \\
&\leftrightarrow \forall \mu : I \to X. \, [(\mu \neq \nabla_{IX}) \to \exists x \dot{\in} X. \, \neg(x \sqsubseteq \mu)] && \{ \text{ contraposition } \} \\
&\to (\nabla_{IX} \neq 0_{IX}) \to \exists x \dot{\in} X. \, \neg(x \sqsubseteq 0_{IX}) && \{ \text{ case of } \mu = 0_{IX} \} \\
&\leftrightarrow (\nabla_{IX} \neq 0_{IX}) \to (\exists x \dot{\in} X). && \{ \, \neg(x \sqsubseteq 0_{IX}) \, \}
\end{aligned}
$$

Noting that

$$
\begin{aligned}
&\rho = \sqcup_{x \dot{\in} \rho} x && \{ \text{ (PA$_*$) } \} \\
&\leftrightarrow \forall \mu : I \to X. \, [\forall x \dot{\in} \rho. \, (x \sqsubseteq \mu) \to (\rho \sqsubseteq \mu)] && \{ \text{ ($\sqcup_*$) } \} \,,
\end{aligned}
$$

(b) is proved similarly to (a). □

Besides the nonemptiness (NE) and (NE$_*$), there are some further axioms related to the point axioms (PA) and (PA$_*$). We now list these axioms:

> (Tot) $\forall X. \, [(\nabla_{IX} \neq 0_{IX}) \to (\mathrm{id}_I = \nabla_{IX} \nabla_{XI})]$
> (Tot$_*$) $\forall \rho : I \to X. \, [(\rho \neq 0_{IX}) \to (\mathrm{id}_I = \rho \rho^\sharp)]$
> (Sub) $\forall \rho : I \to X \, \exists j : S \to X. \, (\rho = \nabla_{IS} j) \wedge (j j^\sharp = \mathrm{id}_S)$
> (Ba) $\forall \alpha : X \to Y. \, (\alpha \sqcup \neg \alpha = \nabla_{XY})$
> (AC) $\forall \alpha : X \to Y. \, [(\mathrm{id}_X \sqsubseteq \alpha \alpha^\sharp) \to \exists f : X \to Y. \, (f \sqsubseteq \alpha)]$

Where $\neg \alpha$ denotes the pseudo complement of α, namely $\neg \alpha = \alpha \Rightarrow 0_{XY}$. A Dedekind category satisfying the axiom (Ba) is a Schröder category. For an object X, S is called *subobject* of X if there exists an injection $j : S \to X$ that is a tfn satisfying $j j^\sharp = \mathrm{id}_S$. Rather informal statements of these axioms are as follows.

> (Tot)　All nonzero universal relations ∇_{IX} are total.
> (Tot$_*$) All nonzero relations $\rho : I \to X$ are total.
> (Sub)　All relations $\rho : I \to X$ are supported by a subobject.
> (Ba)　All relations $\alpha : X \to Y$ have a complement.
> (AC)　All total relations $\alpha : X \to Y$ contain at least one tfn.

So, (Tot) and (Tot$_*$) are called the axioms of totality, (Sub) is called the axiom of subobject, (Ba) is called the axiom of complement, and (AC) is called the (relational) axiom of choice.

The following diagram illustrates the implications between the axioms stated above.

$$(PA) \wedge (Tot_*) \qquad\qquad (PA) \quad\overset{Prop.\ 4}{\Rightarrow}\quad (NE) \quad\rightarrow\quad (Tot)$$

$$\updownarrow (d) \qquad\qquad\qquad \uparrow \qquad\qquad\qquad \uparrow \qquad\qquad \uparrow$$

$$(Ba) \wedge (NE_*) \quad\overset{(d)}{\Leftrightarrow}\quad (PA_*) \quad\overset{Prop.\ 4}{\Rightarrow}\quad (NE_*) \quad\rightarrow\quad (Tot_*)$$

$$\uparrow (b) \qquad\qquad\qquad \uparrow (a) \qquad\qquad \uparrow (b) \qquad\qquad \uparrow (c)$$

$$(Ba) \wedge (NE) \wedge (Sub) \overset{(f)}{\Leftrightarrow} (PA) \wedge (Sub) \quad\rightarrow\quad (NE) \wedge (Sub) \rightarrow (Tot) \wedge (Sub)$$

$$(AC) \wedge (Ba) \wedge (Tot_*) \overset{(e)}{\Leftrightarrow} (AC) \wedge (PA_*)$$

The implications (a) – (f) in the diagram above will be proved in appendix A Proposition 7. The other unspecific implications are trivial.

4 Totality

The next proposition indicates that (Tot_*) is equivalent to the so-called Tarski rule [8].

Proposition 5. *The following four statements are equivalent:*

(a) *All nonzero relations* $\rho : I \rightharpoonup X$ *are total, that is,*
$$\forall \rho : I \rightharpoonup X. \ [(\rho \neq 0_{IX}) \rightarrow (id_I = \rho\rho^\sharp)], \qquad\qquad (Tot_*)$$
(b) *All I-points* $x : I \rightharpoonup X$ *are atomic, that is,*
$$\forall x : I \rightharpoonup X. \ [(\rho \sqsubseteq x) \rightarrow (\rho = 0_{IX}) \vee (\rho = x)],$$
(c) $\mathcal{D}(I, I) = \{0_{II}, id_I\}$, *that is,*
$$\forall u : I \rightharpoonup I. \ (u = 0_{II}) \vee (u = id_I),$$
(d) *All nonzero relations* $\alpha : X \rightharpoonup Y$ *satisfy* $\nabla_{XX}\alpha\nabla_{YY} = \nabla_{XY}$, *that is,*
$$\forall \alpha : X \rightharpoonup Y. \ [(\alpha \neq 0_{XY}) \rightarrow (\nabla_{XX}\alpha\nabla_{YY} = \nabla_{XY})]. \qquad (Tarski\ rule)$$

Proof. (a)→(b) Let $x : I \rightharpoonup X$ be an I-point and $\rho : I \rightharpoonup X$ a relation with $\rho \sqsubseteq x$. If ρ is nonzero, then it is total by (a) and hence $\rho = x$, because ρ and x are functions.

(b)→(c) First recall that the identity relation id_I is an I-point and so by the assumption (b) it is atomic. Let $\rho : I \rightharpoonup I$ be a relation. Then $0_{II} \sqsubseteq \rho \sqsubseteq \nabla_{II} = id_I$ is trivial. But as id_I is atomic, we have $\rho = 0_{II}$ or $\rho = id_I$.

(c)→(d) Let $\alpha : X \rightharpoonup Y$ be a nonzero relation. Then $\nabla_{IX}\alpha\nabla_{YI}$ is also nonzero by the following inclusion.

$$\alpha \sqsubseteq \nabla_{XX}\alpha\nabla_{YY} \qquad\quad \{ \ id_X \sqsubseteq \nabla_{XX} \ \}$$
$$= \nabla_{XI}\nabla_{IX}\alpha\nabla_{YI}\nabla_{IY}. \ \{ \ (Unit) \ \nabla_{XX} = \nabla_{XI}\nabla_{IX} \ \}$$

Hence we have $\nabla_{IX}\alpha\nabla_{YI} = id_I$ by the assumption (c) and so

$$\nabla_{XX}\alpha\nabla_{YY} = \nabla_{XI}\nabla_{IX}\alpha\nabla_{YI}\nabla_{IY} \ \{ \ (Unit) \ \nabla_{XX} = \nabla_{XI}\nabla_{IX} \ \}$$
$$= \nabla_{XI}\nabla_{IY} \qquad\qquad \{ \ \nabla_{IX}\alpha\nabla_{YI} = id_I \ \}$$
$$= \nabla_{XY}. \qquad\qquad\quad \{ \ Proposition\ 1\ (a) \ \}$$

(d)→(a) Let $\rho : I \rightharpoonup X$ be a nonzero relation. Then $\rho\rho^\sharp$ is also nonzero by the basic inclusion $\rho \sqsubseteq \rho\rho^\sharp\rho$. Hence we have

$$
\begin{aligned}
\rho\rho^\sharp &= \mathrm{id}_I\, \rho\rho^\sharp\, \mathrm{id}_I \quad \{ \text{ (id) } \} \\
&= \nabla_{II}\, \rho\rho^\sharp\, \nabla_{II} \; \{ \text{ (Unit) } \mathrm{id}_I = \nabla_{II} \} \\
&= \nabla_{II} \qquad\quad \{ \text{ Tarski rule (d) } \} \\
&= \mathrm{id}_I, \qquad\quad \{ \text{ (Unit) } \mathrm{id}_I = \nabla_{II} \}
\end{aligned}
$$

which means that ρ is total. $\qquad\square$

The following corollary states that under (Tot$_*$) I-points behave like points in set theory.

Corollary 1. *If* (Tot$_*$) *holds, then*

$$\forall x, y : I \to X.\; [(x \neq y) \leftrightarrow (xy^\sharp = 0_{II})].$$

Proof. We will show the contraposition $(x = y) \leftrightarrow (xy^\sharp \neq 0_{II})$.

$$
\begin{aligned}
x = y &\leftrightarrow xy^\sharp = \mathrm{id}_I \; \{ \text{ Proposition 1 (c) } \} \\
&\leftrightarrow xy^\sharp \neq 0_{II}. \; \{ \text{ (Tot$_*$), Proposition 5 (c) } \} \qquad\square
\end{aligned}
$$

(PA$_*$) → (NE$_*$) has been shown in Proposition 4. Also (NE$_*$) → (Tot$_*$) is trivial. Thus we have (PA$_*$) → (Tot$_*$). Next, we show that (Tot$_*$) ↔ (PA$_*$) holds if a Dedekind category with total splittings, relational sums and products satisfies (AC). It is a consequence of Theorem 2 in [10] and (e) of Proposition 7.

A Dedekind category with total splittings, relational sums and product is a Dedekind category satisfying the following three conditions:

- For all equivalence relations $\theta : X \rightharpoonup X$, i.e. $\mathrm{id}_X \sqsubseteq \theta$, $\theta^\sharp \sqsubseteq \theta$ and $\theta\theta \sqsubseteq \theta$, there exists a tfn $s : X \to Q$ such that $s^\sharp s = \mathrm{id}_Q$ and $ss^\sharp = \theta$.
- For all pairs of objects X and Y there exists an object $X + Y$ together with a pair of tfns $i : X \to X + Y$ and $j : Y \to X + Y$ satisfying

$$i^\sharp i \sqcup j^\sharp j = \mathrm{id}_{X+Y}, \quad ii^\sharp = \mathrm{id}_X, \quad jj^\sharp = \mathrm{id}_Y, \quad \text{and} \quad ij^\sharp = 0_{XY}.$$

- For all pairs of objects X and Y there exists an object $X \times Y$ together with a pair of tfns $p : X \times Y \to X$ and $q : X \times Y \to Y$ satisfying

$$pp^\sharp \sqcap qq^\sharp = \mathrm{id}_{X \times Y}, \quad p^\sharp p = \mathrm{id}_X, \quad q^\sharp q = \mathrm{id}_Y, \quad \text{and} \quad p^\sharp q = \nabla_{XY}.$$

The next theorem is immediate from Theorem 2 in [10].

Theorem 1. *Let $\mathcal{D}$ be a Dedekind category with total splitting, relational sums and products. Then* (AC) → (Ba) *in $\mathcal{D}$.* $\qquad\square$

Proposition 7 (e)

$$(\text{AC}) \wedge (\text{Ba}) \wedge (\text{Tot}_*) \leftrightarrow (\text{AC}) \wedge (\text{PA}_*)$$

and Theorem 1 implies the following property.

Corollary 2. *If a Dedekind category $\mathcal{D}$ with total splittings, relational sums and products satisfies* (AC), (Tot$_*$) ↔ (PA$_*$) *in $\mathcal{D}$.* $\qquad\square$

5 Complements

In this section we will prove (PA) $\wedge$ (Tot$_*$) $\to$ (Ba) which is a part of Proposition 7 (d). First we recall a simple sufficient condition for a complete Heyting algebra to be a Boolean algebra.

Lemma 1. *Let A be the set of all atomic elements of a complete Heyting algebra L. If the supremum of A is equal to the greatest element 1 of L, then all elements of L have a complement.*

Proof. Let $h \in L$. First we see $\forall a \in A.\ a \wedge (h \vee \neg h) \neq 0$, where $\neg h$ denotes the pseudo complement $h \Rightarrow 0$ of h. If $a \wedge (h \vee \neg h) = 0$, then $a \wedge h = 0$ and $a \wedge \neg h = 0$, and so, noting that $a \wedge h = 0$ implies $a \leq h \Rightarrow 0 = \neg h$, we have $a = a \wedge \neg h = 0$, which contradicts $a \neq 0$ (since atomic elements are nonzero). Hence one concludes $a \wedge (h \vee \neg h) \neq 0$. As $a \in A$ is atomic by the assumption, we have $a \wedge (h \vee \neg h) = a$ which is equivalent to $a \leq h \vee \neg h$. Therefore $1 = \vee_{a \in A} a \leq h \vee \neg h$. $\square$

Remark 1. Let L be a Heyting algebra. If $a \wedge b = 0$ and $a \vee b = 1$ for $a, b \in L$, then $b = \neg a$. $\square$

The next lemma proves that so-called pairing relations consisting of I-points are atomic if I-points are atomic.

Lemma 2. *Let $x \mathbin{\dot\in} X$ and $y \mathbin{\dot\in} Y$ be I-points. If at least one of x and y is atomic, then so is $x^\sharp y$.*

Proof. First $x^\sharp y \neq 0_{XY}$ follows from $\mathrm{id}_I \sqsubseteq xx^\sharp yy^\sharp$, since I-points are total. Assume that y is atomic. Let $\alpha : X \to Y$ be a nonzero relation with $\alpha \sqsubseteq x^\sharp y$. Then $x\alpha$ is also nonzero from $\alpha = \alpha \sqcap x^\sharp y \sqsubseteq x^\sharp (x\alpha \sqcap y)$. On the other hand the inclusion $x\alpha \sqsubseteq xx^\sharp y = y$ holds by $xx^\sharp = \mathrm{id}_I$. As y is atomic, one concludes $x\alpha = y$. Hence we have $x^\sharp y = x^\sharp x\alpha \sqsubseteq \alpha$, because x is a tfn. This proves $\alpha = x^\sharp y$ and hence $x^\sharp y$ is atomic. $\square$

Theorem 2. (PA) $\wedge$ (Tot$_*$) $\to$ (Ba).

Proof. Assume (PA) $\wedge$ (Tot$_*$). Then all I-points are atomic by Proposition 5 (b) and all pairing relations $x^\sharp y$ for $x \mathbin{\dot\in} X$ and $y \mathbin{\dot\in} Y$ are atomic by Lemma 2. On the other hand it follows from (PA) that

$$
\begin{aligned}
\bigsqcup_{x \dot\in X} \bigsqcup_{y \dot\in Y} x^\sharp y &= \bigsqcup_{x \dot\in X} x^\sharp (\bigsqcup_{y \dot\in Y} y) \quad && \{\ (\cdot_\sqcup)\ \} \\
&= \bigsqcup_{x \dot\in X} x^\sharp \nabla_{IY} && \{\ (\mathrm{PA})\ \} \\
&= (\bigsqcup_{x \dot\in X} x^\sharp) \nabla_{IY} && \{\ (\cdot_\sqcup)\ \} \\
&= (\bigsqcup_{x \dot\in X} x)^\sharp \nabla_{IY} && \{\ (\sqcup^\sharp)\ \} \\
&= \nabla_{IX}^\sharp \nabla_{IY} && \{\ (\mathrm{PA})\ \} \\
&= \nabla_{XI} \nabla_{IY} && \{\ (\nabla^\sharp)\ \} \\
&= \nabla_{XY}. && \{\ \text{Proposition 1 (a)}\ \}
\end{aligned}
$$

Hence the Heyting algebra $\mathcal{D}(X, Y)$ satisfies the sufficient condition of Lemma 1. Therefore all relations $\alpha : X \to Y$ have a complement. $\square$

Suppose that a Dedekind category $\mathcal{D}$ satisfies (PA) and (Tot$_*$). Let $\chi(X)$ be the set of all I-points of X, i.e., $\chi(X) = \{x \mid x \,\dot{\in}\, X\}$. For all relations $\alpha :$ $X \rightharpoonup Y$ define a $\mathcal{D}(I,I)$-relation $\chi(\alpha) : \chi(X) \rightharpoonup \chi(Y)$, namely a mapping $\chi(\alpha) : \chi(X) \times \chi(Y) \to \mathcal{D}(I,I)$ by

$$\chi(\alpha)(x,y) = x\alpha y^\sharp$$

for all $(x,y) \in \chi(X) \times \chi(Y)$. Then, as the cases of [4,7], χ preserves all operations of Dedekind categories. Moreover, since $\mathcal{D}(I,I) = \{0_{II}, \mathrm{id}_I\}$ by Proposition 5, χ determines a full and faithful functor from $\mathcal{D}$ to Rel.

Theorem 3. *A Dedekind category $\mathcal{D}$ satisfying* (PA) $\wedge$ (Tot$_*$) *is representable, that is, there exists a full and faithful functor $\chi : \mathcal{D} \to Rel$.*   □

Remark 2. The last representability theorem gives another proof for Theorem 2: Let $\alpha : X \rightharpoonup Y$ be a relation in a Dedekind category $\mathcal{D}$ which satisfies (PA) $\wedge$ (Tot$_*$). Take a complement $R : \chi(X) \rightharpoonup \chi(Y)$ of $\chi(\alpha)$. Since χ is full, we can choose a relation $\alpha' : X \rightharpoonup Y$ such that $\chi(\alpha') = R$. Then we have

$$\chi(\alpha \sqcup \alpha') = \chi(\alpha) \sqcup \chi(\alpha') = \chi(\alpha) \sqcup R = \nabla_{\chi(X)\chi(Y)} = \chi(\nabla_{XY}), \text{ and}$$
$$\chi(\alpha \sqcap \alpha') = \chi(\alpha) \sqcap \chi(\alpha') = \chi(\alpha) \sqcap R = 0_{\chi(X)\chi(Y)} = \chi(0_{XY}).$$

Hence α' is a complement of α, since χ is faithful.   □

Remark 3. In a Dedekind category, (PA$_*$) $\to$ (Ba) holds since (PA$_*$) implies (PA) and (Tot$_*$).   □

6   L-Relations

Recall that for a complete Heyting algebra L, an L-relation $\alpha : X \rightharpoonup Y$ from a set X into a set Y is defined to be a mapping $\alpha : X \times Y \to L$ (in the ordinary set theory). The ordering $\sqsubseteq$ between L-relations is defined pair-wise, namely for L-relations $\alpha : X \rightharpoonup Y$ and $\alpha' : X \rightharpoonup Y$

$$\alpha \sqsubseteq \alpha' \;\leftrightarrow\; \forall (x,y) \in X \times Y. \, (\alpha(x,y) \le \alpha'(x,y)).$$

The composite $\alpha\beta : X \rightharpoonup Z$ of $\alpha : X \rightharpoonup Y$ followed by $\beta : Y \rightharpoonup Z$ is defined by

$$\forall (x,z) \in X \times Z. \, \alpha\beta(x,z) = \vee_{y \in Y} (\alpha(x,y) \wedge \beta(y,z)).$$

This composition is called sup-min composition.

In the final section we remark some properties of L-relations related to the axioms mentioned in the paper.

Proposition 6. *Let L be a complete Heyting algebra. In the Dedekind category $Rel(L)$ of sets and L-relations the following holds.*

(a) *If* (Tot$_*$) *holds, then $L \cong \{0,1\}$,*
(b) (Sub) $\leftrightarrow$ (Tot$_*$).

Proof. (a) In $Rel(L)$, the unit I is a singleton set.

$$L \cong Rel(L)(I, I) \{ \text{ Def. of } L\text{-relations } \}$$
$$= \{0_{II}, \text{id}_I\}. \quad \{ (\text{Tot}_*), \text{Proposition 5 } \}$$

(b) Recall that the point axiom (PA) always holds in $Rel(L)$.

$$(\text{Sub}) \rightarrow (\text{PA}) \wedge (\text{Sub}) \{ Rel(L) \}$$
$$\rightarrow (\text{PA}_*) \quad \{ \text{ Proposition 7 (a) } \}$$
$$\rightarrow (\text{Tot}_*).$$

Conversely assume (Tot_*). Then by (a) we have $L \cong \{0, 1\}$, and $Rel(L)$ is isomorphic to the Dedekind category Rel of sets and (binary) relations. Hence (Sub) holds in $Rel(L)$. □

It depends on L if the axiom of choice

$$(\text{AC}) \quad \forall \alpha : X \rightarrow Y. \ [(\text{id}_X \sqsubseteq \alpha\alpha^\sharp) \rightarrow \exists f : X \rightarrow Y. \ (f \sqsubseteq \alpha)]$$

holds or not in the Dedekind category $Rel(L)$.

Remark 4. Suppose the axiom of choice in set theory. If L is a complete Boolean algebra, then the axiom of choice (AC) holds in the Dedekind category $Rel(L)$ of sets and L-relations.

Let $\alpha : X \rightarrow Y$ be a total L-relation, namely a mapping $\alpha : X \times Y \rightarrow L$ satisfying $\text{id}_X \sqsubseteq \alpha\alpha^\sharp$. Construct a set $\Gamma = \{\gamma : X \rightarrow Y \mid (\gamma^\sharp\gamma \sqsubseteq \text{id}_Y) \wedge (\gamma \sqsubseteq \alpha)\}$. Then Γ is an inductive set and has a maximal element γ_0 by Zorn's lemma. This γ_0 is a tfn such that $\gamma_0 \sqsubseteq \alpha$. □

Let $[0, 1]$ be the unit interval. The unit interval forms a complete Heyting algebra together with the ordinary inequality relation $\leq$ between real numbers. Since a fuzzy relation $\alpha : X \rightarrow Y$ is a mapping $\alpha : X \times Y \rightarrow [0, 1]$ (in the ordinary set theory), fuzzy relations are $[0, 1]$-relations.

Remark 5. In the Dedekind category *FRel* of sets and fuzzy relations the axiom of choice (AC) fails.

Let $(0, 1)$ be the open real interval between 0 and 1. The unit I of *FRel* is a singleton set $\{*\}$. Consider a fuzzy relation $\rho : I \rightarrow (0, 1)$, that is, a mapping $\rho : I \times (0, 1) \rightarrow [0, 1]$ defined by, for each $t \in (0, 1)$ $\rho(*, t) = t$. Then, ρ is total. On the other hand, for each tfn (I-point) x from I to $(0, 1)$ there exists a unique element $t_x \in (0, 1)$ such that $x(*, t_x) = 1$. This implies $x \not\sqsubseteq \rho$ since $\rho(*, t_x) < x(*, t_x)$. □

In a Dedekind category with (AC) and (Ba), $(\text{Tot}_*) \rightarrow (\text{PA})$ follows from (PA_*) $\rightarrow (\text{PA})$ and Proposition 7 (e) $(\text{AC}) \wedge (\text{PA}_*) \leftrightarrow (\text{AC}) \wedge (\text{Ba}) \wedge (\text{Tot}_*)$. However $(\text{PA}) \rightarrow (\text{Tot}_*)$ need not hold in such a Dedekind category.

Example 1. Let $B = \{0, a, \neg a, 1\}$ be a Boolean algebra with four elements and $Rel(B)$ the Dedekind category. The unit I of $Rel(B)$ is a singleton set $\{*\}$. Consider a B-relation $\rho : I \rightarrow I$ defined by $\rho(*, *) = a$. Then, ρ is nonzero but $\rho\rho^\sharp = \rho \neq \text{id}_I$. This shows that $Rel(B)$ does not satisfy (Tot_*) though it satisfies (AC), (Ba), and (PA). □

7   Conclusion

Dedekind categories may be employed instead of Schröder category as an algebraic framework for relational structure whose members need not have a complement. I-points and the point axioms always connect Dedekind categories with concrete models. In this paper, we have studied relationship between the point axioms and some related axioms in Dedekind categories. It turned out that the usage of point axioms leads to the following restriction within concrete relations:

- A Dedekind category satisfying either (PA) $\wedge$ (Tot$_*$) or (PA$_*$) are suitable only for ordinary relations.
- Assuming either (Tot$_*$) or (Sub), an L-relation becomes an ordinary relations.
- Studying fuzzy relations, (AC) should not be assumed.

Acknowledgement. The authors thank the anonymous reviewers for their suggestions to improve the paper.

References

1. Freyd, P., Scedrov, A.: Categories, allegories. North-Holland, Amsterdam (1990)
2. Goguen, J.A.: L-fuzzy sets. J. Math. Anal. Appl. 18, 145–157 (1967)
3. Ishida, T., Honda, K., Kawahara, Y.: Formal Concepts in Dedekind Categories. In: Berghammer, R., Möller, B., Struth, G. (eds.) RelMiCS/AKA 2008. LNCS, vol. 4988, pp. 221–233. Springer, Heidelberg (2008)
4. Kawahara, Y., Furusawa, H.: An algebraic formalization of fuzzy relations. Fuzzy Sets and Systems 101, 125–135 (1999)
5. MacLane, S.: Categories for the working mathematician, 2nd edn. Springer (1998)
6. Olivier, J.-P., Serrato, D.: Catégories de Dedekind. Morphismes dans les Catégories de Schröder. C. R. Acad. Sci. Paris 260, 939–941 (1980)
7. Schmidt, G., Ströhlein, T.: Relation algebras: Concept of points and representability. Discrete Mathematics 54, 83–92 (1985)
8. Tarski, A.: On the calculus of relations. J. Symbolic Logic 6, 73–89 (1941)
9. Winter, M.: Goguen Categories. A Categorical Approach to L-Fuzzy Relations. Trends in Logic, vol. 25. Springer (2007)
10. Winter, M.: Complements in Distributive Allegories. In: Berghammer, R., Jaoua, A.M., Möller, B. (eds.) RelMiCS/AKA 2009. LNCS, vol. 5827, pp. 337–350. Springer, Heidelberg (2009)
11. Zadeh, L.A.: Fuzzy sets. Information and Control 8, 338–353 (1965)

A   Details of Diagram in Section 3

Proposition 7. (a) $(PA) \wedge (Sub) \rightarrow (PA_*)$,
(b) $(NE) \wedge (Sub) \rightarrow (NE_*)$,
(c) $(Tot) \wedge (Sub) \rightarrow (Tot_*)$,
(d) $(Ba) \wedge (NE_*) \leftrightarrow (PA_*) \leftrightarrow (PA) \wedge (Tot_*)$,
(e) $(AC) \wedge (PA_*) \leftrightarrow (AC) \wedge (Ba) \wedge (Tot_*)$,
(f) $(Ba) \wedge (NE) \wedge (Sub) \leftrightarrow (PA) \wedge (Sub)$.

Proof. For the proofs of (a), (b) and (c) let $\rho : I \rightharpoonup X$ be a relation. The assumption (Sub) claims that

$$\exists j : S \to X. \ (\rho = \nabla_{IS} j) \wedge (jj^\sharp = \mathrm{id}_S).$$

(a) $(PA) \wedge (Sub) \rightarrow (PA_*)$:

$$\rho : I \rightharpoonup X \to \exists j : S \to X. \ \rho = \nabla_{IS} j \ \{ \ (Sub) \ \}$$
$$\to \rho = (\sqcup_{s \,\dot\in\, S} s)j \qquad \{ \ (PA), \ \nabla_{IS} = \sqcup_{s \,\dot\in\, S} s \ \}$$
$$\to \rho = \sqcup_{s \,\dot\in\, S} sj \qquad \{ \ (\cdot\sqcup) \ \}$$
$$\to \rho \sqsubseteq \sqcup_{x \,\dot\in\, \rho} x \qquad \{ \ sj \sqsubseteq \nabla_{IS} j = \rho, \ \{sj \mid s \dot\in S\} \subseteq \{x \mid x \dot\in \rho\} \ \}$$
$$\to \rho = \sqcup_{x \,\dot\in\, \rho} x. \qquad \{ \ (\sqcup_{x \,\dot\in\, \rho} x) \sqsubseteq \rho \ \}$$

(b) $(NE) \wedge (Sub) \rightarrow (NE_*)$:

$$\rho \neq 0_{IX} \to \exists j : S \to X. \ \rho = \nabla_{IS} j \ \{ \ (Sub) \ \}$$
$$\to \nabla_{IS} \neq 0_{IS} \qquad \{ \ \rho = \nabla_{IS} j \neq 0_{IX} \ \}$$
$$\to \exists s \dot\in S \qquad \{ \ (NE) \ \}$$
$$\to sj : I \to X \qquad \{ \ s, j : \mathrm{tfn} \ \}$$
$$\to \exists x \dot\in \rho. \qquad \{ \ sj \sqsubseteq \nabla_{IS} j = \rho \ \}$$

(c) $(Tot) \wedge (Sub) \rightarrow (Tot_*)$:

$$\rho \neq 0_{IX} \to \exists j : S \to X. \ \rho = \nabla_{IS} j \ \{ \ (Sub) \ \}$$
$$\to \mathrm{id}_I = \nabla_{IS} \nabla_{SI} \qquad \{ \ (Tot), \ \nabla_{IS} \neq 0_{IS} \ \}$$
$$\to \mathrm{id}_I = \nabla_{IS} jj^\sharp \nabla_{SI} \qquad \{ \ \mathrm{id}_S = jj^\sharp \ \}$$
$$\to \mathrm{id}_I = \rho\rho^\sharp. \qquad \{ \ \rho = \nabla_{IS} j \ \}$$

(d) (d-1) $(Ba) \wedge (NE_*) \rightarrow (PA_*)$: Set $\rho_0 = \sqcup_{x \,\dot\in\, \rho} x$. Then

$$\rho \not\sqsubseteq \rho_0 \leftrightarrow \neg\rho_0 \sqcap \rho \neq 0_{IX} \ \{ \ (Ba) \ \}$$
$$\to \exists x_0 \dot\in \neg\rho_0 \sqcap \rho \ \{ \ (NE_*) \ \}$$
$$\to x_0 \sqsubseteq \rho_0 \sqcap \neg\rho_0 \ \{ \ \rho_0 = \sqcup_{x \,\dot\in\, \rho} x, \ x_0 \dot\in \rho \ \}$$
$$\to x_0 = 0_{IX} \qquad \{ \ \rho_0 \sqcap \neg\rho_0 = 0_{IX} \ \}$$
$$\to \mathrm{id}_I = 0_{II}, \qquad \{ \ \mathrm{id}_I = x_0 x_0^\sharp \ \}$$

which contradicts $\mathrm{id}_I \neq 0_{II}$. Hence $\rho = \sqcup_{x \,\dot\in\, \rho} x$.
(d-2) $(PA_*) \rightarrow (PA) \wedge (Tot_*)$: It is trivial.

(d-3) $(PA) \wedge (Tot_*) \to (Ba) \wedge (NE_*)$: An implication $(PA) \wedge (Tot_*) \to (NE_*)$ is proved as follows: Let $\rho : I \to X$ be nonzero. Then

$$\rho = \rho \sqcap (\sqcup_{x \,\dot\in\, X} x) \quad \{ (PA) \; \nabla_{IX} = \sqcup_{x \,\dot\in\, X} x \}$$
$$= \sqcup_{x \,\dot\in\, X} (\rho \sqcap x). \quad \{ (\sqcap \sqcup) \}$$

As $\rho \neq 0_{IX}$, there is an I-point $x \,\dot\in\, X$ such that $\rho \sqcap x \neq 0_{IX}$. By (Tot_*) $\rho \sqcap x$ is total and so

$$x \sqsubseteq (\rho \sqcap x)(\rho \sqcap x)^{\sharp} x \; \{ \rho \sqcap x : \text{total} \}$$
$$\sqsubseteq \rho x^{\sharp} x \qquad\qquad \{ (\cdot\sqsubseteq) \}$$
$$\sqsubseteq \rho, \qquad\qquad\quad \{ x : \text{tfn} \}$$

which shows that ρ is nonempty. Another implication $(PA) \wedge (Tot_*) \to (Ba)$ has been shown in Theorem 2.

(e) $(AC) \wedge (PA_*) \to (AC) \wedge (Ba) \wedge (Tot_*)$ is immediate from (d). Since it is trivial that $(AC) \wedge (Tot_*) \to (NE_*)$, $(AC) \wedge (Ba) \wedge (Tot_*) \to (AC) \wedge (PA_*)$ follows from

$$(AC) \wedge (Ba) \wedge (Tot_*) \to (Ba) \wedge (NE_*)$$
$$\to (PA_*) \qquad \{ (d\text{-}1) \}$$

(f) (f-1) $(Ba) \wedge (NE) \wedge (Sub) \to (PA) \wedge (Sub)$:

$$(Ba) \wedge (NE) \wedge (Sub) \to (Ba) \wedge (NE_*) \; \{ (b) \}$$
$$\to (PA). \qquad\quad \{ (d\text{-}1) \}$$

(f-2) $(PA) \wedge (Sub) \to (Ba) \wedge (NE) \wedge (Sub)$:

$$(PA) \wedge (Sub) \to (NE), \qquad\quad \{ \text{Proposition 4} \}$$

$$(PA) \wedge (Sub) \to (PA) \wedge (Tot_*) \; \{ (a), (PA_*) \to (Tot_*) \}$$
$$\to (Ba). \qquad\quad \{ \text{Theorem 2} \} \qquad\qquad \square$$

Two Observations
in Dioid Based Model Refinement

Roland Glück

Institut für Informatik, Universität Augsburg,
D-86135 Augsburg, Germany
glueck@informatik.uni-augsburg.de

Abstract. Dioid based model refinement is an abstract framework for optimality problems on labelled graphs such as the shortest path problem or the maximum capacity problem. Here we show two facts in this area: first, via a language theoretic approach that models with a binary set of edge labels are refinable, and second, the compatibility of linear fixpoint equations with bisimulations. These equations can be used in an algebraic setting for a certain class of optimality problems. Since bisimulations can simplify the complexity of models, they can also reduce the complexity of certain optimality problems.

1 Introduction

In practice one is often confronted with systems containing a large or even infinite number of states and/or transitions, e.g. in control theory, model checking (cf. [1]), automata theory (cf.[8] and [9]) and other cases. If the task consists in ensuring a certain property (optimality, safety, liveness) by refinement of the given system or model, i.e. removing (in practice preventing) transitions, this task can appear to be difficult to solve for the large system. One possible strategy is to reduce the original system to a smaller one using a suitable bisimulation, then to apply a known algorithm to that system, such that a refined subsystem of it fulfils the demanded property, and in a last step to expand that system into a subsystem of the original one. Of course this strategy will not work in all cases. To make sense, the reduction by bisimulation has to decrease the number of states/transitions in a significant way, an algorithm for computing a refined system with the required property has to be known, and the desired property has to be invariant in a certain sense with respect to the chosen bisimulation.

As new material in this paper in contrast to [3] we show the refinability of a special class of optimisation problems (the second step in the above strategy sketch), which will be done via an excursion in language theory. As second new theme we show that linear fixpoint equations over dioids are compatible with suitable bisimulations. In particular, this holds also for the computation of least fixpoints, which can serve to reason about optimal costs in models.

In Section 2 basic definitions concerning models and dioids are given. The announced excursion into language theory takes place in Section 3, and its results are applied to model refinement in Section 4. After some basic facts about

W. Kahl and T.G. Griffin (Eds.): RAMiCS 2012, LNCS 7560, pp. 235–247, 2012.

bisimulations are recapitulated in Section 5. Section 6 deals with fixpoints of linear equations and their compatibility with bisimulations.

2   Definition of Dioids and Target Models

For the readers convenience we recapitulate some definitions and results from [3].

Definition 2.1. A *complete dioid* is a structure $(D, \sum, 0, \cdot, 1)$ such that $(D, \sqsubseteq)$ is a complete lattice with supremum operator $\sum$ and least element 0, where $\sqsubseteq$ is defined by $x \sqsubseteq y \Leftrightarrow \sum\{x, y\} = y$, $(D, \cdot, 1)$ is a monoid and $\cdot$ distributes over $\sum$ from both sides. $\sqsubseteq$ is called the *order* of the complete dioid.

We denote the binary supremum operation by $+$ and often omit the $\cdot$, so $x + y$ stands for $\sum\{x, y\}$ and ab for $a \cdot b$. A complete dioid is called *selective* or shortly an *s-dioid* if $(\sum_{a \in A} a) \in A$ holds for every nonempty finite set $A \subseteq D$. So a dioid is selective iff its order is liner. A dioid with the property $a \sqsubseteq 1$ for all $a \in D$ is called *cumulative*. This property is equivalent to $ab \sqsubseteq a$ and $ba \sqsubseteq a$ for all $a, b \in D$. Moreover, a complete dioid is cumulative iff the implications $a \sqsubseteq b \Rightarrow ac \sqsubseteq b$ and $a \sqsubseteq b \Rightarrow ca \sqsubseteq b$ hold. For further properties we refer the reader to [5] and [3].

Definition 2.2. A model *is a pair* $M = (G, g)$, *where* $G = (V, E)$ *is a directed graph with node set* V *and edge set* E, *and* $g : E \to D$ *is a mapping from the edge set* E *into the carrier set* D *of an s-dioid* $(D, \sum, 0, \cdot, 1)$. *A target model is a pair* $M_T = (M, T)$, *where* $M = ((V, E), g)$ *is a model, and* $T \subseteq V$ *is the so called* target set, *where from every* $v \in V$ *some node* $t \in T$ *is reachable, and no node* $t \in T$ *has an outgoing edge. A model is called* finite *if the associated graph is finite.*

This definition serves to build an abstract theory of a class of optimality problems concerning walks from every node to a node from the target set. The optimisation goal can be e.g. to minimise the length of a walk or to maximise its capacity. For this purpose we give the next definition:

Definition 2.3. Let $M = (G, g)$ with $G = (V, E)$ be a model and $(D, \sum, 0, \cdot, 1)$ the associated s-dioid. Then for a walk $w = x_1 x_2 \ldots x_n$ in G the cost $c(w)$ of w is defined by $c(w) = \prod_{i=1}^{n-1} g(x_i, x_{i+1})$. For two nodes x and y the distance $d(x, y)$ of x and y is defined by $d(x, y) = \sum_{w \in W(x,y)} c(w)$, where $W(x, y)$ denotes the set of all walks in G from x to y. In a target model $M_T = (M, T)$ the target distance $d(x)$ of a node x is defined as $d(x) = \sum_{t \in T} d(x, t)$, where $d(x, t)$ is determined as above in the associated model M. A walk $x_1 x_2 \ldots x_n$ is called optimal if $c(x_1 x_2 \ldots x_n) = d(x_1, x_n)$.

If one is interested in the investigation of shortest walks one will chose the s-dioid $(\mathbb{R}_0^+ \cup \{\infty\}, \inf, \infty, +, 0)$ as codomain of g. The goal, given a target model $M = (((V, E), g), T)$, is to construct an optimal submodel, i.e. a target model $M' = (((V', E'), g'), T')$ with $V' = V$, $E' \subseteq E$, $g' = g|_{E'}$ and $T' = T$, which guarantees optimality. This means that every possible walk w in $G' = (V', E')$ from an arbitrary node $x \in V'$ leads to a node $t \in T'$, and the cost of w equals the target distance $d(x)$ in M. Intuitively this means that an optimal submodel enforces optimal walks. If a model has an optimal submodel it is called *refinable*. Examples can be found again in [3]. Not every model is refinable, consider for example the problem of finding shortest walks in a graph with cycles of negative length.

3   An Excursion into Language Theory

Since the labels along nonempty finite paths in a target starting from the outside of the target set model correspond to words from D^+, and the distance is defined as the supremum over a set of such words (i.e. a nonempty language $L \subseteq D^+$) we consider languages in D^+. The details of this connection will be shown later; first we present the language theoretic matters.

3.1   Definitions

Definition 3.1. *For two words $u = u_1 u_2 \ldots u_n$ and $v = v_1 v_2 \ldots v_m$ we say that u is a skeleton of v if there is a strictly monotone mapping $f : \{1 \ldots n\} \to \{1 \ldots m\}$ with $u_i = v_{f(i)}$.*

Intuitively this means that the symbols of u occur also in v in the original order but (in contrast to a substring) they may be separated by other symbols of v. So abc is a skeleton of $aabbcc$ although it is not a substring of it. As an abbreviation we use $skel(u, v)$ if u is a skeleton of v. Clearly, the relation $skel$ is an order over Σ^+ (but not over Σ^ω, since, e.g., $(aab)^\omega$ is a skeleton of $(ab)^\omega$ and vice versa). Obviously, we have $(xba^n, wba^m) \in skel$ if $(x, w) \in skel \wedge n \le m$ for all $x, w \in \Sigma^*$ and $a, b \in \Sigma$. Note that $(skel \backslash I)^\circ$ (where $^\circ$ denotes the converse of a relation and I the identy relation) is Noetherian, so every language L has a set of minimal elements $min_{skel}(L)$ wrt. $skel$, and $\forall u \in L \exists v \in min_{skel}(L) : skel(v, u)$ holds. A language L is called *skeleton free* if for every pair of distinct words u and v from L holds that u is no skeleton of v. Due to definition for every language L the language $min_{skel}(L)$ is skeleton free.

Definition 3.2. *For a word $u \in \Sigma^*$ with $|u|_a > 0$ ($|u|_a$ denotes the number of occurences of the symbol a in the word u) we define the two (possibly empty) words $u_{\to a}$ and $u_{a \to}$ via the unique decomposition $u = u_{\to a} a u_{a \to}$ with $u_{\to a} \in \Sigma^*$ and $u_{a \to} \in (\Sigma \backslash \{a\})^*$.*

This means that $u_{\to a}$ is the substring of u up to the last occurrence (exclusively) of a, and $u_{a \to}$ is the substring of u after the last occurrence of a. So

$aabbcc_{\to b} = aab$, $aabbcc_{b\to} = cc$, $aabbcc_{\to c} = aabbc$ and $aabbcc_{c\to} = \varepsilon$. We extend this definition pointwise to languages by $L_{\to a} = \{u_{\to a} \mid u \in L \wedge |u|_a > 0\}$ and analogously for $L_{a\to}$.

Definition 3.3. *For an alphabet Σ and a word $u \in \Sigma^+$ we define the span of u over Σ by $span_\Sigma(u) = \{v \in \Sigma^* \mid skel(u,v)\}$. We extend this definition to a language $L \subseteq \Sigma^+$ via $span_\Sigma(L) = \bigcup_{u \in L} span_\Sigma(u)$. A language $L_G \subseteq \Sigma^+$ is called a generator of a language $L \subseteq \Sigma^*$ if $L_G \subseteq L$ and $L \subseteq span_\Sigma(L_G)$.*

With the above definitions it is clear that for every language L the set $min_{skel}(L)$ forms a skeleton free generator of L. Moreover, $min_{skel}(L)$ is a subset of every generator of L.

3.2   Conjectures

We state the following conjectures, which have a surprising connection with model refinability:

Conjecture 3.4. Every language $L \subseteq \Sigma^+$ over a finite alphabet Σ has a finite generator.

This conjecture is equivalent to the following:

Conjecture 3.5. Every skeleton free language $L \subseteq \Sigma^+$ over a finite alphabet Σ is finite.

Of course both conjectures do not hold for infinite alphabets (choose simply $L = \Sigma$ to produce a counterexample). It turns out that these two conjectures are actually the same:

Lemma 3.6. *Conjecture 3.4 and Conjecture 3.5 are equivalent.*

Proof. Conjecture 3.4 $\Rightarrow$ Conjecture 3.5: Let L be a skeleton free language. Since L is skeleton free we have $min_{skel}(L) = L$. Let L_G be a finite generator of L. Because of $min_{skel}(L) \subseteq L_G$ the language $min_{skel}(L)$ is finite, and hence L is finite, too.

Conjecture 3.5 $\Rightarrow$ Conjecture 3.4: Let L be an arbitrary language. Then the set $min_{skel}(L)$ is a generator of L. Moreover, it is skeleton free and hence finite by assumption.

3.3   Proof

Unfortunately, there is no known proof for the conjectures from Subsection 3.2. However, there is a partial result for special alphabets:

Theorem 3.7. *The Conjectures 3.4 resp. 3.5 hold for binary alphabets Σ.*

Proof. W.l.o.g. we assume $\Sigma = \{a, b\}$ and use induction over the length of a shortest word in the langauge L.

Induction base: W.l.o.g. let the shortest word of L be a. Then there can be no other word in L which contains an a. So all other words in L have the form b^m. But there can only be one word of this form: if there were two words b^{m_1} and b^{m_2} in L with $m_1 \neq m_2$ then one of these words would clearly be a skeleton of the other one.

Induction step: W.l.o.g let $u = vb$ be a word of shortest length in L. Then for every $n \in \mathbb{N}$ there can be only finitely many words of the form wba^n in L. Assume there were infinitely many words in $L^n := \{wba^n \in L\}$, and L^n is skeleton free. Then by induction hypothesis there is a finite generator G^n of $L^n_{\to b} \cup \{v\}$. Let wba^n be an arbitrary word in L^n, and $x \in G^n$ be a skeleton of w. But then x or xba^n is a skeleton of wba^n, which is contained in L (the first case occurs if $x = v$, the second case otherwise).

On the other hand there is an upper bound for n such that a word of the form wba^n is an element of L. Therefore let G^b be a finite generator of $L_{\to b}$ (this generator exists due to our induction hypothesis and the fact $v \in L_{\to b}$). For each $w \in G^b$ chose a word $wba^{n_w} \in L$, call the union of these words L^G, and set $n_{max} = max\{n \mid a^n \in L^G_{b \to}\}$. Assume there is a word of the form $wba^n \in L$ with $n > n_{max}$. Then there is a word $xba^{n'} \in L^G$ with $skel(x, w)$ and $n' \leq m$. Hence $xba^{n'}$ is a skeleton of wba^n.

So there are only finitely many $n \in \mathbb{N}$ such that a word of the form wba^n can be in L, and for every $n \in \mathbb{N}$ the set $\{wba^n \mid wba^n \in L\}$ is finite. This means that L is finite, too.

3.4 Remarks

The next step should be to show the conjecture for arbitrary finite alphabets. In the above proof a crucial point is that *skel* is a linear order on $\{a^n \mid n \in \mathbb{N}\}$. If one tries to extend the given proof to arbitrary finite alphabets one could only conclude that there are finitely many words of the form wbu for every $u \in (\Sigma \backslash \{b\})^*$, where vb is a shortest word in L. The second proof ingredient (the upper bound for n) seems not to be so easy to transfer.

4   Refinability

The material from the previous section is connected to model refinement of target models with a cumulative associated s-dioid. In [3] it was shown that finite target models with cumulative associated s-dioid are refinable. If Conjecture 3.4 resp. Conjecture 3.5 hold we have a stronger fact:

Corollary 4.1. *If Conjecture 3.4 resp. Conjecture 3.5 hold every target model $M_T = (((V, E), g), T)$ with cumulative associated s-dioid $(D, \sum, 0, \cdot, 1)$ and finite label set (i.e. $|Im(g)| \leq \infty$) is refinable.*

Proof. First show that for every node $v \in V \backslash T$ there is a walk w from v to some node $t \in T$ with $c(w) = d(v)$ (cf. the definitions from subsection 2). Since we have $d(v) = \sum\limits_{\{w \in W(v,t) \,|\, t \in T\}} c(w)$ we only need to show that there is a finite subset $W' \subseteq W := \{W(v,t) \,|\, t \in T\}$ with $d(v) = \sum\limits_{w \in W'} c(w')$, because due to the linear order of a dioid there will be a concrete element $w' \in W'$ for which $c(w') = d(v)$ holds (note that $W(x,y)$ will be infinite in general, and although $d(v)$ is well-defined there need not to be a walk w with $c(w) = d(v)$). Therefore we map every walk $w = w_1 w_2 \ldots w_n$ to the word $lab(w) := g(w_1, w_2) g(w_2, w_3) \ldots g(w_{n-1}, w_n) \in D^+$, its *label* (remember that we consider only nodes $v \notin T$). Then if for two walks w and w' the relation $skel(w, w')$ holds we have $c(w') \sqsubseteq c(w)$ because of the cumulativity of the associated dioid. Let now $lab(W) := \{lab(w) \,|\, w \in W\}$ be the set of all labels of walks from v to some node in T. By Conjecture 3.4 $lab(W)$ has a finite generator $G_{lab(W)}$. Let now $W' \subseteq W$ be a set of walks from v into T, which contains for every $u \in G_{lab(W)}$ a walk w' with $lab(w') = u$. Then for every $w \in W$ there is a walk $w' \in W'$ with $c(w) \sqsubseteq c(w')$. Hence we have $\sum\limits_{w \in W} c(w) = \sum\limits_{w' \in W'} c(w') = d(v)$.

In order to construct a refinement we have to choose for each node an edge we keep in the refined model. A possible choice is to keep an edge which lies on a shortest optimal path leading into T. By choosing an edge on a shortest optimal path we avoid to get stuck in a cycle, so every path is guaranteed to lead into T, and it is optimal by construction.

Clearly, the idea of this proof can be combined with theorem 3.7 to obtain the following proposition:

Proposition 4.2. *Every target model $M_T = (((V, E), g), T)$ with cumulative associated s-dioid $(D, \sum, 0, \cdot, 1)$ and $|Im(g)| \leq 2$ is refinable.*

Note that this does not mean that every target model with cumulative associated dioid is refinable. Consider for example the model $M_T = (((V, E), g), T)$, given by

- $V = \{s\} \cup \{t\} \cup \{v_i \,|\, i \in \mathbb{N}_0\}$,
- $E = \{(s, v_i) \,|\, i \in \mathbb{N}_0\} \cup \{v_i, t \,|\, i \in \mathbb{N}_0\}$,
- $g((s, v_i)) = g((v_i, t)) = 2^{-i}$ for all $i \in \mathbb{N}_0$ and
- $T = \{t\}$,

with the cumulative associated dioid $(\mathbb{R}_0^+ \cup \{\infty\}, \inf, \infty, +, 0)$. Then there is no optimal (here shortest) walk from s to t, hence the given target model is not refinable.

5   A Short Review of Bisimulations

Since the next result deals with bisimulations we give some important definitions and properties. First we consider edge labelled graphs of the form $G_g = ((V, E), g)$, where $G = (V, E)$ is a graph and $g : E \rightarrow D$ is a labelling of the edges of G with elements drawn from a set D. We write $v_1 \rightarrow_d v_2$ as abbreviation for $(v_1, v_2) \in E \land g((v_1, v_2)) = d$. On labelled graphs we define first the term autobisimulation, where we use the notation R° for the converse of a relation R:

Definition 5.1. *An* autobisimulation *on an edge labelled graph* $G = ((V, E), g)$ *is a right- and left total relation* $B \subseteq V \times V$ *such that for all* $v_1, v_2, v_1' \in V$ *and* $d \in D$ *the implication* $v_1 \rightarrow_d v_2 \land (v_1, v_1') \in B \Rightarrow \exists v_2' : v_1' \rightarrow_d v_2' \land (v_2, v_2') \in B$ *and symmetrically for all* $w_1, w_2, w_1' \in V$ *and* $d \in D$ *the implication* $w_1 \rightarrow_d w_2 \land (w_1, w_1') \in B^\circ \Rightarrow \exists w_2' : w_1' \rightarrow_d w_2' \land (w_2, w_2') \in B^\circ$ *holds.*

An autobisimulation, which is also an equivalence relation is called an *autobisimulation equivalence* or *bisimulation equivalence* for short. Given a partition $V = \bigcup_{i \in I} V_i$ we say that an autobisimulation equivalence B *respects* the partition if for every $i \in I$ the set V_i can be written as the union of suited equivalence classes of B. Since the identity relation is an autobisimulation, and autobisimulations are closed under union, composition and taking the converse, there is for every labelled graph a coarsest autobisimulation which is also an equivalence.

A relational characterisation of bisimulations can be found in [11] and [13].

The purpose of using autobisimulation equivalences is to reduce the number of nodes of a given system without losing too much information about its dynamics. This is done via the following construction, in which v/B denotes the equivalence class of an element v under an equivalence relation B:

Definition 5.2. *Let* $G = ((V, E), g)$ *be an edge labelled graph and* B *an autobisimulation equivalence on* G. *The* quotient G_B *of* G *by* B *is the edge labelled graph* $G_B = ((V_B, E_B), g_B)$ *with*

- $V_B = \{v/B \mid v \in V\}$
- $E_B = \{(v_1/B, v_2/B) \mid (v_1, v_2) \in V\}$
- $g_B((v_1/B, v_2/B)) = g((v_1, v_2))$ *for all* $(v_1, v_2) \in E$

Note that g_B is well defined since B is a bisimulation equivalence. This definition is illustrated in Figure 1.

This construction is also used in the classic papers [8] and [9] for the construction of a minimal finite automaton. [1] uses the quotient to check properties of a given system, and [6] uses a similar construction in the contex of internet pathfinding. The best known algorithm for computing the coarsest bisimulation is given in [10] and has a runtime of $\mathcal{O}(|E| \cdot log(|V|))$.

Usually there is also an expansion operation under consideration, which is a pseudoinverse of the quotient operation (for details see [3] or [4]). Something similar will be done in the following section.

6   Linear Fixpoint Equations

6.1   Basic Definitions

In [5] optimal costs are characterised as minimal solutions of equations of the form $x = Ax + b$ resp. $x = xA + b$ where A denotes a matrix and b a vector.

We will concentrate on such equations of the form $x = Ax + b$; the other type can be handled symmetrically. Those kind of fixpoint equations occur often in mathematics, e.g., the Gauss-Seidel method and the Jacobi method for solving linear equation systems are based on such fixpoint equations (see e.g. [2]). In contrast to traditional linear algebra we will deal also with infinite index sets, so we extend the definition of a matrix and a vector as follows:

Definition 6.1. *Let M be an arbitrary set and $(D, \sum, 0, \cdot, 1)$ an s-dioid. An M-matrix over D is a mapping $A : M \times M \to D$, and an M-vector over D is a mapping $v : M \to D$.*

For brevity we will often use only the terms matrix and vector if the involved sets are clear from the context. In contrast to conventional linear algebra we do not discern between column and row vectors. For a matrix A we write A_{ij} as abbreviation for $A((i,j))$ and analogously for a vector we use v_i as a short form for $v(i)$. We write $D^{M \times M}$ for the set of all M-matrices over $(D, \sum, 0, \cdot, 1)$, and D^M for the set of all M-vectors over $(D, \sum, 0, \cdot, 1)$.

For an M-matrix A and an M-vector v over the same s-dioid $(D, \sum, 0, \cdot, 1)$ we define the product Av as an M-vector over $(D, \sum, 0, \cdot, 1)$, defined by $(Av)_i = \sum_{j \in M} A_{ij} \cdot v_j$. Symmetrically we define the product vA by $(vA)_i = \sum_{j \in M} v_j \cdot A_{ji}$. The sum $v + w$ of two vectors $v, w \in D^M$ is given by $(v + w)_m = v_m + w_m$ for all $m \in M$.

In order to connect labelled graphs with matrices we use the usual approach: if one is interested in shortest walks in a graph with edge labels from $\mathbb{R} \cup \{\infty\}$ every non-existing edge is modelled by a pseudoedge with cost ∞. So for a target model $M_T = (((V, E), g), T)$ or a labelled graph $G = ((V, E), g)$ we define its *adjacency matrix* A by $A_{v_1 v_2} = g((v_1, v_2))$ if $(v_1, v_2) \in E$ and $A_{v_1 v_2} = 0$ otherwise. In this setting the target distances of each node correspond to the minimal solution of the linear fixpoint equation $x = Ax + b$, where $b_i = 0$ for all $i \notin T$ and $b_i = 1$ for all $i \in T$ (of course the concrete values of 0 and 1 and the order is given by the associated dioid of M_T). So we define on D^M for an ordered set $(D, \sqsubseteq)$ the ordering $\sqsubseteq_M$ by $x \sqsubseteq_M y \Leftrightarrow \forall m \in M : x_m \sqsubseteq y_m$. In this case suprema are built componentwise, i.e. $(\sum_{v^M \in E^M} v^M)_m = \sum_{v^M \in E^M} v_m^M$ for arbitrary $E^M \subseteq D^M$ and all $m \in M$.

6.2   Application of Bisimulations

To make use of bisimulations for solving linear fixpoint equations we need some more definitions concerning the interplay between adjacency matrices, vectors, autobisimulation equivalences and quotients.

Definition 6.2. *Let $G = ((V, E), g)$ be a labelled graph with adjacency matrix A and B an autobisimulation equivalence for G. The quotient matrix A_B of A by B is defined as the adjacency matrix of the quotient G/B. The partition induced by b for a vector $b \in Im(g)^V$ is the partition of V into the equivalence classes* *of the equivalence relation $R := \{(v_1, v_2) \mid b_{v_1} = b_{v_2}\}$. For an autobisimulation equivalence on G respecting the partition induced by b we define the vector $b_B \in Im(g)^{V_B}$ by $b_{(v/B)} = b_v$ for all $v \in V$. Given a vector $y_B \in Im(g)^{V_B}$ with indices in V_B the expansion $y_B \backslash B \in Im(g)^V$ of y_B by B is defined by $(y_B \backslash B)_v = (y_B)_{(v/B)}$.*

We will illustrate this in the depiction from Figure 1, which is inspired by the fourth chapter of [5].

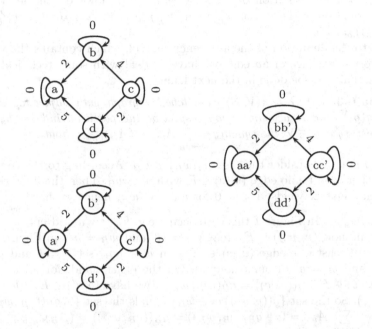

Fig. 1. A labelled graph (left) and its coarsest quotient (right)

On the left side a labelled graph with nodeset $V = \{a, b, c, d, a', b', c', d'\}$ is given. We assume that the edge labels are drawn from the s-dioid $(\mathbb{R}_0^+ \cup \infty, min, \infty, +, 0)$. The order $\sqsubseteq$ of this dioid corresponds to the relation $\geq$ on $\mathbb{R} \cup \infty$ with least element ∞. The adjacency matrix A of the left side graph is given by $A_{aa} = 0$, $A_{ba} = 2$, $A_{ab} = \infty$ (note that here the zero corresponds to ∞!) etc. Its coarsest quotient with adjacency matrix A_B is depicted on the right side. For better readability the set parentheses in the nodes' captions of the quotient are omitted.

Let us consider the vector v, given by $v_a = v_{a'} = 0$ and $v_b = v_c = v_d = v_{b'} = v_{c'} = v_{d'} = \infty$. A solution of the equation $x = Ax + v$ is given by the vector y with $y_a = 0$, $y_b = y_c = y_d = 1$, $y_{a'} = 0$, $y_{b'} = 2$, $y_{c'} = 6$ and $y_{d'} = 5$. For example we have

$$- \; y_a = 0 = min\{0+0, \infty+1, 0\} = 0$$
$$- \; y_b = 1 = min\{2+0, 0+1, \infty+1, \infty\} = 1 \text{ and}$$
$$- \; y_{c'} = 6 = min\{\infty+0, \infty+1, 4+2, 0+6, 2+5, 6\} = 6.$$

Also the vector y' with $y'_a = y'_{a'} = 0$, $y'_b = y'_{b'} = 2$, $y'_c = y'_{c'} = 6$ and $y'_d = y'_{d'} = 5$ is a solution of $x = Ax + v$. On contrast to y it is the least solution of $x = Ax + v$ (remember that the order corresponds to $\geq$), since it describes the target distances to a reps. a', according to the choice of v.

Analogously, the vector w_B with $(w_B)_{aa'} = 0$ and $(w_B)_{bb'} = (w_B)_{cc'} = (w_B)_{dd'} = 1$ is a solution of $x_B = A_B x_B + v_B$. However, the least solution of this equation is the vector w'_B with $(w'_B)_{aa'} = 0$, $(w'_B)_{bb'} = 2$, $(w'_B)_{cc'} = 6$ and $(w'_B)_{dd'} = 5$.

Due to the definition of the adjacency matrix, which contains the value 0 to indicate the absence of an edge we have to deal with some technical inconveniences. This will be done in the next lemma.

Lemma 6.3. *Let $G = ((V, E), g)$ a labelled graph with adjacency matrix A, $y \in Im(g)^V$ and B a bisimulation respecting the partition induced by y. Then for arbitrary $v, w \in V$ the equality $\sum_{w \in w_B} A_{vw} = (A_B)_{v_B w_B}$ holds.*

Proof. First we consider the case $(v_B w_B) \notin E_B$. According to the construction of G_B there can be no edge $(v, w) \in E$ with $w \in w_B$. Then the definition of the adjacency matrix yields $A_{vw} = 0$ for all $w \in w_B$, hence we have $\sum_{w \in w_B} A_{vw} = 0 = (A_B)_{v_B w_B}$. In Figure 1 this case occurs e.g. for $v = a$ and $w = c'$.

Assume now $(v_B w_B) \in E_B$ (e.g. $v_B = bb'$ and $w_B = aa'$ in Figure 1). Then there is at least one edge $(v, w) \in E$ with $w \in w_B$ (set $v = b$ and $w = a$ or $v = b'$ and $w = a'$ in our example). On the other hand, for all $w \in w_B$ we have $(v, w) \notin E \vee g((v, w)) = g_B((v_B, w_B))$ (we have $(b, a') \notin E \wedge (b', a) \notin E$ in Figure 1). So the set $\{g((v, w)) \mid w \in w_B\}$ equals the set $\{0, g_B((v_B, w_B))\}$, and we have $\sum_{w \in w_B} A_{vw} = 0 + g_B((v_B, w_B)) = g_B((v_B, w_B)) = (A_B)_{v_B, w_B}$.

For given A, b and B it is easy to construct A_B and b_B. Conversely, it is also easy to construct $b_B \backslash B$ if b_B and B are given. If the involved elements are finite this is possible in linear time; in the infinite case one has to work with symbolic methods. So if one has to solve a linear fixpoint equation it could be advantageous to try to solve it on a quotient and to construct a solution of the original equation from a solution on the quotient. This approach is correct, as is stated in the next theorem.

Theorem 6.4. *Let $G = ((V, E), g)$ be a labelled graph with adjacency matrix A and consider the linear fixpoint equation $x = Ax + b$. If B is an autobisimulation equivalence on G respecting the partition induced by b and y_B is a solution of the linear fixpoint equation $x_B = A_B x_B + b_B$ then $y_B \backslash B$ is a solution of $x = Ax + b$.*

Proof. Unfolding the matrix multiplication yields that we have to show the equality $(y_B \backslash B)_v = \sum_{w \in V} (A_{vw} \cdot (y_B \backslash B)_w) + b_v$ for all $v \in V$ under the assumption $(y_B)_{v_B} = \sum_{w_B \in V_B} ((A_B)_{v_B w_B} \cdot (y_B)_{w_B}) + (b_B)_{v_B}$ for all $v_B \in V_B$. For an arbitrary $v \in V$ we have, due to the definition of the expansion operator, the equality $(y_B \backslash B)_v = (y_B)_{v_B}$, so we are done if we succeed to show that $\sum\limits_{w \in V} (A_{vw} \cdot (y_B \backslash B)_w) + b_v = \sum\limits_{w_B \in V_B} ((A_B)_{v_B w_B} \cdot (y_B)_{w_B}) + (b_B)_{v_B}$ holds for all $v \in V$. Because B is by assumption a bisimulation equivalence respecting the partition induced by b we have $b_v = (b_B)_{v_B}$, so the claim reduced to the equality $\sum\limits_{w \in V} (A_{vw} \cdot (y_B \backslash B)_w) = \sum\limits_{w_B \in V_B} ((A_B)_{v_B w_B} \cdot (y_B)_{w_B})$. According to the definition of the expansion operation the left side is identical to $\sum\limits_{w \in V} (A_{vw} \cdot (y_B)_{w_B})$. Then we can split the sum using the equivalence classes of B (which are the elements of V_B) and obtain the expression $\sum\limits_{w_B \in V_B} \sum\limits_{w \in w_B} (A_{vw} \cdot (y_B)_{w_B})$. Using Lemma 6.3 and distributivity yields the desired term $\sum\limits_{w_B \in V_B} ((A_B)_{v_B w_B} \cdot (y_B)_{w_B})$.

This theorem show only how fixpoints of linear equations can be constructed from fixpoints of the quotient. The question arises whether least fixpoints with respect to $\sqsubseteq_M$ are also compatible with bisimulations in a sense similar as in Theorem 6.4. There is a positive answer, as stated in the next theorem:

Theorem 6.5. *Let $G = ((V, E), g)$ be a labelled graph with adjacency matrix A and B an autobisimulation equivalence on G respecting the partition induced by the vector $b \in D^V$. Let y be the least fixpoint of the function $f(x) = Ax + b$ with respect to $\sqsubseteq_V$, and let y_B be the least fixpoint of the function $f_B(x_B) = A_B x_B + b_B$ with respect to $\sqsubseteq_{V_B}$. Then the equality $y = y_B \backslash B$ holds.*

Proof. Since the edge labels are drawn from a complete ordered set (an s-dioid) and both the functions f and f_B preserve suprema (this follows from the fact that in every s-dioid multiplication distributes over arbitrary suprema) the least fixpoint of these functions can be determined via the famous fixpoint iteration theorem of Kleene. (This theorem appears first in [7] as "recursion theorem" in the context of partial recursive functions. A version using lattices was given in [12]. The intellectual priority remains open.)

So the least fixpoint ν_f of f equals $\sum\limits_{i \geq 0} f^i(0^V)$, where $f^i(0^V)$ denotes the i-fold application of f to the vector 0^V, defined by $0_v^V = 0$ for all $v \in V$. With analogous notation we have the equation $\nu_{f_B} = \sum\limits_{i \geq 0} f_B^i(0^{V_B})$ for the least fixpoint ν_{f_B} of f_B. The sums above are the suprema with respect to the orders $\sqsubseteq_V$ and $\sqsubseteq_{V_B}$ resp.

The idea is to show that for all $i \in \mathbb{N}$ the equality $f^i(0^{V_B}) \backslash B = f^i(0^V)$ holds. Then the claim follows because the suprema with respect to $\sqsubseteq_V$ and $\sqsubseteq_{V_B}$ are the pointwise suprema of $\sqsubseteq$.

The claimed equality can be shown by simple induction:

Induction base: The equation holds for $i = 0$ since $0^{V_B} \backslash B = 0^V$ is obviously true.

Induction step: Assume $f^i(0^{V_B}) \backslash B = f^i(0^V)$ and fix an arbitrary $v \in V$. Due to the definition of $\backslash B$ we have to show that $A(f^i(0^V))_v + b_v = A_B(f^i(0^{V_B}))_{v_B} + (b_B)_{v_B}$ holds. B was supposed to be compatible with the partition induced by b;

so $b_v = (b_B)_{v_B}$ holds trivially; it remains to show $A(f^i(0^V))_v = A_B(f^i(0^{V_B}))_{v_B}$.
Therefore we calculate:

$A(f^i(0^V))_v =$
  { Definition of multiplication of matrix and vector }
$\sum_{w \in V} A_{vw} \cdot (f^i(0^V))_w =$
  { rearranging the sum along the equivalence classes of B }
$\sum_{w_B \in V_B} \sum_{w \in w_B} A_{vw} \cdot (f^i(0^V))_w =$
  { induction hypothesis }
$\sum_{w_B \in V_B} \sum_{w \in w_B} A_{vw} \cdot (f^i(0^{V_B}))_{w_B} =$
  { Lemma 6.3, distributivity }
$\sum_{w_B \in V_B} (A_B)_{v_B w_B} \cdot (f^i(0^{V_B}))_{w_B} =$
  { Definition of multiplication of matrix and vector }
$A_B(f^i(0^{V_B}))_{v_B}$

Intuitively this proof relies on the fact that the computation of f^i and f_B^i can be imagined as parallel labelings of the nodes in V and V_B during the execution of the Kleene fixpoint iteration. Thereby the values of a node $v \in V$ and of its equivalence class $v_B \in V_B$ always coincide in each step.

6.3 Connection with Model Refinement

Whereas Theorem 6.4 has more an aesthetic value and no direct application in model refinement, Theorem 6.5 has a deeper connection to optimality problems. As already mentioned above, the optimal values of nodes can be identified using the least fixpoint of a linear fixpoint equation. So Theorem 6.5 shows that the approach via a bisimulation quotient yields correct results for this problem class. However, it does not give an answer how to refine the model. In general, refinement is impossible, for instance in the presence of cycles of negative length. If one is interested in determining the optimal cost values via the least fixpoint method the application of bisimulations can lead to a speed up since in general the quotient is smaller than the original graph. Of course, it depends on how much the graph is reduced, see also the considerations in [3] and [4]. In contrast to the previous work in [3] the proof is done via algebraic calculation and not by point/nodewise considerations.

7 Conclusion and Outlook

We presented two independent observations which arose during the study of model refinement and bisimulations. The future work in the direction of the first observation is clear: it will be about the proof of Conjecture 3.4 and Conjecture 3.5. This would show the refinability of every target model with a finite associated cumulative s-dioid.

The second observation (especially Theorem 6.4) shows only how linear fix-points can be determined from fixpoints of the quotient. It is open whether the other way round, i.e. constructing a fixpoint on the quotient equation from an arbitrary fixpoint on the original equation, is also possible.

Acknowledgements. I am grateful to Bernhard Möller and the anonymus referees for valuable remarks and discussions, and to Evgeny Kissin for his piano recital in Munich during which I figured out parts of the proof of Theorem 3.7.

References

1. Baier, C., Katoen, J.P.: Principles of Model Checking. MIT Press (2008)
2. Ciarlet, P.G.: Introduction to Numerical Linear Algebra and Optimisation. Cambridge University Press (1989)
3. Glück, R.: Using Bisimulations for Optimality Problems in Model Refinement. In: de Swart, H. (ed.) RAMICS 2011. LNCS, vol. 6663, pp. 164–179. Springer, Heidelberg (2011)
4. Glück, R., Möller, B., Sintzoff, M.: Model Refinement Using Bisimulation Quotients. In: Johnson, M., Pavlovic, D. (eds.) AMAST 2010. LNCS, vol. 6486, pp. 76–91. Springer, Heidelberg (2011)
5. Gondran, M., Minoux, M.: Graphs, Dioids and Semirings. Springer (2008)
6. Gurney, A.J.T., Griffin, T.G.: Pathfinding through Congruences. In: de Swart, H. (ed.) RAMICS 2011. LNCS, vol. 6663, pp. 180–195. Springer, Heidelberg (2011)
7. Kleene, S.: Introduction to Metamathematics. Wolters-Noordhoff - Groningen (1952)
8. Myhill, J.: Finite automata and the representation of events. WADD TR-57-624, pp. 112–137 (1957)
9. Nerode, A.: Linear automaton transformations. Proceedings of the American Mathematical Society 9(4), 541–544 (1958)
10. Paige, R., Tarjan, R.: Three partition refinement algorithms. SIAM Journal for Computing 16(6), 973–989 (1987)
11. Schmidt, G., Ströhlein, T.: Relations and Graphs: Discrete Mathematics for Computer Scientists. Springer (1993)
12. Tarski, A.: A lattice-theoretical fixpoint theorem and its applications. Pacific Journal of Mathematics 5(2), 285–309 (1955)
13. Winter, M.: A relation-algebraic theory of bisimulations. Fundam. Inf. 83(4), 429–449 (2008)

Relation Algebras, Matrices, and Multi-valued Decision Diagrams

Francis Atampore and Michael Winter*

Department of Computer Science
Brock University
St. Catharines, ON, Canada
{fa10qv,mwinter}@brocku.ca

Abstract. In this paper we want to further investigate the usage of matrices as a representation of relations within arbitrary heterogeneous relation algebras. First, we want to show that splittings do exist in matrix algebras assuming that the underlying algebra of the coefficients provides this operation. Second, we want to outline an implementation of matrix algebras using reduced ordered multi-valued decision diagrams. This implementation combines the efficiency of operations based on those data structures with the general matrix approach to arbitrary relation algebras.

1 Introduction

Relation algebras, and category of relations, in particular, have been extremely useful as a formal system in various areas of mathematics and computer science. Applications range from logic [15, 20], fuzzy relations [25], program development [7], program semantics [16, 26], and graph theory [4, 16] to social choice theory [6][1]. Visualizing relations and computing expression in the language of relations can be very helpful while working within abstract theory. The RelView system [3] was developed for exactly this purpose. The system visualizes relations as Boolean matrices. Internally it uses reduced ordered binary decision diagrams (ROBDDs) in order to provide very efficient implementations of the operations on relations.

The RelView system is based on the standard model of relation algebras, i.e., Boolean matrices or, equivalently, sets of pairs. Therefore, this system cannot visualize computations in non-standard models of the abstract theory of heterogeneous relations. This is particularly important if one considers properties that are true in the standard model but not in all models. An example of such a property is given by the composition of two (heterogeneous) universal relations, i.e., two relations that relate every pair of elements between different sets. In the

---

* The author gratefully acknowledges support from the Natural Sciences and Engineering Research Council of Canada.

[1] This list of topics, papers, and books is not meant to be exhaustive. It is supposed to serve as starting point for further research.

W. Kahl and T.G. Griffin (Eds.): RAMiCS 2012, LNCS 7560, pp. 248–263, 2012.

standard model one will always obtain the universal relation, while this might not be the case in some non-standard models. Another example is given by the relationship between the power set of a disjoint union of two sets A and B and the product of the power set of A and the power set of B. In the standard model both constructions lead to isomorphic objects, while this might not be the case in certain non-standard models.

In [22, 23] it has been shown that relations from arbitrary relation algebras can be represented by matrices. Instead of the Boolean values one has to deal with more general coefficients. As a consequence one obtains that all standard operations on relations correspond to known matrix operations. In this paper we want to extend the general matrix approach to additional operations within relation algebras such as splittings and relational powers. It has been shown by multiple examples [5] that splittings are an important construction in the application of relational methods. Having this construction available shows once more that it is sufficient to use matrices as a representation for arbitrary relation algebras. Furthermore, we want to outline an implementation of matrix algebras using reduced ordered multi-valued decision diagrams. Since this implementation resorts to a data structure similar to that of RelView it combines an efficient computation of the operations on relations with the general matrix approach for arbitrary relation algebras.

The remainder of this paper is organized as follows. In Section 2 we recall the basic theory of heterogeneous relation algebras. After recalling the pseudo-representation theorem using matrices we will show in Section 3 that splittings in matrix algebras do exist if the underlying algebra of the coefficients provides this kind of operation. Finally, we will outline the implementation of matrix algebras using multi-valued decision diagrams in Section 4.

2   Heterogeneous Relation Algebras

In this section we recall some fundamentals on heterogeneous relation algebras. Heterogeneous relation algebras are a categorical version of Tarski's relation algebras with the additional requirements that the underlying Boolean algebras are complete and atomic. For further details we refer to [11, 16, 18].

We will denote the collection of objects and the collection of morphisms of a category $\mathcal{C}$ by $\mathrm{Obj}_\mathcal{C}$ and $\mathrm{Mor}_\mathcal{C}$, respectively. Composition is written as "$;$", which has to be read from left to right, i.e., $f; g$ means "f first, then g". For a morphism f in a category $\mathcal{C}$ with source A and target B we use $f \in \mathcal{C}[A, B]$ and $f : A \to B$ interchangeably. Finally, the identity morphism in $\mathcal{C}[A, A]$ is denoted by $\mathbb{I}_A$.

Definition 1. *A (heterogeneous abstract) relation algebra is a locally small category $\mathcal{R}$. The morphisms are usually called relations. In addition, there is a totally defined unary operation $\breve{}_{AB} : \mathcal{R}[A, B] \longrightarrow \mathcal{R}[B, A]$ between the sets of morphisms, called conversion. The operations satisfy the following rules:*

1. *Every set $\mathcal{R}[A,B]$ carries the structure of a complete atomic Boolean algebra with operations $\sqcup_{AB}, \sqcap_{AB}, \overline{}_{AB}$, zero element $\bot\!\!\!\bot_{AB}$, universal element $\top\!\!\!\top_{AB}$, and inclusion ordering $\sqsubseteq_{AB}$.*
2. *The Schröder equivalences*

$$Q;R \sqsubseteq_{AC} S \iff Q^{\smile};\overline{S} \sqsubseteq_{BC} \overline{R} \iff \overline{S};R^{\smile} \sqsubseteq_{AB} \overline{Q}$$

hold for relations $Q : A \to B, R : B \to C$ and $S : A \to C$.

As usual we omit all indices of elements and operations for brevity if they are not important or clear from the context.

The standard example of a relation algebra is the category **Rel** of sets and binary relations, i.e., sets of ordered pairs, with the usual operations. We will use this example frequently in order to motivate or illustrate definitions and properties of relations. Notice that those relations can also be represented by Boolean matrices as shown in [16–18] and the RelView system [3].

In the following lemma we have summarized several standard properties of relations. We will use them as well as other basic properties such as $\mathbb{I}_A^{\smile} = \mathbb{I}_A$ throughout this paper without mentioning. Proofs can be found in any of the following [11, 16–18].

Lemma 1. *Let $\mathcal{R}$ be a relation algebra and $Q : A \to B$, $R, R_1, R_2 : B \to C$ and $S : A \to C$ be relations. Then we have:*

1. $Q;(R_1 \sqcap R_2) \sqsubseteq Q;R_1 \sqcap Q;R_2,$
2. $Q;R \sqcap S \sqsubseteq (Q \sqcap S;R^{\smile});(R \sqcap Q^{\smile};S)$ *(Dedekind formula),*
3. $Q \sqsubseteq Q;Q^{\smile};Q.$

The inclusion $Q;X \sqsubseteq R$ for two relations $Q : A \to B$ and $R : A \to C$ has a greatest solution $Q\backslash R = \overline{Q^{\smile};\overline{R}}$ in X called the right residual of Q and R. Similarly, the inclusion $Y;S \sqsubseteq R$ has a greatest solution $R/S = \overline{\overline{R};S^{\smile}}$ in Y called the left residual of S and R. In **Rel** the two constructions are given by the relations $Q\backslash R = \{(x,y) \mid \forall z : (z,x) \in Q$ implies $(z,y) \in R\}$ and $R/S = \{(z,u) \mid \forall y : (u,y) \in S$ implies $(z,y) \in R\}$. Using both constructions together we obtain the definition of a symmetric quotient $\mathrm{syQ}(Q,R) = Q\backslash R \sqcap Q^{\smile}/R^{\smile}$. The characteristic property of the symmetric quotient can be stated as

$$X \sqsubseteq \mathrm{syQ}(Q,R) \iff Q;X \sqsubseteq R \text{ and } X;R^{\smile} \sqsubseteq Q^{\smile}.$$

Notice that in **Rel** we have $\mathrm{syQ}(Q,R) = \{(x,y) \mid \forall z : (z,x) \in Q$ iff $(z,y) \in R\}$.

We define the notion of a homomorphism between relation algebras as usual, i.e., as a functor that preserves the additional structure. A pair of homomorphisms $F : \mathcal{R} \to \mathcal{S}, G : \mathcal{S} \to \mathcal{R}$ is called an equivalence iff $F \circ G$ and $G \circ F$ are naturally isomorphic to the corresponding identity functors.

The relational description of disjoint unions of sets is the relational sum [18, 26]. This construction corresponds to the categorical biproduct.

Definition 2. *Let $\{A_i \mid i \in I\}$ be a set of objects indexed by a set I. An object $\sum_{i \in I} A_i$ together with relations $\iota_j \in \mathcal{R}[A_j, \sum_{i \in I} A_i]$ for all $j \in I$ is called a relational sum of $\{A_i \mid i \in I\}$ iff for all $i, j \in I$ with $i \neq j$ the following holds*

$$\iota_i ; \breve{\iota_i} = \mathbb{I}_{A_i}, \qquad \iota_i ; \breve{\iota_j} = \perp\!\!\!\perp_{A_i A_j}, \qquad \bigsqcup_{i \in I} \breve{\iota_i} ; \iota_i = \mathbb{I}_{\sum_{i \in I} A_i}.$$

$\mathcal{R}$ *has relational sums iff for every set of objects the relational sum does exist.*

For a set of two objects $\{A, B\}$ this definition corresponds to usual definition of the relational sum. As known categorical biproducts and hence relational sums are unique up to isomorphism.

A partial equivalence relation Ξ on a set A is an equivalence relation that does not need to be totally defined. Algebraically, it satisfies $\Xi^{\smile} = \Xi$ (symmetric) and $\Xi; \Xi = \Xi$ (idempotent). Because of the defining properties partial equivalence relations are also called symmetric idempotent relations, or sid's for short.

We will omit the proof of the following lemma because of lack of space. The proof can be found in the long version of this paper [1].

Lemma 2. *If $\Xi, \Theta : A \to A$ are partial equivalence relations so that $\Xi; \Theta \sqsubseteq \Theta$, then $\Xi \sqcap \overline{\Theta}$ is a partial equivalence relation with $(\Xi \sqcap \overline{\Theta}); \Xi = \Xi \sqcap \overline{\Theta}$.*

Given a partial equivalence relation Ξ in **Rel** one can compute the set of all equivalence classes whenever Ξ is defined. This concept is called a splitting [11].

Definition 3. *Let $\Xi \in \mathcal{R}[A, A]$ be a partial equivalence relation. An object B together with a relation $\psi \in \mathcal{R}[B, A]$ is called a splitting of Ξ iff*

$$\psi; \psi^{\smile} = \mathbb{I}_B, \qquad \psi^{\smile}; \psi = \Xi.$$

A relation algebra has splittings iff for all partial equivalence relations a splitting exists.

Notice that splittings are unique up to isomorphism, that ψ is total and injective, and that the relation $\psi^{\smile}$ is a partial function. Furthermore, we may distinguish two special cases of splittings. If Ξ is also reflexive, i.e., an equivalence relation, then $\psi^{\smile}$ is total. In that case the splitting corresponds to the construction of the set of equivalence classes with the canonical epimorphism $\psi^{\smile}$ mapping elements to their equivalence class. If $\Xi \sqsubseteq \mathbb{I}_A$, then Ξ corresponds to a subset of A. In this case the splitting becomes the subobject induced by Ξ with ψ the corresponding injection.

The last construction we want to introduce in this section is the abstract counterpart of a power set, called the relational power. The definition is based on the fact that every relation $R : A \to B$ in **Rel** can be transformed into a function $f_R : A \to \mathcal{P}(B)$ where $\mathcal{P}(B)$ denotes the power set of B.

Definition 4. *Let $\mathcal{R}$ be a relation algebra. An object $\mathcal{P}(A)$, together with a relation $\varepsilon_A : A \to \mathcal{P}(A)$ is called a relational power of A iff*

$$\mathrm{syQ}(\varepsilon_A, \varepsilon_A) \sqsubseteq \mathbb{I}_{\mathcal{P}(A)} \quad and \quad \mathrm{syQ}(R, \varepsilon_A) \text{ is total}$$

for all relations $R : B \to A$. $\mathcal{R}$ has relational powers iff the relational power for any object exists.

Notice that in the case of the relation algebra **Rel** and a relation $R : A \to B$ the construction $\mathrm{syQ}(R^{\smile}, \varepsilon_B) : A \to \mathcal{P}(B)$ is the function f_R mentioned above.

For technical reasons we follow [11] and call an object B a pre-power of A if there is a relation $T : A \to B$ so that $\mathrm{syQ}(R, T)$ is total for all relations $R : C \to A$, i.e., a relational power is a pre-power with the additional requirement $\mathrm{syQ}(T, T) \sqsubseteq \mathbb{I}$. If $\mathcal{R}$ has splittings, then we obtain a relational power of A from a pre-power by splitting the equivalence relation $\mathrm{syQ}(T, T)$. This fact indicates once more that splittings are an important construction in the theory of relations.

3   Matrix Algebras

Given a heterogeneous relation algebra $\mathcal{R}$, an algebra of matrices with coefficients from $\mathcal{R}$ may be defined.

Definition 5. *Let $\mathcal{R}$ be a relation algebra. The algebra $\mathcal{R}^+$ of matrices with coefficients from $\mathcal{R}$ is defined by:*

1. *An object of $\mathcal{R}^+$ is a function from an arbitrary index set I to $\mathrm{Obj}_{\mathcal{R}}$.*
2. *For every pair $f : I \to \mathrm{Obj}_{\mathcal{R}}, g : J \to \mathrm{Obj}_{\mathcal{R}}$ of objects from $\mathcal{R}^+$, the set of morphisms $\mathcal{R}^+[f, g]$ is the set of all functions $R : I \times J \to \mathrm{Mor}_{\mathcal{R}}$ so that $R(i, j) \in \mathcal{R}[f(i), g(j)]$ holds.*
3. *For $R \in \mathcal{R}^+[f, g]$ and $S \in \mathcal{R}^+[g, h]$ composition is defined by*

$$(R; S)(i, k) := \bigsqcup_{j \in J} R(i, j); S(j, k).$$

4. *For $R \in \mathcal{R}^+[f, g]$ conversion and negation are defined by*

$$R^{\smile}(j, i) := (R(i, j))^{\smile}, \qquad \overline{R}(i, j) := \overline{R(i, j)}.$$

5. *For $R, S \in \mathcal{R}^+[f, g]$ union and intersection are defined by*

$$(R \sqcup S)(i, j) := R(i, j) \sqcup S(i, j), \qquad (R \sqcap S)(i, j) := R(i, j) \sqcap S(i, j).$$

6. *The identity, zero and universal elements are defined by*

$$\mathbb{I}_f(i_1, i_2) := \begin{cases} \amalg_{f(i_1) f(i_2)} : i_1 \neq i_2 \\ \mathbb{I}_{f(i_1)} : i_1 = i_2, \end{cases}$$

$$\amalg_{fg}(i, j) := \amalg_{f(i) g(j)}, \qquad \mathbb{T}_{fg}(i, j) := \mathbb{T}_{f(i) g(j)}.$$

Obviously, an object in $\mathcal{R}^+$ may be seen as a (in general non-finite) sequence of objects from $\mathcal{R}$, and a morphism in $\mathcal{R}^+$ may be seen as a (in general non-finite) matrix indexed by objects from $\mathcal{R}$. Notice that any Boolean algebra forms a relation algebra if we define composition as the meet operation and converse is the identity function. We obtain Boolean matrices in $\mathbb{B}^+$ if we choose the Boolean algebra with two values $\mathbb{B} = \{0, 1\}$. These matrices are a natural representation of the relations in **Rel**.

The proof of the following result is an easy exercise and is, therefore, omitted.

Lemma 3. $\mathcal{R}^+$ *is a relation algebra.*

Furthermore, the possibility to build disjoint unions of arbitrary sets indexed by a set gives us the following. A detailed proof can be found in [11].

Theorem 1. $\mathcal{R}^+$ *has relational sums.*

In addition to relational sums, matrix algebras also provide the essential part of relational powers. Again, a detailed proof can be found in [11].

Theorem 2. *If $\mathcal{R}$ is small, then $\mathcal{R}^+$ has pre-powers.*

In [11] it was shown that every relation algebra $\mathcal{R}$ can be embedded into an algebra $\mathcal{R}_{sid}$ that has splittings. Furthermore, if $\mathcal{R}$ has relational sums, so does $\mathcal{R}_{sid}$. As a special case we obtain that every matrix algebra can be embedded into an algebra with splittings. However, this new algebra does not need to be a matrix algebra again. Altogether, these constructions do not provide any hint when the matrix algebra itself already provides splittings. Such a characterization is important if we want to consider matrix algebras as a general representation and/or visualization of relation algebras. We will come back to this problem later in this section.

Following the notion used in algebra, we call an object A integral if there are no zero divisors within the algebra $\mathcal{R}[A, A]$. The class of integral objects will define the basis of $\mathcal{R}$.

Definition 6. *An object A of a relation algebra $\mathcal{R}$ is called integral iff for all $Q, R \in \mathcal{R}[A, A]$ the equation $Q; R = \perp\!\!\!\perp_{AA}$ implies $Q = \perp\!\!\!\perp_{AA}$ or $R = \perp\!\!\!\perp_{AA}$. $\mathcal{R}$ is called integral iff all objects of $\mathcal{R}$ are integral. The basis $\mathcal{B}_{\mathcal{R}}$ of $\mathcal{R}$ is defined as the full subcategory given by the class of all integral objects.*

As usual, we omit the index $\mathcal{R}$ in $\mathcal{B}_{\mathcal{R}}$ when its meaning is clear from the context. Notice that the basis is normally a lot "smaller" than the original relation algebra. In particular, if $\mathcal{R}$ has relational sums, then all objects in $\mathcal{B}$ are irreducible objects with respect to the sum construction.

The following theorem is proved in [22, 23]. It can be seen as a pseudo-representation theorem indicating that it is completely sufficient to consider matrix algebras over integral relations algebras when considering the standard operations on relations.

Theorem 3. *Let $\mathcal{R}$ be a relation algebra with relational sums and subobjects and $\mathcal{B}$ the basis of $\mathcal{R}$. Then $\mathcal{R}$ and $\mathcal{B}^+$ are equivalent.*

Since matrix algebras also provide relational sums (Theorem 1) and pre-powers (Theorem 2), these algebras also cover both additional constructions. As mentioned already above it has not been shown that we can perform splittings in matrix algebras. In order to prove such a theorem we will use the following conventions. Suppose $M : f \to f$ is a square matrix where $f : I \to \text{Obj}_{\mathcal{R}}$ is an object of $\mathcal{R}^+$. The set I can be well-ordered by the axiom of the choice, and, hence, is isomorphic to some ordinal number α. For simplicity we identify I and α in the rest of this section and call M a matrix of size α. In addition, we will denote by M_β with $\beta \le \alpha$ the submatrix of M of size β, i.e., $M_\beta : f_\beta \to f_\beta$ with $f_\beta(\gamma) = f(\gamma)$ and $M_\beta(\gamma, \delta) = M(\gamma, \delta)$ for all $\gamma, \delta \le \beta$. Obviously we have $M_\alpha = M$ and $(M_\beta)_\gamma = M_\gamma$ for $\gamma \le \beta \le \alpha$.

In order to explain intuitively how to compute the splitting of a partial equivalence relation in matrix form we first consider the case of Boolean matrices. Notice that the Boolean values seen as a relation algebra form a category with one object that does not provide splittings. We have to add a so-called zero object, i.e., an object 0 for which $\bot\!\bot_{00} = \top\!\top_{00}$. Now, each row of a partial equivalence relation written as a Boolean matrix corresponds to one equivalence class. A 1 (or true) entry indicates that the element of the corresponding column is in the equivalence class, and a 0 (or false) indicates that the element is not in the class. It is easy to verify that keeping one instance of each different row gives the splitting. Instead of removing multiple copies of a row, we could also replace the row by a row of $\bot\!\bot$'s with source 0. This works because a zero object is a unit for the relational sum, i.e., $A + 0$ is isomorphic to A for all objects A. Notice that the latter approach has the advantage that the matrix of the splitting has the same size than the original matrix. If we want to generalize this procedure, we first notice that each element on the diagonal of a partial equivalence relation is itself a partial equivalence relation. Those relations are not necessarily independent of each other. The relation $M(\beta, \gamma)$ relates the two partial equivalence relations $M(\beta, \beta)$ and $M(\gamma, \gamma)$. If we split $M(\beta, \beta)$, then we only have to split the remaining part of $M(\gamma, \gamma)$ that was not already covered by the splitting of $M(\beta, \beta)$, i.e., the relation $M(\gamma, \gamma) \sqcap \overline{M(\gamma, \beta); M(\beta, \gamma)}$. Notice that the relationship between the diagonal elements is trivial in the Boolean case. We will illustrate the general splitting process by an example at the end of this section.

In order to establish a theorem about splittings in matrix algebras we first need to prove some basic properties of partial equivalence relations in matrix form. We will omit the proofs of the lemma below, which can be found in [1].

Lemma 4. *If $M : f \to f$ is a partial equivalence relation in $\mathcal{R}^+$ of size α, then we have for all β, γ, δ and relations R:*

1. $M(\beta, \gamma); M(\gamma, \delta) \sqsubseteq M(\beta, \delta)$,
2. $M(\beta, \beta); M(\beta, \gamma) = M(\beta, \gamma)$,
3. $M(\beta, \gamma); \overline{M(\gamma, \delta); R} \sqsubseteq \overline{M(\beta, \delta); R}$.

The next lemma allows a stepwise definition of partial equivalence relations based on the diagonal elements of M.

Lemma 5. *If $M : f \to f$ is a partial equivalence relation in $\mathcal{R}^+$ of size α, then the relations $\Theta_\beta := \bigsqcup_{\gamma < \beta} M(\beta, \gamma); M(\gamma, \beta)$ are partial equivalence relations with $M(\beta, \beta); \Theta_\beta \sqsubseteq \Theta_\beta$. Furthermore, the relations*

$$\Xi_\beta := M(\beta,\beta) \sqcap \overline{\Theta_\beta} = M(\beta,\beta) \sqcap \prod_{\gamma<\beta} \overline{M(\beta,\gamma); M(\gamma,\beta)}$$

for $1 \le \beta \le \alpha$ are also partial equivalence relations.

Our final lemma states some properties about the partial equivalence relations Ξ_β and their splittings.

Lemma 6. *Let $M : f \to f$ be a partial equivalence relation in $\mathcal{R}^+$ of size α, and Ξ_β for $1 \le \beta \le \alpha$ be the partial equivalence relations defined in Lemma 5. Furthermore, suppose $R_\beta : A_\beta \to f(\beta)$ splits Ξ_β in $\mathcal{R}$. Then we have:*

1. $M(\gamma,\beta); R_\beta^\smile = \perp\!\!\!\perp_{f(\gamma)A_\beta}$ *for all $\gamma < \beta$.*
2. $M(\beta,\gamma); \Xi_\gamma; M(\gamma,\beta) = M(\beta,\gamma); M(\gamma,\beta) \sqcap \prod_{\delta<\gamma} \overline{M(\beta,\delta); M(\delta,\beta)}.$
3. $\bigsqcup_{\gamma\le\delta} M(\beta,\gamma); \Xi_\gamma; M(\gamma,\beta) = \bigsqcup_{\gamma\le\delta} M(\beta,\gamma); M(\gamma,\beta)$ *for all $1 \le \delta \le \beta$.*

We are now ready to prove our main theorem of this section.

Theorem 4. *Let $\mathcal{R}$ be relation algebra with splittings. Then $\mathcal{R}^+$ has splittings.*

Proof. Let $M : f \to f$ be a partial equivalence relation in $\mathcal{R}^+$ of size α, and let $R_\beta : A_\beta \to f(\beta)$ be a splitting of Ξ_β in $\mathcal{R}$. Define $g : \alpha \to \mathrm{Obj}_\mathcal{R}$ and $N : g \to f$ by

$$g(\beta) = A_\beta,$$

$$N(\beta,\gamma) = \begin{cases} R_\beta; M(\beta,\gamma) & \text{iff } \beta \le \gamma, \\ \perp\!\!\!\perp_{A_\beta f(\gamma)} & \text{otherwise,} \end{cases}$$

with $\beta, \gamma \le \alpha$. We have

$$(N; N^\smile)(\beta,\delta) = \bigsqcup_{\gamma\le\alpha} N(\beta,\gamma); N^\smile(\gamma,\delta)$$

$$= \bigsqcup_{\gamma\le\alpha} N(\beta,\gamma); N(\delta,\gamma)^\smile$$

$$= \bigsqcup_{\max(\beta,\delta)\le\gamma\le\alpha} N(\beta,\gamma); N(\delta,\gamma)^\smile \qquad \text{Def. of } N$$

$$= \bigsqcup_{\max(\beta,\delta)\le\gamma\le\alpha} R_\beta; M(\beta,\gamma); (R_\delta; M(\delta,\gamma))^\smile$$

$$= \bigsqcup_{\max(\beta,\delta)\le\gamma\le\alpha} R_\beta; M(\beta,\gamma); M(\gamma,\delta); R_\delta^\smile \qquad M \text{ symmetric}$$

$$= R_\beta; \left(\bigsqcup_{\max(\beta,\delta)\le\gamma\le\alpha} M(\beta,\gamma); M(\gamma,\delta) \right); R_\delta^\smile$$

If $\beta \neq \delta$, then we obtain

$$R_\beta; \left(\bigsqcup_{\max(\beta,\delta) \leq \gamma \leq \alpha} M(\beta,\gamma); M(\gamma,\delta) \right); R_\delta^{\smallsmile}$$

$$\sqsubseteq R_\beta; \left(\bigsqcup_{\gamma \leq \alpha} M(\beta,\gamma); M(\gamma,\delta) \right); R_\delta^{\smallsmile}$$

$$= R_\beta; (M; M)(\beta,\delta); R_\delta^{\smallsmile}$$

$$= R_\beta; M(\beta,\delta); R_\delta^{\smallsmile} \qquad\qquad M \text{ idempotent}$$

$$= \amalg_{A_\beta A_\delta},$$

where the last equality follows from Lemma 6(1) since we have either $\beta < \delta$ or $\delta < \beta$. If $\beta = \delta$, then we compute

$$R_\beta; \left(\bigsqcup_{\beta \leq \gamma \leq \alpha} M(\beta,\gamma); M(\gamma,\beta) \right); R_\beta^{\smallsmile}$$

$$= R_\beta; M(\beta,\beta); R_\beta^{\smallsmile} \qquad\qquad \text{Lemma 4(2) for } \gamma = \beta \text{ and (1) otherwise}$$

$$= R_\beta; R_\beta^{\smallsmile}; R_\beta; M(\beta,\beta); R_\beta^{\smallsmile} \qquad R_\beta; R_\beta^{\smallsmile} = \mathbb{I}_{A_\beta}$$

$$= R_\beta; \Xi_\beta; M(\beta,\beta); R_\beta^{\smallsmile} \qquad R_\beta^{\smallsmile}; R_\beta = \Xi_\beta$$

$$= R_\beta; \Xi_\beta; R_\beta^{\smallsmile} \qquad\qquad \text{Lemma 5 and Lemma 2}$$

$$= R_\beta; R_\beta^{\smallsmile}; R_\beta; R_\beta^{\smallsmile} \qquad R_\beta^{\smallsmile}; R_\beta = \Xi_\beta$$

$$= \mathbb{I}_{A_\beta},$$

i.e., we have just shown that $N; N^{\smallsmile} = \mathbb{I}_g$. In order to verify that $N^{\smallsmile}; N = M$ consider

$$(N^{\smallsmile}; N)(\beta,\delta) = \bigsqcup_{\gamma \leq \alpha} N^{\smallsmile}(\beta,\gamma); N(\gamma,\delta)$$

$$= \bigsqcup_{\gamma \leq \alpha} N(\gamma,\beta)^{\smallsmile}; N(\gamma,\delta)$$

$$= \bigsqcup_{\gamma \leq \min(\beta,\gamma)} N(\gamma,\beta)^{\smallsmile}; N(\gamma,\delta) \qquad \text{Def. } N$$

$$= \bigsqcup_{\gamma \leq \min(\beta,\gamma)} M(\beta,\gamma); R_\gamma^{\smallsmile}; R_\gamma; M(\gamma,\delta) \qquad \text{Def. } N$$

$$= \bigsqcup_{\gamma \leq \min(\beta,\gamma)} M(\beta,\gamma); \Xi_\gamma; M(\gamma,\delta).$$

We immediately obtain from Lemma 4(1)

$$M(\beta,\gamma); \Xi_\gamma; M(\gamma,\delta) \sqsubseteq M(\beta,\gamma); M(\gamma,\gamma); M(\gamma,\delta) \sqsubseteq M(\beta,\delta),$$

and, hence, $(N^{\smile}; N)(\beta, \delta) \sqsubseteq M(\beta, \delta)$. For the converse inclusion assume $\beta \leq \delta$. Then we have

$$
\begin{aligned}
&M(\beta, \delta) \\
&= M(\beta, \beta); M(\beta, \delta) && \text{Lemma 4(2)} \\
&= \left(\bigsqcup_{\gamma \leq \beta} M(\beta, \gamma); M(\gamma, \beta) \right); M(\beta, \delta) && \text{Lemma 4(1) and (2) for } \gamma = \beta \\
&= \left(\bigsqcup_{\gamma \leq \beta} M(\beta, \gamma); \Xi_\gamma; M(\gamma, \beta) \right); M(\beta, \delta) && \text{Lemma 6(3)} \\
&= \bigsqcup_{\gamma \leq \beta} M(\beta, \gamma); \Xi_\gamma; M(\gamma, \beta); M(\beta, \delta) \\
&\sqsubseteq \bigsqcup_{\gamma \leq \beta} M(\beta, \gamma); \Xi_\gamma; M(\gamma, \delta). && \text{Lemma 4(3)}
\end{aligned}
$$

The case $\delta \leq \beta$ is shown analogously. This completes the proof. $\qquad\square$

We want to illustrate the previous theorem by an example. In this example we use $\mathcal{B}$-fuzzy relations where $\mathcal{B}$ is a Boolean algebra. This is a special case of so-called $\mathcal{L}$-fuzzy relations where $\mathcal{L}$ is a Heyting algebra. For further details on these kind of fuzzy relations we refer to [25]. As already mentioned above every Boolean algebra is also a relation algebra where composition is given by the meet operation and converse is the identity. Let B_{abc} be the Boolean algebra with the three atoms a, b, c. We will denote arbitrary elements of B_{abc} by the sequence of atoms below that element, e.g., ab or bc or abc, or 0 for the least element. In order to create a relation algebra based on B_{abc} that has splittings we need to consider also the Boolean algebras of all elements smaller or equal a given element x of B_{abc}. We will denote this Boolean algebra by B_x. Now, the objects of the relation algebra are the Boolean algebras B_x for every $x \in B_{abc}$, and the morphisms between B_x and B_y are given by the Boolean algebra $B_{x \cap y}$, where $x \cap y$ is the intersection of the two sets of atoms x and y. For our example we consider the object $[B_{abc}, B_{abc}, B_{abc}, B_{abc}]$ and following partial equivalence relation in matrix form:

$$
M = \begin{array}{c} \\ B_{abc} \\ B_{abc} \\ B_{abc} \\ B_{abc} \end{array} \begin{pmatrix} \overset{\displaystyle B_{abc}}{ab} & \overset{\displaystyle B_{abc}}{0} & \overset{\displaystyle B_{abc}}{b} & \overset{\displaystyle B_{abc}}{0} \\ 0 & ab & 0 & a \\ b & 0 & bc & 0 \\ 0 & a & 0 & a \end{pmatrix}
$$

Following the proof of the theorem above we obtain the following four partial equivalence relations $\Xi : B_{abc} \to B_{abc}$.

$$\Xi_1 := ab,$$

$$\Xi_2 := ab \sqcap \overline{0 \sqcap 0} = ab \sqcap abc = ab,$$

$$\Xi_3 := bc \sqcap \overline{b} \sqcap b \sqcap \overline{0 \sqcap 0} = bc \sqcap ac \sqcap abc = c,$$

$$\Xi_4 := a \sqcap \overline{0 \sqcap 0} \sqcap \overline{a} \sqcap a \sqcap \overline{0 \sqcap 0} = a \sqcap abc \sqcap bc \sqcap abc = 0.$$

The splittings for each of those relations is given by

$$R_1 = R_2 := ab : B_{ab} \to B_{abc}, \quad R_3 := c : B_c \to B_{abc}, \quad R_4 := 0 : B_0 \to B_{abc}.$$

Notice that the source object of each of those relations is different from B_{abc}. We obtain the matrix N as:

$$N = \begin{array}{c c} & \begin{array}{c c c c} B_{abc} & B_{abc} & B_{abc} & B_{abc} \end{array} \\ \begin{array}{c} B_{ab} \\ B_{ab} \\ B_c \\ B_0 \end{array} & \left(\begin{array}{c c c c} ab & 0 & b & 0 \\ 0 & ab & 0 & a \\ 0 & 0 & c & 0 \\ 0 & 0 & 0 & 0 \end{array} \right) \end{array}$$

An easy computation shows that N indeed splits M. Moreover, since B_0 is the trivial Boolean algebra with $0 = 1$ the last row of the matrix can actually be dropped from N. In the abstract language of relation algebras this corresponds to move from an object $A + 0$ to A where 0 is a null object, i.e., an object that is neutral (up to isomorphism) with respect to the relational sum.

4   Matrix Algebras and Multi-valued Decision Diagrams

In this section we want to introduce multi-valued Decision Diagrams (MDDs) and how they can be used to implement heterogeneous relation using the matrix algebra approach. Decision diagrams are one of the contemporary symbolic data structures used to represent logic functions. A multiple-valued decision diagram is a natural extension of reduced ordered decision diagrams (ROBDD) [9] to the multi-valued case. MDDs are considered to be more efficient, and they perform better than ROBDDs with respect to memory size and path length [13].

Let V be a set of finite size r. An r-valued function f is a function mapping V^n for some n to V. We will identify the n input values of f using a set of variables $X = \{x_0, x_1, ..., x_{n-1}\}$. Each x_i as well as $f(X)$ is r-valued, i.e., it represents an element from V. The function f can be represented by a multi-valued decision diagram. Such a decision diagram is directed acyclic graph (DAG). Each terminal node is labeled by a distinct value from V, and every non-terminal node is labeled by an input variable x_i and has r outgoing edges [12].

An MDD is ordered (OMDD) if there is an order on the set of variables X so that for every path from the root to a leave node all variables appear at most once in that order. Furthermore, a MDD is called reduced if the graph does not contain isomorphic subgraphs and no nodes for which all r children are isomorphic. Notice that a reduced MDD will have at most r leaves, one for each value of V. A MDD that is ordered and reduced is called a reduced ordered multi-valued decision diagram (ROMDD). Both ROBDDs and ROMDDs have been widely studied. Most of the techniques used when implementing a package for the creation and manipulation of ROMDDs are those already known from the binary case. These techniques includes edge negation, adjacent level interchange, operator nodes and logical operation [14].

MDDs are usually traversed in one of the following three ways:

1. A depth-first traversal starting at the top node and moving along the edges from each node to the descendants or child nodes. This technique is a very well-known conventional graph traversal.
2. ROMDDs can be traversed horizontally by moving from one node to another of all nodes labeled by the same variable. This corresponds to a specific breath-first traversal.
3. ROMDDs can also be traversed by applying both techniques described above at the same time.

We want to illustrate by an example how relations can be implemented using MDDs. We will assume that relations are given as matrices. The elements of the matrices become the leaf or terminal nodes of the MDD after encoding the domain and the range of the relation by a suitable set of variables. In our example we want to use so-called $\mathcal{L}$-fuzzy relations as already mentioned in the previous section. Now consider the following Heyting algebra $\mathcal{L}$:

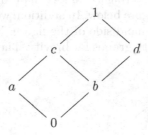

From the figure above we can derive operation tables for the meet and the join operation of the lattice $\mathcal{L}$. For example, we have $c \sqcap d = b$, $a \sqcap b = 0$, $a \sqcap c = a$, $1 \sqcup b = 1$, and $a \sqcup b = c$.

Now suppose that R and S are two $\mathcal{L}$-fuzzy relations represented by the following matrices, i.e., elements from $\mathcal{L}^+$:

$$R = \begin{pmatrix} a & b & a \\ 0 & 1 & c \end{pmatrix} \quad S = \begin{pmatrix} a & a & 1 \\ 0 & b & d \end{pmatrix}$$

Using the meet operation of the underlying lattice $\mathcal{L}$ we can compute the intersection (or meet) of R and S as:

$$T := R \sqcap S = \begin{pmatrix} a & 0 & a \\ 0 & b & b \end{pmatrix}$$

In order to implement these relations using MDDs we have to encode the corresponding matrices first as functions and then as graphs. We will adopt the method used in the RelView system for the binary case to MDDs. First we have to encode the row and column indices by variables ranging over the lattice $\mathcal{L}$. Since $\mathcal{L}$ has 6 elements, we need only one variable for the rows and one for the columns of R, and similarly for S. However, in order to obtain a totally defined function we have to enlarge the matrices so that its size is a power of the number of elements, i.e., the corresponding function is defined for every possible input for each variable. We will fill the new entries with 0's. The figure below shows R enlarged to a proper size with its rows and columns labeled by values, i.e., by elements of the lattice $\mathcal{L}$, for the row variable u and the column variable v. In addition, the figure shows the corresponding function f_R encoding R where _ stands for an arbitrary parameter not listed before:

$$
\begin{array}{c}
\begin{array}{ccccccc}
 & 0 & a & b & c & d & 1 \\
0 & a & b & a & 0 & 0 & 0 \\
a & 0 & 1 & c & 0 & 0 & 0 \\
b & 0 & 0 & 0 & 0 & 0 & 0 \\
c & 0 & 0 & 0 & 0 & 0 & 0 \\
d & 0 & 0 & 0 & 0 & 0 & 0 \\
1 & 0 & 0 & 0 & 0 & 0 & 0
\end{array}
\end{array}
\qquad
\begin{array}{ccc}
u & v & f_R(u,v) \\
\hline
0 & 0 & a \\
0 & a & b \\
0 & b & a \\
a & 0 & 0 \\
a & a & 1 \\
a & b & c \\
\hline
- & - & 0
\end{array}
$$

Now based on the variable ordering $u < v$ we can produce a MDD representing f_R as shown on the left in the figure below. In addition, we present the corresponding reduced graph on the right-hand side of the figure. Notice that we only display the essential part of R in both graphs for brevity. The additional 0's are skipped:

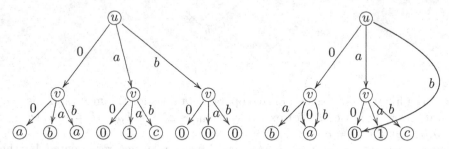

Similarly, we construct the ROMDD for S shown below:

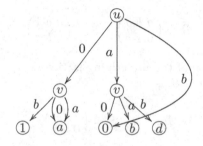

In order to compute the ROMDD for T from the ROMDDs for R and S we use an operation called `apply` on ROMDDs. This operation applies a function to a leaf node of R and the corresponding leaf node of S. Corresponding leaf nodes are determined by the same path from the root to the leaf in both graphs, i.e., the same spot in each matrix. The function that is applied to the leaf nodes is given by a table as mentioned above. For example, if we follow the edges labeled 0a (in that order) in the graph for R we get to a leaf labeled with b. In the graph of S we obtain a. The meet operation of $\mathcal{L}$ that is passed as parameter of `apply` will give 0, the leaf node of the path $0a$ in the graph of T.

The operation `apply` can be used to implement several operations of relations on their ROMDD representation. Another version of the `apply` operation applies an unary function to a each leaf node of a single ROMDD. This version can be used to implement relation algebraic operations such as complement and transpose. In addition to these apply operations we also use a so-called abstraction operation. This operation applies a function to all elements of an entire row or column of a matrix. It behave similarly to a `fold` in a functional programming language, and it reduces the size of the matrix. For instance, if T is a $n \times m$ matrix and we apply the abstraction operation with the abstraction variable set to the entire set of row encoding variables, then the result of abstraction is a row vector, that is $1 \times m$ matrix. The product of two matrices can be computed by a combination of the apply and abstraction operations. We refer to [2, 14] for more details on operations on MDDs.

The splitting of a partial equivalence relation $M : f \to f$ can be computed using the function `apply` only. First of all, obtaining the Ξ_β's, the A_β's, and the R_β's does not use any matrix operation accept accessing the coefficients of the matrix. Given these components we generate a new MDD representing the following matrix:

$$
R = \begin{pmatrix}
R_1 & R_1 & R_1 \cdots \\
\amalg_{A_2 f(1)} & R_2 & R_2 \cdots \\
\amalg_{A_3 f(1)} & \amalg_{A_3 f(2)} & R_3 \cdots \\
\vdots & \vdots & \vdots & \ddots
\end{pmatrix}
$$

If we now use `apply` with R, M and the composition operation of the underlying algebra of the coefficients we will obtain the matrix N of Theorem 4. Notice, however, that this application of `apply` does not correspond to any operation of a relation algebra.

4.1 RelMDD - A Library for Manipulating Relations Based on MDDs

The RelView system [3] implements relations in **Rel** using Boolean matrices represented as binary decision diagrams (BDDs) [8]. Our library RelMDD implements arbitrary heterogeneous relation algebras using the matrix algebra approach represented by ROMDDs. RelMDD is a library written in the programming language C. It is a package that can be imported by other programs and/or languages such as Java and Haskell when programming or manipulating arbitrary relations. The implementation is currently restricted to the basic operations of relation algebras, i.e., union, intersection, composition, converse, and complement. By design the package is capable of manipulating relations from both classes of models, standard models and non-standard models of relation algebras. In our implementation MDDs were implemented using algebraic decision diagrams [2]. By taking this approach we were able to use a well-known package for these diagrams called CUDD [19].

5   Conclusion and Future Work

In this paper we have shown that splittings do exist in matrix algebras assuming that the underlying algebra of the coefficients provides this operation. This shows once more that it is sufficient to use matrices as a representation for arbitrary relation algebras. In addition, we have outlined an implementation of matrix algebras using ROMDDs. This implementation combines two major advantages over a regular array implementation of matrices. First of all, it is suitable for arbitrary relation algebras and is not restricted to the standard model **Rel**. In addition, it uses an advanced data structure that is known to work more efficiently. The RelView system, in particular, has proven that an implementation of relations using decision diagrams is of great benefit.

The package implements all standard operations on relations. A future project will add further operations such as sums and splittings. The latter will then also allow to compute relational powers and so-called weak relational products [24]. Another project will be a suitable module for the programming language Haskell that makes the RelMDD package available in this language.

References

1. Atampore, F., Winter, M.: Relation Algebras, Matrices, and Multi-Valued Decision Diagrams. Brock University, Dep. of Computer Science Report CS-12-03 (2012), http://www.cosc.brocku.ca/research/reports
2. Bahar, I.R., Frohm, E.A., Gaona, C.M., Hachtel, G.D., Macii, E., Pardo, A., Somenzi, F.: Algebraic Decision Diagrams amd their Applications. In: Proceedings of the International Conference on Computer-Aided Design, pp. 188–191. IEEE (1993)
3. Berghammer, R., Neumann, F.: RELVIEW – An OBDD-Based Computer Algebra System for Relations. In: Ganzha, V.G., Mayr, E.W., Vorozhtsov, E.V. (eds.) CASC 2005. LNCS, vol. 3718, pp. 40–51. Springer, Heidelberg (2005)
4. Berghammer, R.: Applying Relation Algebra and RelView to solve Problems on Orders and Lattices. Acta Informatica 45(3), 211–236 (2008)
5. Berghammer, R., Winter, M.: Embedding Mappings and Aplittings with Applications. Acta Informatica 47(2), 77–110 (2010)
6. Berghammer, R., Rusinowska, A., de Swart, H.C.M.: Applying Relation Algebra and RelView to Measures in a Social Network. European Journal of Operational Research 202, 182–195 (2010)
7. Bird, R., de Moor, O.: Algebra of Programming. Prentice Hall (1997)

8. Brace, K.S., Rudell, L.R., Bryant, E.R.: Efficient Implementation of a BDD Package. In: ACM/IEEE Design Automation Conference, pp. 40–42. IEEE (1990)
9. Bryant, R.E.: Graph-based Algorithms for Boolean Function Manipulation. IEEE Trans. on Computers 35(8), 677–691 (1986)
10. Chin, L.H., Tarski, A.: Distributive and Modular Laws in the Arithmetic of Relation Algebras. University of California Press, Berkley and Los Angeles (1951)
11. Freyd, P., Scedrov, A.: Categories, Allegories. North-Holland (1990)
12. Miller, D.M., Drechsler, R.: Implementing a Multiple-Valued Decision Diagram Package. In: Proceedings of the 28th International Symposium on Multiple-Valued Logic (ISMVL 1998), pp. 52–57. IEEE (1998)
13. Nagayama, S., Sasao, T.: Code Generation for Embedded Systems using Heterogeneous MDDs. In: 12th Workshop on Synthesis and System Integration of Mixed Information Technologies (SASIMI 2003), pp. 258–264 (2003)
14. Nagayama, S., Sasao, T.: Compact Representations of Logic Functions using Heterogeneous MDDs. In: Proceedings of the 33rd International Symposium on Multiple-Valued Logic (ISML 2003), pp. 3168–3175. IEEE (2003)
15. Balbiani, P., Orlowska, E.: A Hierarchy of Modal Logics with Relative Accessibility Relations. Journal of Applied Non-Classical Logics 9(2-3) (1999)
16. Schmidt, G., Ströhlein, T.: Relationen und Graphen. Springer (1989); English version: Relations and Graphs. Discrete Mathematics for Computer Scientists. EATCS Monographs on Theoret. Comput. Sci. Springer (1993)
17. Schmidt, G.: Relational Mathematics. Encyclopedia of Mathematics and its Applications, vol. 132. Cambridge University Press (2010)
18. Schmidt, G., Hattensperger, C., Winter, M.: Heterogeneous Relation Algebras. In: Brink, C., Kahl, W., Schmidt, G. (eds.) Relational Methods in Computer Science. Advances in Computer Science. Springer, Vienna (1997)
19. Somenzi, F.: CUDD: CU Decision Diagram Package; Release 2.5.0. Department of Electrical, Computer, and Energy Engineering, University of Colorado (2012), http://vlsi.colorado.edu/~fabio/CUDD/cuddIntro.html
20. Tarski, A., Givant, S.: A Formalization of Set Theory without Variables. Amer. Math. Soc. Colloq. Publ.41 (1987)
21. Winter, M.: Strukturtheorie Heterogener Relationenalgebren mit Anwendung auf Nichtdetermismus in Programmiersprachen. Dissertationsverlag NG Kopierladen GmbH, München (1998)
22. Winter, M.: Relation Algebras are Matrix Algebras over a suitable Basis. University of the Federal Armed Forces Munich, Report Nr. 1998-05 (1998)
23. Winter, M.: A Pseudo Representation Theorem for various Categories of Relations. TAC Theory and Applications of Categories 7(2), 23–37 (2000)
24. Winter, M.: Weak Relational Products. In: Schmidt, R.A. (ed.) RelMiCS/AKA 2006. LNCS, vol. 4136, pp. 417–431. Springer, Heidelberg (2006)
25. Winter, M.: Goguen Categories - A Categorical Approach to L-fuzzy Relations. Trends in Logic, vol. 25 (2007)
26. Zierer, H.: Relation Algebraic Domain Constructions. Theoret. Comput. Sci. 87, 163–188 (1991)

Incremental Pseudo Rectangular Organization of Information Relative to a Domain

Sahar Ismail and Ali Jaoua

Computer Science and Engineering Department, Qatar University
saharisl@yahoo.com, jaoua@qu.edu.qa

Abstract. Information resources in today's cyber communities over the World Wide Web are increasingly growing in size with an ever increasing pace of change. As information demand increases, more knowledge management and retrieval applications need to exhibit a degree of resilience towards information change, and must be able to handle incremental changes in a reasonable time. In this paper we are defining a new system that utilizes new conceptual methods using the notion of pseudo maximal rectangles (i.e. the union of all non enlargeable rectangles containing a pair (a, b) of a binary relation) for managing incremental information organization and structuring in a dynamic environment. The research work in hand focuses on managing changes in an information store relevant to a specific domain of knowledge attempted through addition and deletion of information. The incremental methods developed in this work should support scalability in change-prone information stores and be capable of producing updates to end users in an efficient time. The paper will also discuss some algorithmic aspects and evaluation results concerning the new methods.

Keywords: pseudo maximal rectangle, incremental rectangular decomposition, incremental structuring, information organization.

1 Introduction

The change in technology and information-sharing culture over the various forms of media have introduced new ways of doing business and imposed high demand on change-prone applications. Business enterprises are relying more on their information processing capabilities to gain a competitive market edge and executives are relying more on various forms of decision support systems that act as a central processing hub for knowledge extraction and organization of resources fed from various channels to support real-time decision making and day-to-day business. Such systems capable of performing efficient knowledge extraction, summarization and organization are not new. However, the amount of information handled, the pace of change in that information (volatility), and the response time required are all stress factors requiring a new way of thinking and new design approaches.

Plenty of techniques in information science, information retrieval, data mining, and machine learning have been extensively used to address the problem

W. Kahl and T.G. Griffin (Eds.): RAMiCS 2012, LNCS 7560, pp. 264–277, 2012.

of efficient incremental information management. Using Formal Concept Analysis (FCA) [5] and relational algebra for information analysis is advantageous over statistical data analysis techniques since the former capitalizes on structural similarities in the information processed for recognizing and generalizing knowledge [3]. The Galois lattice structure is one common and argumentatively comprehensive information representation model for structured information. To address the problem of efficient incremental update of the structured information spaces using FCA and relational algebra methods, an increasing number of research work have been done on developing algorithms that handle the Galois lattice more efficiently without necessarily re-computing the whole lattice structure on the emergence of new objects or attributes in an information store. Other methods have approached the problem differently by replacing the problematic size of the Galois lattice with an approximate information structure holding a minimal number of maximal rectangles covering the entire space [10]. Although finding a minimal coverage of maximal rectangles for an information context is an NP-complete problem [2], it has the advantage of reducing rectangular representation by reducing the number of rectangles considerably compared to the Galois lattice structure. However, methods based on minimal decomposition into maximal rectangles do not exhibit the same degree of accuracy in terms of the resulting rectangular representation of a space of information since they represent an approximation [8]. Although lots of optimization algorithms address the problem of minimal coverage extraction from a space of information, aspects related to incremental information organization and maintenance are still a hot research topic.

In this paper we will present a new method for managing information stores incrementally using a new approximation technique. The new method employs pseudo maximal rectangles or unions of maximal rectangles to structure information incrementally in an effective fashion.

2   Related Work

The problem of incremental information structuring and organization using FCA and relational algebra methods has been the subject of research for a long time. The main challenge faced by most of these methods is to implement them in a reasonably efficient way suitable for real-time systems. Methods proposed for handling incremental information organization based on conceptual analysis methods can be categorized into few types. One type discussed the problem of incremental information organization in an information space organized as a Galois lattice. Such methods include the work of Godin et al. [4] which describes a set of algorithms for updating a lattice structure using the cardinality of maximal rectangles. This method proposed addition update operations of domain objects only converging in quadratic time in terms of the number of maximal rectangles in a Galois lattice which has a theoretical exponential growth. Other methods based on the Galois lattice employ the local lattice structure for updating the Galois lattice with newly arriving data incrementally as described by Carpineto and Romano [1]. This method proposed addition and deletion update operations of both objects and attributes in a Galois lattice converging in quadratic time in terms of the number of maximal rectangles in a Galois lattice for the addition operation while in linear time for the deletion operation. The work of Valtchev et al. [13] approaches the problem differently through one of the most common practices in computer science. It is based on the divide-and-conquer philosophy by partitioning the binary context into fragments for faster matrix manipulation rather than processing the whole context at once.

Research has also been done on incremental conceptual methods based on minimal rectangular coverage of an information store that models maximal rectangles in non lattice-like structures. Such approaches focus mainly on optimizing information extraction, organization and update time. The work of Jaoua et al. [7] addresses this problem and proposes a solution for managing the minimal coverage set incrementally through addition and deletion operations of objects and attributes converging in quadratic time in terms of the number of maximal rectangles in the minimal coverage set. It is worth noting that since these methods rely on the Galois lattice or on minimal coverage sets, they show exponential growth in terms of the addition and deletion operations since the number of maximal rectangles has a theoretical exponential growth.Significant work has also been done in the area of supervised machine learning to address the incremental rectangular updates as in the work of Fisher [9], Hunt et al. [10] and Reinke and Michalski [2].

3 Background

To understand the newly proposed methods in the context of FCA and relational algebra we present here basic definitions and formal background bridging the gap with relational algebra.

3.1 Binary Relations

A binary relation R between two finite sets G and M is a subset of the Cartesian product $G \times M$. An element in R is denoted by (a, b), where b denotes an image of a by R [8]. For a binary relation R, the following subsets are associated:

1. The set of images of a defined by: $a.R = \{b \mid (a, b) \in R\}$.
2. The set of antecedents of b defined by: $R.b = \{a \mid (a, b) \in R\}$.
3. The domain of R is defined by: $Dom(R) = \{a \mid \exists b : (a, b) \in R\}$.
4. The range of R is defined by: $Ran(R) = \{b \mid \exists a$ such that $(a, b) \in R\}$.
5. The cardinality of R defined by: $Card(R) =$ number of pairs in R.
6. Let R and R' be two binary relations on $A \times C$ and $C \times B$, respectively. We define the relative product (or composition) of R and R' as the relation on $A \times B$ given by $R; R' = \{(a, b) \mid \exists t$ such that $(a, t) \in R$ and $(t, b) \in R'\}$,
7. The converse relation of R on $M \times G$ is given by: $R^{-1} = \{(a, b) \mid (b, a) \in R\}$.

8. For a in G, we define the partial identity relation
   $I(a.R) = \{(b, b) \in M \times M \mid (a, b) \in R\}$.
   Similarly, for b in M, we define the partial identity relation
   $I(R.b) = \{(a, a) \in G \times G \mid (a, b) \in R\}$.
9. The complement of the binary relation R with respect to its typing $G \times M$
   is the relation on $G \times M$ given by $\overline{R} := \{(a, b) \mid (a, b) \notin R\}$.

3.2 Rectangles and Maximal Rectangles

Rectangle

Let R be a binary relation defined between two sets G and M. Given sets $A \subseteq G$ and $B \subseteq M$, a rectangle of R is a pair (A, B) such that $A \times B \subseteq R$ where A is the domain of the rectangle and B its range. The rectangle closure $R*$ of a binary relation is defined by the Cartesian product: $R* = Dom(R) \times Ran(R)$ [7]. In information retrieval system a rectangle (A, B) composed of $Card(A) \times Card(B)$ elements is represented in an economical way in terms of memory space by only $Card(A) + Card(B)$ elements by linking set A to set B through only one intermediate element representing the rectangle instead of linking each element of A to each element of B. The difference $Card(A) \times Card(B) - (Card(A) + Card(B))$ is the economy of the rectangle (A, B).

Maximal Rectangle

Let R be a binary relation defined between two sets G and M, a rectangle (A, B) of R is maximal if for all A' and B', $A \times B \subseteq A' \times B' \subseteq R \rightarrow A = A'$ and $B = B'$ [7].

3.3 Formal Context in FCA

Formal Concept Analysis (FCA) is a set of mathematical theories for data analysis and rectangular structures. It uses formal contexts and Galois lattices for information modelling and hierarchical visualization. FCA is increasingly applied in rectangular decomposition for data clustering, data analysis and information retrieval. In FCA, a formal context K is defined as the triplet structure $K = (G, M, R)$ where G and M represent the sets of the objects and attributes respectively and R is a binary relation defined over the two sets G and M, i.e. $R \subseteq G \times M$ [12].

3.4 Galois Operators

A Galois connection is a conceptual learning structure used to extract precise knowledge from an existing formal context. The Galois connection represents a duality in the formal context defined using the two Galois operators f and g.

The first operator f defines the set of shared antecedents or objects for some set $A \subseteq G$ as defined in Equation (1)

$$f(A) = \{b \in M \mid \forall a \in A, (a, b) \in R\} \tag{1}$$

The g operator; on the other hand, defines the set of images or properties shared by some set $B \subseteq M$ as defined in Equation (2)

$$g(B) = \{a \in G \mid \forall b \in B, (a, b) \in R\} \tag{2}$$

We can also define the **closure**$(A) = g(f(A)) = A'$, and the **closure**$(B) = f(g(B)) = B'$ [8].

In FCA terms, a non-enlargeable rectangle or maximal rectangle is called a formal concept of a formal context $K = (G, M, R)$ to be a pair (A, B) such that $A \subseteq G, B \subseteq M, f(A) = B$, and $g(B) = A$ [4]. A non-enlargeable rectangle has been previously referred to as a maximal rectangle [7]. An order relation "$<$" can be defined between two maximal rectangles (A_1, B_1) and (A_2, B_2) where (A_1, B_1) is a *sub maximal rectangle* of (A_2, B_2) iff $A_1 \subseteq A_2$ and $B_2 \subseteq B_1$. Based on this hierarchical order relation of the maximal rectangles, we can consider a Galois Lattice as the set of all maximal rectangles defined over the context $K = (G, M, R)$ ordered hierarchically as the **Galois Lattice** of the context K [1].

3.5   Gain or Weight of a Relation

The gain $W(R)$ of a binary relation R is given by Equation (3) as:

$$W(R) = \left(\frac{r}{d \times c}\right)(r - (d + c)) \tag{3}$$

where, r is the cardinality of R (i.e. the number of pairs in the binary relation R), d is the cardinality of the domain of R, and c is the cardinality of the range of R [6]. We may notice here that the ratio r over $(d \times c)$ is a measure of the density of the relation assessing information precision associated to the replacement of R by its rectangular closure $R*$. It is maximal (i.e. equal to value 1) when R is a rectangle. In that case without loss of information, we may replace $d \times c$ pairs by one pair linking directly the set of d elements of the domain of R to the set of c elements of its range. The difference of r with the sum $(d + c)$ is a measure of information economy obtained by the replacement of the relation by its rectangular closure $R*$. So the product $W(R)$ is an estimate of the weight of a relation as the product of information precision by information economy assessment.

3.6   Pseudo Rectangular Coverage

A pseudo maximal rectangle is a sub-relation that is associated with a pair in a binary relation. A pseudo maximal rectangle $PS(a, b)$ associated with a pair

(a, b) in a binary relation R is the union of all maximal rectangles containing the pair (a, b) [11] and is defined as:

$$PS(a, b) = I(R.b); \ R \ ; \ I(a.R) \tag{4}$$

$PS(a, b)$ is simultaneously the pre-restriction of R by the antecedents of b and its post-restriction by the images of a. Formally the pre-restriction (respectively the post-restriction) is realized by the left composition (respectively the right composition) of R by the partial identity relation $I(R.b)$ (respectively the partial identity relation $I(a.R)$). A pseudo maximal rectangle is an elementary relation offering an interesting conceptual construct that can be used to analyse the hidden features of a binary relation. The strength of a pair is one such criterion defined using the properties of the pseudo maximal rectangle representing that pair. The pair is called the pivot of the pseudo maximal rectangle. Consider the formal context $K = (G, M, R)$ used to model a set of documents (macro context). The relation R would be then associating or indexing documents in the domain with words in the codomain of R. If a word w is indexing a document d then w is weakly associated with all words w contained in document d represented by the binary context R, with strength s defined as:

$$s(d, w) = |d.R| \times |w.R^{-1}| - (|d.R| + |w.R^{-1}|) \tag{5}$$

To calculate the strength $s(d, w)$ of a pair (d, w) given by $W(w.R^{-1} \times d.R)$ in Equation (5), we need to calculate the pseudo maximal rectangle PS and calculate the weight of the resulting relation as defined in Equation (6)

$$W(PS(d, w)) = W(I(R.w); \ R \ ; \ I(d.R)) \tag{6}$$

This definition was optimized in [11] to avoid computing the pseudo maximal rectangle and calculate its economy by function W, defined in Equation (7) to be

$$W(PS) = Strength(d, w) = s(d, w) \tag{7}$$

Comparing the definition of W given by Equation (6) and that given in Equation (7), we find that the computational overhead of Equation (7) is far less than that of Equation (6) which costs $(n \times m)^2$ operations (where n is the cardinality of the domain of R and m is the cardinality of the range) compared to a linear time process required for Equation (7) in terms of the number of pairs in R [11]. The significance of the strength measure of a pair $s(d, w)$ in the binary relation R is that it reflects the importance of the relationship it summarizes that links semantically a set of documents and a set of words and at the same time enabling us to save a number of pairs equal to $(|d.R| \times |w.R^{-1}|) - (|d.R| + |w.R^{-1}|)$. This measure was used in [11] to produce micro and macro structures of a document space and will be employed in the newly proposed algorithms as part of this research.

4   Algorithms

The newly proposed methods can be seen as a system of three types of algorithms:

1. Algorithms to handle initial pseudo rectangular decomposition of an information store
2. Algorithms to handle incremental changes to information via addition operation
3. Algorithms to handle incremental changes to information via deletion operation

4.1   Non-incremental Algorithm

The main objective of this algorithm is to create a structure of pseudo maximal rectangles that covers the information store in a minimal way. The pseudo maximal rectangles are organized in a number of max-heaps. Each of these heaps has as root the domain category with which it mostly overlaps. The information store is modeled as a binary relation associating documents with words for finding a macro-structure that relates documents to each other's. The strategy the algorithm uses to find the minimal pseudo rectangular coverage for a relation efficiently is through careful selection of pairs for expansion into pseudo maximal rectangles. Each of the pairs selected is considered a pivot to the corresponding pseudo maximal rectangle, which is calculated and positioned in the correct heap location under the most relevant category. The strategy for selecting the next pairs representing a pseudo maximal rectangle pivot is based on two factors: the familiarity of the pair and the weight of that pair. The familiarity measure gives a priority to pairs having words common to some of the domain categories, while the weight measure gives a priority to the pairs with the highest information gain or the largest of pseudo maximal rectangles. The weight of a pair represents the strength of the bridge which the pair creates between a number of domain elements and a number of codomain elements. The higher this weight is the more information is gained by considering this pseudo maximal rectangle. The weight of a pair (a, b) with respect to a binary relation $R : G \times M$ is calculated using the following equation $w(a, b) = (|b.R^{-1}| \times |a.R|) - (|b.R^{-1}| + |a.R|)$

After a pseudo maximal rectangle is calculated, all its pairs are marked as covered so that the two previously mentioned selection criteria for pseudo rectangle pivots apply only to pairs in the relation that haven't been yet covered by any pseudo maximal rectangle. When the next best pair found has an index of -1, this is an indication that the entire relation has been already covered which forms a termination criterion to the algorithm.

If we would like to examine the time complexity of the non-incremental algorithm in terms of the total number of pairs of 1's (densities of 1's) in a relation N. The worst case scenario for the algorithm is to decompose the relation into N pseudo maximal rectangles where each pair represents an isolated cluster. The algorithm will be in the worst case $O(N^2)$. The best case scenario for this algorithm, on the other hand, is to be entirely covering R with 1 pseudo maximal rectangle. In this case, the algorithm will be in the best case $O(N \log N)$.

4.2 Incremental Addition Algorithm

The incremental addition algorithms handle updates to the information store in the form of added domain or codomain elements besides the addition of associations between existing domain and codomain elements. Each of the addition cases above has its own specificity in terms of update strategy; however, they all share one generic algorithmic skeleton for the addition that can be summarized as follows:

1. Identify parts of R to update on the addition of pairs, domain or codomain elements.
2. Identify pseudo maximal rectangles to be updated with the new information.
3. Structure the information which is identified as irrelevant to all existing pseudo rectangles in a similar fashion to the non-incremental algorithm

On the event of added domain, codomain or association elements to a binary relation, certain parts of the solution elements need to be updated. For example, in a context of documents representing domain elements and words representing codomain elements, the addition of a new document with an entirely new sets of words will have a similar execution flow to that of the non-incremental algorithm except that structuring will be only required for the newly added pairs as opposed to the entire space as in the non-incremental algorithm.

On the other hand, the addition of new document(s) associated with old words in the context will require a different update strategy. First, the weights of the old columns representing the images of the newly added documents will be recalculated to reflect the new significance or weight of the corresponding pairs. Then, each pseudo maximal rectangle having its pivot or any of its pairs sharing an image or an antecedent with any of the new pairs will be updated by adding all the new pairs. These pseudo maximal rectangles have as well updated weights already calculated in the first step; the update of weights might necessitate the repositioning of the pseudo maximal rectangles in the heap. Inversely, the event of adding new word(s) associated with existing documents will require updating the weights of all the rows representing the antecedents of the new words. After recalculating these weights, the pseudo maximal rectangles having their pivots or pairs sharing an extremity (i.e. image or antecedent) with the new pairs will be updated and their positions in the respective heaps will be reconsidered.

Adding a new pair to the binary relation for an existing domain and codomain elements like updating a document with a word already in the binary context will require updating the weights of rows and columns that will be affected by the addition of the new pairs. All pairs in the rows representing the set of antecedents of the new pairs will have their weights recalculated. In addition, all pairs in the columns representing the images of the new pairs will have their weights recalculated. After that, each pseudo rectangle with a pivot or a pair sharing an image or an antecedent with the new pairs will be updated by adding the new pairs and might need to be repositioned in the heap.

Analyzing the performance for the incremental add algorithms we notice that most of the execution time is consumed by the weight recalculation of the related columns and rows between the new pairs and the old pairs. This assumption is supported by the fact that the size of the pair increments is assumed to be trivial compared to the size of the entire relation and due to the fact that practically the minimal pseudo rectangular coverage can be achieved with a far less number of pseudo maximal rectangles than the size of the entire relation. For that we revisited the incremental add algorithms to identify then the need for weight recalculation. The information about the weight of a pair is used for two main purposes. First, in the process of initial discovery and structuring, the pivot of the next pseudo maximal rectangle is selected based on the weight criterion. Second, the weight of the pivot identifies the location of the pseudo maximal rectangle in the respective category heap. The purpose of keeping the weights of all pairs in the information store at all times updated, is to ensure an up-to-date information store ready for changes as they happen. However, if we relax this condition and update the weights only when necessary and as required, we can improve the execution time dramatically especially when the information store is updated with small increments or decrements of information.

4.3   Performance of the Incremental Addition Algorithm

Studying the time complexity of the incremental_add algorithm in terms of the number of the number of pairs to be added which is n, we find that in the worst case scenario, we will have a structure of N pseudo maximal rectangles all having common extremities with the n new pairs. In such case, the algorithm will be in the worst case in $O(n \log n + N \log N)$. In the best case scenario however, we can assume that none of the pseudo maximal rectangles will have a common extremity with the newly added maximal rectangles in which case, only structuring effort is required that converges in $O(n \log n)$ in the best case.

Further examination of the performance of the incremental add and delete operations in practice showed promising scalability results when compared with the structuring time obtained using the non-incremental version of the pseudo rectangular coverage algorithm. For the purpose of this evaluation, we have used part of the NSF Research Award Abstracts 1990–2003 as the testing Data Set which is provided by the UC Irvine Machine Learning Database Repository[1]. The reason for selecting that data set is for its rich textual contents. The NSF data set provides as a bag-of-word the entire vocabulary of the documents in use which is another interesting aspect to test while structuring with the proposed method.

When comparing the runtime obtained by running the non-incremental method against the incremental-add method. Both methods will start at 1000 documents which generate a context of around 60,000 pairs to process. In the non-incremental version, in order to add a document, a restructuring of the entire space will be required. Starting from a context of size 61,970 (1000 documents) and running non-incrementally until 1018 documents are structured. The

---

[1] http://archive.ics.uci.edu/ml/datasets/
NSF+Research+Award+Abstracts+1990-2003

incremental-add method on the other hand, will start from the benchmark structure of 1000 documents obtained incrementally or non-incrementally, and then add one document at a time until 1018 documents are entirely structured forming a context of 62981 pairs. The experimental results show that the time required to structure 1001 documents with over 62 thousand pairs non-incrementally would be around 63 seconds at all times. However if we have a context with the 1000 documents already structured it is merely enough to add the document using the incremental-add method to achieve the required task in around 0.67 seconds. As shown in Figure 1: Runtime compared (incremental-add and non-incremental method) the time difference between the two approaches is considerable although both demonstrate steady growth as the number of pairs in the context increase. The run time for the incremental approach is cumulative.

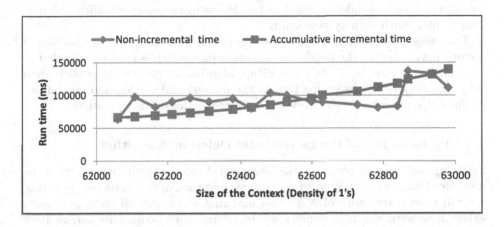

Fig. 1. Runtime compared (incremental-add and non-incremental method)

4.4   Incremental Deletion Algorithm

Similar to the incremental addition, the incremental deletion scenarios for domain elements, codomain elements or associations have the following algorithmic skeleton:

1. Identify parts of R that will be affected by the deletion of a pair or a domain or codomain element
2. Identify pseudo maximal rectangles to be updated by removing some of their pairs
3. Identify pseudo maximal rectangles to be deleted due to deletion of their pivots
4. Structure pairs belonging to deleted pseudo maximal rectangles using the addition algorithms

Similar to the addition cases, on the event of deleting a domain element all the weights of the columns representing the images of each of the documents to be deleted will be recalculated. Then for each pseudo maximal rectangles having one of its pairs (except the pivot) in the deleted pairs list, the pseudo maximal rectangle will be updated by removing the pairs to be deleted and the pseudo maximal rectangle is repositioned in the heap if required.

On the other hand, for each pseudo maximal rectangle having its pivot in the list of deleted pairs, the deleted pairs are removed along with the pivot, the label of the pseudo maximal rectangle is cleared from the list of used labels and the remaining pairs of the pseudo maximal rectangle are re-organized using the addition of associations' algorithm. In some scenarios, the deletion in the first step will be major that none of the existing pseudo maximal rectangles can be found relevant to the pairs to structure. In this case these pairs are used to construct new pseudo maximal rectangles as in the case of addition of new document(s) with entirely new words.

The cases of deletion of words or associations will be similar to the deletion of documents except for the weight recalculation part which is analogous to that of the addition scenarios. As in the addition algorithms, the weight recalculation takes the majority of execution time in the incremental version and for that an optimized version has been developed for incremental delete operations.

4.5   Performance of the Incremental Deletion Algorithm

Studying the time complexity of the incremental_delete algorithm in terms of the number of pairs to be deleted which is n, we find that in the worst case scenario, we will have a structure of N pseudo maximal rectangles all having common extremities with the n new pairs where all pseudo maximal rectangles have been broken and their pairs require restructuring. In such case, the algorithm converges in the worst case in $O(N^2)$.

In the best case scenario however, we can assume that none of the pseudo maximal rectangles will have a pivot with a common extremity with the newly added maximal rectangles in which case, only recalculation of a pivot pair which has a constant time and repositioning in heap is required. In that case, the algorithm also converges in $O(N^2)$.

Similarly, comparing the runtime obtained by running the non-incremental method at varying levels of documents and the incremental method for incremental_delete starting at one thousand and eighteen documents. In the non-incremental version, starting with 1018 documents (context of size 62936), we will need to restructure the entire space every time we need to delete a document. The incremental-delete method on the other hand, will start from the benchmark structure of 1018 documents obtained incrementally or non-incrementally, and then delete one document at a time until 1000 documents are entirely structured forming a context of 62981 pairs. The experimentation results plotted in Figure 2: Runtime compared (incremental_delete and non-incremental method)

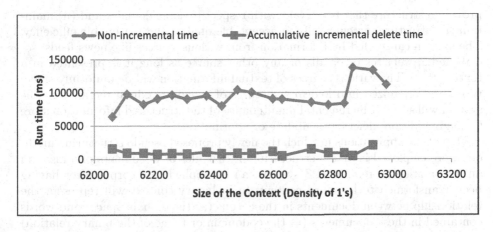

Fig. 2. Runtime compared (incremental-delete and non-incremental method)

demonstrates the pattern of the run time compared to the non-incremental structuring. We can conclude from the graph that the run time required to delete a document from a context consisting of 1018 documents (approximately 6.4 sec) is far less than structuring 1017 documents from scratch in a non-incremental fashion (approximately 133 sec).

The quality of the derived structure for the incremental methods was also evaluated in terms of two classic information evaluation metrics: (i) cohesion: measuring the inner tightness in a pseudo maximal rectangle and (ii) separation: measuring the degree of overlapping among the pseudo maximal rectangles. In both cases, the incremental methods were maintaining reasonable quality in the derived structures compared to the non-incremental addition or deletion methods.

5   IPS System Overview

The objective of the new incremental solution is to employ new methods based on Formal Concept Analysis that can be used for efficient incremental information organization. These new methods are based on the notion of pseudo maximal rectangles (or unions of maximal rectangles) offering an efficient incremental update with reasonable quality and improved scalability. The efficiency of the incremental methods will be measured by means of time complexity analysis and the quality will be measured using cohesion and separation metrics.

The new methods will also make advantage of domain specific information modelled as bags of words each representing a category in the respective domain. Incremental update aspects of an information store are the main focus of this work while changes in the domain of knowledge are briefly addressed. The architecture of the new system; named Incremental pseudo rectangular Structuring System (IPS System) supports handling two main functions: (i) finding a minimal pseudo rectangular coverage for information non-incrementally and provide a structure that is relevant with respect to some domain, and (ii) maintain the structure incrementally in acceptable quality and reasonable efficiency. The system can be fed by information from various sources like news feeds, social media, websites or emails or any other source as long it is presented in a textual form. The various sources of textual information will be uniformly structured and organized into logical structures of pseudo maximal rectangles. The system will support incremental maintenance of the structured information store through addition, deletion or update of information.

One of the applications in which the newly proposed pseudo structuring methods were employed is macro-structuring of text files or documents organized in an information store. The IPS system can be applied to a corpus after having been transformed to a binary context. Such binary context will represent the relationship between documents in the corpus (as the domain space) and words contained in these documents (as the codomain or range of the binary relation) in a binary matrix. The transformation however to a binary context entails some natural language processing aspects that should be used to enrich the semantics of the derived binary context and reduce the gap between the discrete form enforced by the binary representation compared to the original representation of the documents.

Another interesting application in the IPS System is incremental text summarization and documents features extraction where micro level structuring starts from the creation of a binary context representing a single document to be processed by associating its sentences to its contained words.

6   Conclusion

As the information space is increasingly growing in size and in a relatively fast pace, most applications nowadays need to exhibit efficient incremental information management capabilities. In this paper, we presented a new structuring method based on relational algebra and FCA intended to address the problem of incremental information organization and structuring. This method uses minimal pseudo rectangular decomposition as a structuring strategy for obtaining a scalable and robust structure for a change-friendly information store. The newly proposed methods perform well compared to other incremental management techniques that are based on Galois lattice or minimal rectangular decomposition as presented in the Related work section since they support high scalability resulting from the fact that the number of pseudo maximal rectangles is far less than that of maximal rectangles covering a binary relation. The experimental results for the addition and deletion algorithms were implemented in a new system (IPS system) which supports a number of applications like corpus organization, text summarization and feature extraction.

Testing results have highlighted some of the advantages and disadvantages of the incremental strategies adopted by these methods. The experiments showed that the incremental methods scale well in terms of the number of pseudo maximal rectangles to maintain incrementally; however, noticeable impact on the quality of the pseudo maximal rectangles structures was detected and could be subject to further improvement. Measures of quality structures quality like cohesion and separation were used to evaluate the quality of derived structures. Furthermore, experimentation results of the new methods showed as well the need for identifying the optimal update frequency and mechanism for an information store as a function of its update frequency and update amount.

Acknowledgement. This research work is supported by the FINANCIAL WATCH project which was made possible through a grant from the Qatar National Research Fund QNRF (NPRP 085831-101). The contents of this publication are solely the responsibility of the authors and do not necessarily represent the official views of the QNRF. Authors thank the anonymous reviewers for their valuable comments. Thanks are also due to Professor Marcelo Frias for his precious advice during the preparation of the last version.

References

1. Carpineto, C., Romano, G.: A lattice conceptual clustering system and its application to browsing retrieval. Machine Learning 24, 95–122 (1996)
2. Garey, R., Johnson, D.: Computers and Intractability: A Guide to the Theory of NP Completeness. W. H. Freeman and Company (1979)
3. Godin, R.: L'utilisation de Treillis pour l'Accès aux Systèmes d'Information. PhD thesis, University of Montreal (1986)
4. Godin, R., Missaoui, R., Alaoui, H.: Incremental concept formation algorithms based on galois (concept) lattices. Computational Intelligence 11, 246–267 (1995)
5. Hunt, E., Martin, J., Stone, P.: Experiments in Induction. Academic Press (1966)
6. Jaoua, A.: Pseudo-conceptual text and web structuring. In: Proceedings of The Third Conceptual Structures Tool Interoperability Workshop (CS-TIW2008). CEUR Workshop Proceedings, vol. 352, pp. 22–32 (2008)
7. Jaoua, A., Duwairi, R., Elloumi, S., Yahia, S.B.: Data Mining, Reasoning and Incremental Information Retrieval through Non Enlargeable Rectangular Relation Coverage. In: Berghammer, R., Jaoua, A.M., Möller, B. (eds.) RelMiCS/AKA 2009. LNCS, vol. 5827, pp. 199–210. Springer, Heidelberg (2009)
8. Khcherif, R., Gammoudi, M., Jaoua, A.: Using difunctional relations in information organization. Inf. Sci. Appl. 125(1-4), 153–166 (2000)
9. Li, K., Du, Y., Xiang, D., Chen, H., Liao, Z.: A Method for Building Concept Lattice Based on Matrix Operation. In: Huang, D.-S., Heutte, L., Loog, M. (eds.) ICIC 2007. LNCS (LNAI), vol. 4682, pp. 350–359. Springer, Heidelberg (2007)
10. Meghini, C., Spyratos, N.: Computing Intensions of Digital Library Collections. In: Kuznetsov, S.O., Schmidt, S. (eds.) ICFCA 2007. LNCS (LNAI), vol. 4390, pp. 66–81. Springer, Heidelberg (2007)
11. Neapolitan, R., Naimipour, K.: Foundations of Algorithms using JAVA Pseudocode. Jones and Barlett (2004)
12. Reinke, R., Michalski, R.: Incremental learning of concept descriptions: A method and experimental results. In: Hayes, J.E., Michie, D., Richards, J. (eds.) Machine Intelligence, vol. 11, pp. 263–288. Oxford University Press, Inc., New York (1988)
13. Valtchev, P., Missaoui, R., Lebrun, P.: A partition-based approach towards constructing Galois (concept) lattices. Discrete Math. 256(3), 801–829 (2002)

Relational Concepts in Social Choice

Gunther Schmidt

Fakultät für Informatik, Universität der Bundeswehr München
85577 Neubiberg, Germany
gunther.schmidt@unibw.de

Abstract. What now is called social choice theory has ever since attracted mathematicians — not least several Nobel laureates — who try to capture the comparison relations expressed and to aggregate them. Their results are often referred to nowadays. The purpose of this paper is to make point-free relation-algebraic mathematics available as a tool for the study of social choice. Thus, we provide simplification, additional systematics, more compact relation-algebraic proofs and also an access to solving such problems with programs in the language TITUREL — at least for the medium sized cases.

Keywords: relation, social choice, rationalization, revealed preference.

1 Introduction

Social choice is concerned with sets of decisions expressed by individuals and tries to aggregate these to a collective decision relation. Much of this paper is some sort of a translation of the respective theory to a point-free relation-algebraic form. However, it is not simply translated, but also in a non-trivial way transferred to a shorthand form. This in turn provides new insights and enables more compact algebraic proofs. It is, however, also a valuable scientific step that may help understanding the highly involved concepts. Much of the discussion runs along [Sen70, Suz83, Wri85].

2 Relation-Algebraic Preliminaries

This section is inserted to make the paper more or less self-contained, giving [SS89, SS93, Sch11] as a general reference. We write $R : V \longrightarrow W$ if R is a relation with source V and target W, often conceived as a subset of $V \times W$. If the sets V and W are finite of size m and n, respectively, and ordered, we may consider R as a Boolean matrix with m rows and n columns; called a *homogeneous* relation when $m = n$.

We assume the reader to be familiar with the basic operations on relations, namely R^{T} (*converse*), $\overline{R}$ (*negation*), $R \cup S$ (*union*), $R \cap S$ (*intersection*), and $R \,;\, S$ (*composition*), the predicate $R \subseteq S$ (*containment*), and the special relations[1] $\perp\!\!\!\perp$ (*empty relation*), $\top$ (*universal relation*), and $\mathbb{I}$ (*identity relation*).

---

[1] Suppressing indices here.

W. Kahl and T.G. Griffin (Eds.): RAMiCS 2012, LNCS 7560, pp. 278–293, 2012.
© Springer-Verlag Berlin Heidelberg 2012

A *heterogeneous relation algebra* is a structure that

— is a category with respect to composition " $;$ " and identities $\mathbb{I}$,
— has complete atomic Boolean lattices with $\cup, \cap, ^-, \bot\!\!\bot, \top\!\!\top, \subseteq$ as morphism sets,
— obeys rules for transposition in connection with the category and the lattice aspect just mentioned that may be stated in either one of the following two ways:

Dedekind $R;S \cap Q \subseteq (R \cap Q;S^\mathsf{T});(S \cap R^\mathsf{T};Q)$ or

Schröder $R;S \subseteq Q \iff R^\mathsf{T};\overline{Q} \subseteq \overline{S} \iff \overline{Q};S^\mathsf{T} \subseteq \overline{R}.$

Residuals are often introduced via $A;B \subseteq C \iff A \subseteq \overline{\overline{C};B^\mathsf{T}} =: C/B$, where B is divided from C on the right side. Intersecting such residuals in $\mathsf{syq}(R,S) :=$ $\overline{R^\mathsf{T};\overline{S}} \cap \overline{\overline{R}^\mathsf{T};S}$, the *symmetric quotient* $\mathsf{syq}(R,S) : W \longrightarrow Z$ of two relations $R : V \longrightarrow W$ and $S : V \longrightarrow Z$ is introduced. Symmetric quotients serve the purpose of 'column comparison': $\big[\mathsf{syq}(R,S)\big]_{wz} = \forall v \in V : R_{vw} \leftrightarrow S_{vz}$.

The symmetric quotient is not least applied to introduce *membership relations* $\varepsilon : X \longrightarrow \mathcal{P}(X)$ between a set X and its powerset $\mathcal{P}(X)$ or $\mathbf{2}^X$. These can be characterized algebraically up to isomorphism demanding $\mathsf{syq}(\varepsilon,\varepsilon) \subseteq \mathbb{I}$ and surjectivity of $\mathsf{syq}(\varepsilon,R)$ for all R. With a membership the powerset ordering is easily described as $\Omega = \overline{\varepsilon^\mathsf{T};\overline{\varepsilon}}$.

There is another point to observe, namely the transition from a *subset* $V \subseteq X$, conceived as a relation $V : X \longrightarrow \mathbb{1}$, to its counterpart *element* $e_V = \mathsf{syq}(\varepsilon,V) \subseteq \mathbf{2}^X$. It often helps if one makes this difference explicit, using membership ε in

$$
\begin{array}{c}
\varepsilon = \begin{array}{c} a \\ b \\ c \end{array}
\begin{pmatrix}
0 & 1 & 0 & 1 & 0 & 1 & 0 & 1 \\
0 & 0 & 1 & 1 & 0 & 0 & 1 & 1 \\
0 & 0 & 0 & 0 & 1 & 1 & 1 & 1
\end{pmatrix}
\qquad
\begin{pmatrix} 0 \\ 1 \\ 1 \end{pmatrix} = \varepsilon;e_V = V \\
e_V^\mathsf{T} = (0 \; 0 \; 0 \; 0 \; 0 \; 0 \; 1 \; 0)
\end{array}
$$

with column labels $\{\}, \{a\}, \{b\}, \{a,b\}, \{c\}, \{a,c\}, \{b,c\}, \{a,b,c\}$.

3 Order versus Preference

Orderings are generalized to preference structures as they have developed over the years and partly *after* the tremendous success of the work by Kenneth Arrow and Amartya Sen. Historically, orders E or strictorders C have more or less been used at free will with the possibility in mind that with $E = C \cup \mathbb{I}$ and $C = E \cap \overline{\mathbb{I}}$ everything may be freely converted from one form to the other.

However, this is not really true; with orderings E one often looses all the consequences of the Ferrers property. A (possibly heterogeneous) relation R has the Ferrers property if $R;\overline{R}^\mathsf{T};R \subseteq R$, which expresses that one may find from any situation R_{ij} and R_{km} that either R_{kj} or R_{im}; an absolutely useful condition giving rise to a plethora of consequences concerning thresholds, semiorders, and intervalorders.

Orderings do not comfortably fit into the hierarchy of order concepts (see [Sch11] Prop. 12.1) in contrast to preorders, i.e., reflexive and transitive relations. This together with other indications has persuaded us to prefer the irreflexive form — not least that irreflexive Ferrers orderings are the slightly more general concept since E Ferrers implies C Ferrers, but not vice versa.

A next problem came up when researchers started investigating preference structures as a generalization of orderings. The by now standard way is to consider a so-called weak preference relation $R = $ 'is not worse than' and derive from it strict preference P, indifference I, and incomparability J. We have collected in [Sch11] Prop. 13.9 much of the dispersed information on how these concepts are interrelated. The bijective mutual transitions $\alpha : R \mapsto (P, I, J)$ and $\beta : (P, I, J) \mapsto R$ can be given explicitly as

$$\alpha(R) := (R \cap \overline{R}^\mathsf{T}, R \cap R^\mathsf{T}, \overline{R} \cap \overline{R}^\mathsf{T}) \qquad \text{and} \qquad \beta(P, I, J) := P \cup I.$$

Prop. 3.1.ii justifies the 'is not worse'-idea. Because one feels that indifference should be reflexive, it gives reason to demand already R to be reflexive.

Proposition 3.1. i) $P \subseteq \overline{\mathbb{I}}$ for every R.

ii) R reflexive $\implies$ $\mathbb{I} \subseteq I$.

Proof: i) $R \cap \mathbb{I} = (R \cap \mathbb{I})^\mathsf{T} \subseteq R^\mathsf{T}$ implies $\mathbb{T} = \overline{R} \cup \overline{\mathbb{I}} \cup R^\mathsf{T}$ and $R \cap \overline{R}^\mathsf{T} \subseteq \overline{\mathbb{I}}$.

ii) $\mathbb{I} \subseteq R \implies \mathbb{I} \subseteq R \cap R^\mathsf{T} = I$. $\square$

One obtains always the partition $P \cup P^\mathsf{T} \cup I \cup J = \mathbb{T}$ and observes that P is asymmetric, I is reflexive and symmetric, and J is irreflexive and symmetric.

Then several other concepts are defined, mentioned not least in [Suz83]. The following shows their translation into a point-free — and thus shorthand — form. As defined above, we will always have a relation R for which its asymmetric part is defined as

$$P := P(R) := R \cap \overline{R}^\mathsf{T}.$$

When R is agreed upon, we will use the respective shorter version. Since R is in general not an ordering, one has to investigate the following concepts anew from scratch that are concerned with cycle avoidance.

Definition 3.2. We consider the relation R and use its asymmetric part P.

i) R **quasi-transitive** $:\Longleftrightarrow$ P transitive
ii) R **acyclic** $:\Longleftrightarrow$ $P^+ \subseteq \overline{P}^\mathsf{T}$
iii) R **acyclic**$_{\text{Sen}}$ $:\Longleftrightarrow$ $P^+ \subseteq R$

iv) R **consistent** $\quad:\Longleftrightarrow\quad P_{;}R^{*}\subseteq\overline{R}^{\mathsf{T}}$

v) P **progressively finite** $\quad:\Longleftrightarrow\quad \varepsilon\subseteq\mathbb{T}_{;}(\varepsilon\cap\overline{P_{;}\varepsilon})$ $\qquad\qquad\square$

Being progressively finite is the adequate relation-algebraic formulation that excludes an infinite run over ever new points in the same way as running into a circuit; cf. [SS93], p. 121. The condition is easily understood interpreting the right side as looking for elements of the subset from which one cannot proceed according to P to another point inside it: $\overline{P_{;}\varepsilon}$.

In the following example, the two non-empty sets $\{3,4\}$ and $\{1,3,4\}$ do not have a maximal element so that the corresponding columns in $\varepsilon\cap\overline{P_{;}\varepsilon}$ vanish.

$$P = \begin{array}{c}1\\2\\3\\4\end{array}\begin{pmatrix}0&1&0&1\\0&0&0&0\\0&0&0&1\\0&0&1&0\end{pmatrix}\qquad \varepsilon\cap\overline{P_{;}\varepsilon} = \begin{array}{c}1\\2\\3\\4\end{array}\begin{pmatrix}0&1&0&0&0&1&0&0&0&0&0&0&0&0&0\\0&0&1&1&0&0&1&1&0&0&1&1&0&0&1&1\\0&0&0&0&1&1&1&1&0&0&0&0&0&0&0&0\\0&0&0&0&0&0&0&0&1&1&1&1&0&0&0&0\end{pmatrix}$$

Fig. 1. Illustrating the condition of being progressively finite; P is not

Several interdependencies follow immediately. One is in particular interested in consistent preference; that is, one does not like iterated preference with indifferences in between to result in preference in reverse direction. A lot of literature has appeared how to avoid problems of this kind.

Proposition 3.3. Let be given the situation of the preceding definition.

i) R transitive $\qquad\Longrightarrow\qquad R$ quasi-transitive, i.e., P transitive

ii) R transitive $\qquad\Longrightarrow\qquad R$ consistent

iii) R consistent $\qquad\Longrightarrow\qquad R$ acyclic

iv) R quasi-transitive $\qquad\Longrightarrow\qquad R$ acyclic

v) R acyclic$_{\text{Sen}}$ $\qquad\Longrightarrow\qquad R$ acyclic

vi) R acyclic$_{\text{Sen}}$ $\qquad\not\Longleftarrow\qquad R$ acyclic

Proof: i) The proof of $P_{;}P\subseteq P$ decomposes into two parts:

$P_{;}P\subseteq R_{;}R\subseteq R$ since R is assumed to be transitive

$P_{;}P = (R\cap\overline{R}^{\mathsf{T}})_{;}(R\cap\overline{R}^{\mathsf{T}})\subseteq\overline{R}^{\mathsf{T}}$, where the latter follows via the Schröder rule and transitivity from $(R^{\mathsf{T}}\cap\overline{R})_{;}R^{\mathsf{T}}\subseteq\overline{R}\cup R^{\mathsf{T}}$.

ii) $P_{;}R^{*}\subseteq\overline{R}^{\mathsf{T}}\iff R^{\mathsf{T}}_{;}R^{*\mathsf{T}}\subseteq\overline{P}=\overline{R}\cup R^{\mathsf{T}}$, which holds due to transitivity.

iii) $P^{+}\subseteq R^{+} = R^{*}_{;}R\subseteq\overline{P}^{\mathsf{T}}$, the last step uses consistency in Schröderized form: $R^{\mathsf{T}}_{;}R^{*\mathsf{T}}\subseteq\overline{P}$.

iv) If P is transitive, acyclicity reads $P\subseteq\overline{P}^{\mathsf{T}}$. This, however, is trivially satisfied in view of the definition of P:

$$P = R\cap\overline{R}^{\mathsf{T}}\subseteq\overline{R}^{\mathsf{T}}\cup R = \overline{R\cap\overline{R}^{\mathsf{T}}}^{\mathsf{T}} = \overline{P}^{\mathsf{T}}.$$

v) since $P^+ \subseteq R \subseteq \overline{R}^\mathsf{T} \cup R = \overline{P}^\mathsf{T}$

vi) $R = \begin{matrix} 1 \\ 2 \\ 3 \end{matrix} \begin{pmatrix} 0 & 1 & 0 \\ 0 & 0 & 1 \\ 0 & 0 & 0 \end{pmatrix}$ provides a counter-example with non-reflexive R.   □

4   The Mechanics of Being Greatest

Since $R : X \longrightarrow X$ need not be an ordering, we must be very careful and avoid any informal reasoning, because much — but not all — stays the same. With $\mathsf{ubd}_R(\varepsilon) = \overline{\overline{R}^\mathsf{T}{}_{;}\varepsilon} : X \longrightarrow 2^X$, we obtain the set of upper bounds of all subsets in one hit. Upper bound points may exist, or not, and there may be one or many.

Given a relation $R : X \longrightarrow X$, we also introduce $\max_R : X \longrightarrow 2^X$ as assigning *the set* of maximal elements; in the finite case, this set will always be non-empty for a non-empty set. Executing this simultaneously,

$$\max{}_R(\varepsilon) := \varepsilon \cap \overline{(R \cap \overline{R}^\mathsf{T}){}_{;}\varepsilon} = \varepsilon \cap \overline{P{}_{;}\varepsilon}$$

describes columnwise those elements that belong to the set and for which it is not the case that they are in relation $R \cap \overline{R}^\mathsf{T}$ — i.e. strictly R-below — to any element of the set.

In much a similar way, we here conceive the `gre` to deliver *always* a result; however, the result may correspond to the empty set indicating that there is no greatest *element*. In contrast to the classical case, there may occur several greatest elements for a relation R which is not an ordering. That is, given a relation $R : X \longrightarrow X$, we type this function as $\mathsf{gre}_R : X \longrightarrow 2^X$. One has to intersect sets with their upper bound sets,

$$\mathsf{gre}{}_R(\varepsilon) = \varepsilon \cap \mathsf{ubd}{}_R(\varepsilon),$$

to get greatest element sets for all subsets 'columnwise' simultaneously. For the set $\{1, 4\}$ in Fig. 2, we get the result $\{4\}$, e.g. This $\{4\}$ is a subset $\subseteq X$, for which we will now consider the corresponding element in 2^X; and this executed simultaneously for all greatest element sets, resulting (see the end of Sect. 2) in a relation G, so that:

$$\mathsf{gre}{}_R(\varepsilon) = \varepsilon{}_{;}G \qquad \text{and} \qquad G := \mathsf{syq}(\varepsilon, \mathsf{gre}{}_R(\varepsilon)).$$

With this highly compact notation, we will now generalize a result best known for orderings to arbitrary R. Concerning (ii) in Prop. 4.1, one often says that for R a finite preorder in every nonempty subset S a maximal element exists.

Proposition 4.1.   Let an arbitrary homogeneous relation R be given.

i) $\mathsf{gre}_R(\varepsilon) \subseteq \max_R(\varepsilon)$

ii) R finite preorder $\implies$ $\varepsilon \subseteq \mathbb{T}{}_{;}\max_R(\varepsilon)$

iii) R preorder $\implies$ $\mathsf{gre}_R(\varepsilon) = \max_R(\varepsilon) \cap \mathbb{T}{}_{;}\mathsf{gre}_R(\varepsilon)$.

Proof: i) We have to prove $\varepsilon \cap \overline{R^{\mathsf{T}}}_{;}\varepsilon \subseteq \varepsilon \cap \overline{(R \cap \overline{R}^{\mathsf{T}})_{;}\varepsilon}$, but this is obvious.

ii) The asymmetric part P of a finite preorder R is certainly progressively finite (does not admit cycling), so that with Def. 3.2.v $\varepsilon \subseteq \mathbb{T}_{;}(\varepsilon \cap \overline{P_{;}\varepsilon}) = \mathbb{T}_{;}\mathbf{max}_R(\varepsilon)$.

iii) In view of (i), only $\mathbf{max}_R(S) \subseteq \mathbf{gre}_R(S)$ needs a proof. Assume a point $x \subseteq \mathbf{gre}_R(S) = S \cap \overline{R^{\mathsf{T}}}_{;}S$ to exist, which is equivalent to $x \subseteq S \subseteq R_{;}x$.

Now we consider an arbitrary point $z \subseteq \mathbf{max}_R(S) = S \cap \overline{(R \cap \overline{R}^{\mathsf{T}})_{;}S}$, which implies $z \subseteq S \subseteq (R \cup \overline{R}^{\mathsf{T}})_{;}z$.

Combining all this crosswise, $z \subseteq S \subseteq R_{;}x$ and $x \subseteq S \subseteq (R \cup \overline{R}^{\mathsf{T}})_{;}z$ where the latter implies $z \subseteq (R^{\mathsf{T}} \cup \overline{R})_{;}x$. From both follows $z \subseteq [R \cap (\overline{R} \cup R^{\mathsf{T}})]_{;}x = [(R \cap \overline{R}) \cup (R \cap R^{\mathsf{T}})]_{;}x = (R \cap R^{\mathsf{T}})_{;}x = I_{;}x$. Shunting and transposing gives $x \subseteq I_{;}z$, so that in total

$$z \subseteq S \subseteq R_{;}x \subseteq R_{;}I_{;}z \subseteq R_{;}R_{;}z \subseteq R_{;}z$$

due to transitivity of a preorder. This means $z \subseteq \mathbf{gre}_R(S)$.   □

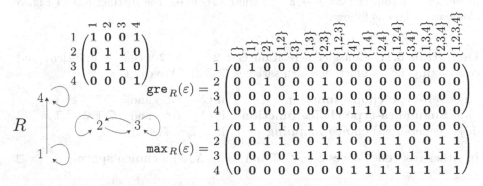

Fig. 2. $\mathbf{gre}_R(\varepsilon)$ and $\mathbf{max}_R(\varepsilon)$ using the membership relation ε

There hold further interesting formulae in case of greatest element sets.

Proposition 4.2. For every homogeneous relation R

i) $\mathsf{syq}(\mathbf{gre}_R(\varepsilon), \varepsilon) \subseteq \Omega^{\mathsf{T}}$,

ii) $\mathbf{gre}_R(\varepsilon)_{;}\Omega^{\mathsf{T}} \cap \varepsilon = \mathbf{gre}_R(\varepsilon)$.

Proof: i) $\overline{\Omega} = \varepsilon^{\mathsf{T}}_{;}\overline{\varepsilon} \subseteq \varepsilon^{\mathsf{T}}_{;}(\overline{\varepsilon} \cup \overline{R^{\mathsf{T}}}_{;}\varepsilon) = \varepsilon^{\mathsf{T}}_{;}\overline{\mathbf{gre}_R(\varepsilon)}$
$\subseteq \varepsilon^{\mathsf{T}}_{;}\overline{\mathbf{gre}_R(\varepsilon)} \cup \overline{\varepsilon}^{\mathsf{T}}_{;}\mathbf{gre}_R(\varepsilon) = \overline{\mathsf{syq}(\varepsilon, \mathbf{gre}_R(\varepsilon))}$.

ii) This means by definition $(\varepsilon \cap \overline{R^{\mathsf{T}}}_{;}\varepsilon)_{;}\Omega^{\mathsf{T}} \cap \varepsilon = \varepsilon \cap \overline{R^{\mathsf{T}}}_{;}\varepsilon$. We will use $\varepsilon_{;}\Omega = \varepsilon$. Direction $\supseteq$ is clear because Ω is reflexive. For $\subseteq$, we may restrict ourselves to showing

$$\overline{R^{\mathsf{T}}}_{;}\varepsilon_{;}\Omega^{\mathsf{T}} \subseteq \overline{R^{\mathsf{T}}}_{;}\varepsilon \quad \Longleftrightarrow \quad \overline{R^{\mathsf{T}}}_{;}\varepsilon_{;}\Omega \subseteq \overline{R^{\mathsf{T}}}_{;}\varepsilon. \qquad \square$$

Although R is not an ordering, the interpretation is not very far from the ordering case: Stepping down from some greatest element of a set via the powerset ordering Ω^T, but staying inside that set, one will remain in the set of greatest elements. (In case R is an order, the greatest element set would be an at most 1-element set.)

5   Preferences versus Choice Functions

Choice is considered in powersets, where one indicates the — often strictly smaller — subsets of a subset from which elements *may be chosen*. When looking at definitions in [Sen70], e.g., one will find that the author is careful in demanding non-empty argument sets to which non-empty choice sets are assigned. Suzumura [Suz83] (page 27) discussed this in detail and decided for Sen's way. In an appendix of Chapt. 2, however, he also discusses slightly more general variants.

We go here even further and start from a set X of so-called **conceivable states** of which we intend to form subsets $\varepsilon : X \longrightarrow 2^X$ and consider the powerset ordering $\Omega : 2^X \longrightarrow 2^X$ of these. To make the distinctions in Fig. 3 clear, we define as follows:

Definition 5.1.   Consider a relation $C : 2^X \longrightarrow 2^X$ that is univalent and contracting, i.e., a function which satisfies $C \subseteq \Omega^\mathsf{T}$. We call C a

  i) **Sen-type choice function**   if $C \subseteq \mathbb{T}{;}\varepsilon$ and $C{;}\mathbb{T} = \varepsilon^\mathsf{T}{;}\mathbb{T}$,
  ii) **Suzumura-type choice function** if $C \subseteq \mathbb{T}{;}\varepsilon$ and $C{;}\mathbb{T} \subseteq \varepsilon^\mathsf{T}{;}\mathbb{T}$,
 iii) (generalized) **choice mapping**   if $C{;}\mathbb{T} = \mathbb{T}$.

In either case, one defines $\mathcal{S} := C{;}\mathbb{T}$ and calls $(X, \mathcal{S})$ a **choice space**.   □

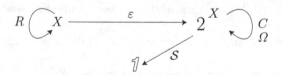

Fig. 3.   Typing choice functions C as opposed to weak preferences R

In this way, results in (i,ii) are assigned only to non-empty sets and results are always non-empty subsets of the argument since $C \subseteq \mathbb{T}{;}\varepsilon$. The C in (iii) will, due to contraction, assign the empty set to the empty set.

Every Sen-type choice function is obviously a Suzumura choice function. None of the two can ever be a choice mapping which is totally defined by definition. However, both — in particular the more general Suzumura version — are easily converted to the generalized mapping.

Proposition 5.2. i) For C a choice function, $C_{\mathrm{Gen}} := C \cup \overline{C;\mathbb{T}} ; \overline{\varepsilon^{\mathsf{T}};\mathbb{T}}^{\mathsf{T}}$ is a generalized choice mapping.

ii) Given any choice mapping C, we obtain $C_{\mathrm{Suz}} := C \cap \mathbb{T};\varepsilon$ as a Suzumura choice function.

Proof: The proof is obvious when looking at Fig. 4.   □

Researchers have always been very careful to execute all the case distinctions that arise when admitting an empty choice, be it from a non-empty subset, or of the empty subset. Being an empty choice might, however, smoothly be interpreted as an abstention. We will see that when proceeding to point-free relation-algebraic handling these problems disappear and results obtain a more uniform shape.

$$
C_{\mathrm{Suz}} = \begin{array}{c} \\ \{\} \\ \{1\} \\ \{2\} \\ \{1,2\} \\ \{3\} \\ \{1,3\} \\ \{2,3\} \\ \{1,2,3\} \end{array}
\begin{pmatrix}
0 & 0 & 0 & 0 & 0 & 0 & 0 & 0 \\
0 & 0 & 0 & 0 & 0 & 0 & 0 & 0 \\
0 & 0 & 0 & 0 & 0 & 0 & 0 & 0 \\
0 & 0 & 0 & 1 & 0 & 0 & 0 & 0 \\
0 & 0 & 0 & 0 & 0 & 0 & 0 & 0 \\
0 & 0 & 0 & 0 & 1 & 0 & 0 & 0 \\
0 & 0 & 0 & 0 & 0 & 0 & 1 & 0 \\
0 & 0 & 0 & 0 & 0 & 0 & 0 & 0
\end{pmatrix}
\qquad
C_{\mathrm{Gen}} = \begin{array}{c} \\ \{\} \\ \{1\} \\ \{2\} \\ \{1,2\} \\ \{3\} \\ \{1,3\} \\ \{2,3\} \\ \{1,2,3\} \end{array}
\begin{pmatrix}
1 & 0 & 0 & 0 & 0 & 0 & 0 & 0 \\
1 & 0 & 0 & 0 & 0 & 0 & 0 & 0 \\
1 & 0 & 0 & 0 & 0 & 0 & 0 & 0 \\
0 & 0 & 0 & 1 & 0 & 0 & 0 & 0 \\
1 & 0 & 0 & 0 & 0 & 0 & 0 & 0 \\
0 & 0 & 0 & 0 & 1 & 0 & 0 & 0 \\
0 & 0 & 0 & 0 & 0 & 0 & 1 & 0 \\
1 & 0 & 0 & 0 & 0 & 0 & 0 & 0
\end{pmatrix}
$$

Fig. 4. Toggling between Suzumura choice function and choice mapping

6 Generating Choice Functions from Preferences

Now we look at possibilities how to obtain choice functions or mappings. The frequently applied idea is to start from any relation R on a set X and let C map every subset of X to the subset of its R-greatest elements — recall that the definition of the functional $\mathbf{gre}_R(u)$ above has already sailed free from the requirement that R be an ordering.

Definition 6.1. Given any homogeneous relation R, not necessarily an order or a preorder, we call $C :- \mathbf{syq}(\mathbf{gro}_R(\varepsilon), \varepsilon)$ its **corresponding choice mapping** and speak of the corresponding choice function F of

i) **Suzumura-type**   if $F = C \cap \mathbb{T};\varepsilon$,
ii) **Sen-type**   if $F = C \cap \mathbb{T};\varepsilon$ and in addition $F;\mathbb{T} = \varepsilon^{\mathsf{T}};\mathbb{T}$.   □

The claim for C to be a mapping needs a proof which is given below as Prop. 6.2.i,ii. The side conditions in Def. 6.1.i,ii seem slightly artificial.

The typical investigation is now to look at R and try to guarantee certain favourable properties of C; that it generates a Sen-type choice function, e.g. Such work has a great tradition, and we cannot report much of it; in particular, because we have changed part of the foundation in moving to choice mappings.

$$
\begin{array}{c}
1\ 2\ 3 \\
1 \\ 2 \\ 3
\end{array}
\begin{pmatrix}
1 & 1 & 1 \\
1 & 1 & 1 \\
0 & 0 & 1
\end{pmatrix}
\qquad
\begin{array}{c}
\{\} \\ \{1\} \\ \{2\} \\ \{1,2\} \\ \{3\} \\ \{1,3\} \\ \{2,3\} \\ \{1,2,3\}
\end{array}
\begin{pmatrix}
0 & 0 & 0 & 0 & 0 & 0 & 0 & 0 \\
0 & 1 & 0 & 0 & 0 & 0 & 0 & 0 \\
0 & 0 & 1 & 0 & 0 & 0 & 0 & 0 \\
0 & 0 & 0 & 1 & 0 & 0 & 0 & 0 \\
0 & 0 & 0 & 0 & 1 & 0 & 0 & 0 \\
0 & 0 & 0 & 0 & 1 & 0 & 0 & 0 \\
0 & 0 & 0 & 0 & 1 & 0 & 0 & 0 \\
0 & 0 & 0 & 0 & 1 & 0 & 0 & 0
\end{pmatrix}
\qquad
\begin{array}{c}
\{\} \\ \{1\} \\ \{2\} \\ \{1,2\} \\ \{3\} \\ \{1,3\} \\ \{2,3\} \\ \{1,2,3\}
\end{array}
\begin{pmatrix}
1 & 0 & 0 & 0 & 0 & 0 & 0 & 0 \\
0 & 1 & 0 & 0 & 0 & 0 & 0 & 0 \\
0 & 0 & 1 & 0 & 0 & 0 & 0 & 0 \\
0 & 0 & 0 & 1 & 0 & 0 & 0 & 0 \\
0 & 0 & 0 & 0 & 1 & 0 & 0 & 0 \\
0 & 0 & 0 & 0 & 1 & 0 & 0 & 0 \\
0 & 0 & 0 & 0 & 1 & 0 & 0 & 0 \\
0 & 0 & 0 & 0 & 1 & 0 & 0 & 0
\end{pmatrix}
$$

with column headers $\{\}\ \{1\}\ \{2\}\ \{1,2\}\ \{3\}\ \{1,3\}\ \{2,3\}\ \{1,2,3\}$

Fig. 5. R, its corresponding Sen-type choice function, and choice mapping

We will soon see that we have dropped conditions for reasons of simplicity and uniformity. The latter idea is very much supported by the following proposition. By the way, (iv) of Prop. 6.2 has in [Sen70] been termed 'property α'.

Proposition 6.2.   Let be given a homogeneous relation R and its corresponding choice mapping $C = \mathsf{syq}(\mathsf{gre}_R(\varepsilon), \varepsilon)$. Then

i) C is indeed a mapping, i.e., total and univalent,
ii) $C \subseteq \Omega^{\mathsf{T}}$,
iii) $\varepsilon{;}C^{\mathsf{T}} = \mathsf{gre}_R(\varepsilon)$,
iv) $\varepsilon{;}C^{\mathsf{T}} = \varepsilon{;}C^{\mathsf{T}}{;}\Omega^{\mathsf{T}} \cap \varepsilon$.

Proof: i) C is a mapping by definition; cf. [Sch11] Def. 7.13.

ii) is the statement of Prop. 4.2.i.

iii) $\varepsilon{;}C^{\mathsf{T}} = \varepsilon{;}\big[\mathsf{syq}(\mathsf{gre}_R(\varepsilon), \varepsilon)\big]^{\mathsf{T}} = \varepsilon{;}\mathsf{syq}(\varepsilon, \mathsf{gre}_R(\varepsilon)) = \mathsf{gre}_R(\varepsilon)$ according to [Sch11] Prop. 7.14.

iv) We start with $\subseteq$: The first containment is trivial since Ω is reflexive, while the second is a consequence of (ii). For $\supseteq$, we start with

$$
\varepsilon{;}\Omega = \varepsilon, \quad \text{which implies} \quad (\varepsilon \cap \overline{R}^{\mathsf{T}}{;}\varepsilon){;}\Omega \subseteq \overline{R}^{\mathsf{T}}{;}\varepsilon{;}\Omega = \overline{R}^{\mathsf{T}}{;}\varepsilon = \overline{R^{\mathsf{T}}{;}\varepsilon \cup \overline{\varepsilon}}
$$
$$
\Longleftrightarrow \quad (\varepsilon \cap \overline{\overline{R}^{\mathsf{T}}{;}\varepsilon}){;}\Omega^{\mathsf{T}} \subseteq \overline{\varepsilon} \cup \overline{\overline{R}^{\mathsf{T}}{;}\varepsilon}
$$
$$
\Longleftrightarrow \quad \mathsf{gre}_R(\varepsilon){;}\Omega^{\mathsf{T}} \cap \varepsilon = (\varepsilon \cap \overline{\overline{R}^{\mathsf{T}}{;}\varepsilon}){;}\Omega^{\mathsf{T}} \cap \varepsilon \subseteq \overline{\overline{R}^{\mathsf{T}}{;}\varepsilon}
$$
$$
\Longrightarrow \quad \varepsilon{;}C^{\mathsf{T}}{;}\Omega^{\mathsf{T}} \cap \varepsilon = \mathsf{gre}_R(\varepsilon){;}\Omega^{\mathsf{T}} \cap \varepsilon \subseteq \varepsilon \cap \overline{\overline{R}^{\mathsf{T}}{;}\varepsilon} = \mathsf{gre}_R(\varepsilon) = \varepsilon{;}C^{\mathsf{T}} \qquad \square
$$

Normally, several groups of conditions are assembled and then the proof is given that R defines a Sen-type choice function. We will here proceed the other way round and first formulate the condition on C aimed at.

Proposition 6.3. A choice mapping C corresponding to R will have a corresponding Sen-type choice function precisely when the following condition on R is satisfied

$$\varepsilon \subseteq \mathbb{T}\,;(\varepsilon \cap \overline{\overline{R}^{\mathsf{T}}\,;\varepsilon}).$$

Proof: The Sen condition on C is that it assigns non-empty subsets to non-empty argument sets, i.e., $C^{\mathsf{T}} \cap \mathbb{T}\,;\varepsilon \subseteq \varepsilon^{\mathsf{T}}\,;\mathbb{T}$, which one will verify looking at Fig. 6, derived from Fig. 5.

The condition slightly modified is $C^{\mathsf{T}}\,;(\mathbb{I} \cap \mathbb{T}\,;\varepsilon) = C^{\mathsf{T}} \cap \mathbb{T}\,;\varepsilon \subseteq \varepsilon^{\mathsf{T}}\,;\mathbb{T}$. Using the Schröder rule, since C is a mapping, and using Prop. 6.2.iii, we get

$$C\,;\overline{\varepsilon^{\mathsf{T}}\,;\mathbb{T}} = \overline{C\,;\varepsilon^{\mathsf{T}}\,;\mathbb{T}} = \overline{[\mathbf{gre}_R(\varepsilon)]^{\mathsf{T}}\,;\mathbb{T}} \subseteq \overline{\mathbb{I} \cap \mathbb{T}\,;\varepsilon}.$$

Negating, transposing, and expanding **gre** gives

$$\mathbb{I} \cap \mathbb{T}\,;\varepsilon \subseteq \mathbb{T}\,;(\varepsilon \cap \overline{\overline{R}^{\mathsf{T}}\,;\varepsilon}),$$

from which we obtain the final result as

$$\varepsilon \subseteq \mathbb{T}\,;\varepsilon = \mathbb{T}\,;(\mathbb{I} \cap \mathbb{T}\,;\varepsilon) \subseteq \mathbb{T}\,;\mathbb{T}\,;(\varepsilon \cap \overline{\overline{R}^{\mathsf{T}}\,;\varepsilon}) = \mathbb{T}\,;(\varepsilon \cap \overline{\overline{R}^{\mathsf{T}}\,;\varepsilon}). \qquad \square$$

	{}	{1}	{2}	{1,2}	{3}	{1,3}	{2,3}	{1,2,3}
{}	1	0	0	0	0	0	0	0
{1}	0	1	0	0	0	0	0	0
{2}	0	0	1	0	0	0	0	0
{1,2}	0	0	0	1	0	0	0	0
{3}	0	0	0	0	1	1	1	1
{1,3}	0	0	0	0	0	0	0	0
{2,3}	0	0	0	0	0	0	0	0
{1,2,3}	0	0	0	0	0	0	0	0

$$C^{\mathsf{T}}$$

	{}	{1}	{2}	{1,2}	{3}	{1,3}	{2,3}	{1,2,3}
{}	0	0	0	0	0	0	0	0
{1}	0	1	0	0	0	0	0	0
{2}	0	0	1	0	0	0	0	0
{1,2}	0	0	0	1	0	0	0	0
{3}	0	0	0	0	1	0	0	0
{1,3}	0	0	0	0	0	1	0	0
{2,3}	0	0	0	0	0	0	1	0
{1,2,3}	0	0	0	0	0	0	0	1

$$\mathbb{I} \cap \mathbb{T}\,;\varepsilon = \mathbb{I} \cap \varepsilon^{\mathsf{T}}\,;\mathbb{T}$$

	{}	{1}	{2}	{1,2}	{3}	{1,3}	{2,3}	{1,2,3}
{}	0	0	0	0	0	0	0	0
{1}	1	1	1	1	1	1	1	1
{2}	1	1	1	1	1	1	1	1
{1,2}	1	1	1	1	1	1	1	1
{3}	1	1	1	1	1	1	1	1
{1,3}	1	1	1	1	1	1	1	1
{2,3}	1	1	1	1	1	1	1	1
{1,2,3}	1	1	1	1	1	1	1	1

$$\varepsilon^{\mathsf{T}}\,;\mathbb{T}$$

Fig. 6. Condition on a choice mapping to lead to a (Sen) choice function

Once we are in this position, we may look for combinations of the widely known conceivable properties of R that satisfy this requirement; e.g., being reflexive and/or connex, and/or transitive etc. The homogeneous relation R will be called **connex** provided $\mathbb{T} = R \cup R^{\mathsf{T}}$; it is thus reflexive and complete, the latter meaning $\overline{\mathbb{I}} = R \cup R^{\mathsf{T}}$.

Proposition 6.4. Whenever R is connex, and quasi-transitive on a finite set, the corresponding choice mapping $C := \mathsf{syq}(\mathbf{gre}_R(\varepsilon), \varepsilon)$ will give rise to a Sen-type choice function.

Proof: Following Prop. 6.3, we have to prove

$$R \cup R^{\mathsf{T}} = \mathbb{T}, \quad P\,;P \subseteq P \quad \Longrightarrow \quad \varepsilon \subseteq \mathbb{T}\,;(\varepsilon \cap \overline{\overline{R}^{\mathsf{T}}\,;\varepsilon}).$$

Since $\overline{R}^{\mathsf{T}} \subseteq R$ and thus $P = R \cap \overline{R}^{\mathsf{T}} = \overline{R}^{\mathsf{T}}$, this means $\varepsilon \subseteq \mathbb{T}_i(\varepsilon \cap \overline{P_i\varepsilon})$. However, this is the condition for being progressively finite according to Def. 3.2; and indeed, as a transitive and by construction asymmetric relation on a finite set, P is a strictorder, and thus progressively finite. □

Traditionally, many more such results are proved, usually with page-long free-style proofs. The one above written in full, in contrast, may be proof-checked.

Often the criterion is acyclicity.

Proposition 6.5. Let R be a finite connex relation. Then the corresponding choice mapping $C := \mathtt{syq}(\mathtt{gre}_R(\varepsilon), \varepsilon)$ will give rise to a Sen-type choice function provided R is acyclic$_{\mathrm{Sen}}$.

Proof: As in the preceding proof, we get $\overline{R}^{\mathsf{T}} \subseteq R$ from connexity, and thus $P = R \cap \overline{R}^{\mathsf{T}} = \overline{R}^{\mathsf{T}}$, so that we have to prove $\varepsilon \subseteq \mathbb{T}_i(\varepsilon \cap \overline{P_i\varepsilon})$. We use that being progressively finite is equivalent with being circuit-free $P^+ \subseteq \overline{\mathbb{I}}$ in case of finiteness; cf. [SS93] Prop. 6.3.2.

Now we proceed assuming P not to be circuit-free. Then there exists a finite at least 2-element sequence of points $x_1, x_2, \ldots x_{n+1} = x_1$ such that $x_i \subseteq P_i x_{i+1}$, counting the indices cyclically modulo n. With Sen-acyclicity $P^+ \subseteq R$, we obtain that they are all mutually related $x_i \subseteq R_i x_j$ for $i, j = 1, \ldots n$; and therefore also $x_i \subseteq I_i x_j$. This is a contradiction, because P, P^{T}, I, J form a disjunction. □

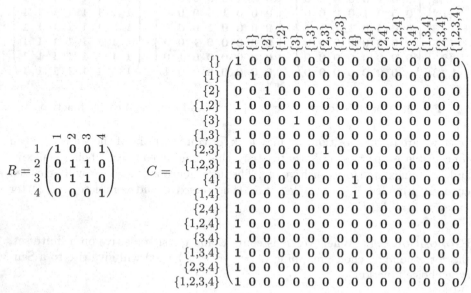

Fig. 7. A homogeneous relation R determining a generalized choice mapping C

The above matrices visualize forming the choice mapping. (One should remember that Sen ususally presents the matrix of an ordering with the greatest element down to the least.) Obviously, R is not an ordering. One will recognize that there is no greatest element in the set $\{1, 3\}$ resulting in assigning the empty set via C. On the other hand, the set $\{4\}$ is at the same time the set of greatest elements of $\{1, 4\}$ and $\{4\}$.

7   Rationalization Conceived as a Galois Correspondence

Since it is always a promising situation when one finds some Galois correspondence, we mention here the following result. So far, however, we have not had the opportunity to look for all its possible consequences.

Proposition 7.1. There exists a Galois correspondence between the R- and the C-side. It concerns arbitrary relations R and C, the latter contained in Ω^{T}, and looks as follows

$$\pi(C) \subseteq R \quad \Longleftrightarrow \quad C \subseteq \sigma(R)$$

with $\quad \sigma(R) := \overline{\mathbf{gre}_R(\varepsilon)^{\mathsf{T}}{}_;\varepsilon} \quad$ and $\quad \pi(C) := \varepsilon_;C_;\varepsilon^{\mathsf{T}}$.

Proof: We will use that $C_;\varepsilon^{\mathsf{T}} \subseteq \varepsilon^{\mathsf{T}}$, which is trivial because we have $C \subseteq \Omega^{\mathsf{T}}$.

$$C \subseteq \sigma(R) = \overline{\mathbf{gre}_R(\varepsilon)^{\mathsf{T}}{}_;\varepsilon}$$
$$\Longleftrightarrow \quad \overline{\mathbf{gre}_R(\varepsilon)^{\mathsf{T}}{}_;\varepsilon} \subseteq \overline{C}$$
$$\Longleftrightarrow \quad C_;\varepsilon^{\mathsf{T}} \subseteq \left[\mathbf{gre}_R(\varepsilon)\right]^{\mathsf{T}} = \varepsilon^{\mathsf{T}} \cap \overline{\varepsilon^{\mathsf{T}}{}_;\overline{R}}$$
$$\Longleftrightarrow \quad C_;\varepsilon^{\mathsf{T}} \subseteq \overline{\varepsilon^{\mathsf{T}}{}_;\overline{R}}$$
$$\Longleftrightarrow \quad \varepsilon^{\mathsf{T}}{}_;\overline{R}{}_;\varepsilon \subseteq \overline{C}$$
$$\Longleftrightarrow \quad \overline{R}^{\mathsf{T}}{}_;\varepsilon_;C \subseteq \overline{\varepsilon}$$
$$\Longleftrightarrow \quad \varepsilon_;C^{\mathsf{T}}{}_;\varepsilon^{\mathsf{T}} \subseteq R^{\mathsf{T}}$$
$$\Longleftrightarrow \quad \varepsilon_;C_;\varepsilon^{\mathsf{T}} \subseteq R \qquad\qquad\qquad \square$$

This correspondence seems to be related with rationalization.

Definition 7.2. We consider some choice function $C : 2^X \longrightarrow 2^X$. A relation $R : X \longrightarrow X$ is said to **rationalize** C if $\varepsilon_;C^{\mathsf{T}} = \mathbf{gre}_R(\varepsilon)$. If such an R exists, C is called a **rational** choice. If this R is in addition an ordering, C is called a **fully rational** choice. $\qquad\qquad\qquad\square$

Should the choice C have been constructed starting from some relation R, this underlying R will obviously rationalize C, since, according to [Sch11] Prop. 7.14, $X = \varepsilon_;\mathbf{syq}(\varepsilon, X)$ for every X. But there may exist other rationalizing relations, not least via the above Galois mechanism. There are more Cs than Rs, so that one may hope for an adjunction.

8 Revealing a Preference Out of a Choice Function

Rationalization asks whether an executed choice C has followed some 'rational' criterion R. While we have so far defined a choice function starting from an arbitrary relation R, we will now go in reverse direction and try to reveal (i.e., extract) such a relation R from an arbitrary choice function C.

Definition 8.1. For every choice function $C : 2^X \longrightarrow 2^X$, we define the following $R : X \longrightarrow X$, calling it the

i) **revealed preference** $R_C := \varepsilon{;}C{;}\varepsilon^{\mathsf{T}}$,

ii) **revealed strict preference** $R_C^* := (\varepsilon{;}C \cap \bar{\varepsilon}){;}\varepsilon^{\mathsf{T}}$.   □

In [Suz83], (ii) is written as $R_C^* = \bigcup_{S \in \mathcal{S}} [C(S) \times \{S \setminus C(S)\}]$, and explained with *x is R_C^*-preferred to y if and only if x is chosen and y could have been chosen but was actually rejected from some $S \in \mathcal{S}$.* (Order reversed!)

It is certainly an important case when the revealed R can somehow re-determine the C one has been starting from.

The following is mentioned in order to demonstrate that the construct of a choice *mapping* — as opposed to the choice *functions* — is indeed a profitable idea.

Proposition 8.2. Def. 8.1.i delivers the same relation, regardless of whether formed of a choice mapping, its corresponding Sen-type choice function, or its corresponding Suzumura-type choice function, i.e.,

$$\varepsilon{;}C{;}\varepsilon^{\mathsf{T}} = \varepsilon{;}(C \cap \mathbb{T}{;}\varepsilon){;}\varepsilon^{\mathsf{T}}.$$

Proof: For the Suzumura-case (as well as for the Sen-case which has an additional condition on C), we apply two times obvious matrix product formulae, which say, e.g., that annihilating columns of the second factor is equivalent to annihilating these columns in a product:

$$\varepsilon{;}(C \cap \mathbb{T}{;}\varepsilon){;}\varepsilon^{\mathsf{T}} = (\varepsilon{;}C \cap \mathbb{T}{;}\varepsilon){;}\varepsilon^{\mathsf{T}} = \varepsilon{;}C{;}(\varepsilon^{\mathsf{T}} \cap [\mathbb{T}{;}\varepsilon]^{\mathsf{T}}) = \varepsilon{;}C{;}\varepsilon^{\mathsf{T}}$$ □

We are, thus, again enabled to go back and forth between relations $R : X \longrightarrow X$ and relations $C : 2^X \longrightarrow 2^X$ with Def. 6.1 and Def. 8.1. The question immediately arises, to which extent a revealed R_C obtained from a C which is obtained from R resembles the original relation. We have indicated this idea with the Galois correspondence above. For reasons of time and manpower, it has not yet been made a central point of our investigation. In any case, the main question is, to what extent going forth and back again comes close to an identity. It is answered below.

Proposition 8.3. i) For any R, the R_C obtained from its corresponding choice mapping satisfies $R_C \subseteq R$.

ii) In addition: R reflexive implies equality $R_C = R$.

Proof: i) $R_C = \varepsilon_i C_i \varepsilon^\mathsf{T} = \varepsilon_i [\mathbf{gre}_R(\varepsilon)]^\mathsf{T} = \varepsilon_i (\varepsilon^\mathsf{T} \cap \overline{\varepsilon^\mathsf{T} {;} \overline{R}}) \subseteq \varepsilon_i \overline{\varepsilon^\mathsf{T} {;} \overline{R}} \subseteq R.$

ii) This proof, which we omit, seems to need pointwise consideration. □

Fig. 8 gives an example for being unequal when R is not reflexive: Not even C resembles R in an adequate way.

$$R = \begin{matrix} 1 \\ 2 \\ 3 \end{matrix}\begin{pmatrix} 1 & 0 & 1 \\ 0 & 1 & 0 \\ 0 & 0 & 0 \end{pmatrix} \qquad C = \begin{matrix} \{\} \\ \{1\} \\ \{2\} \\ \{1,2\} \\ \{3\} \\ \{1,3\} \\ \{2,3\} \\ \{1,2,3\} \end{matrix}\begin{pmatrix} 1 & 0 & 0 & 0 & 0 & 0 & 0 & 0 \\ 0 & 1 & 0 & 0 & 0 & 0 & 0 & 0 \\ 0 & 0 & 1 & 0 & 0 & 0 & 0 & 0 \\ 1 & 0 & 0 & 0 & 0 & 0 & 0 & 0 \\ 1 & 0 & 0 & 0 & 0 & 0 & 0 & 0 \\ 1 & 0 & 0 & 0 & 0 & 0 & 0 & 0 \\ 1 & 0 & 0 & 0 & 0 & 0 & 0 & 0 \\ 1 & 0 & 0 & 0 & 0 & 0 & 0 & 0 \end{pmatrix} \qquad R_C = \begin{matrix} 1 \\ 2 \\ 3 \end{matrix}\begin{pmatrix} 1 & 0 & 0 \\ 0 & 1 & 0 \\ 0 & 0 & 0 \end{pmatrix}$$

Fig. 8. $R_C \not\subseteq R$

9 Axiomatization of Choice

Once choice functions are established, researchers usually proceed to the characterization of desirable properties of choice. Many famous people have contributed to this idea and the interdependency of all these conceivable axioms has widely been investigated.

One usually starts with certain intuitively clear and appealing postulates and looks in which way these may be satisfied or not. Impossibility theorems are well known that destroy any hope for choice mechanisms that follow simple axiomatizations. It seems that highly complicated ones are necessary.

Cycles of preference are counter-intuitive. Demanding transitivity, they are excluded, but this is often considered too hard a condition; so indifference is admitted. We recall postulates that are intended to prohibit cycles. See, e.g., Prop. 6.5.

Definition 9.1. Assume a choice function C and revealed preferences thereof.

i) An H-**cycle** from some point x to x is given when $\left[R^*_{C}{;}(R_C)^+\right]_{xx}$.

ii) An SH-**cycle** from some point x to x is given when $\left[R_{C}{;}(R^*_C)^+\right]_{xx}$. □

The following axioms for the revealed R are often demanded to avoid cycles.

Definition 9.2. We consider the revealed preferences of some choice function.

i) **Houthakker's axiom of revealed preference** (HOA) demands that there be no H-cycle, i.e., $(R_C)^+ \subseteq \overline{R_C^*}^\mathsf{T}$.

ii) The **strong axiom of revealed preference** (SA) demands that there be no SH-cycle, i.e., $(R_C^*)^+ \subseteq \overline{R_C}^\mathsf{T}$.

iii) The **weak axiom of revealed preference** (WA) demands that there be no 2-step cycle, i.e., $R_C^* \subseteq \overline{R_C}^\mathsf{T}$. □

Corresponding axioms for the choice functions themselves have also been formulated and the interrelationship has been discussed.

Definition 9.3. We consider the choice function C as well as its revealed preference together with the membership relation. We will speak of the

i) **strong congruence axiom** SCA if $\varepsilon\,;C \cap R_C^{+\mathsf{T}}\,;\varepsilon \subseteq \varepsilon$,

ii) **weak congruence axiom** WCA if $\varepsilon\,;C \cap R_C^\mathsf{T}\,;\varepsilon \subseteq \varepsilon$. □

We have seen on several occasions that we need not explicitly mention $\mathcal{S} := C\,;\mathbb{T}$ every time. Not least Prop. 8.2 has shown that the empty rows of C or those that are non-empty, but assign an empty choice may be neglected without affecting the overall structure. Having this in mind, we consider, e.g., the weak congruence axiom (WCA). In [Suz83], it is presented as

$$\forall S \in \mathcal{S} : [x \in S \,\&\, \{\exists y \in C(S) : (x, y) \in R_C\}] \to x \in C(S).$$

Firstly, quantification over x is not mentioned. Another typical flaw of such considerations is that, in this case, the $S \in \mathcal{S}$ appears — without making this visible — as a subset that may contain elements and also as an element over which quantification may run. Let us denote the element in the powerset corresponding to S as e. (We also remember that our ordering is transposed compared with [Suz83].)

$$\forall x : \forall e : \left[\varepsilon_{xe} \wedge \{\exists y : (\varepsilon\,;C^\mathsf{T})_{ye} \wedge (R_C)_{yx}\}\right] \to (\varepsilon\,;C^\mathsf{T})_{xe}$$

$$\forall x : \forall e : \left[\overline{\varepsilon}_{xe} \vee \overline{\exists y : (\varepsilon\,;C^\mathsf{T})_{ye} \wedge (R_C)_{yx}}\right] \vee (\varepsilon\,;C^\mathsf{T})_{xe}$$

$$\forall x : \forall e : \left[\overline{\varepsilon} \cup \overline{R_C^\mathsf{T}\,;\varepsilon\,;C^\mathsf{T}}\right]_{xe} \vee (\varepsilon\,;C^\mathsf{T})_{xe}$$

$$\varepsilon \cap R_C^\mathsf{T}\,;\varepsilon\,;C^\mathsf{T} \subseteq \varepsilon\,;C^\mathsf{T}$$

$$\varepsilon\,;C \cap R_C^\mathsf{T}\,;\varepsilon \subseteq \varepsilon$$

At last, the function C has been multiplied from the right side, using a standard formula. In analogy follows the strong congruence axiom (SCA).

We mention the following well-known implications without giving full proofs.

Proposition 9.4

i) HOA $\iff$ SCA

ii) HOA $\implies$ SA $\implies$ WA

iii) WA $\iff$ WCA

Proof: ii) is trivial since $R_C^* \subseteq R_C$.

iii) Condition WA demands for the revealed strict preference $R_C^* = (\varepsilon_i C \cap \bar{\varepsilon})_i \varepsilon^\mathsf{T}$

$$R_C^* = (\varepsilon_i C \cap \bar{\varepsilon})_i \varepsilon^\mathsf{T} \subseteq \overline{R_C}^\mathsf{T}$$

$$\Longleftrightarrow R_C^\mathsf{T}{}_i \varepsilon \subseteq \overline{\varepsilon_i C} \cup \varepsilon$$

$$\Longleftrightarrow \varepsilon_i C \cap R_C^\mathsf{T}{}_i \varepsilon \subseteq \varepsilon, \quad \text{i.e., WCA} \qquad\qquad \square$$

10   Concluding Remark

This text is certainly just a first step directed towards a study of social choice using relations and towards computational social choice. Lifting to a point-free relation-algebraic treatment, we have achieved several goals. Firstly, this is a shorthand notation that facilitates work at least for the initiated. Secondly, we got rid of many case distinctions necessary in Sen's or Suzumura's approach; not least are relational proofs more easily computer-checkable. Scientific progress by this article may also be found in the unification of the choice concepts and in relating them to formally manipulable formulae such as being progressively finite, etc. Finally, writing all this down — as it has indeed been done — in the relational reference language TITUREL, an immediate execution on a computer became possible, at least for moderately sized tasks.

Many more attempts allow a relational approach, not least centered around the Gibbard paradox with its standard rights rules; cf. [Wri85]. One may study the Arrow or the Chernoff Axiom relationally and many more as well as a lot of Pareto modelling.

Acknowledgment. The author gratefully acknowledges the detailed comments and suggestions of the unknown referees.

References

[Sch11] Schmidt, G.: Relational Mathematics. Encyclopedia of Mathematics and its Applications, vol. 132. 584 pages. Cambridge University Press (2011) ISBN 978-0-521-76268-7

[Sen70] Sen, A.K.: Collective choice and social welfare. Mathematical Economics Texts, vol. 5. Holden-Day, San Francisco (1970)

[SS89] Schmidt, G., Ströhlein, T.: Relationen und Graphen. Mathematik für Informatiker. Springer (1989) ISBN 3-540-50304-8, ISBN 0-387-50304-8

[SS93] Schmidt, G., Ströhlein, T.: Relations and Graphs — Discrete Mathematics for Computer Scientists. EATCS Monographs on Theoretical Computer Science. Springer (1993) ISBN 3-540-56254-0, ISBN 0-387-56254-0

[Suz83] Suzumura, K.: Rational choice, collective decisions, and social welfare. Cambridge University Press (1983), Reprinted 2009

[Wri85] Wriglesworth, J.L.: Libertarian conflicts in social choice. Cambridge University Press (1985)

An Algebra of Layered Complex Preferences

Bernhard Möller and Patrick Roocks

Institut für Informatik, Universität Augsburg, D-86135 Augsburg, Germany
{moeller,roocks}@informatik.uni-augsburg.de

Abstract. Preferences allow more flexible and personalised queries in database systems. Evaluation of such a query means to select the maximal elements from the respective database w.r.t. to the preference, which is a partial strict-order. Often one requires the additional property of negative transitivity; such a *strict weak order* induces equivalence classes of "equally good" tuples, arranged in layers of the order. We extend our recent algebraic, point-free, calculus of database preferences to cope with weak orders. Since the approach is completely first-order, off-the-shelf automated provers can be used to show theorems concerning the evaluation algorithms for preference-based queries and their optimisation. We use the calculus to transform arbitrary preferences into layered ones and present a new kind of Pareto preference as an application.

Keywords: relational algebra, complex preferences, preference algebra.

1 Introduction

Personalised database systems are designed to regard user wishes more comprehensively to support queries with an exact specification of the user's preferences. For example a consumer wants to buy a new car, and the two most important criteria are high power and low fuel consumption. These goals are *conflicting*, because cars with high power tend to have a higher fuel consumption. To get the optimal results according to both of these equally important goals from a database, the concept of *skyline queries* [BKS01] is used: A car belongs to the result set if there is no other car which is better in both criteria, i.e. has a lower fuel consumption and a higher power. In a 2-dimensional diagram for both criteria the result set looks like a "skyline", viewed from the origin.

Example 1.1. As an introductory example, consider the data set in table 1. The skyline query for minimal fuel consumption and maximal power returns the "BMW 5" and "Mercedes E", because each of these is better than the other by one criterion. The "Audi 6" is not returned, as it is worse by both criteria. □

Imagine that a large database, for example a catalogue containing all the cars for the European market, returns a quite large result for the above skyline query. Assume that the consumer has even more wishes, for example prefers cars with a specific colour, but this is *less important* than the preference for low fuel and high power.

W. Kahl and T.G. Griffin (Eds.): RAMiCS 2012, LNCS 7560, pp. 294–309, 2012.

Table 1. Example of a data set of cars

Model	Fuel	Power	Color
BMW 5	11.4	230	silver
Mercedes E	12.1	275	black
Audi 6	12.7	225	red

Our preference model is not limited to skyline queries but based on the more comprehensive approach of PREFERENCE SQL [Kie05] and its current implementation [KEW11]. There, such hierarchies of user wishes can be handled by *complex preferences*. The constructs of the *Pareto* and the *Prioritisation* preference are used to model equal and more/less important user preferences. Complex preferences are constructions where *base preferences* like "lowest fuel consumption" are at the bottom of the hierarchy. We exemplify the formal notation.

Example 1.2. In the abstract notation of Preference SQL, the preference for "Lowest fuel consumption and (equally important) highest power, *both more important than* a preference for black cars" can be expressed by

$$P = (\text{LOWEST}(fuel) \otimes \text{HIGHEST}(power)) \,\&\, \text{POS}(color, \{\text{black}\}) \,,$$

where LOWEST and HIGHEST induce the "<" and >" orders on their respective numerical domains, while POS creates a preference for values contained in the given set on a discrete domain. Pareto-composition and Prioritisation are denoted by $\otimes$ and $\&$ and are explained formally later on. □

In PREFERENCE SQL many base preferences have the nice property of being strict weak orders, and prioritisation constructs of strict weak orders have this property again. In the scope of preferences we also call such relations *layered preferences*. This allows fast algorithms and a very intuitive way to define *equally good* results: The incomparability relation w.r.t. a layered preference is an equivalence relation. Unfortunately the Pareto preference constructor does not preserve layered preferences. This is the technical reason for an counterintuitive effect which occurs in Preference SQL, shown in the following example.

Example 1.3. The best objects according to P from Example 1.2 in the data set of table 1 are again "BMW 5" and "Mercedes E". This is quite counterintuitive, because the preference for black cars should decide for the Mercedes only. □

With the support of our algebraic calculus of database preferences [MRE12] we define a transformation of the Pareto preference to a layered one which then avoids the counterintuitive effect of the previous example. We show the well-definedness and some other interesting properties algebraically using our calculus. The complex preferences are represented in an abstract relation algebra embedded into a *join algebra* which allows reasoning about complex preferences in a point-free fashion.

2   An Algebraic Calculus for Complex Preferences

Preferences are homogeneous binary relations which additionally are strict partial orders. They are associated with a set of attributes, which label the columns in a database table. The type domain of a preference is induced by the data types occurring in the attribute set, e.g., for the data type "float" the type domain is isomorphic to the floating-point numbers.

In our algebraic approach we abstract from the case of concrete relations and largely work in the setting of idempotent semirings. For some theorems in this paper, however, it becomes necessary to introduce converse and complement, i.e., to work with the special case of abstract relation algebras.

Next to a gain in generality, passing to the more abstract setting has another advantage: searching for appropriate *algebraic characterisations* of relation-like elements, typing, join, etc. forces one to think about the truly essential properties of these objects and operations. On the other hand, the resulting axiomatisation should be simple and small, but still "efficiently usable" to allow the deployment of automatic theorem provers. In our case study [MR12] we have illustrated how a non-trivial law of preference algebra can be shown in PROVER9.

2.1   The Join Algebra for Preferences

We will now recapitulate some definitions from [MRE12]. While that paper also presented a typing mechanism and a join operator that allow treating the case of several attributes, we will not fully use these parts in the present paper.

Definition 2.1. An *idempotent semiring* consists of a set S of elements together with binary operations $+$ of *choice* and $\cdot$ of *composition*. Both are required to be associative, choice also to be commutative and idempotent. Moreover, composition has to distribute over choice in both arguments. Finally there have to be units 0 for choice and 1 for composition.

Homogeneous binary relations over a set form an idempotent semiring with choice $\cup$ and composition ";", which have $\emptyset$ and the identity relation as their respective units.

Hence when we talk about semiring elements $a, b, c, \ldots$ one should always keep in mind that in the concrete instance of preferences they represent homogeneous binary relations between database tuples.

Definition 2.2. Every idempotent semiring induces a *subsumption order* by $x \leq y \Leftrightarrow x + y = y$. A *test* is an element $x \leq 1$ that has a complement $\neg x$ relative to 1, i.e., which satisfies $x + \neg x = 1$ and $x \cdot \neg x = 0 = \neg x \cdot x$.

Tests, sometimes also called *coreflexives*, are algebraic counterparts of subsets of values on which the preference relations "operate". In the relational semiring they are sub-identities, i.e., subsets of the identity relation. The largest test is 1, the smallest 0, standing for the set of all values and the empty set. Sum and composition of tests correspond to union and intersection of the represented sets.

We will use small letters $a, b, c, ...$ at the beginning of the alphabet to denote arbitrary semiring elements and $p, q, r, ...$ to denote tests.

A special test is produced by the inverse image operator $\langle a \rangle\, q$:

Definition 2.3. Following [DMS06], the *(forward) diamond* is axiomatised by the universal property $\langle a \rangle\, q \leq p \Leftrightarrow a \cdot q \leq p \cdot a$. This also leads to the *domain* of a, representing the set of objects having an a-successor, namely $\ulcorner a =_{df} \langle a \rangle\, 1$.

The inverse image has many useful algebraic properties, e.g. isotony, distributivity over "+" and strictness w.r.t. 0 (all in both arguments).

The semiring elements are intended to represent database preferences, which are associated with *attributes*, i.e. database columns. To model this we introduce typed elements as in [MRE12].

Let $\mathcal{A}$ be a set of *attribute names*. A *type* is a subset $T \subseteq \mathcal{A}$ of attribute names. The *join* $T_1 \bowtie T_2$ of two types T_1, T_2 is their union $T_1 \cup T_2$. For every type T we assume a test 1_T representing the type domain of the attribute set T.

By convention, attribute names $A, B, ... \in \mathcal{A}$ are also used for the "simple" (i.e. single element set) types $\{A\}, \{B\}, ...$ We will use them in the examples, whereas in the theorems we use "general" types (i.e. subsets of $\mathcal{A}$ named by $T, T_1, T_a, ...$).

We define a *type assertion*

$$a :: T^2 \quad \Leftrightarrow_{df} \quad a \leq 1_T \cdot a \cdot 1_T \ .$$

A special case is formed by the tests, i.e., sub-identities:

$$p :: T \quad \Leftrightarrow_{df} \quad p \leq 1_T \ .$$

In the concrete relational instances we use in the examples, the type domain for a type T is denoted by the set D_T and the type assertions mean:

$$a :: T^2 \quad \Leftrightarrow \quad a \subseteq D_T \times D_T, \quad p :: T \quad \Leftrightarrow \quad p \subseteq D_T \ .$$

This explains the use of T^2 and T: While general elements are interpreted as homogeneous binary relations, tests are interpreted as sets. We use $a :: T^{(2)}$ to denote "$a :: T$ or $a :: T^2$".

For each type T we require a smallest element 0_T and a largest element $\top_T$; they correspond to the empty set and the full relation in the concrete instances. For every $x :: T^{(2)}$ we have $0_T \leq x \leq \top_T$.

Additionally to [MRE12] we introduce *sub-types*. Let $r :: T$ be a test. The r-*induced sub-type of* T, formally $T[r]$, is defined by the new identity $1_{T[r]} =_{df} r$. For the type assertions this implies

$$p :: T[r] \quad \Leftrightarrow \quad p \leq r, \quad a :: T[r]^2 \quad \Leftrightarrow \quad a \leq r \cdot a \cdot r \ .$$

For the greatest element of this type we have $\top_{T[r]} = r \cdot \top_T \cdot r$ while the smallest element $0_{T[r]} = 0_T$ remains unchanged.

For sake of readability we use the following abbreviations for $x :: T^2$:

$$0_x =_{df} 0_T \ , \quad 1_x =_{df} 1_T \ , \quad \top_x =_{df} \top_T \ .$$

The following definitions are the formal foundations for our preference calculus. First, we define abstract relation algebras using the axiomatisation in [Mad97]:

Definition 2.4 (Abstract relation algebra). An *abstract relation algebra* is an idempotent semiring with additional operators $(...)^{-1}, \overline{(...)})$ for converse and complement, axiomatised by the Schröder equivalences and Huntington's axiom:

$$x \cdot y \leq z \Leftrightarrow x^{-1} \cdot \overline{z} \leq \overline{y} \Leftrightarrow \overline{z} \cdot y^{-1} \leq \overline{x} , \qquad x = \overline{\overline{x} + y} + \overline{\overline{x} + \overline{y}} .$$

For our applications, we additionally stipulate the Tarski rule

$$a \neq 0_a \Rightarrow \mathsf{T}_a \cdot a \cdot \mathsf{T}_a = \mathsf{T}_a ,$$

where $\mathsf{T}_a = \overline{0_a}$.

We assume that our underlying semiring is an abstract relation algebra and each type domain is closed under converse and complement, i.e. for $x :: T^{(2)}$ we have also $x^{-1}, \overline{x} :: T^{(2)}$. Note that this also holds for sub-types $T[r]$, where the complement is relative to r.

For an easier notation, we introduce the meet operation and the difference between two elements as follows:

$$x \sqcap y =_{df} \overline{\overline{x} + \overline{y}} , \qquad x - y =_{df} x \sqcap \overline{y} .$$

For relations these correspond to intersection and set difference. For tests $p, q \leq 1$ they are equal to composition and relative complement:

$$p \sqcap q = p \cdot q , \qquad p - q = p \cdot \neg q .$$

To work with relational compositions of preferences, which we need for the *complex preferences* in the following sections, we introduce the join operator $\bowtie$ for general elements. If $a :: T_a^{(2)}, b :: T_b^{(2)}$ then $a \bowtie b :: (T_a \bowtie T_b)^{(2)}$. In [MRE12] we introduced a *join algebra* where we required associativity, commutativity of $\bowtie$ and distributivity of $\bowtie$ over $+$. Additionally the diamond operator distributes over $\bowtie$ and we have an exchange law for $\bowtie$ and ".". We need these laws for some technical steps and will not go into detail here; for the main results in this paper, we just need the existence of joins for the construction of complex preferences. For disjoint types T_a and T_b, in the concrete relational instances the type domain of $T_a \bowtie T_b$ is the Cartesian product $D_{T_a} \times D_{T_b}$.

The *maximal* objects w.r.t. element $a :: T^2$ and test $p :: T$ are represented by the test

$$a \rhd p =_{df} p - \langle a \rangle p .$$

With the abstract relation algebra and the join algebra we conclude our formal framework for point-free reasoning about preference relations.

Although we use an abstract relation algebra, in the examples one may always think of a concrete representation, where general elements $a, b, c, ...$ are relations and tests $p, q, r, ...$ are sets. To make our examples easy to follow, we sometimes use a point-wise notation: For a relation $a :: T^2$ the expression $t \, a \, t'$ for tuples $t, t' \in D_T$ means that the tuple t is related to t' via a. Analogously we use $\neg(t \, a \, t')$ if t is not a-related to t'.

2.2   Preferences and SV Relations

Definition 2.5 ((Layered) preferences). A relation $a :: T^2$ is a *preference* if and only if it is irreflexive and transitive, i.e.

1. $a \sqcap 1_a = 0_a$,
2. $a \cdot a \leq a$.

a is a *layered preference* if additionally *negative transitivity* $\overline{a} \cdot \overline{a} \leq \overline{a}$ holds.

Strict partial orders satisfying negative transitivity are sometimes called *strict weak orders*. In the scope of preferences, i.e. in [Kie05] they are called *weak order preferences*. In this paper we will only use the term *layered preferences*. The reason for this is that such relations induce a "layered structure", i.e. there is always a function $f : D_A \to \mathbb{N}$ s.t. $t_1 \, a \, t_2 \Leftrightarrow f(t_1) < f(t_2)$, which is shown in [Fis70], Thm. 2.2.

To allow a more precise modelling of user preferences, complex combinations of elementary preference relations may be formed. One example is given by the *prioritisation preference* $a \,\&\, b :: (T_a \bowtie T_b)^2$ for preferences $a :: T_a^2, b :: T_b^2$, which is also known as the *lexicographic order*. The intuitive meaning of $a \,\&\, b$ is: "Better w.r.t. a in the T_a-part, or equal there w.r.t. a and better in the T_b-part w.r.t. b". In the preference algebra of [MRE12] this can be formalised as

$$a \,\&\, b = a \bowtie T_b + 1_a \bowtie b .$$

So "equal w.r.t. a" is represented by the identity 1_a, while T_b represents that one does not care about the T_b-part when the T_a-part already decides about the overall relation.

However, it is questionable whether this always meets the user's expectation. If a is a layered preference, then the incomparability relation $\overline{(a + a^{-1})}$ is an equivalence relation. Hence, if two tuples t_1, t_2 are incomparable w.r.t. a layered preference a but t_2 is better than t_1 w.r.t. b, it is quite intuitive to say that t_2 is better than t_1 (and not incomparable with t_1) in $a \,\&\, b$.

Formally this is reflected by the *SV-semantics* of [Kie05]. "SV" stands for "substitutable values" and means that a comparison between two tuples t_1, t_2 with respect to a remains unchanged if t_1 is substituted for an SV-related t'_1. In relational notation, with s_a being the SV relation and "$\equiv$" meaning logical equivalence of formulas, we have

$$\forall t_1, t'_1, t_2 : \quad t_1 \, s_a \, t'_1 \implies t_1 \, a \, t_2 \equiv t'_1 \, a \, t_2 \,\wedge\, t_2 \, a \, t_1 \equiv t_2 \, a \, t'_1 . \tag{1}$$

We give an algebraic characterisation of SV relations.

Definition 2.6 (SV relation). For a preference $a :: T_a^2$ we call $s_a :: T_a^2$ an *SV relation for* a, if s_a fulfils the following properties:

1. The relation s_a is an equivalence relation, i.e. s_a is reflexive ($1_a \leq s_a$), symmetric ($s_a^{-1} = s_a$) and transitive ($s_a \cdot s_a \leq s_a$).

2. s_a is compatible with a:
  (a) $s_a \sqcap a = 0_a$,
  (b) $s_a \cdot a \leq a$,
  (c) $a \cdot s_a \leq a$.

If the SV relation is not stated explicitly, then it is assumed to be the identity, i.e. we set $s_a = 1_a$.

Using this concept we will below adapt our definition of prioritisation. Next to the prioritisation preference there is another important complex preference constructor called *Pareto composition* which combines two preferences $a :: T_a^2, b :: T_b^2$. It is denoted as $a \otimes b :: (T_a \bowtie T_b)^2$ and has the intuitive meaning: "Better w.r.t. to a in the T_a-part or b in the T_b-part and not worse (i.e. equal or better) in the other part".

Using SV relations where the "equal w.r.t to a" is formalised by "s_a-related" we define the following complex preferences:

Definition 2.7 (Prioritisation and Pareto composition with SV). Let $a :: T_a^2$ and $b :: T_b^2$ be preferences with associated SV relations $s_a :: T_a^2$ and $s_b :: T_b^2$. The prioritisation is given by:

$$a \,\&\, b :: (T_a \bowtie T_b)^2 \,,$$
$$a \,\&\, b = a \bowtie T_b + s_a \bowtie b \,,$$

whereas the Pareto composition is defined as

$$a \otimes b :: (T_a \bowtie T_b)^2 \,,$$
$$a \otimes b = a \bowtie (s_b + b) + (s_a + a) \bowtie b \,.$$

We say that $a \,\&\, b$ or $a \otimes b$ is *SV-preserving* if $s_{a\&b} = s_a \bowtie s_b$ or $s_{a\otimes b} = s_a \bowtie s_b$, respectively. Note that one may always define SV relations other than 1 for complex preferences, as long as they fulfil the conditions of Definition 2.6.

Corollary 2.8. *The above notions are well-defined, i.e. $s_a \bowtie s_b$ is indeed a valid SV relation for $a \,\&\, b$ and $a \otimes b$.*

Proof. Straightforward from distributivity and isotony of join, the exchange laws and some other standard laws, cf. [MRE12]. □

Now we consider SV relations larger than 1_a. For weak orders a typical SV relation is the incomparability relation, which we state in the following lemma:

Lemma 2.9. *If $a :: T^2$ is a layered preference then $s_a = \overline{a + a^{-1}}$ is an SV relation.*

Proof. The equivalence property for this relation is well known and $s_a \sqcap a = 0_a$ is clear. We show $s_a \cdot a \leq a$. By definition of s_a, the exchange law for complement and converse, and finally the infimum property we infer:

$$s_a \cdot a = \left(\overline{a} \sqcap \overline{a^{-1}}\right) \cdot a = \left(\overline{a} \sqcap (\overline{a})^{-1}\right) \cdot a \leq (\overline{a})^{-1} \cdot a \,.$$

We still have to show that $(\bar{a})^{-1}{\cdot}a \leq a$. By the Schröder equivalences, this is equivalent to $\bar{a}{\cdot}\bar{a} \leq \bar{a}$, which is true by negative transitivity of a. For $a{\cdot}s_a \leq a$ an analogous argument holds; hence s_a is compatible with a.   □

Note, that this does not hold if a is not a layered preference, which we show with the following example:

Example 2.10. Assume preferences $a :: A^2, b :: B^2$ on attributes A, B with $D_A = D_B = \{0, 1, 2\}$. The preferences a and b both are the $<$-relation on the natural numbers, i.e. we have $(0\,x\,1), (1\,x\,2)$ and $(0\,x\,2)$ for $x \in \{a, b\}$. Assume tuples $t_1 = (1, 2)$ and $t_2 = (2, 1)$. Then, by definition, t_1 and t_2 are not related w.r.t. to $(a{\otimes}b)$ or $1_A \Join 1_B$, i.e. they are incomparable. Now we consider the incomparability relation $s_{\text{inc}} =_{df} \overline{(a \otimes b) + (a \otimes b)^{-1}}$. We have, by definition of $a \otimes b$,

$$(2, 0)\,(a \otimes b)\,(2, 1), \quad \neg((2, 0)\,(a \otimes b)^k\,(1, 2)), \quad \neg((2, 1)\,(a \otimes b)^k\,(1, 2)),$$

where $k \in \{-1, 1\}$. By definition of s_{inc} this implies

$$\neg((2, 0)\,s_{\text{inc}}\,(2, 1)), \quad (2, 0)\,s_{\text{inc}}\,(1, 2), \quad (2, 1)\,s_{\text{inc}}\,(1, 2),$$

which means that s_{inc} is not transitive, hence it is no equivalence relation and therefore no SV relation for a.

3   The Pareto-regular Preference

Unfortunately, layered preferences are not closed under the Pareto operator. Therefore the question arises whether there is a similar operator under which layered preferences are closed. The answer is yes and our strategy to construct such a preference is as follows:

- Let r be a test representing the *basic set*, i.e. the data set. We take the maxima of r w.r.t. $(a \otimes b)$, i.e. $(a \otimes b) \triangleright r$, and call them layer-0 elements.
- We remove them from the basic set, i.e. we define $r_1 = r - (a \otimes b) \triangleright r$, take their maxima and call these layer-1 elements.
- We iterate this process to obtain the layer-n elements for $n = 2, 3, \ldots$.
- We define a new preference induced by ordering the elements according to their layers, placing layer 0 at the top and the layer with the largest number at the bottom. This yields a layered preference by construction.

We will call the new preference the *Pareto-regular* preference. This name stems from the implementation of preferences in PREFERENCE SQL [KEW11]: The keyword `regular` after a layered preference means that the SV relation from Lemma 2.9 is applied. As the Pareto preference need not be a layered preference, we call "Pareto-regular" the result of transforming the Pareto preference into a layered preference and then applying Lemma 2.9.

The process of successive removal of maximal elements corresponds to the repeated removal of sinks in a classical algorithm for cycle detection in directed graphs [Mar59, SS93]. That algorithm terminates when the set of maxima/sinks to be removed becomes empty. The original graph contained a cycle iff the remaining set is nonempty. Since our preference relations, as strict partial orders, correspond to acyclic graphs, the iteration necessarily will reach the empty set, which, however, is not counted as a separate layer. We will make this more precise in the next section.

3.1   Computing Layer-i Elements

The concept of layer-i elements was originally introduced under the term "Iterated preferences" in [Cho03], where the maximum operator is called "winnow operator". Here we give an algebraic definition, and prove some properties. In particular, we show that the induced order is indeed a layered preference.

Definition 3.1 (Layer-i Elements). Let $a :: T_a^2$ be a preference, and $r :: T_a$ a basic set. For $i = 0, 1, 2, \ldots$ we define the tests q_i and r_i characterising the layer-i elements and the remainders, resp.:

$$q_i =_{df} a \triangleright r_i \text{ where } r_i =_{df} r - \sum_{j=0}^{i-1} q_j .$$

By convention, the empty sum is 0_a, hence we have $r_0 = r$.

A mnemonic for the q_i is that the letter "b" for "best", rotated by 180° becomes a "q". This matches our convention that $a, b, c, \ldots$ are used for general elements and p, q, r for tests.

Using our algebra we deduce a closed formula for the r_i. In this, we write ra short for $r \cdot a$. The powers x^k are defined by $x^k = x^{k-1} \cdot x$ and $x^0 = 1$. The proof of the following lemma can be found in A.1.

Lemma 3.2 (Closed formula for layer-i elements). *For $i \in \mathbb{N}$ we have the following properties:*

1. *$(ra)^{i+1} \leq (ra)^i$ provided $i > 0$,*
2. *$\langle (ra)^{i+1} \rangle r \leq \langle (ra)^i \rangle r$,*
3. *$r_i = \langle (ra)^i \rangle r$.*

With this lemma we have a compact representation of layer-n elements. This helps us to show some interesting properties of r_i and q_i which we will need for the construction of the Pareto-regular preference later on. These properties are stated in the following lemma, which is proved in A.2.

Lemma 3.3. *Assume q_i, r_j as in Definition 3.1. We have:*

1. *The r_i are decreasing in i, i.e., $r_0 \geq r_1 \geq r_2 \geq \ldots$.*
2. *The q_i are pairwise disjoint, i.e., for $i \neq j$ we have $q_i \cdot q_j = 0_a$.*
3. *Let r be finite, i.e., assume that there do not exist infinitely many disjoint $p_i \neq 0$ with $\sum_i p_i = r$. Then the calculation of the r_i becomes stationary, i.e. there exists an $N \in \mathbb{N}$ with $N = \max\{k \in \mathbb{N} \mid r_k \neq 0_a\}$.*
4. *The q_i cover r, i.e., $\sum_{i=0}^{N} q_i = r$.*
5. *For $i \leq j$ we have $q_i \cdot a \cdot q_j = 0_a$.*

3.2   The Induced Layered Preference

Now we will construct the preference induced by the layer-i elements and the corresponding induced SV relation.

Definition 3.4 (Induced layered preference). Let $a :: T_a^2$ be a preference and $r :: T_a$ a basic set. Consider the corresponding layer-i elements $q_i = a \triangleright \langle (ra)^i \rangle r$ with $i \in [1, N]$ and $N = \max\{k \in \mathbb{N} \mid r_k \neq 0_a\}$ (see Lemma 3.3.3). We define relations b_{ij} $(i, j \in [1, N])$ by $b_{ij} = q_i \cdot T_a \cdot q_j$. In the concrete model these represent universal relations between the sets q_i and q_j. With their help, the *induced layered preference* $m(a, r) :: T_a[r]^2$ is defined as

$$m(a, r) =_{df} \sum_{i>j} b_{ij} \, ,$$

where $T_a[r]$ is the sub-type of T_a with identity r and greatest element $r \cdot T_a \cdot r$.

By the summation over $i > j$ the less preferred elements w.r.t. to a (with higher layer numbers) are $m(a, r)$-related to the more preferred elements (with lower layer numbers).

A corresponding SV relation $s_{m(a,r)} :: T_a[r]^2$ is defined as

$$s_{m(a,r)} =_{df} \sum_{i} b_{ii} \, .$$

We note an important property of the relations b_{ij}: by disjointness of the q_i and the Tarski rule we have

$$b_{ij} \cdot b_{kl} = \begin{cases} b_{il} & \text{if } j = k, \\ 0_a & \text{otherwise.} \end{cases} \tag{2}$$

Our goal is to construct the Pareto-regular preference, and we are close to this by constructing $m(a, r)$ for a Pareto preference $a = a_1 \otimes a_2$. But is the resulting relation fulfilling the (layered) preference properties, i.e. would such a preference be well-defined? We show this in the next lemma.

Lemma 3.5

1. *The relation $m(a, r)$ from the previous definition is a layered preference.*
2. *$s_{m(a,r)}$ is an SV relation for $m(a, r)$.*

Proof

1. Transitivity follows from the definition of $m(a, r)$ and eq. (2). Again by definition of $m(a, r)$ and disjointness of q_i (Lemma 3.3.2) we have irreflexivity. It remains to show that negative transitivity holds. Note that due to the type $T_a[r]^2$ of $m(a, r)$ and $s_{m(a,r)}$ the complement $\overline{(...)}$ is relative to r. With this we infer $(\overline{m(a, r)})^2 \leq \overline{m(a, r)}$:

$$\left(\overline{m(a,r)} \right)^2 = \left(\sum_{i \leq j} b_{ij} \right) \cdot \left(\sum_{k \leq l} b_{kl} \right) = \sum_{i \leq j \leq l} b_{ij} \cdot b_{jl} \leq \sum_{i \leq l} b_{il} = \overline{m(a,r)} \, .$$

2. We infer that

$$\overline{m(a,r) + m(a,r)^{-1}} = \overline{\sum_{i>j} b_{ij} + \sum_{i<j} b_{ij}} = \overline{\sum_{i \neq j} b_{ij}} = \sum_i b_{ii} = s_{m(a,r)}.$$

Together with Lemma 2.9 this shows the claim.   □

We give two further useful properties of the induced layered preference:

- The original preference is still contained in the induced layered preference.
- The induced SV relation is part of the incomparability relation of the original preference.

Formally this is stated in the following lemma. There again we restrict the "original preference" and the "incomparability relation" to r on both sides, because the induced layered preference and the induced SV relation is only defined on the basic set r.

Lemma 3.6. *Let $a :: T_a^2$ be a preference and $r :: T_a$ a basic set. We have:*

1. $r \cdot a \cdot r \leq m(a,r)$.
2. $s_{m(a,r)} \leq r \cdot \overline{(a + a^{-1})} \cdot r$.

Proof

1. By Lemma 3.3.5 we get $q_i \cdot a \cdot q_j = 0$ for $i \leq j$. This implies:

$$\sum_{i \leq j} q_i \cdot a \cdot q_j = 0_a .\tag{3}$$

We use this in the following deduction:

> TRUE
>
> $\Leftrightarrow$ {| definition T_a |}
>
> $a \leq T_a$
>
> $\Rightarrow$ {| $q_j \cdot (...), (...) \cdot q_i$, summation over $i > j$ |}
>
> $\sum_{i>j} q_i \cdot a \cdot q_j \leq \sum_{i>j} q_i \cdot T_a \cdot q_j$
>
> $\Leftrightarrow$ {| Eq. (3) (additional term is 0), def. of b_{ij} |}
>
> $\sum_{i>j} q_i \cdot a \cdot q_j + \sum_{i \leq j} q_i \cdot a \cdot q_j \leq \sum_{i>j} b_{ij}$
>
> $\Leftrightarrow$ {| re-indexing of sum, def. of $m(a,r)$ |}
>
> $\sum_{i,j} q_i \cdot a \cdot q_j \leq m(a,r)$
>
> $\Leftrightarrow$ {| distributivity and $\sum_i q_i = r$ (Lemma 3.3.4) |}
>
> $r \cdot a \cdot r \leq m(a,r)$.

2. The claim is equivalent to

$$s_{m(a,r)} \sqcap (r \cdot a^k \cdot r) = 0_a \text{ for } k \in \{-1, 1\} .$$

From Part 1 we obtain $r \cdot a^k \cdot r \leq m(a^k, r)$ for $k \in \{-1, 1\}$, because a^{-1} is again a preference, hence the same argument holds for it. Thus it is sufficient to prove:

$$s_{m(a,r)} \sqcap m(a^k, r) = 0_a \text{ for } k \in \{-1, 1\} .$$

This follows from the definitions of $s_{m(a,r)}$ and $m(a^k, r)$ and the disjointness of the q_i (Lemma 3.3.2). $\qquad\qquad\square$

Remark 3.7. The inequations in the previous lemma are equations if a is already a layered preference.

Now we have everything ready to define the Pareto-regular preference. All we have to do is to apply $m(...)$ to the classic Pareto preference, which yields a well defined result by the previous lemma.

Definition 3.8 (Pareto-regular preference). For preferences $a :: T_a^2$, $b :: T_b^2$ and a basic set $r :: T_a \bowtie T_b$ the Pareto-regular preference and its SV relation are defined as:

$$a \otimes_{\text{reg}} b :: (T_a \bowtie T_b)^2 ,$$
$$a \otimes_{\text{reg}} b = m(a \otimes b, r) ,$$
$$s_{a \otimes_{\text{reg}} b} = s_{m(a \otimes b, r)} .$$

Note that the Pareto-regular preference depends on the concrete basic set, i.e. the data stored in the database table. This is a fundamental difference to the classical preferences, which are independent of the data. But Lemma 3.6 tells us that this is kind of "harmless".

3.3 Application: Pareto(-regular) and Prioritisation

With the Pareto-regular preference we have a layered preference which is quite similar to the classic Pareto preference (which is not layered in general). But this is primarily a technical feature, which at first sight changes nothing, because the maxima set of a Pareto preference is the same as that for the associated Pareto-regular preference. For preferences $a :: T_a^2$, $b :: T_b^2$ and a basic set $r :: T_a \bowtie T_b$ we get immediately from the definitions that $(a \otimes b) \triangleright r = (a \otimes_{\text{reg}} b) \triangleright r$.

So where is the *practical* difference between $\otimes$ and $\otimes_{\text{reg}}$? The effect of the latter becomes evident in combination with other complex preferences, especially if the Pareto-regular preference is the first part of a prioritisation. We consider the following example:

Example 3.9. Let $a :: A^2$, $b :: B^2$, $c :: C^2$ be preferences on attributes A, B, C. Let their type domains be $D_A = D_B = D_C = \{1, 2\}$. Let a, b, c each be the $<$-order, i.e. we have $(1 \, x \, 2)$ for $x \in \{a, b, c\}$. Consider the basic set $r :: A \bowtie B \bowtie C$ given by $t_1 =_{df} (1, 2, 1)$, $t_2 =_{df} (2, 1, 2)$, $r =_{df} t_1 + t_2$, and the preferences

$$d_1 =_{df} (a \otimes_{\text{reg}} b) \& c, \quad d_2 =_{df} (a \otimes b) \& c .$$

For $(a \otimes b)$ we have only one maxima set $q_0 = (a \otimes b) \triangleright r = r$. Hence we get:

$$t_1 \, d_1 \, t_2 \, , \quad \neg(t_1 \, d_2 \, t_2) \, .$$

This means that the preference c decides about the maxima of d_i only if the previous Pareto preference $a \otimes b$ has no incomparable elements. This incomparability is avoided by construction in the Pareto-regular preference. Probably the preference d_1 is what the user expected, or at least expected more than d_2. □

This is the abstract formulation of the introductory example on a car database (Examples 1.1–1.3). The same results are obtained as follows: Define $a =_{df}$ LOWEST(*fuel*), $b =_{df}$ HIGHEST(*power*), $c =_{df}$ POS(*color*, {black}) and set the tuples t_1 and t_2 to the fuel/power/color values of "BMW 5" (t_1) and "Mercedes E" (t_2) as denoted in table 1 on page 295; then the result is that "Mercedes E" is the best object according d_1.

Remark 3.10. Note that applying $m(...)$ after the prioritisation in the previous example does not result in Pareto-regular behaviour, i.e., for the preference $d_3 =_{df} m((a \otimes b) \& c, r)$ we get $\neg(t_1 \, d_3 \, t_2)$.

4   Conclusion and Outlook

The search for another kind of Pareto preference was originally motivated by a past project where Preference SQL was used for context-aware suggestions in a hiking-tour recommender. In [REM+12] the context model is described and some sample queries are given; these are very complex preference constructs, where base preferences or Pareto compositions are put into long prioritisation chains. Within the project we noticed that the less-prioritized preferences, like "c" in Example 3.9, are not decisive for the maxima set, because a Pareto preference at the beginning of the term (like "$a \otimes b$" in the example) generates incomparable elements, hence the set of maxima cannot be reduced by c. With the Pareto-regular preference these prioritisation chains become a chain of "filters" where the set of maxima can be reduced by adding less-prioritized preferences at the end of the chain. By calculating not only the maximal elements for the Pareto preference but the layer-i elements we are also able to answer *TOP-k* queries, i.e. a query like "What are the 10 best elements according to preference a".

For this work we have extended our algebraic calculus with layered preferences and SV relations. Interesting connections and properties can be stated and proved algebraically, i.e., in a point-free fashion. These methods are important for a better understanding of preferences and constructing algorithms for preference evaluation.

For future research, there is a large number of theorems about preferences which have only been proved in a point-wise fashion which makes the proofs hardly readable. These theorems are important for optimizing the algorithms for evaluating preference queries. In addition, we are also working on methods for parallel computation of the maxima according to preferences; hopefully algebraic methods will support us, e.g., the methods of concurrent Kleene Algebra.

References

[BKS01]   Börzsönyi, S., Kossmann, D., Stocker, K.: The Skyline Operator. In: Data Engineering (ICDE 2001), pp. 421–430 (2001)

[Cho03]   Chomicki, J.: Preference Formulas in Relational Queries. ACM Transactions on Database Systems 28(4), 427–466 (2003)

[DMS06]   Desharnais, J., Möller, B., Struth, G.: Kleene Algebra With Domain. ACM Transactions on Computational Logic 7, 798–833 (2006)

[Fis70]   Fishburn, P.C.: Utility Theory for Decision Making. Wiley, New York (1970)

[KEW11]   Kießling, W., Endres, M., Wenzel, F.: The Preference SQL System – An Overview. IEEE Data Eng. Bull. 34(2), 11–18 (2011)

[Kie05]   Kießling, W.: Preference Queries with SV-Semantics. In: International Conference on Management of Data (COMAD 2005), pp. 15–26 (2005)

[Mad97]   Maddux, R.: Relation Algebras. In: Brink, C., Kahl, W., Schmidt, G. (eds.) Relational Methods in Computer Science. Advances in Comp. Sci. Springer (1997)

[Mar59]   Marimont, R.: A New Method of Checking the Consistency of Precedence Matrices. Journal of the ACM 6(2), 164–171 (1959)

[MR12]   Möller, B., Roocks, P.: Proof of the Distributive Law for Prioritisation and Pareto Composition, http://tinyurl.com/c79wdrt

[MRE12]   Möller, B., Roocks, P., Endres, M.: An Algebraic Calculus of Database Preferences. In: Gibbons, J., Nogueira, P. (eds.) MPC 2012. LNCS, vol. 7342, pp. 241–262. Springer, Heidelberg (2012)

[REM+12]   Roocks, P., Endres, M., Mandl, S., Kießling, W.: Composition and Efficient Evaluation of Context-Aware Preference Queries. In: Lee, S.-G., Peng, Z., Zhou, X., Moon, Y.-S., Unland, R., Yoo, J. (eds.) DASFAA 2012, Part II. LNCS, vol. 7239, pp. 81–95. Springer, Heidelberg (2012)

[SS93]   Schmidt, G., Ströhlein, S.: Relations and Graphs — Discrete Mathematics for Computer Scientists. Springer (1993)

A   Proofs of Section 3.1

A.1   Proof of Lemma 3.2

1. By transitivity of a we have

$$(ra)^2 = r{\cdot}a{\cdot}r{\cdot}a \leq r{\cdot}a{\cdot}a \leq r{\cdot}a = (ra)^1$$

which implies transitivity of (ra). Iterated application of transitivity shows the claim.

2. For $i = 0$ we obtain by a diamond property

$$\langle (ra)^1 \rangle\, r = \langle ra \rangle\, r = r{\cdot} \langle a \rangle\, r \leq r = \langle (ra)^0 \rangle\, r$$

For $i > 0$ the claim is immediate from Part 1 and isotony of diamond.

3. We perform again an induction on i.
   - $i = 0$: $\langle (ra)^0 \rangle\, r = \langle 1 \rangle\, r = r$.
   - $i \to i + 1$: Assume $r_i = \langle (ra)^i \rangle\, r$.

$$r_{i+1}$$
$$= \quad \{\!\!\{ \text{ definitions } \}\!\!\}$$
$$r - \sum_{j=0}^{i} q_j$$
$$= \quad \{\!\!\{ \text{ splitting the sum and definitions } \}\!\!\}$$
$$r_i - q_i$$
$$= \quad \{\!\!\{ \text{ definition } q_i \}\!\!\}$$
$$r_i - a \triangleright r_i$$
$$= \quad \{\!\!\{ \text{ definition } \triangleright \}\!\!\}$$
$$r_i - (r_i - \langle a \rangle \, r_i)$$
$$= \quad \{\!\!\{ \text{ definition of } -, \text{ De Morgan } \}\!\!\}$$
$$r_i \cdot (\neg r_i + \langle a \rangle \, r_i)$$
$$= \quad \{\!\!\{ \text{ distributivity, } p \cdot \neg p = 0 \}\!\!\}$$
$$r_i \cdot \langle a \rangle \, r_i$$
$$= \quad \{\!\!\{ \ r_i \leq r \text{ by definition } \}\!\!\}$$
$$r_i \cdot r \cdot \langle a \rangle \, r_i$$
$$= \quad \{\!\!\{ \text{ diamond property } \}\!\!\}$$
$$r_i \cdot \langle ra \rangle \, r_i$$
$$= \quad \{\!\!\{ \text{ induction hypothesis } \}\!\!\}$$
$$(\langle (ra)^i \rangle \, r) \cdot (\langle ra \rangle \, \langle (ra)^i \rangle \, r)$$
$$= \quad \{\!\!\{ \text{ diamond property, definition of powers } \}\!\!\}$$
$$(\langle (ra)^i \rangle \, r) \cdot (\langle (ra)^{i+1} \rangle \, r)$$
$$= \quad \{\!\!\{ \text{ Part 2 } \}\!\!\}$$
$$\langle (ra)^{i+1} \rangle \, r$$

A.2 Proof of Lemma 3.3

1. Immediate from Lemma 3.2.
2. Let w.l.o.g. $j \geq i+1$. It follows:

$$q_i \cdot q_j$$
$$= \quad \{\!\!\{ \text{ definition of } r_i \text{ and } \triangleright \}\!\!\}$$
$$(r_i - \langle a \rangle \, r_i) \cdot (r_j - \langle a \rangle \, r_j)$$
$$= \quad \{\!\!\{ \text{ Boolean algebra } \}\!\!\}$$
$$r_i \cdot r_j - (\langle a \rangle \, r_i + \langle a \rangle \, r_j)$$
$$= \quad \{\!\!\{ \ r_j \leq r_i \text{ by (1) and } j \geq i+1, \text{ isotony of diamond } \}\!\!\}$$
$$r_j - \langle a \rangle \, r_i$$
$$= \quad \{\!\!\{ \text{ Lemma 3.2 for } r_i, r_j \}\!\!\}$$
$$\langle (ra)^j \rangle \, r - \langle a \rangle \, \langle (ra)^i \rangle \, r$$

$$\leq \quad \{\!| \ r \leq 1_A \ |\!\}$$
$$\langle (ra)^j \rangle \, r - r \cdot \langle a \rangle \langle (ra)^i \rangle \, r$$
$$= \quad \{\!| \text{ diamond properties } |\!\}$$
$$\langle (ra)^j \rangle \, r - \langle (ra)^{i+1} \rangle \, r$$
$$= \quad \{\!| \ (ra)^j \leq (ra)^{i+1} \text{ by Lemma 3.2.1 and } j \geq i+1 \ |\!\}$$
$$0_a$$

3. By transitivity and irreflexivity of a there are always maximal elements in non-empty sets, i.e. we have $r \neq 0_a \Rightarrow a \triangleright r \neq 0_a$. Hence $r_i \neq 0_a$ implies $q_i \neq 0_a$. Additionally the q_i are pairwise disjoint by Part 2, hence r_{i+1} is strictly less (i.e. $r_{i+1} \leq r_i \ \wedge \ r_{i+1} \neq r_i$) than r_i. Induction shows that the sequence r_i is strictly decreasing for $i = 0, ..., (N+1)$ and equals 0_a for $i = (N+1), ..., \infty$.

4. Immediate from Part 3 and the definition of the q_i, since the definition of N implies $r_{N+1} = 0_a$.

5. First, we have

$$q_i \cdot a \cdot q_j = 0_a$$
$$\Leftrightarrow \quad \{\!| \text{ domain is strict w.r.t. } 0_a \ |\!\}$$
$$\ulcorner (q_i \cdot a \cdot q_j) = 0_a$$
$$\Leftrightarrow \quad \{\!| \text{ definition of diamond } |\!\}$$
$$\langle q_i \cdot a \rangle \, q_j = 0_a$$
$$\Leftrightarrow \quad \{\!| \text{ property of diamond } |\!\}$$
$$q_i \cdot \langle a \rangle \, q_j = 0_a$$

Now,

$$q_i \cdot \langle a \rangle \, q_j$$
$$= \quad \{\!| \text{ definition of } q_i, q_j \ |\!\}$$
$$(r_i - \langle a \rangle \, r_i) \cdot \langle a \rangle \, (r_j - \langle a \rangle \, r_j)$$
$$\leq \quad \{\!| \text{ isotony of diamond } |\!\}$$
$$(r_i - \langle a \rangle \, r_i) \cdot \langle a \rangle \, r_j$$
$$\leq \quad \{\!| \ i \leq j, \text{ hence } r_j \leq r_i \text{ by Part 1 } |\!\}$$
$$(r_i - \langle a \rangle \, r_i) \cdot \langle a \rangle \, r_i$$
$$= \quad \{\!| \text{ Boolean algebra } |\!\}$$
$$0_a$$

Continuous Relations
and Richardson's Theorem*

Hitoshi Furusawa[1], Toshikazu Ishida[2], and Yasuo Kawahara

[1] Department of Mathematics and Computer Science, Kagoshima University,
Kagoshima, Japan
[2] Center for Fundamental Education, Kyushu Sangyo University, Fukuoka, Japan

Abstract. The paper intends to seek a definition of continuous relations
with relational methods and gives another proof of Richardson's theorem
on nondeterministic cellular automata.

1 Introduction

The theory of cellular automata [2,8] (CA, for short) has been studied as a model
of complex systems connecting mathematics, physics and other natural sciences.
In 1972 Richardson [6] initially introduced a topology on the set of configurations of CA (tessellations) and proved a theorem, which we call Richardson's
theorem, to characterize parallel transition relations of nondeterministic CA using a notion of continuous relations. His theorem characterized (parallel) transition relations of CA by the following three properties: (a) the transition relation
is commutative with (parallel) shifts, (b) the states of each cell after transition are independently defined, and (c) the transition relation is continuous and
closed. Richardson's theorem was applied to solve some problems concerned with
Garden-of-Eden configurations. He independently but indirectly defined topological notions such as the compactness and the continuity of relations via points
of accumulation. Also his forward continuity of relations was comprised of the
continuity (which will be defined in section 3) and the closedness of transition
relations. These unusual topological arguments made his paper difficult to read.
Although Richardson's theorem asserted for nondeterministic CA, almost all
applications of the main theorem have been limited to the deterministic case.

On the other hand Brattka and Hertling [1] and Ziegler [9] focused on the
continuity of relations related to computability. The continuity of relations has
rarely been treated in the field of relation algebras. It is historically known that
Choquet [3] studied the continuity of relations.

Despite there are many natural phenomena with nondeterministic behaviors,
we could meet very few researches on nondeterministic CA. Such a situation may
cause from a lack of mathematical tools for analyzing nondeterministic CA. This
is a reason why it is important to give algebraic formalisms for nondeterministic CA. To this end the paper intends to apply relational methods to study CA.

---

* This work was supported in part by Grants-in-Aid for Scientific Research (C)
22500016 from Japan Society for the Promotion of Science (JSPS).

W. Kahl and T.G. Griffin (Eds.): RAMiCS 2012, LNCS 7560, pp. 310–325, 2012.

In particular we improve the continuity of relations due to [1] with relational terms and develop a more formal treatment of the continuity. Though almost all computations of relations in the paper are manipulated by basic rules of relation algebras, some proofs concerned with the compactness of topological spaces have to use pointwise arguments related to so-called the point axiom. Also we remark a helpful fact that the product topology of a finite discrete space has a uniform-like basis consisting of equivalence classes made by projections. Finally we will give a modern proof of Richardson's theorem combined topological and relational devices.

The paper organized as follows. In Section 2 we briefly explain basic notions and notations on binary relations between sets. Here we define a total function (simply, tfn) to be a univalent and total relation, but it is just an ordinary function (or a mapping). Also we recall the domain (relation) of a relation and a rectangular relation and state their basic properties. The domains of relations will be used in the definition of continuous relations, and open relations between topological spaces are define using the notion of rectangular relations. In Section 3 we reformulate the definition of the continuity of relations between topological spaces with relational notations. Intuitively speaking a relation α is continuous if the converse relation $\alpha^\sharp$ transforms all open relations into open relations in the domain of α. Thus if α is totally defined then the continuity of α may be very similar to the continuity of ordinary functions. Also some fundamental results in [1] will be restated with relational language. In Section 4 we will discuss compact and Hausdorff spaces which are elementary in the theory of topology, but we try to define them with relational terms. The compactness of transition relations for CA topologically represents the finiteness of local relations (rules) of CA. As the preliminary for the later section we extend basic theorems that continuous relations preserve the compactness, and that the inverse function of a continuous bijection from a compact space onto a Hausdorff space is continuous. In Section 5 we recall the topology on the configuration space of CA, just the product topology of the discrete space of finite states. It is known that in some cases the configuration space with the topology is compact, totally disconnected, perfect metric space and is homeomorphic to the Cantor discontinuum. To treat the product topology with relations we give a useful basis of the topology consisting of equivalence classes made by projections. Note that an equivalence class of the projection equivalence is approximately equal to an open ball for the Baire metric on the configuration space. In Section 6 we will give a relational definition of nondeterministic CA and prove Richardson's theorem with relational notation. We should note that the definition of transition relations of CA using intersection of relations is quite different from the ordinary definition for deterministic CA and often difficult to handle, because the sharpness [4] for (infinite) products might be hidden. Finally we show a new theorem for nondeterministic CA by applying the compactness of configuration spaces: a transition relation of CA is surjective iff all transition relations restricted to finite cell sets are surjective.

2   Algebras of Binary Relations

In this section we give a brief of notions and notations on binary relations, needed in the later discussion. For further details on relational issues the readers should refer to [5,7]. Remark that all relations treated in the paper are not abstract ones but concrete relations between sets.

A (binary) relation α from a set X into a set Y, written $\alpha : X \rightharpoonup Y$, is a subset $\alpha \subseteq X \times Y$. The set of all set relations from X into Y will be denoted by $Rel(X, Y)$. Clearly $Rel(X, Y)$ is identical with the power set $\wp(X \times Y)$ and forms a boolean algebra. The boolean operations for relations will be denoted by usual symbols: $\sqsubseteq$ (inclusion), $\sqcup$ (union), $\sqcap$ (intersection) and $^-$ (complement). The zero relation $0_{XY} : X \rightharpoonup Y$ and the universal relation $\nabla_{XY} : X \rightharpoonup Y$ are relations with $0_{XY} = \emptyset$ and $\nabla_{XY} = X \times Y$, respectively. It is trivial that $0_{XY} \sqsubseteq \alpha \sqsubseteq \nabla_{XY}$ for all relations $\alpha : X \rightharpoonup Y$. The identity relation $\mathrm{id}_X : X \rightharpoonup X$ is a relation such that $(x, x') \in \mathrm{id}_X \leftrightarrow x = x'$. The composite $\alpha\beta : X \rightharpoonup Z$ of a relation $\alpha : X \rightharpoonup Y$ followed by a relation $\beta : Y \rightharpoonup Z$ is defined by $(x, z) \in \alpha\beta \leftrightarrow \exists y \in Y. [(x, y) \in \alpha \wedge (y, z) \in \beta]$. The converse (or transpose) $\alpha^\sharp : Y \rightharpoonup X$ of α is defined by $(y, x) \in \alpha^\sharp \leftrightarrow (x, y) \in \alpha$. For relations $\alpha : X \rightharpoonup Y, \beta : Y \rightharpoonup Z$ and $\gamma : X \rightharpoonup Z$ the Dedekind formula (DF, for short) $\alpha\beta \sqcap \gamma \sqsubseteq \alpha(\beta \sqcap \alpha^\sharp \gamma)$ holds. It is distinctive and useful in algebras of relations.

A relation $\alpha : X \rightharpoonup Y$ is called *univalent* if $\alpha^\sharp \alpha \sqsubseteq \mathrm{id}_Y$, *total* if $\mathrm{id}_X \sqsubseteq \alpha\alpha^\sharp$, and *surjective* if $\mathrm{id}_Y \sqsubseteq \alpha^\sharp \alpha$. A univalent and total relation is called a *tfn* (total function, or mapping), and a tfn $\alpha : X \rightharpoonup Y$ will be introduced by $\alpha : X \rightarrow Y$. The *domain* (relation) $\lfloor \alpha \rfloor : X \rightharpoonup X$ of a relation $\alpha : X \rightharpoonup Y$ is defined by $\lfloor \alpha \rfloor = \alpha\alpha^\sharp \sqcap \mathrm{id}_X$.

Proposition 1. *Let* $\alpha, \alpha' : X \rightharpoonup Y$ *and* $\beta : Y \rightharpoonup Z$ *be relations. Then*

(a) $\lfloor \alpha \rfloor^\sharp = \lfloor \alpha \rfloor$, $\quad \lfloor \alpha \rfloor \alpha = \alpha$, $\quad \alpha \sqsubseteq \alpha' \rightarrow \lfloor \alpha \rfloor \sqsubseteq \lfloor \alpha' \rfloor$,

(b) $\alpha : total \leftrightarrow \lfloor \alpha \rfloor = \mathrm{id}_X$, $\quad u \sqsubseteq \mathrm{id}_X \leftrightarrow \lfloor u \rfloor = u$,

(c) $\lfloor \alpha\beta \rfloor \sqsubseteq \lfloor \alpha \rfloor$, $\quad \beta : total \rightarrow \lfloor \alpha\beta \rfloor = \lfloor \alpha \rfloor$,

(d) $\lfloor \alpha \sqcap \alpha' \rfloor = \alpha\alpha'^\sharp \sqcap \mathrm{id}_X$, $\quad \lfloor \alpha \sqcup \alpha' \rfloor = \lfloor \alpha \rfloor \sqcup \lfloor \alpha' \rfloor$,

(e) $\alpha : univalent \rightarrow \alpha \lfloor \beta \rfloor = \lfloor \alpha\beta \rfloor \alpha$,

(f) $\nabla_{WY} \lfloor \beta \rfloor = \nabla_{WZ} \beta^\sharp$, $\quad \alpha \lfloor \beta \rfloor = \alpha \sqcap \nabla_{XZ} \beta^\sharp$,

(g) $\beta^\sharp : univalent \rightarrow \alpha\beta = (\alpha^- \beta)^- \lfloor \beta^\sharp \rfloor$. $\qquad\qquad\qquad\qquad\qquad$ $\square$

A relation $\alpha : X \rightharpoonup Y$ is called *rectangular* if $\alpha\nabla_{YX}\alpha \sqsubseteq \alpha$. The rectangular relation is a universal relation between subsets of X and Y. It is trivial that α is rectangular iff $\alpha^\sharp$ is rectangular. Note that $\alpha \sqsubseteq \alpha\nabla_{YX}\alpha$ always holds since $\alpha = \alpha \sqcap \alpha \sqsubseteq \alpha(\mathrm{id}_Y \sqcap \alpha^\sharp \alpha) \sqsubseteq \alpha\alpha^\sharp \alpha \sqsubseteq \alpha\nabla_{YX}\alpha$ by using DF. Also it follows from $\nabla_{XY} = \nabla_{XI}\nabla_{IY}$ that α is rectangular iff $\alpha = (\nabla_{IY}\alpha^\sharp)^\sharp \nabla_{IX}\alpha$.

Proposition 2. (a) *All meets of rectangular relations are also rectangular.*
(b) *All composites of rectangular relations are also rectangular.*

Proof. (a) Let $\alpha_j : X \rightharpoonup Y$ be rectangular for all j. Then

$$(\sqcap_j \alpha_j) \nabla_{YX} (\sqcap_j \alpha_j) \sqsubseteq \sqcap_j \alpha_j \nabla_{YX} \alpha_j$$
$$\sqsubseteq \sqcap_j \alpha_j. \qquad \{ \alpha_j : \text{rectangular} \}$$

(b) Let $\alpha : X \rightharpoonup Y$ and $\beta : Y \rightharpoonup Z$ be rectangular. Then

$$\alpha\beta \nabla_{ZX} \alpha\beta \sqsubseteq \alpha \nabla_{YX} \alpha\beta \nabla_{ZX} \alpha\beta \ \{ \alpha \sqsubseteq \alpha \nabla_{YX} \alpha \}$$
$$\sqsubseteq \alpha \nabla_{YX} \alpha\beta \nabla_{ZY} \beta \quad \{ \nabla_{ZX} \alpha \sqsubseteq \nabla_{ZY} \}$$
$$\sqsubseteq \alpha\beta. \qquad \{ \alpha, \beta : \text{rectangular} \}$$

This completes the proof. □

Let I be a singleton set. An element x of a set X (i.e. $x \in X$) will be identi-fied with a tfn $x : I \to X$. A subset S of X will be identified with a relation $\rho : I \rightharpoonup X$ or a subidentity $u \sqsubseteq \text{id}_X$. Also the empty subset of X and the whole set X are represented by the zero relation 0_{IX} and the universal relation ∇_{IX}, respectively. The algebraic framework *Rel* of all relations between sets forms a category satisfies the point axiom (PA$_*$) and the totality (Tot$_*$):

(PA$_*$) $\forall \rho : I \rightharpoonup X. \ (\sqcup_{x \in \rho} x = \rho)$,
(Tot$_*$) $\forall \rho : I \rightharpoonup X. \ (\rho \neq 0_{IX}) \to (\rho\rho^\sharp = \text{id}_I)$.

3 Continuous Relations

In this section we will state a relational treatment of topologies and continuous relations, and prove some basic properties of topology though they are basic and essentially the same as usual.

A *topology* $\mathcal{O}(X)$ on a set X is a subset of *Rel*(I, X) closed under arbitrary union and finite intersection. Of course, $0_{IX} \in \mathcal{O}(X)$ and $\nabla_{IX} \in \mathcal{O}(X)$. A *topological space* is a pair of a set X and a specified topology $\mathcal{O}(X)$ on X. Let X and Y be topological spaces. A rectangular relation $\alpha : X \rightharpoonup Y$ is called *open* if $\nabla_{IY} \alpha^\sharp \in \mathcal{O}(X)$ and $\nabla_{IX} \alpha \in \mathcal{O}(Y)$. All composites of open rectangular relations are also open: Let $\alpha : X \rightharpoonup Y$ and $\beta : Y \rightharpoonup Z$ be open rectangular relations. If $\alpha\beta \neq 0_{XY}$ then $\nabla_{IX} \alpha\beta = \nabla_{IX} \alpha\beta \nabla_{ZY} \beta = \nabla_{IY} \beta \in \mathcal{O}(Z)$ by Tarski rule. Otherwise $\nabla_{IX} \alpha\beta = 0_{IZ} \in \mathcal{O}(Z)$ is clear. Moreover, a relation $\alpha : X \rightharpoonup Y$ is called *open* if it is a union of open rectangular relations. The set of all open relations from X into Y will be denoted by $\mathcal{O}(X, Y)$. It is trivial that $\mathcal{O}(X, Y) \subseteq Rel(X, Y)$. Also $\mathcal{O}(X, Y)$ is closed under arbitrary union and finite intersection and isomorphic to the usual product topology on $X \times Y$. In particular, $\mathcal{O}(I, X) = \mathcal{O}(X)$ holds because $\mathcal{O}(I) = \{0_{II}, \text{id}_I\} = Rel(I, I)$. Note that $\alpha \in \mathcal{O}(X, Y)$ iff $\alpha^\sharp \in \mathcal{O}(Y, X)$. All composites of open relations are also open.

According to Brattka and Hertling [1] we give a relational definition of the continuity of relations between topological spaces as follows:

Definition 1. Let X and Y be topological spaces. A relation $\alpha : X \rightharpoonup Y$ is *continuous* if for all $\rho \in \mathcal{O}(Y)$ there exists $\mu \in \mathcal{O}(X)$ such that $\rho\alpha^\sharp = \mu\lfloor\alpha\rfloor$. □

Notation. Let $\beta : Y \rightarrow Z$ be a relation and $\mathcal{U}$ a subset of $Rel(X, Y)$. A subset $\mathcal{U}\beta$ of $Rel(X, Z)$ is defined by $\mathcal{U}\beta = \{\alpha\beta \mid \alpha \in \mathcal{U}\}$.

Using the above notation a relation $\alpha : X \rightarrow Y$ is continuous iff $\mathcal{O}(Y)\alpha^\sharp \subseteq \mathcal{O}(X)\lfloor\alpha\rfloor$. Also a total relation $\alpha : X \rightarrow Y$ is continuous iff $\mathcal{O}(Y)\alpha^\sharp \subseteq \mathcal{O}(X)$. Thus the continuity for relations between topological spaces given in Definition 1 can be understood a natural extension of the ordinary continuity for tfns.

Proposition 3. (a) *All open relations are continuous.*
(b) *All subidentities $u \sqsubseteq \mathrm{id}_X$ are continuous.*
(c) *All rectangular relations are continuous.*
(d) *All unions of continuous tfns are continuous.*
(e) *For a tfn $f : X \rightarrow Y$ with $\mathcal{O}(X) = \mathcal{O}(Y)f^\sharp$ both of f and $f^\sharp$ are continuous.*
(f) *A relation $\alpha : X \rightarrow Y$ is continuous iff $\forall Z.\ \mathcal{O}(Z, Y)\alpha^\sharp \subseteq \mathcal{O}(Z, X)\lfloor\alpha\rfloor$.*

Proof. (a) If α is open, then $\mathcal{O}(Y)\alpha^\sharp = \mathcal{O}(Y)\alpha^\sharp\lfloor\alpha\rfloor \subseteq \mathcal{O}(X)\lfloor\alpha\rfloor$, since composites of open relations are open. (Thus all zero relations 0_{XY} and all universal relations ∇_{XY} are continuous.)
(b) Let $u \sqsubseteq \mathrm{id}_X$. Then $\mathcal{O}(X)u^\sharp = \mathcal{O}(X)\lfloor u\rfloor$, since $u^\sharp = u = \lfloor u\rfloor$.
(c) Let $\alpha : X \rightarrow Y$ be rectangular and $\rho \in \mathcal{O}(Y)$. If $\rho\alpha^\sharp = 0_{IX}$ then $\rho\alpha^\sharp = 0_{IX}\lfloor\alpha\rfloor$ and $0_{IX} \in \mathcal{O}(X)$. If $\rho\alpha^\sharp \neq 0_{IX}$ then $\rho\alpha^\sharp$ is total by (Tot_*) and so

$$\begin{aligned}
\rho\alpha^\sharp &= \rho\alpha^\sharp\nabla_{XI}\nabla_{IY}\alpha^\sharp && \{\ \alpha : \text{rectangular}\ \} \\
&= \nabla_{IY}\alpha^\sharp && \{\ \rho\alpha^\sharp\nabla_{XI} = \mathrm{id}_I \leftarrow \rho\alpha^\sharp : \text{total}\ \} \\
&= \nabla_{IX}\lfloor\alpha\rfloor. && \{\ \nabla_{IX} \in \mathcal{O}(X)\ \}
\end{aligned}$$

This proves that α is continuous.
(d) Let $\{f_j : X \rightarrow Y \mid j \in J\}$ be a nonempty set of continuous tfns. Then for $\rho \in \mathcal{O}(Y)$ we have

$$\rho(\sqcup_{j\in J}f_j)^\sharp = (\sqcup_{j\in J}\rho f_j^\sharp)\lfloor\sqcup_{j\in J}f_j\rfloor.\ \{\ \lfloor\sqcup_{j\in J}f_j\rfloor = \mathrm{id}_X\ \}$$

Hence $\sqcup_{j\in J}f_j$ is continuous, since $\rho f_j^\sharp \in \mathcal{O}(X)$.
(e) Let $f : X \rightarrow Y$ be a tfn with $\mathcal{O}(X) = \mathcal{O}(Y)f^\sharp$. The continuity of f is direct from $\mathcal{O}(Y)f^\sharp \subseteq \mathcal{O}(X)$. On the other hand $f^\sharp$ is continuous, because

$$\forall\rho \in \mathcal{O}(X)\ \exists\mu \in \mathcal{O}(Y).\ \rho = \mu f^\sharp \qquad \{\ \mathcal{O}(X) \subseteq \mathcal{O}(Y)f^\sharp\ \}$$

and

$$\begin{aligned}
\rho(f^\sharp)^\sharp &= \rho f \\
&= \mu f^\sharp f && \{\ \rho = \mu f^\sharp\ \} \\
&= \mu\lfloor f^\sharp\rfloor. && \{\ f^\sharp f \sqsubseteq \mathrm{id}_Y\ \}
\end{aligned}$$

(f) ($\leftarrow$) Take $Z = I$. Then we have $\mathcal{O}(Y)\alpha^\sharp \subseteq \mathcal{O}(X)\lfloor\alpha\rfloor$, since $\mathcal{O}(X) = \mathcal{O}(I, X)$.
($\rightarrow$) Let $\xi \in \mathcal{O}(Z, Y)$. Then

$$\begin{aligned}
\xi\alpha^\sharp &= (\sqcup_j\xi_j)\alpha^\sharp && \{\ \exists\xi_j : \text{open rectangular}\ \} \\
&= \sqcup_j\xi_j\nabla_{YI}\nabla_{IZ}\xi_j\alpha^\sharp && \{\ \xi_j = \xi_j\nabla_{YI}\nabla_{IZ}\xi_j\ \} \\
&= \sqcup_j\xi_j\nabla_{YI}\tau_j\lfloor\alpha\rfloor && \{\ \exists\tau_j \in \mathcal{O}(I, X) \leftarrow \alpha : \text{continuous}\ \} \\
&= (\sqcup_j\xi_j\nabla_{YI}\tau_j)\lfloor\alpha\rfloor && \\
&\in \mathcal{O}(Z, X)\lfloor\alpha\rfloor. && \{\ \tau_j, \xi_j\nabla_{YI} : \text{open}\ \}
\end{aligned}$$

This completes the proof. $\qquad\qquad\qquad\qquad\qquad\qquad\qquad\qquad\qquad\qquad\qquad$ $\square$

As usual a relation α is *closed* if the complement α^- of α is open.

Corollary 1. *Let $f : X \to A$ and $g : Y \to B$ be continuous tfns. Then for all open (closed) relations $\xi : A \to B$ the composite $f\xi g^\sharp$ is open (closed).*   □

Unions of continuous relations are not always continuous. Take two distinct reals $r, s \in \mathbb{R}$ and consider a relation $\alpha = r^\sharp \nabla_{I\mathbb{R}} \sqcup \nabla_{RI} s : \mathbb{R} \to \mathbb{R}$. Remark that $(x, y) \in \alpha \leftrightarrow (x = r) \vee (y = s)$. Both $\nabla_{RI} s$ and $r^\sharp \nabla_{I\mathbb{R}}$ are rectangular and so continuous. Also α is total since $\nabla_{RI} s$ is total. On the other hand the complement r^- of r is open since points are closed, and we have

$$r^- \alpha^\sharp = r^- (r^\sharp \nabla_{I\mathbb{R}} \sqcup \nabla_{RI} s)$$
$$= s \notin \mathcal{O}(\mathbb{R}), \qquad \{\ r^- r^\sharp = 0_{II},\ r^- \nabla_{RI} = \mathrm{id}_I\ \}$$

which means that α is not continuous.

Remark. The composite of continuous relations is not always continuous. A counter example is given in [1].

Proposition 4. *Let $\alpha : X \to Y$ and $\beta : Y \to Z$ be continuous relations.*

(a) *If α is univalent, then $\alpha\beta$ is continuous.*
(b) *$\alpha \sqcap \mu^\sharp \nabla_{IY}$ is continuous for all relations $\mu : I \to X$.*
(c) *If $\nabla_{IZ} \beta^\sharp : I \to Y$ is open, then $\alpha\beta$ is continuous.*

Proof. (a) Let $\sigma \in \mathcal{O}(Z)$. Then

$$\sigma(\alpha\beta)^\sharp = \sigma\beta^\sharp\alpha^\sharp$$
$$= \rho\lfloor\beta\rfloor\alpha^\sharp \quad \{\ \exists\rho \in \mathcal{O}(Y).\ \sigma\beta^\sharp = \rho\lfloor\beta\rfloor\ \}$$
$$= \rho\alpha^\sharp\lfloor\alpha\beta\rfloor \quad \{\ \alpha\lfloor\beta\rfloor = \lfloor\alpha\beta\rfloor\alpha \leftarrow \alpha^\sharp\alpha \sqsubseteq \mathrm{id}_Y\ \}$$
$$= \mu\lfloor\alpha\rfloor\lfloor\alpha\beta\rfloor \quad \{\ \exists\mu \in \mathcal{O}(X).\ \rho\alpha^\sharp = \mu\lfloor\alpha\rfloor\ \}$$
$$= \mu\lfloor\alpha\beta\rfloor. \quad \{\ \lfloor\alpha\beta\rfloor \sqsubseteq \lfloor\alpha\rfloor\ \}$$

(b) As $\lfloor\mu^\sharp\rfloor$ is a subidentity it is univalent and continuous by Proposition 3 (b). Hence $\alpha \sqcap \mu^\sharp \nabla_{IY} = \lfloor\mu^\sharp\rfloor\alpha$ is continuous by (a).
(c) Let $\sigma \in \mathcal{O}(Z)$. Then

$$\sigma\beta^\sharp = \rho\lfloor\beta\rfloor \qquad \{\ \exists\rho \in \mathcal{O}(Y).\ \sigma\beta^\sharp = \rho\lfloor\beta\rfloor \leftarrow \beta : \text{continuous}\ \}$$
$$= \rho \sqcap \nabla_{IZ}\beta^\sharp \in \mathcal{O}(Y). \{\ \nabla_{IZ}\beta^\sharp \in \mathcal{O}(Y)\ \}$$

Since α is continuous there exists a relation $\mu \in \mathcal{O}(X)$ such that $\sigma\beta^\sharp\alpha^\sharp = \mu\lfloor\alpha\rfloor$. Hence

$$\sigma(\alpha\beta)^\sharp = \sigma(\alpha\beta)^\sharp\lfloor\alpha\beta\rfloor \quad \{\ \lfloor\alpha\rfloor\alpha = \alpha\ \}$$
$$= \sigma\beta^\sharp\alpha^\sharp\lfloor\alpha\beta\rfloor$$
$$= \mu\lfloor\alpha\rfloor\lfloor\alpha\beta\rfloor \quad \{\ \sigma\beta^\sharp\alpha^\sharp = \mu\lfloor\alpha\rfloor\ \}$$
$$= \mu(\lfloor\alpha\rfloor \sqcap \lfloor\alpha\beta\rfloor)$$
$$= \mu\lfloor\alpha\beta\rfloor. \qquad \{\ \lfloor\alpha\beta\rfloor \sqsubseteq \lfloor\alpha\rfloor\ \}$$

This completes the proof.   □

4 Compact Relations

In this section we will generalize a basic theorem that the inverse function of a continuous bijection from a compact space onto a Hausdorff space is continuous.

An open cover $\mathcal{U}$ of a relation $\alpha : X \rightharpoonup Y$ is a subset of $\mathcal{O}(X,Y)$ such that $\alpha \sqsubseteq \sqcup \mathcal{U}$. A relation $\alpha : X \rightharpoonup Y$ is *compact* if for all open cover $\mathcal{U}$ of α there exists a finite subset $\mathcal{U}'$ of $\mathcal{U}$ which also covers α. A topological space X is *compact* if ∇_{IX} is compact.

Proposition 5. *Let $\alpha : X \rightharpoonup Y$ and $\beta : Y \rightharpoonup Z$ be relations. If $\alpha \lfloor \beta \rfloor$ is compact and β is univalent and continuous, then $\alpha\beta$ is compact.*

Proof. Assume that $\alpha\beta \sqsubseteq \sqcup_{j \in J} \delta_j$ for $\delta_j \in \mathcal{O}(X,Z)$. As β is continuous, by Proposition 3 (f) there exists $\alpha_j \in \mathcal{O}(X,Y)$ such that $\delta_j\beta^\sharp = \alpha_j\lfloor\beta\rfloor$. Remark that $\alpha_j\beta = \alpha_j\lfloor\beta\rfloor\beta = \delta_j\beta^\sharp\beta \sqsubseteq \delta_j$ since β is univalent. Then

$$
\begin{aligned}
\alpha\lfloor\beta\rfloor &\sqsubseteq \alpha\beta\beta^\sharp && \{\ \lfloor\beta\rfloor \sqsubseteq \beta\beta^\sharp\ \} \\
&\sqsubseteq (\sqcup_{j \in J}\delta_j)\beta^\sharp && \{\ \alpha\beta \sqsubseteq \sqcup_{j \in J}\delta_j\ \} \\
&\sqsubseteq \sqcup_{j \in J}\alpha_j. && \{\ \delta_j\beta^\sharp = \alpha_j\lfloor\beta\rfloor \sqsubseteq \alpha_j\ \}
\end{aligned}
$$

By the compactness of $\alpha\lfloor\beta\rfloor$, there exists a finite subset F of J such that $\alpha\lfloor\beta\rfloor \sqsubseteq \sqcup_{j \in F}\alpha_j$. Hence we have

$$
\begin{aligned}
\alpha\beta &= \alpha\lfloor\beta\rfloor\beta && \{\ \beta = \lfloor\beta\rfloor\beta\ \} \\
&\sqsubseteq (\sqcup_{j \in F}\alpha_j)\beta && \{\ \alpha\lfloor\beta\rfloor \sqsubseteq \sqcup_{j \in F}\alpha_j\ \} \\
&\sqsubseteq \sqcup_{j \in F}\delta_j. && \{\ \alpha_j\beta \sqsubseteq \delta_j\ \}
\end{aligned}
$$

This proves that $\alpha\beta$ is compact.   □

Corollary 2. *The composite of a compact relation $\alpha : X \rightharpoonup Y$ followed by a continuous tfn $f : Y \to Z$ is compact.*   □

A topological space X is *Hausdorff* if $\mathrm{id}_X^- \in \mathcal{O}(X,X)$.

Proposition 6. (a) *If ∇_{XY} is compact and $\alpha : X \rightharpoonup Y$ is closed, then α is compact.*
(b) *If X is Hausdorff and $\rho : I \rightharpoonup X$ is compact, then ρ is closed.*

Proof. (a) Let $\mathcal{U}$ be an open cover of α. Then $\mathcal{U} \cup \{\alpha^-\}$ is an open cover of ∇_{XY}. By the compactness of ∇_{XY} there exists a finite subcover $\mathcal{U}' \subseteq \mathcal{U} \cup \{\alpha^-\}$ of ∇_{XY}. Then $\mathcal{U}' \cap \mathcal{U}$ is finite and covers α.
(b) As X is Hausdorff $\mathrm{id}_X^- = \sqcup_{j \in J}\xi_j$ holds for open rectangular relations $\xi_j : X \rightharpoonup X$. Also we have $\nabla_{IX}\xi_j^\sharp \sqsubseteq (\nabla_{IX}\xi_j)^-$ for all $j \in J$:

$$
\begin{aligned}
\nabla_{IX}\xi_j \sqcap \nabla_{IX}\xi_j^\sharp &\sqsubseteq \nabla_{IX}\xi_j^\sharp(\xi_j\nabla_{XI}\nabla_{IX}\xi_j \sqcap \mathrm{id}_X) && \{\ \mathrm{DF}\ \} \\
&\sqsubseteq \nabla_{IX}\xi_j^\sharp(\xi_j \sqcap \mathrm{id}_X) && \{\ \xi_j : \text{rectangular}\ \} \\
&= 0_{IX}. && \{\ \xi_j \sqsubseteq \mathrm{id}_X^-\ \}
\end{aligned}
$$

Let $x \in \rho^-$ and define a subset J' of J by $j \in J' \leftrightarrow x^\sharp \nabla_{IX} \sqcap \xi_j \neq 0_{XX}$. For $j \in J'$ we have $x \sqcap \nabla_{IX} \xi_j^\sharp \neq 0_{IX}$ by DF and hence $x \sqsubseteq \nabla_{IX} \xi_j^\sharp$, because points are atomic. On the other hand $x^\sharp \rho \sqsubseteq \mathrm{id}_X^- = \sqcup_{j \in J} \xi_j$ implies $x^\sharp \rho \sqsubseteq \sqcup_{j \in J'} \xi_j$ and $\rho = x x^\sharp \rho \sqsubseteq \sqcup_{j \in J'} \nabla_{IX} \xi_j$. Thus $\{\nabla_{IX} \xi_j \mid j \in J'\}$ is an open cover of ρ. As ρ is compact, there is a finite subset F of J' such that $\rho \sqsubseteq \sqcap_{j \in F} \nabla_{IX} \xi_j$. Hence we have $x \sqsubseteq \sqcap_{j \in F} \nabla_{IX} \xi_j^\sharp \sqsubseteq \sqcap_{j \in F} (\nabla_{IX} \xi_j)^- \sqsubseteq \rho^-$, which proves $\rho^- \in \mathcal{O}(X)$ because $\sqcap_{j \in F} \nabla_{IX} \xi_j^\sharp \in \mathcal{O}(X)$. $\qquad\square$

A relation $\alpha : X \rightarrow Y$ is a *partial bijection* if $\alpha^\sharp \alpha \sqsubseteq \mathrm{id}_Y$ and $\alpha \alpha^\sharp \sqsubseteq \mathrm{id}_X$.

Proposition 7. *Let $\alpha : X \rightarrow Y$ be a partial bijection from a compact space X into a Hausdorff space Y. If α is continuous and $\nabla_{IY} \alpha^\sharp$ is closed, then $\alpha^\sharp$ is continuous.*

Proof

$$
\begin{aligned}
\rho \in \mathcal{O}(X) &\rightarrow \rho^- : \text{closed} \\
&\rightarrow \rho^- \lfloor \alpha \rfloor = \rho^- \sqcap \nabla_{IY} \alpha^\sharp : \text{closed} \quad \{ \nabla_{IY} \alpha^\sharp : \text{closed} \} \\
&\rightarrow \rho^- \lfloor \alpha \rfloor : \text{compact} \quad\quad\quad\quad \{ \text{Proposition 6 (a)} \} \\
&\rightarrow \rho^- \alpha : \text{compact} \quad\quad\quad\quad\quad \{ \text{Proposition 5} \} \\
&\rightarrow \rho^- \alpha : \text{closed} \quad\quad\quad\quad\quad\;\; \{ \text{Proposition 6 (b)} \} \\
&\rightarrow (\rho^- \alpha)^- \in \mathcal{O}(Y).
\end{aligned}
$$

As $\rho \alpha = (\rho^- \alpha)^- \lfloor \alpha^\sharp \rfloor$ holds by Proposition 1 (g), $\alpha^\sharp$ is continuous. $\qquad\square$

5 Product Topologies

In this section we recall some basic properties of product topologies.

For a pair of nonempty sets Q and X the set of all tfns $c : X \rightarrow Q$ is denoted by Q^X. For a subset U of X the notation $p_U : Q^X \rightarrow Q^U$ denotes the projection (or may be the restriction of coordinates) from Q^X onto Q^U. In particular, when $U = \{x\}$, p_U will be written as $p_x : Q^X \rightarrow Q$. Also define an equivalence relation $\theta[U] : Q^X \rightarrow Q^X$ by $\theta[U] = p_U p_U^\sharp$.

Proposition 8. *Let U and V be subsets of X. Then*

(a) $\theta[U] = \sqcap_{x \in U} p_x p_x^\sharp$, $\theta[X] = \mathrm{id}_{Q^X}$ *and* $\theta[\emptyset] = \nabla_{Q^X Q^X}$,

(b) $\theta[U \cup V] = \theta[U] \sqcap \theta[V]$ *and* $\theta[U \cap V] = \theta[U] \theta[V]$. $\qquad\square$

Let Q be a finite set with the discrete topology $\wp(Q)$. Of course Q is a Hausdorff compact space. The *product topology* $\mathcal{O}(Q^X)$ on Q^X is the least topology such that for all $x \in X$ the projection $p_x : Q^X \rightarrow Q$ is continuous. It is known that in some cases the product space Q^X is compact, totally disconnected, perfect metric space and is homeomorphic to the Cantor discontinuum. In what follows we assume that every product space Q^X has the compact Hausdorff topology $\mathcal{O}(Q^X)$. If U is a finite set then $\mathcal{O}(Q^U)$ is discrete, that is, $\mathcal{O}(Q^U) = \wp(Q^U)$.

Notation. $U \subseteq_* X$ denotes that U is a *finite* subset of X.

The following proposition indicates that the product space Q^X has a kind of uniform topology.

Proposition 9. (a) $\forall c \in Q^X \; \forall U \subseteq_* X. \; c\theta[U] \in \mathcal{O}(Q^X)$,
(b) $\forall \mu \in \mathcal{O}(Q^X) \forall c \in \mu \; \exists W_c \subseteq_* X. \; \mu = \sqcup_{c \in \mu} c\theta[W_c]$.

Proof. (a) Recall $c\theta[U] = \sqcap_{x \in U} c p_x p_x^\sharp$, $x p_x \in \wp(Q) = \mathcal{O}(Q)$ and p_x is continuous. Hence $c\theta[U]$ is open since U is finite.
(b) Let $\mu : I \to Q^X$ be an open subset and $c \in \mu$. As the product topology $\mathcal{O}(Q^X)$ consists of all unions of finite meets $\sqcap_{j \in J} \rho_j p_{x_j}^\sharp$, where $\rho_j : I \to Q$ (open subset) and $x_j \in X$ for all $j \in J$. Thus there exists a finite meet $\sqcap_{j \in J} \rho_j p_{x_j}^\sharp$ such that $c \sqsubseteq \sqcap_{j \in J} \rho_j p_{x_j}^\sharp \sqsubseteq \mu$. Set $W_c = \{x_j \mid j \in J\}$. Then

$$
\begin{aligned}
c\theta[W_c] &\sqsubseteq (\sqcap_{j \in J} \rho_j p_{x_j}^\sharp)\theta[W_c] \\
&\sqsubseteq \sqcap_{j \in J} \rho_j p_{x_j}^\sharp \theta[x_j] \quad \{\; x_j \in W_c \;\} \\
&= \sqcap_{j \in J} \rho_j p_{x_j}^\sharp \quad \{\; \theta[x_j] = p_{x_j} p_{x_j}^\sharp \;\} \\
&\sqsubseteq \mu.
\end{aligned}
$$

Hence we have $\mu = \sqcup_{c \in \mu} c\theta[W_c]$, since $\theta[W_c]$ is an equivalence relation. □

Proposition 10. (a) *A tfn $f : A \to Q^X$ is continuous iff for all $x \in X$ the composite $f p_x$ is continuous.*
(b) *For $U \subseteq X$ the projection $p_U : Q^X \to Q^U$ is continuous.*
(c) *A relation $\alpha : A \rightharpoonup Q^X$ is continuous iff for all $W \subseteq_* X$ the composite $\alpha p_W : A \rightharpoonup Q^W$ is continuous.*

Proof. (a) ($\to$) It is trivial. ($\leftarrow$) Assume that for all $x \in X$ the composite $f p_x$ is continuous. We have to show that $\mu f^\sharp$ is open for all $\mu \in \mathcal{O}(Q^X)$. By Proposition 9 (b) for all $c \in \mu$ there is a finite subset $W_c \subseteq_* X$ such that $\mu = \sqcup_{c \in \mu} c\theta[W_c]$. Then

$$
\begin{aligned}
c\theta[W_c] f^\sharp &= c(\sqcap_{x \in W_c} p_x p_x^\sharp) f^\sharp \\
&= \sqcap_{x \in W_c} c p_x p_x^\sharp f^\sharp \quad \{\; c, f : \text{tfn} \;\} \\
&= \sqcap_{x \in W_c} c p_x (f p_x)^\sharp.
\end{aligned}
$$

As $c p_x : I \to Q$ is open and $f p_x$ is continuous, $c\theta[W_c] f^\sharp$ is a finite meet of open sets and so it is open. Hence $\mu f^\sharp = \sqcup_{c \in \mu} c\theta[W_c] f^\sharp$ is open.
(b) It is a corollary of (a).
(c) ($\to$) As p_U is a continuous tfn, αp_U is continuous by Proposition 4 (b).
($\leftarrow$) Assume that for all $W \subseteq_* X$ the composite αp_W is continuous. Let $\mu \in \mathcal{O}(Q^X)$. Again by Proposition 9 (b) μ can be written as $\mu = \sqcup_{c \in \mu} c\theta[W_c]$, where $W_c \subseteq_* X$. Then

$$
\begin{aligned}
\rho\alpha^\sharp &= \sqcup_{c \in \mu} c p_{W_c} (\alpha p_{W_c})^\sharp \quad \{\; \theta[W] = p_W p_W^\sharp \;\} \\
&= \sqcup_{c \in \mu} \xi_c \lfloor \alpha p_{W_c} \rfloor \quad \{\; \exists \xi_c \in \mathcal{O}(Q^X). \; c p_{W_c} (\alpha p_{W_c})^\sharp = \xi_c \lfloor \alpha p_{W_c} \rfloor \;\} \\
&= \sqcup_{c \in \mu} \xi_c \lfloor \alpha \rfloor \quad \{\; p_{W_c} : \text{total} \;\} \\
&= (\sqcup_{c \in \mu} \xi_c) \lfloor \alpha \rfloor,
\end{aligned}
$$

which proves that α is continuous. □

Let G be a group with a unit element e. The inverse of $x \in G$ will be denoted by x^{-1}. For subsets W, V of G the subset $\{wv \mid w \in W, v \in V\}$ of G is denoted by WV. If $W = \{x\}$ is a singleton set, $\{x\}V$ will be written as xV for short. For all $x \in G$ the shift tfn $t_x : Q^G \to Q^G$ is defined by

$$\forall a \in G.\ t_x p_a = p_{x^{-1}a}.$$

All shifts t_x $(x \in G)$ are continuous.

Proposition 11. *Let $x, y \in G$ and $V \subseteq G$. Then*

(a) $t_e = \mathrm{id}_{Q^G}$,
(b) $t_x t_y = t_{yx}$,
(c) $\theta[xV] = t_{x^{-1}}\theta[V]t_x$.

Proof. (a) $t_e p_a = p_{e^{-1}a} = \mathrm{id}_{Q^G}p_a$.
(b) $t_x t_y p_a = t_x p_{y^{-1}a} = p_{x^{-1}y^{-1}a} = p_{(yx)^{-1}a} = t_{yx} p_a$.
(c) $\theta[xV] = p_{xV}p_{xV}^\sharp = t_{x^{-1}}p_V (t_{x^{-1}}p_V)^\sharp = t_{x^{-1}}\theta[V]t_x$.

This completes the proof.   □

6   Richardson's Theorem

In this section we will give a relational definition of transition relations for non-deterministic CA and prove a relational version of Richardson's theorem to characterize the transition relations.

Let Q be a nonempty finite state set and G a group with a unit element e. For a finite subset N of G a relation $\lambda : Q^N \rightharpoonup Q$ is called a local relation or a local rule. The transition relation $\tau_\lambda : Q^G \rightharpoonup Q^G$ for λ is defined by

$$\tau_\lambda = \sqcap_{x \in G}t_{x^{-1}}p_N \lambda p_x^\sharp.$$

The above definition extends deterministic CA on groups [2]. It is clear that if a local relation $\lambda : Q^N \rightharpoonup Q$ is a tfn, then the transition relation $\tau_\lambda : Q^G \rightharpoonup Q^G$ is also a tfn. Of course, τ_λ is total iff λ is total.

Now we will show the first part of Theorem 1 in [6] with relational methods.

Theorem 1 (Richardson 1972). *Let $\lambda : Q^N \rightharpoonup Q$ be a local relation. The transition relation $\tau_\lambda : Q^G \rightharpoonup Q^G$ satisfies the following conditions:*

(a) $\forall x \in G.\ t_x \tau_\lambda = \tau_\lambda t_x$,
(b) $\sqcap_{x \in G}\tau_\lambda p_x p_x^\sharp = \tau_\lambda$,
(c) τ_λ *is closed and continuous.*   □

Proof. (a) $\forall a \in G.\ \tau_\lambda = t_{a^{-1}}\tau_\lambda t_a$:

$$
\begin{aligned}
t_{a^{-1}}\tau_\lambda t_a &= t_{a^{-1}}(\sqcap_{x \in G}t_{x^{-1}}p_N \lambda p_x^\sharp)t_a && \{\ \tau_\lambda\ \} \\
&= \sqcap_{x \in G}t_{x^{-1}a^{-1}}p_N \lambda p_x^\sharp t_a && \{\ t_a : \text{bijection}\ \} \\
&= \sqcap_{x \in G}t_{(ax)^{-1}}p_N \lambda p_{ax}^\sharp && \{\ t_{a^{-1}}p_x = p_{ax}\ \} \\
&= \sqcap_{x \in G}t_{x^{-1}}p_N \lambda p_x^\sharp && \{\ aG = G\ \} \\
&= \tau_\lambda.
\end{aligned}
$$

(b) $\sqcap_{x \in G} \tau_\lambda p_x p_x^\sharp = \tau_\lambda$:

$$\tau_\lambda \sqsubseteq \sqcap_{x \in G} \tau_\lambda p_x p_x^\sharp, \qquad \{\, p_x : \text{tfn} \,\}$$

$$\begin{aligned}
\sqcap_{x \in G} \tau_\lambda p_x p_x^\sharp &\sqsubseteq \sqcap_{x \in G} t_{x^{-1}} p_N \lambda p_x^\sharp p_x p_x^\sharp \quad \{\, \tau_\lambda \sqsubseteq t_{x^{-1}} p_N \lambda p_x^\sharp \,\} \\
&= \sqcap_{x \in G} t_{x^{-1}} p_N \lambda p_x^\sharp \qquad\qquad \{\, p_x : \text{tfn} \,\} \\
&= \tau_\lambda.
\end{aligned}$$

(c-1) τ_λ is closed:

$$\begin{aligned}
&\to \forall x \in G.\ t_{x^{-1}} p_N \lambda p_x^\sharp : \text{ closed} \quad \{\, \text{Corollary 1} \,\} \\
&\to \tau_\lambda = \sqcap_{x \in G} t_{x^{-1}} p_N \lambda p_x^\sharp : \text{ closed.}
\end{aligned}$$

(c-2) τ_λ is continuous:
We have to show that

$$\forall \rho \in \mathcal{O}(Q^G)\, \exists \mu \in \mathcal{O}(Q^G).\ \rho \tau_\lambda^\sharp = \mu \lfloor \tau_\lambda \rfloor.$$

By Proposition 9 (b) $\rho \in \mathcal{O}(Q^G)$ can be written as $\rho = \sqcup_{c \in \rho} c\theta[W_c]$ for some $W_c \subseteq_* G$. Then we have

$$\begin{aligned}
\rho \tau_\lambda^\sharp &= \sqcup_{c \in \rho} c\theta[W_c] \tau_\lambda^\sharp \\
&= \sqcup_{c \in \rho} c\theta[W_c] \tau_\lambda^\sharp \theta[W_c N] \lfloor \tau_\lambda \rfloor \quad \{\, \text{Lemma 1 (c) below} \,\} \\
&= (\sqcup_{c \in \rho} c\theta[W_c] \tau_\lambda^\sharp \theta[W_c N]) \lfloor \tau_\lambda \rfloor.
\end{aligned}$$

Set $\mu = \sqcup_{c \in \rho} c\theta[W_c] \tau_\lambda^\sharp \theta[W_c N]$. Then $\mu \in \mathcal{O}(Q^G)$ by Proposition 12 below. This completes the proof. $\qquad\qquad\square$

Proposition 12. $\forall U \subseteq_* G\, \forall \eta : I \rightarrowtail Q^G.\ \eta\theta[U] :$ *closed and open.*

Proof. Recall $\eta\theta[U] = \eta p_U p_U^\sharp$. As Q^U is discrete, ηp_U is closed and open in Q^U and so $\eta p_U p_U^\sharp$ is also closed and open in Q^G by the continuity of p_U. $\qquad\square$

Lemma 1. *Let* $\lambda : Q^N \rightarrow Q$ *be a local relation and* $\tau_\lambda : Q^G \rightarrowtail Q^G$ *the transition relation. Then for all* $W \subseteq G$ *the following holds.*

(a) $\theta[WN] \tau_\lambda \theta[W] \sqcap \tau_\lambda \theta[W^-] \sqsubseteq \tau_\lambda$,
(b) $\tau_\lambda \theta[W] = \lfloor \tau_\lambda \rfloor \theta[WN] \tau_\lambda \theta[W]$,
(c) $\theta[W] \tau_\lambda^\sharp = \theta[W] \tau_\lambda^\sharp \theta[WN] \lfloor \tau_\lambda \rfloor$.

Proof. (a) The inclusion follows from

$$\begin{aligned}
\theta[WN] \tau_\lambda \theta[W] &\sqsubseteq \sqcap_{x \in W} \theta[WN] t_{x^{-1}} p_N \lambda p_x^\sharp \theta[W] \quad \{\, \tau_\lambda \,\} \\
&\sqsubseteq \sqcap_{x \in W} \theta[xN] t_{x^{-1}} p_N \lambda p_x^\sharp \theta[x] \qquad \{\, x \in W \,\} \\
&= \sqcap_{x \in W} \theta[xN] t_{x^{-1}} p_N \lambda p_x^\sharp \qquad\quad \{\, p_x^\sharp \theta[x] = p_x^\sharp \,\} \\
&= \sqcap_{x \in W} t_{x^{-1}} \theta[N] t_x t_{x^{-1}} p_N \lambda p_x^\sharp \quad \{\, \theta[xN] = t_{x^{-1}} \theta[N] t_x \,\} \\
&= \sqcap_{x \in W} t_{x^{-1}} p_N \lambda p_x^\sharp, \qquad\qquad\quad \{\, \theta[N] p_N = p_N \,\}
\end{aligned}$$

and

$$\tau_\lambda \theta[W^-] \sqsubseteq \sqcap_{x \in W^-} t_{x^{-1}} p_N \lambda p_x^\sharp \theta[W^-] \,\{\, \tau_\lambda \,\}$$
$$\sqsubseteq \sqcap_{x \in W^-} t_{x^{-1}} p_N \lambda p_x^\sharp \theta[x] \quad \{\, x \in W^- \,\}$$
$$= \sqcap_{x \in W^-} t_{x^{-1}} p_N \lambda p_x^\sharp.$$

(b) The inclusion $\tau_\lambda \theta[W] \sqsubseteq \lfloor \tau_\lambda \rfloor \theta[WN] \tau_\lambda \theta[W]$ is obvious:

$$\tau_\lambda \theta[W] = \lfloor \tau_\lambda \rfloor \tau_\lambda \theta[W] \qquad \{\, \alpha = \lfloor \alpha \rfloor \alpha \,\}$$
$$\sqsubseteq \lfloor \tau_\lambda \rfloor \theta[WN] \tau_\lambda \theta[W]. \,\{\, \mathrm{id}_{Q^G} \sqsubseteq \theta[WN] \,\}$$

Note that $\theta[W^-]\theta[W] = \theta[W^- \cap W] = \theta[\emptyset] = \nabla_{Q^G Q^G}$ by Proposition 8. The converse inclusion $\lfloor \tau_\lambda \rfloor \theta[WN] \tau_\lambda \theta[W] \sqsubseteq \tau_\lambda \theta[W]$ follows from

$$\lfloor \tau_\lambda \rfloor \theta[WN] \tau_\lambda \theta[W] \sqsubseteq \theta[WN] \tau_\lambda \theta[W] \sqcap \tau_\lambda \nabla_{Q^G Q^G} \quad \{\, \lfloor \tau_\lambda \rfloor = \mathrm{id}_{Q^G} \sqcap \tau_\lambda \nabla_{Q^G Q^G} \,\}$$
$$= \theta[WN] \tau_\lambda \theta[W] \sqcap \tau_\lambda \theta[W^-]\theta[W] \quad \{\, \nabla_{Q^G Q^G} = \theta[W^-]\theta[W] \,\}$$
$$\sqsubseteq (\theta[WN] \tau_\lambda \theta[W] \sqcap \tau_\lambda \theta[W^-])\theta[W] \,\{\, (\mathrm{DF}),\, \theta[W] : \text{equiv. rel} \,\}$$
$$\sqsubseteq \tau_\lambda \theta[W]. \qquad\qquad\qquad\qquad \{\, (a) \,\}$$

(c) It is only the converse of (b). □

Corollary 3. *If the transition relation τ_λ is a partial bijection, then $\tau_\lambda^\sharp$ is also continuous.*

Proof. It is easy to check that $\nabla_{IQ^G} \tau_\lambda^\sharp = \sqcap_{x \in G} \nabla_{IQ} \lambda^\sharp (t_{x^{-1}} p_N)^\sharp$. Thus $\nabla_{IQ^G} \tau_\lambda^\sharp$ is closed and Proposition 7 yields the result. □

Corollary 4. *Let $\xi : Q \rightharpoonup Q$ be a relation. The relation $\xi_* : Q^X \rightharpoonup Q^X$ defined by $\xi_* = \sqcap_{x \in X} p_x \xi p_x^\sharp$ is continuous.*

Proof. Set $N = \{e\}$. Then we have $\xi_* = \tau_\xi$ and so the assertion is direct from Theorem 1 (c). □

Next we will show the second part of Theorem 1 in [6].

Theorem 2. *If a relation $\delta : Q^G \rightharpoonup Q^G$ satisfies the conditions*

(a) $\forall x \in G.\ t_x \delta = \delta t_x,$
(b) $\sqcap_{x \in G} \delta p_x p_x^\sharp = \delta,$
(c) δ *is closed and continuous,*

then there exists a local relation $\lambda : Q^N \rightharpoonup Q$ such that

$$\delta = \lfloor \delta \rfloor \tau_\lambda.$$

Proof. The composite $\gamma = \delta p_e$ is continuous by Proposition 4 (b) and is compact by Corollary 2, because δ is closed in a compact relation $\nabla_{Q^G Q^G}$ and p_e is continuous. Thus, by Lemma 2 below it holds that

$$\exists N \subseteq_* G.\ \gamma = \lfloor \gamma \rfloor \theta[N] \gamma.$$

Set $\lambda = p_N^\sharp \gamma$. Then

$$
\begin{aligned}
\lfloor \delta \rfloor \tau_\lambda &= \lfloor \delta \rfloor (\sqcap_{x \in G} t_{x^{-1}} p_N p_N^\sharp \gamma p_x^\sharp) & \{ \lambda = p_N^\sharp \gamma \} \\
&= \sqcap_{x \in G} \lfloor \delta \rfloor t_{x^{-1}} \theta[N] \gamma p_x^\sharp & \{ \lfloor \delta \rfloor \sqsubseteq \mathrm{id}_{Q^G} \} \\
&= \sqcap_{x \in G} t_{x^{-1}} \lfloor \delta \rfloor \theta[N] \gamma p_x^\sharp & \{ (a) \} \\
&= \sqcap_{x \in G} t_{x^{-1}} \lfloor \gamma \rfloor \theta[N] \gamma p_x^\sharp & \{ \lfloor \delta \rfloor = \lfloor \gamma \rfloor \} \\
&= \sqcap_{x \in G} t_{x^{-1}} \delta p_e p_x^\sharp & \{ \lfloor \gamma \rfloor \theta[N] \gamma = \gamma = \delta p_e \} \\
&= \sqcap_{x \in G} \delta t_{x^{-1}} p_e p_x^\sharp & \{ (a) \} \\
&= \sqcap_{x \in G} \delta p_x p_x^\sharp & \{ t_{x^{-1}} p_e = p_x \} \\
&= \delta. & \{ (b) \} \qquad \square
\end{aligned}
$$

Lemma 2. *If a relation* $\gamma : Q^G \rightharpoonup Q$ *is continuous and compact, then there exists a finite subset* $N \subseteq_* G$ *such that*

$$
\gamma = \lfloor \gamma \rfloor \theta[N] \gamma.
$$

Proof. First note that the inclusion $\gamma \sqsubseteq \lfloor \gamma \rfloor \theta[N] \gamma$ holds for all $N \subseteq G$.

$$
\begin{aligned}
\gamma &= \lfloor \gamma \rfloor \gamma \\
&\sqsubseteq \lfloor \gamma \rfloor \theta[N] \gamma. \quad \{ \mathrm{id}_{Q^G} \sqsubseteq \theta[N] \}
\end{aligned}
$$

Thus it suffices to show that

$$
\exists N \subseteq_* G. \ \lfloor \gamma \rfloor \theta[N] \gamma \sqsubseteq \gamma.
$$

Since γ is continuous and every $q \in Q$ is open, for each $(c, q) \in \gamma$ there exists a finite subset $W_{(c,q)}$ of G such that

$$
c\theta[W_{(c,q)}] \lfloor \gamma \rfloor \sqsubseteq q\gamma^\sharp,
$$

which is equivalent to

$$
\lfloor \gamma \rfloor \theta[W_{(c,q)}] c^\sharp q \sqsubseteq \gamma. \tag{A}
$$

A set $\{ \theta[W_{(c,q)}] c^\sharp q \mid (c, q) \sqsubseteq \gamma \}$ is an open cover of γ, because $\theta[W_{(c,q)}] c^\sharp q$ is an open rectangular relation and $c^\sharp q \sqsubseteq \theta[W_{(c,q)}] c^\sharp q$. As γ is compact, the open cover has a finite subcover, that is,

$$
\exists F \subseteq_* J. \ \gamma \sqsubseteq \sqcup_{j \in F} \theta[W_{(c_j, q_j)}] c_j^\sharp q_j. \tag{B}
$$

Set $N = \cup_{j \in F} W_{(c_j, q_j)}$. Then we have $\lfloor \gamma \rfloor \theta[N] \gamma \sqsubseteq \gamma$.

$$
\begin{aligned}
\lfloor \gamma \rfloor \theta[N] \gamma &\sqsubseteq \lfloor \gamma \rfloor \theta[N] (\sqcup_{j \in F} \theta[W_{(c_j, q_j)}] c_j^\sharp q_j) \ \{ (B) \} \\
&= \sqcup_{j \in F} \lfloor \gamma \rfloor \theta[N] \theta[W_{(c_j, q_j)}] c_j^\sharp q_j \\
&\sqsubseteq \sqcup_{j \in F} \lfloor \gamma \rfloor \theta[W_{(c_j, q_j)}] c_j^\sharp q_j & \{ (C) \ \text{below} \} \\
&\sqsubseteq \gamma. & \{ (A) \}
\end{aligned}
$$

This completes the proof. $\qquad \square$

Remark

(A) $\rho\lfloor\gamma\rfloor \sqsubseteq q\gamma^{\sharp} \leftrightarrow \lfloor\gamma\rfloor\rho^{\sharp}q \sqsubseteq \gamma$:

$$\rho\lfloor\gamma\rfloor \sqsubseteq q\gamma^{\sharp} \leftrightarrow q^{\sharp}\rho\lfloor\gamma\rfloor \sqsubseteq \gamma^{\sharp} \ \{\ q : \text{tfn}\ \}$$
$$\leftrightarrow \lfloor\gamma\rfloor\rho^{\sharp}q \sqsubseteq \gamma.$$

(C) $W \subseteq N \rightarrow \theta[N]\theta[W] \sqsubseteq \theta[W]$:

$$\theta[N]\theta[W] \sqsubseteq \theta[W]\theta[W] \ \{\ \theta[N] \sqsubseteq \theta[W] \leftarrow W \subseteq N\ \}$$
$$\sqsubseteq \theta[W]. \qquad \{\ \theta[W] : \text{equiv. rel.}\ \} \hspace{2cm} \square$$

For a (nonempty) subset S of G the restricted transition relation $\tau_S : Q^{SN} \rightharpoonup Q^S$ is defined by

$$\tau_S = \sqcap_{x\in S}p_{SN}^{\sharp}t_{x^{-1}}p_N\lambda p_x^{\sharp}p_S.$$

Lemma 3. (a) $\tau_\lambda p_S \sqsubseteq p_{SN}\tau_S$,
(b) If τ_λ is surjective, then so is τ_S for all $S \subseteq G$,
(c) If λ is total, then $\tau_\lambda p_S = p_{SN}\tau_S$.

Proof. (a)

$$\tau_\lambda p_S \sqsubseteq (\sqcap_{x\in S}t_{x^{-1}}p_N\lambda p_x^{\sharp})p_S \qquad \{\ S \subseteq G\ \}$$
$$= (\sqcap_{x\in S}\theta[SN]t_{x^{-1}}p_N\lambda p_x^{\sharp}\theta[S])p_S \quad \{\ \text{Proposition 8, 11}\ \}$$
$$= (\sqcap_{x\in S}p_{SN}p_{SN}^{\sharp}t_{x^{-1}}p_N\lambda p_x^{\sharp}p_S p_S^{\sharp})p_S \ \{\ \theta[S] = p_S p_S^{\sharp}\ \}$$
$$= p_{SN}(\sqcap_{x\in S}p_{SN}^{\sharp}t_{x^{-1}}p_N\lambda p_x^{\sharp}p_S)p_S^{\sharp}p_S \ \{\ p_{SN},\ p_S : \text{tfn}\ \}$$
$$= p_{SN}\tau_S. \qquad \{\ p_S^{\sharp}p_S = \mathrm{id}_{Q^S}\ \}$$

(b) Assume that τ_λ is surjective. Then

$$\mathrm{id}_{Q^S} = p_S^{\sharp}p_S$$
$$\sqsubseteq p_S^{\sharp}\tau_\lambda^{\sharp}\tau_\lambda p_S \quad \{\ \tau_\lambda : \text{surjective}\ \}$$
$$\sqsubseteq \tau_S^{\sharp}p_{SN}^{\sharp}p_{SN}\tau_S \ \{\ \text{(a)}\ \tau_\lambda p_S \sqsubseteq p_{SN}\tau_S\ \}$$
$$= \tau_S^{\sharp}\tau_S. \qquad \{\ p_{SN}^{\sharp}p_{SN} = \mathrm{id}_{Q^{SN}}\ \}$$

(d) Assume that $\lambda : Q^N \rightharpoonup Q$ be a total relation and define relations $\alpha_x : Q^G \rightharpoonup Q$ and $\beta_x : Q \rightharpoonup Q^G$ for all $x \in G$ by

$$\alpha_x = \nabla_{Q^G Q}, \quad \beta_x = p_x^{\sharp} \qquad \text{for } x \in S,$$
$$\alpha_x = t_{x^{-1}}p_N\lambda, \quad \beta_x = \nabla_{QQ^G} \text{ for } x \in S^-.$$

Then (0) $\forall x \in G.\ \alpha_x\beta_x = \nabla_{Q^G Q^G}$ and (1) $\nabla_{Q^G Q^G} = (\sqcap_{x\in S^-}\alpha_x p_x^{\sharp})\theta[S]$ hold:

(0-1) $x \in S \quad \rightarrow \alpha_x\beta_x = \nabla_{Q^G Q}p_x^{\sharp} = \nabla_{Q^G Q^G}, \{\ p_x : \text{total}\ \}$
(0-2) $x \in S^- \rightarrow \alpha_x\beta_x = \alpha_x\nabla_{QQ^G} = \nabla_{Q^G Q^G}, \{\ \alpha_x : \text{total} \leftarrow \lambda : \text{total}\ \}$
(0-3) $x \in S^- \rightarrow p_x\beta_x = p_x\nabla_{QQ^G} = \nabla_{Q^G Q^G}. \{\ p_x : \text{total}\ \}$

$$(1)\ \nabla_{Q^G Q^G} = \sqcap_{x \in G} \alpha_x \beta_x \qquad\qquad\quad \{\ (0)\ \}$$
$$= (\sqcap_{x \in G} \alpha_x p_x^\sharp)(\sqcap_{x \in G} p_x \beta_x)\ \{\ \text{Axiom of choice}\ \}$$
$$= (\sqcap_{x \in S^-} \alpha_x p_x^\sharp)\theta[S]. \qquad \{\ (0\text{-}3)\ \}$$

Therefore the identity $p_{SN}\tau_S = \tau_\lambda p_S$ follows from

$$p_{SN}\tau_S = (\sqcap_{x \in S} t_{x^{-1}} p_N \lambda p_x^\sharp) p_S \qquad\qquad\qquad \{\ \text{Proof of (a)}\ \}$$
$$= \{(\sqcap_{x \in S} t_{x^{-1}} p_N \lambda p_x^\sharp) \sqcap (\sqcap_{x \in S^-} \alpha_x p_x^\sharp)\theta[S]\} p_S\ \{\ (1)\ \}$$
$$= \{(\sqcap_{x \in S} t_{x^{-1}} p_N \lambda p_x^\sharp)\theta[S] \sqcap (\sqcap_{x \in S^-} \alpha_x p_x^\sharp)\} p_S\ \{\ (\mathrm{DF})\ \}$$
$$= \{(\sqcap_{x \in S} t_{x^{-1}} p_N \lambda p_x^\sharp) \sqcap (\sqcap_{x \in S^-} \alpha_x p_x^\sharp)\} p_S$$
$$= \tau_\lambda p_S. \qquad\qquad\qquad\qquad\qquad\qquad\qquad\qquad \square$$

Finally we prove a nondeterministic version of a part of Theorem 3 in [6].

Theorem 3. *If a local relation $\lambda : Q^N \rightharpoonup Q$ is total, then*

$$(\forall S \subseteq_* G.\ [\tau_S : surjective]) \to [\tau_\lambda : surjective].$$

Proof. Let $\lambda : Q^N \rightharpoonup Q$ be a total relation. We will show the contraposition

$$\neg[\tau_\lambda : surjective] \ \to\ (\exists S \subseteq_* G.\ \neg[\tau_S : surjective]).$$

Assume that τ_λ is not surjective. Then $\exists c \in Q^G.\ \nabla_{Q^G I}\, c \sqcap \tau_\lambda = 0_{Q^G Q^G}$. Since τ_λ is closed by Theorem 1 and $Q^G \times Q^G$ is compact Hausdorff, there exists a finite subset $S \subseteq_* G$ such that $\nabla_{Q^G I}\, c\theta[S] \sqcap \tau_\lambda = 0_{Q^G Q^G}$. The following picture illustrates this situation.

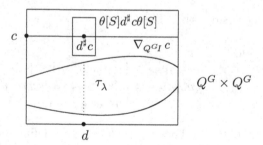

Then we have

$$\nabla_{Q^{SN} I}\, c p_S \sqcap \tau_S$$
$$= \nabla_{Q^{SN} I}\, c p_S \sqcap p_{SN}^\sharp p_{SN} \tau_S \qquad \{\ p_{SN}^\sharp p_{SN} = \mathrm{id}_{Q^{SN}}\ \}$$
$$= \nabla_{Q^{SN} I}\, c p_S \sqcap p_{SN}^\sharp \tau_\lambda p_S \qquad \{\ \text{Lemma 3 (c)}\ \}$$
$$\sqsubseteq p_{SN}^\sharp(p_{SN} \nabla_{Q^{SN} I}\, c p_S p_S^\sharp \sqcap \tau_\lambda) p_S\ \{\ (\mathrm{DF})\ \}$$
$$= p_{SN}^\sharp(\nabla_{Q^G I}\, c\theta[S] \sqcap \tau_\lambda) p_S$$
$$= 0_{Q^{SN} Q^S}, \qquad\qquad\qquad \{\ \nabla_{Q^G I}\, c\theta[S] \sqcap \tau_\lambda = 0_{Q^G Q^G}\ \}$$

which proves that τ_S is not surjective. $\qquad\qquad\qquad\qquad\qquad \square$

7 Conclusion

We investigated the continuity of relations between topological spaces using the relational notation, and developed some fundamental properties of continuous relations, including a modern proof of Richardson's theorem for nondeterministic cellular automata. The contribution might be expected to lead to further applications of nondeterministic CA.

Acknowledgment. The authors appreciate anonymous referees for helpful comments and suggestions.

References

1. Brattka, V., Hertling, P.: Continuity and computability of relations. Informatik Berichte, vol. 164. FernUniversität in Hagen (1994)
2. Ceccherini-Silberstein, T., Coornaert, M.: Cellular automata and groups. Springer (2010)
3. Choquet, G.: Convergence. Annales de l'université de Grenoble 23, 57–112 (1947-1948)
4. Desharnais, J.: Monomorphic characterization of n-ary direct products. Information Sciences 119(3-4), 275–288 (1999)
5. Freyd, P., Scedrov, A.: Categories, allegories. North-Holland, Amsterdam (1990)
6. Richardson, D.: Tessellations with local transformations. J. Computer and System Sciences 6, 373–388 (1972)
7. Schmidt, G., Ströhlein, T.: Relations and graphs, Discrete Mathematics for Computer Scientists. EATCS-Monographs on Theoret. Comput. Sci. Springer, Berlin (1993)
8. Wolfram, S.: A new kind of science. Wolfram Media (2002)
9. Ziegler, M.: Relative computability and uniform continuity of relations. Logic seminar 2011, Technische Universität Darmstadt (2011)

Relations on Hypergraphs

John G. Stell

School of Computing, University of Leeds, Leeds, U.K.

Abstract. A relation on a hypergraph is a binary relation on the set consisting of the nodes and hyperedges together, and which satisfies a constraint involving the incidence structure of the hypergraph. These relations correspond to join-preserving mappings on the lattice of sub-hypergraphs. This paper studies the algebra of these relations, in particular the analogues of the familiar operations of complement and converse of relations. When generalizing from relations on a set to relations on a hypergraph we find that the Boolean algebra of relations is replaced by a weaker structure: a pair of isomorphic bi-Heyting algebras, one of which arises from the relations on the dual hypergraph. The paper also considers the representation of sub-hypergraphs as relations and applies the results obtained to mathematical morphology for hypergraphs.

1 Introduction

The study of relations on a set [Sch11] is a well-established area both in terms of theory and applications to computing and other disciplines. This paper studies the algebraic properties of relations on a set which carries additional structure with the elements being the edges and nodes of a hypergraph. The relations we define are required to be compatible with this additional structure.

There are a number of practical applications that motivate the study of relations on hypergraphs or more simply on graphs. In conceptual modelling the value of relations on sets is well-known, but consider a case such as a network of railway stations (nodes) linked by tracks (edges). In this case there may be relationships between nodes and edges: a section of track and a station might be maintained by the same company, or some attribute of the track (e.g. level of security risk) might be dependent on the same attribute for the station. Similarly we might want to model relationships between nodes or between edges.

Another application concerns the need to model networks at a variety of levels of detail. A simple approach to granularity for sets comes from rough set theory [Paw82]. An equivalence relation on a set U can give a coarsened view of the set – the individual elements of the set are clumped together into equivalence classes which can be thought of as 'elements' of the set but at a lower level of detail. Given an arbitrary subset $X \subseteq U$, rough set theory considers two ways in which X can be approximated by sets of equivalence classes. This process can be generalized to arbitrary relations on a set which can be seen as instruments providing approximations of subsets [Ste10]. A particular case is mathematical morphology [Ser82, Ste07, BHR07], where the 'structuring elements', described [Ser82] as 'probes' on digital images, define relations on pixels.

W. Kahl and T.G. Griffin (Eds.): RAMiCS 2012, LNCS 7560, pp. 326–341, 2012.
© Springer-Verlag Berlin Heidelberg 2012

A binary (black and white) image is just a subset of these pixels and the operations of erosion, dilation, opening and closing used in mathematical morphology are then ways of approximating such images, usually described as filters on the images. All these techniques operate on a set of pixels, but there have been some initial investigations on extending mathematical morphology to graphs and hypergraphs [HV93, Ste07, CNS09, BB11]. It appears a reasonable conjecture that these investigations will be applicable to approximations of graphs.

One possible approach to the extension of mathematical morphology from sets to graphs or hypergraphs, initiated in [Ste07], is to base the investigation on a suitable generalization of relations; considering relations on a hypergraph or graph rather than relations on a set. However, the set of all relations on a given hypergraph does not admit all the relational operations familiar from the set case, and an account of the operations that do exist does not appear to have been given. In particular, the operations of the converse of a relation and the complement of a relation, are only present in weaker forms when dealing with relations on a hypergraph.

In this paper we present new results on the properties of operations for relations on a hypergraph. This significantly extends some earlier work [Ste07, Ste10] which only used composition of relations with just one of the two converse operations revealed here. In particular these earlier papers did not consider the two generalizations of complement (a pseudocomplement and its dual) introduced in Section 4 below, nor the role of the operations of relative addition, left and right residuation and the combination of converse and complement. These operations are demonstrated to be important for mathematical morphology on hypergraphs and we use them to generalize some well known identities from sets to hypergraphs.

Section 2 starts by reviewing the type of hypergraphs used in the paper and some well-known facts about the lattice of sub-hypergraphs (which we generally just call subgraphs). The following section defines the central notion of the paper, that of a relation on a hypergraph, and establishes some basic properties and some of the operations available on these relations. Section 4 then shows that although relations on a hypergraph have neither converses nor complements, there are four weaker operations which generalize these. While relations on a set form a Boolean algebra, we see that relations on a hypergraph only form a bi-Heyting algebra which is isomorphic to its opposite. The applications of the relational operations to mathematical morphology for hypergraphs are presented in Section 5, and the final section presents conclusions and potential future work.

Notation Used in This Paper

The converse and the complement of R are denoted $\smile R$ and $-R$ respectively. The choice of this notation instead of the common $\check{R}$ and $\overline{R}$ is dictated by the properties of the two weaker forms of converse and the two weaker forms of complement that we meet later. None of these four operations is an involution and since we need to consider several of them applied to a single relation it proved unwieldy to stack several symbols vertically. The combination of complement and converse, $-\smile R = \smile - R$, will be denoted $\frown R$.

For relations on a set U which is clear from the context, $1'$ denotes the identity relation on U, the universal relation $U \times U$ is denoted 1, and the empty relation is denoted 0.

Some knowledge of adjunctions on posets (or Galois connections) is assumed; details can be found in [Tay99]. The notation $f \dashv g$ will be used when f is left adjoint to g. The idea [Tay99, p152], of viewing $\dashv$ as an arrow (with the horizontal dash as the shaft of the arrow, and the vertical dash as the head of the arrow) proceeding from the left adjoint to the right adjoint is also adopted in diagrams where $\dashv$ may appear rotated as an arrow between two other arrows.

For relations R and S on a set U, their composition is denoted $R \,;\, S$. Unary operators will be assumed to have higher precedence than binary ones. Thus. for example, $\frown R \,;\, \smile S \,;\, \frown R$ will mean $(\frown R) \,;\, (\smile S) \,;\, (\frown R)$.

The operation $S \mapsto R \,;\, S$ will be denoted $R \,;\, \_$ and similarly for occurrences of $\_$ in other expressions. The operations $R \,;\, \_$ and $\_ \,;\, R$ have right adjoints $R \backslash \_$ and $\_ / R$ respectively. The binary operations $\backslash$ and $/$ are called left and right residuation and satisfy $R \backslash S = -(\smile R \,;\, -S)$ and $R \,/\, S = -(-R \,;\, \smile S)$. The operation $\dagger$ defined by $R \dagger S = -(-R \,;\, -S)$ is called relative addition.

For a poset, $P = (X, \leqslant)$, its opposite, $(X, \geqslant)$, will be denoted by $(P)^{\mathrm{op}}$. The term 'isomorphism of posets' is used to mean an order-preserving bijection, and $\cong$ will denote an isomorphism.

2   Hypergraphs

Hypergraphs can be defined as consisting of a set N of nodes and a set E of edges (or hyperedges) together with an incidence function $i : E \to \mathscr{P}N$ from E to the powerset of N. This approach allows several edges to be incident with the same set of nodes, and also allows edges incident with the empty set of nodes. Variations exist but this approach is adopted here. An example is shown on the left in Figure 1 (adapted from [Ste10]) where the edges are drawn as curves enclosing the nodes with which they are incident.

When studying relations on these structures it is more convenient to use an equivalent definition, in which there is a single set comprising both the edges and nodes together. This has been used in [Ste07, Ste10] and is based on using a similar approach to graphs in [BMSW06].

Definition 1. *A **hypergraph** consists of a set U and a relation $\varphi \subseteq U \times U$ such that for all $x, y, z \in U$,*

1. *if $(x, y) \in \varphi$ then $(y, y) \in \varphi$, and*
2. *if $(x, y) \in \varphi$ and $(y, z) \in \varphi$ then $y = z$.*

From a hypergraph described in this way we can obtain the edges as those $u \in U$ for which $(u, u) \notin \varphi$, whereas the nodes satisfy $(u, u) \in \varphi$. Figure 1 shows an example. A **sub-hypergraph** of (U, φ) is defined as a subset $K \subseteq U$ for which

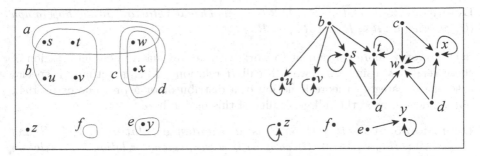

Fig. 1. Left hand side: hypergraph with edge set $E = \{a, b, c, d, e, f\}$ and node set $N = \{s, t, u, v, w, x, y, z\}$. On the right: the corresponding relation φ on $N \cup E$.

$k \in K$ and $(k, u) \in \varphi$ imply $u \in K$. The **dual hypergraph** of (U, φ) is (U, φ^*), where $\varphi^* = -\varphi \cap (1' \cup \smile\varphi)$. In the dual the edges and nodes are interchanged.

The sub-hypergraphs of a given hypergraph (U, φ) form a complete lattice which is a sub-lattice of the Boolean algebra $\mathscr{P}U$. This sub-lattice is not a Boolean algebra as the complement of a sub-hypergraph need not be a sub-hypergraph. There is a more general structure, a bi-Heyting algebra, which also appears when we consider the lattice of relations on a hypergraph.

Definition 2. *Let $\mathscr{L}$ be a lattice with top element 1 and bottom element 0. A* **pseudocomplement** *of $a \in \mathscr{L}$ is an element $m \in \mathscr{L}$ such that for all x in $\mathscr{L}$, $a \wedge x = 0$ iff $x \leqslant m$. A* **dual pseudocomplement** *of a is an element $m \in \mathscr{L}$ such that for all x in $\mathscr{L}$, $a \vee x = 1$ iff $m \leqslant x$.*

These operations generalize the usual complement of a Boolean algebra. The lattice of subgraphs of a graph is well-known [Law86, Law91] to have both pseudocomplements and dual pseudocomplements of all elements, and also to support more general operations as follows.

Definition 3. *A binary operation $\Rightarrow: \mathscr{L} \times \mathscr{L} \to \mathscr{L}$ is a* **relative pseudocomplement**, *if for all $x, y, z \in \mathscr{L}$ we have $x \leqslant (y \Rightarrow z)$ iff $x \wedge y \leqslant z$.*
A **dual relative pseudocomplement** *is a binary operation $\smallsetminus$ on $\mathscr{L}$ such that $x \leqslant (y \vee z)$ iff $(x \smallsetminus y) \leqslant z$.*

A **Heyting algebra** is a lattice $\mathscr{L}$ with 0 and 1 and equipped with a relative pseudocomplement, if it also has a dual relative pseudocomplement then it is a **bi-Heyting algebra**.

3   Relations on a Hypergraph

3.1   Definition of H-Relations

For any hypergraph (U, φ) the relations we need can be defined as follows.

Definition 4. *A relation $R \subseteq U \times U$ is **graphical relation** on the hypergraph (U, φ) when $\varphi ; R \subseteq R$ and $R ; \varphi \subseteq R$.*

It is, however, more convenient to work with an alternative characterization of these relations, using what we will call H-relations. This is actually a special case of the category-theoretic notion of a distributor or a pro-functor [Bor94], but we do not need the full generality of this notion here.

Definition 5. *A relation H on U is an **incidence relation** if $1' \subseteq H$ and $(H \cap -1') ; (H \cap -1') = 0$. Given a set U with incidence relation H, a relation R on U is an **H-relation** if $R = H ; R ; H$.*

It is straightforward to check that when (U, φ) is a hypergraph the reflexive closure, φ°, of φ is an incidence relation on U. Further, any incidence relation H arises in this way since defining $\varphi = H \cap ((H \cap -1') ; (\smile H \cap -1'))$, gives a hypergraph (U, φ) with $\varphi^\circ = H$.

Proposition 1. *Let (U, φ) be a hypergraph and R a relation on U. Then R is a graphical relation on (U, φ) iff R is an H-relation where $H = \varphi^\circ$.*   □

Distinct hypergraphs (U, φ_1) and (U, φ_2) can have $(U, \varphi_1^\circ) = (U, \varphi_2^\circ)$ when, for example, an element $u \in U$ appears as an isolated node in (U, φ_1) but as an empty edge in (U, φ_2). For a hypergraph (U, φ), the dual hypergraph (U, φ^*) is related to the converse of the corresponding incidence relation. Clearly $(\varphi^*)^\circ = \smile(\varphi^\circ)$ and H is an incidence relation iff $\smile H$ is an incidence relation.

In the remainder of this paper we use H to denote an incidence relation on a fixed but arbitrary set U. If we refer to an arbitrary relation without further explanation we will mean a binary relation on the set U. We will use the term **subgraph** in this setting for a subset $K \subseteq U$ such that $k \in K$ and $(k, u) \in H$ imply $u \in K$.

3.2   Basic Properties of H-Relations

Lemma 2. *Let R and S be H-relations, and let A be any relation on U.*

1. *H is an H-relation, and $R ; H = R = H ; R$.*
2. *$A \subseteq H ; A ; H$.*
3. *The relations $1 = U \times U$ and $0 = \varnothing$ are both H-relations.*
4. *$R \cup S$ is an H-relation.*
5. *$R \cap S$ is an H-relation.*
6. *$R ; S$ is an H-relation.*

Proof. We use basic properties of relations and the fact that $H ; H = H$, which immediately establishes part 1.

2. Since $1' \subseteq H$, we have $1' ; A ; 1' \subseteq H ; A ; H$.
3. Since $1' \subseteq H$, we have $1 = H ; 1 ; H$. Clearly $0 = H ; 0 ; H$.
4. $H ; (R \cup S) ; H = (H ; R ; H) \cup (H ; S ; H) = R \cup S$.

5. $H ; (R \cap S) ; H \subseteq (H ; R ; H) \cap (H ; S ; H) = R \cap S$. The reverse inclusion comes from part 2.

6. Using $H ; H = H$ and the fact that R and S are H-relations we have
$H ; (R ; S) ; H = H ; (H ; R ; H) ; (H ; S ; H) ; H = R ; S$.   □

We use the notation H-Rel for the set of all H-relations. The special case of $H = 1'$ includes all relations on U as H-relations. The set of all relations on U will be denoted Rel. The above results show that H-Rel is closed under the operations of union, intersection and composition in Rel, that H is the identity element for composition in H-Rel, and that the greatest and least elements of H-Rel are those in Rel. However, H-Rel is neither closed under complementation nor under taking converses, as can be seen from the example in Figure 2.

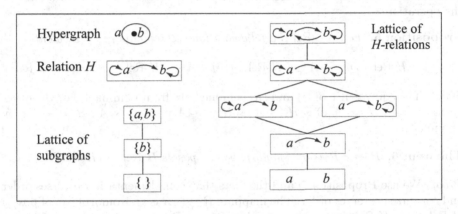

Fig. 2. Hypergraph with edge a and node b showing the lattice of H-relations

Various alternative definitions of a relation on a hypergraph can be given. An obvious possibility is to take two relations, one on the edges and one on the nodes, with a requirement that related edges have incident nodes which are related in some way. The choice taken here instead generalizes the characterization of arbitrary relations on the set U as sup-preserving (union-preserving) mappings on the powerset $\mathscr{P}U$. This well-known characterization is significant in mathematical morphology [BHR07], and we return to this in Section 5 below.

Let H-Sub denote the lattice of all subgraphs. Given any H-relation R, and any $K \in H$-Sub, we can check that defining $R(K) = \{u \in U : \exists k \in K \,((k, u) \in R)\}$ makes the assignment $K \mapsto R(K)$ a sup-preserving operations on H-Sub. The appropriate algebraic setting for these operations is a quantale [Ros90] and it is straightforward to check that we have the following result [Ste10].

Theorem 3. *The H-relations in H-Rel form a quantale under composition of relations, with unit H, which is isomorphic to the quantale of sup-preserving mappings on the lattice H-Sub.*   □

3.3 The Lattice H-Rel

The inclusion of H-Rel in Rel preserves all meets and joins. Hence this inclusion has both a left adjoint and a right adjoint. To determine these adjoints we start with a lemma.

Lemma 4. *Let A be any relation. Then $H \mathbin{;} A \mathbin{;} H = A$ iff $\smile H \mathbin{;} -A \mathbin{;} \smile H = -A$.*

Proof. The operation $H \mathbin{;} \_\_ \mathbin{;} H : \text{Rel} \to \text{Rel}$ is left adjoint to $H \setminus \_\_ / H$. Hence $A \subseteq H \setminus A / H$ if and only if $H \mathbin{;} A \mathbin{;} H \subseteq A$. Writing $H \setminus A / H$ as $-(\smile H \mathbin{;} -A \mathbin{;} \smile H)$ gives $H \mathbin{;} A \mathbin{;} H \subseteq A$ if and only if $\smile H \mathbin{;} -A \mathbin{;} \smile H \subseteq -A$. The result follows since $A \subseteq H \mathbin{;} A \mathbin{;} H$ and $-A \subseteq \smile H \mathbin{;} -A \mathbin{;} \smile H$ by Lemma 2. □

An immediate corollary is the following. Although H-relations are closed neither under complement nor under converse, they are closed under their combination: the operation $\frown$.

Proposition 5. *For $A \in \text{Rel}$ the following four statements are equivalent*

(i) $A \in H\text{-Rel}$ (ii) $-A \in \smile H\text{-Rel}$ (iii) $\smile A \in \smile H\text{-Rel}$ (iv) $\frown A \in H\text{-Rel}$.

Proof. The equivalence of (i) and (ii) is immediate from Lemma 4. For the other parts, $H \mathbin{;} A \mathbin{;} H = A$ iff $\smile H \mathbin{;} \smile A \mathbin{;} \smile H = \smile A$ iff $H \mathbin{;} -\smile A \mathbin{;} H = -\smile A$ by Lemma 4. □

Theorem 6. *$R \mapsto \frown R$ is an isomorphism of posets $H\text{-Rel} \to (H\text{-Rel})^{\text{op}}$.*

Proof. We use Proposition 5 and the facts that complementation reverses order and converse preserves order. The mapping $R \mapsto -R$ is an isomorphism of posets $H\text{-Rel} \to (\smile H\text{-Rel})^{\text{op}}$ and the mapping $R \mapsto \smile R$ is an isomorphism of posets $(\smile H\text{-Rel})^{\text{op}} \to (H\text{-Rel})^{\text{op}}$ so their composite has the required property. □

Proposition 7. *Let R and S be H-relations. Then $R \setminus S$, R / S, and $R \dagger S$ are also H-relations.*

Proof. To show that $R \setminus S$ is an H-relation, use that fact that $R \setminus S = -(\smile R \mathbin{;} -S)$. Then $R \setminus S$ is an H-relation iff $\smile R \mathbin{;} -S$ is an $(\smile H)$-relation by Lemma 4. But both $\smile R$ and $-S$ are $(\smile H)$-relations, so their composite is by Lemma 2. The cases of R / S and $R \dagger S$ are similar. □

Proposition 8. *For any relation A, and any H-relation R,*

1. *$H \setminus A / H$ is an H-relation, and*
2. *$H \setminus R / H = R$.*

Proof. 1. By Lemma 4, $H \setminus A/H$ is an H-relation iff $-(H \setminus A / H)$ is a $(\smile H)$-relation. But $-(H \setminus A / H) = \smile H \mathbin{;} -A \mathbin{;} \smile H$ which is a $(\smile H)$-relation by Lemma 2.

2. By Lemma 4, $-R = \smile H \mathbin{;} -R \mathbin{;} \smile H$, so $R = -(\smile H \mathbin{;} -R \mathbin{;} \smile H) = H \setminus R / H$. □

Theorem 9. *The inclusion H-Rel $\subseteq$ Rel has adjoints as follows.*

$$\text{Rel} \xrightleftharpoons[\bot]{\;H\;;\;\_\;;\;H\;} H\text{-Rel} \xrightleftharpoons[H\setminus\_/H]{\bot} \text{Rel}$$

Proof. To show that $H\;;\;\_\;;\;H$ is left adjoint to the inclusion of H-Rel in Rel we need to show that for $A \in$ Rel and $R \in H$-Rel, $H\;;\;A\;;\;H \subseteq R$ iff $A \subseteq R$. This is straightforward using $A \subseteq H\;;\;A\;;\;H$ and that $H\;;\;\_\;;\;H$ is order-preserving.

The second adjunction can be obtained from the opposite of the first since we can write $H\setminus R/H = \frown(H\;;\frown R;H)$ and compose the three adjunctions below.

$$H\text{-Rel} \xrightleftharpoons[\frown]{\cong} H\text{-Rel}^{\mathrm{op}} \xrightleftharpoons[H\;;\;\_\;;\;H]{\bot} \text{Rel}^{\mathrm{op}} \xrightleftharpoons[\frown]{\cong} \text{Rel}$$

$\square$

4 Generalized Complement and Converse

We have seen from the example in Figure 2 that the lattice of H-relations is not a Boolean algebra. Theorem 6 shows that this lattice is isomorphic to its opposite. This symmetry is not shared by H-Sub in general, as can be seen by considering the 5 element lattice of subgraphs of a graph with a single edge incident with two nodes. We next show that H-Rel is a bi-Heyting algebra; this reveals the two operations that are essentially weaker versions of the usual complement in Rel.

4.1 H-Rel is a Bi-heyting Algebra

Proposition 10. *The lattice H-Rel has a dual pseudocomplement given by $\lnot R = H\;;\;-R\;;\;H$, and a pseudocomplement given by $\neg R = H\setminus -R/H$.*

Proof. This follows from the fact that these H-relations are respectively the least H-relation containing $-R$ and the greatest H-relation contained within $-R$. $\square$

The converse complement interchanges the pseudocomplement with its dual in the following sense.

Proposition 11. *For any $R \in H$-Rel, $\frown\lnot R = \neg\frown R$ and $\frown\neg R = \lnot\frown R$*

Proof. Using the identity $H\setminus R/H = \frown(H\;;\frown R;H)$ this is a direct calculation from the definitions of $\neg$ and $\lnot$. $\square$

It follows from well-known properties of pseudocomplements and dual pseudo-complements (see for example [BD74, LM92]) that the operations $\neg$ and $\lnot$ on H-Rel have properties including the following for any $R, S \in H$-Rel.

Proposition 12

1. $\neg(R \cup S) = \neg R \cap \neg S$

2. $\neg(R \cap S) = \neg\neg(\neg R \cup \neg S)$

3. $R \cap \neg R = 0$

4. $\neg\neg\neg R = \neg R$

5. $R \cup \neg\neg R = \neg\neg R$
   (i.e. $R \subseteq \neg\neg R$)

6. $\lrcorner(R \cup S) = \lrcorner\lrcorner(\lrcorner R \cap \lrcorner S)$

7. $\lrcorner(R \cap S) = \lrcorner R \cup \lrcorner S$

8. $R \cup \lrcorner R = 1$

9. $\lrcorner\lrcorner\lrcorner R = \lrcorner R$

10. $R \cap \lrcorner\lrcorner R = \lrcorner\lrcorner R$
    (i.e. $\lrcorner\lrcorner R \subseteq R$)   □

The lattice H-Rel supports two more general operations from which the pseudocomplement and its dual can be obtained. The following result follows by verifying that the operations satisfy Definition 3.

Proposition 13. *In H-Rel there is a relative pseudocomplement, $\Rightarrow$, and a dual relative pseudocomplement, $\diagdown$, defined as follows for H-relations R and S.*

$$R \Rightarrow S = H \setminus (-R \cup S) / H \qquad S \diagdown R = H\,;(S \cap -R)\,; H$$

Proof. For the relative pseudocomplement we need that $T \subseteq (R \Rightarrow S)$ iff $T \cap R \subseteq S$ for all H-relations R, S, T. Now $T \cap R \subseteq S$ is equivalent to $T \subseteq -R \cup S$ in Rel although $-R \cup S$ may not be an H-relation. The result follows from Theorem 9 because for any $Q \in$ Rel we have $T \subseteq H \setminus Q / H$ iff $T \subseteq Q$.

The dual relative pseudocomplement can be justified similarly using the other adjunction from Theorem 9.   □

Note that $1 \diagdown R = \lrcorner R$ and $\neg R = R \Rightarrow 0$. The lattice H-Rel is complete because Rel is, so from Theorem 6 and Proposition 13 we have:

Theorem 14. *The lattice H-Rel is a complete bi-Heyting algebra which is isomorphic to its opposite and is also isomorphic to the lattice of $\smallsmile H$-relations.*   □

4.2   Converse Operations on H-Rel

H-relations are not closed under converse. In a similar way to the complement there are two weaker operations on H-Rel that can play the role of the converse.

Definition 6. *The **right converse** $\smallsmile R$ and the **left converse** $\mathbin{\smallsmile} R$ of $R \in H$-Rel are defined:*

$$\smallsmile R = H \setminus \smallsmile R / H \qquad and \qquad \mathbin{\smallsmile} R = H\,;\smallsmile R\,; H.$$

Three basic properties of converse of relations on a set are $\smallsmile\smallsmile R = R$, and $\smallsmile(R\,;S) = \smallsmile S\,;\smallsmile R$, and $\smallsmile 1' = 1'$, however, the analogues of these equations do not hold for the weaker forms of converse available in H-Rel.

We can construct an example of relation R where not only is the iterated left converse $\smallsmile^2 R = \smallsmile\smallsmile R$ strictly greater than R, but repeatedly applying this operation produces an infinite sequence of successively strictly greater relations. Consider the graph with edges $\{2n+1 : n \in \mathbb{Z}\}$ and nodes $\{2n : n \in \mathbb{Z}\}$ and where edge $2n+1$ is incident only with nodes $2n$ and $2n+2$. This means we have $U = \mathbb{Z}$ and H is the relation $1' \cup \{(2n+1, 2n) : n \in \mathbb{Z}\} \cup \{(2n+1, 2n+2) : n \in \mathbb{Z}\}$. Define R to be the relation $\{(1,2)\}$. It can be checked that $(-1, 4) \in \smile^2 R$, that $(-3, 6) \in \smile^4 R$. In general $\smile^{n+2} R$ is strictly larger than $\smile^n R$ for all n. Likewise, the right-converse can be repeated on a suitable infinite relation on the same graph to demonstrate that even powers of this operation can become successively strictly smaller. These examples do not really indicate that the right and left converse have very weak properties, because neither $\smile\smile R$ nor $\frown\smile R$ is the appropriate analogue of $\smile\frown R$, for which we need to consider $\smile\frown R$ and $\frown\smile R$ in view of the following result.

Proposition 15. *The left and right converses are adjoints.* $\text{H-Rel} \overset{\smile}{\underset{\frown}{\rightleftarrows}} \perp \text{ H-Rel}$

Proof. It is straightforward to check that both operations are order-preserving. We also need to check for any H-relation R that $\smile\frown R \subseteq R \subseteq \frown\smile R$. From $\frown R \subseteq \smile R$ we get that $\smile\frown R \subseteq R$, but since $\smile\frown R$ is the least H-relation containing $\smile\frown R$ we have $\smile\frown R \subseteq R$. Similarly for the other inclusion. □

Proposition 16. *The following identities hold for all $R \in H\text{-Rel}$.*

$$\neg R = \frown\smile R \qquad \neg R = \smile\frown R \qquad \llcorner R = \frown\smile R \qquad \llcorner R = \smile\frown R$$

$$\smile R = \frown\neg R \qquad \smile R = \neg\frown R \qquad \smile R = \frown\llcorner R \qquad \smile R = \llcorner\frown R$$

$$\llcorner\neg R = \smile\smile R \qquad \neg\llcorner R = \frown\frown R \qquad \neg\neg R = \smile\frown R \qquad \llcorner\llcorner R = \frown\smile R$$

Proof. Using the identity $H \backslash R / H = \frown(H; \frown R; H)$ we can establish $\neg R = \frown\smile R$ and $\llcorner R = \frown\smile R$ straightforwardly from the definitions. By replacing R by $\frown R$ in $\neg R = \frown\smile R$ we get $\neg\frown R = \frown\smile\frown R$ so by Proposition 11 we get $\frown\llcorner R = \frown\smile\frown R$ and thus $\llcorner R = \smile\frown R$. Similarly $\neg R = \smile\frown R$. This gives us the first row of four identities in the proposition. The remaining identities follow easily from these using the fact that $\frown$ is an involution. □

Proposition 17. *For all $R, S \in H\text{-Rel}$,*

(i) $\smile(R \cup S) = \smile R \cup \smile S,$ (iii) $\smile(R \cap S) = \llcorner\llcorner(\smile R \cap \smile S),$

(ii) $\frown(R \cap S) = \smile R \cap \frown S,$ (iv) $\frown(R \cup S) = \neg\neg(\frown R \cup \frown S).$

Proof. Parts (i) and (ii) are immediate because $\smile$ is a left adjoint (which preserves joins) and $\frown$ is a right adjoint (which preserves meets). For part (iii), we have $\smile(R \cap S) = \llcorner\frown(R \cap S) = \llcorner(\frown R \cup \frown S)$. By Proposition 12 part 6, this is $\llcorner\llcorner(\llcorner\frown R \cap \llcorner\frown S)$ and the result follows from Proposition 16. Part (iv) is similar. □

For the interaction of the converses with composition we have:

Proposition 18. $\smallsmile S\,;\smallsmile R \subseteq \smallsmile(R;S) \subseteq \smallfrown(R;S) \subseteq \smallfrown(R;S) \subseteq \smallfrown S\,;\smallfrown R.$

Proof. Since $\smallsmile(R\,;S)$ is the maximal H-relation inside $\smallfrown(R\,;S)$, we have that $\smallsmile(R\,;S) \subseteq \smallfrown(R\,;S)$. But from $\smallsmile S \subseteq \smallfrown S$ and $\smallsmile R \subseteq \smallfrown R$ we get $\smallsmile S\,;\smallsmile R \subseteq \smallfrown S\,;\smallfrown R$ so that $\smallsmile S\,;\smallsmile R \subseteq \smallfrown(R\,;S)$ and hence $\smallsmile S\,;\smallsmile R \subseteq \smallsmile(R\,;S)$ by the maximality of $\smallsmile(R\,;S)$. Dually for the parts involving $\smallfrown$. $\qquad\square$

4.3   Lonely Elements and the Dual Pseudocomplement

Since H is the identity element for composition in H-Rel it is natural to consider how the other operations interact with this particular relation. Here we consider the action of the double dual pseudocomplement. We start by defining $u \in U$ to be **lonely** if $(u,v) \in H$ implies $u = v$. If u is lonely with respect to the incidence relation H and (U,φ) is any hypergraph where $H = \varphi^\circ$ then in the hypergraph u is either an isolated node (not incident with any edges) or an empty edge (not incident with any nodes).

Theorem 19. *(i)* $\lnot\lnot H \subseteq 1'$, *and*

(ii) for any $u \in U$, $(u,u) \in \lnot\lnot H$ iff u is lonely.

Proof. Suppose first that $(u,v) \in H$ with $u \neq v$. Since $(u,v) \in H$, and $(v,u) \in -H$, and $(u,u) \in H$ we have $(u,u) \in H\,;-H\,;H = \lnot H$. Similarly both (u,v) and (v,v) belong to $\lnot H$.

For any $w,z \in U$, if $(w,z) \in \lnot\lnot H$ there must exist x and y such that $(w,x),(y,z) \in H$ and $(x,y) \in -\lnot H \subseteq H$, and when this happens the set $\{w,x,y,z\}$ can contain at most two distinct elements because of the properties of H.

If $(u,u) \in \lnot\lnot H$ there must be $x,y \in U$ with $(u,x),(y,u) \in H$ and $(x,y) \in -\lnot H \subseteq H$. But the set $\{u,x,y\}$ contains at most two distinct elements so either $u = x$ or $u = y$. If $u = x$ then both (u,y) and (y,u) are in H as $-\lnot H \subseteq H$, so that $u = y$ and $(u,u) = (x,y) \notin \lnot H$. But we have shown above that $(u,u) \in \lnot H$, so that $u = x$ cannot hold. If $u \neq x$ and $u \neq y$ then $x = y$ and we have a contradiction from $(u,x),(x,u) \in H$. The only remaining possibility is that $u \neq x$ and $u = y$ but this implies that $(u,x) \in H$ and $(x,u) \in -\lnot H \subseteq H$ which is again impossible. Hence $(u,u) \notin \lnot\lnot H$.

Similar arguments show that $(u,v),(v,v) \notin \lnot\lnot H$. So we have established that when $(u,v) \in H$ with $u \neq v$ then none of $(u,u),(u,v),(v,v)$ is in $\lnot\lnot H$. Since $\lnot\lnot H \subseteq H$ we conclude $\lnot\lnot H \subseteq 1'$, and $(u,u) \notin \lnot\lnot H$ when u is not lonely.

It remains to show that $(u,u) \in \lnot\lnot H$ for any lonely u. Expanding $\lnot\lnot H$ to $H\,;-(H\,;-H\,;H)\,;H$, it is sufficient to show that for u lonely, $(u,u) \notin H\,;-H\,;H$ and this is easily established. $\qquad\square$

5   Mathematical Morphology for Hypergraphs

Mathematical morphology [Ser82, BHR07] is used in image processing and has an algebraic foundation in lattice theory. This section begins by recalling some basic operations in mathematical morphology. These operations can be described as operations on subsets, $X \subseteq U$, induced by a relation on U. In sections 5.2 and 5.3 we go on to show how these operations can be generalized to operations on subgraphs of a hypergraph which are induced by an H-relation. The definition of erosion and dilation on subgraphs by means of an H-relation is very straightforward. However, the properties that these operations have for subgraphs are often significantly weaker than the properties of the corresponding operations on subsets. Determining these properties is less straightforward, and we begin to determine these at the end of section 5.3.

5.1   Mathematical Morphology for Sets

Given an arbitrary relation $A \subseteq U \times U$, two operations are defined on the powerset $\mathscr{P}U$.

Definition 7. *For $X \subseteq U$ the **dilation**, $X \oplus A$ is $\{u \in U : \exists x \in X \, ((x,u) \in A)\}$. The **erosion** $A \ominus X$ is $\{u \in U : \forall v \in U \, ((u,v) \in A \text{ implies } v \in X)\}$.*

The motivation for writing the subset on the left in dilation and on the right in erosion comes from the identities $X \oplus (A\,;B) = (X \oplus A) \oplus B$ and $(A\,;B) \ominus X = A \ominus (B \ominus X)$. These operations are well known and appear in many contexts other than mathematical morphology.

Erosion and dilation by a relation A are operations on subsets of U. However, subsets can be regarded as relations (subset X being $1' \cap X \times X$) and then erosion and dilation can easily be extended to operations on arbitrary relations. In this context they appear simply as composition with A and right residuation by A. The well-known property in mathematical morphology that $X \oplus \smile A = -(A \ominus -X)$ is then nothing more than $B\,; \smile A = -(-B\,/\,A)$. However, this more general context is not the appropriate setting for considering erosion and dilation on subgraphs. This point is justified in section 5.3 where we see that for graphs one analogue of $X \oplus \smile A = -(A \ominus -X)$ does not extend to a statement about operations on H-relations, so it is necessary to consider erosion and dilation as operations on subgraphs.

For any relation A we can define $\mathsf{Range}\,A = \{y \in U : \exists x \, (x,y) \in A\}$, and $\mathsf{CoRange}\,A = \{y \in U : \forall x \, (x,y) \in A\}$. Erosion and dilation can be characterized relationally:

Proposition 20. $U \times (X \oplus A) = (U \times X)\,;A$ *and* $U \times (A \ominus X) = (U \times X)\,/\,A$.

Proof. It is straightforward to check we have adjoints as follows.

$$\mathscr{P}U \quad \underset{\mathsf{CoRange}}{\overset{U \times \_}{\underset{\perp}{\rightleftarrows}}} \quad \mathsf{Rel} \quad \underset{\_\,/\,A}{\overset{\_\,;A}{\underset{\perp}{\rightleftarrows}}} \quad \mathsf{Rel} \quad \underset{U \times \_}{\overset{\mathsf{Range}}{\underset{\perp}{\rightleftarrows}}} \quad \mathscr{P}U$$

From the definitions of Range and $\oplus$ we get $\mathsf{Range}((U \times X)\,;A) = X \oplus A$, and since erosion is right adjoint to dilation $\mathsf{CoRange}((U \times X)\,/\,A) = A \ominus X$.

Now $(U \times X)\,;A$ is of the form $U \times Y$ for some $Y \subseteq U$, so we deduce that $U \times (\mathsf{Range}((U \times X)\,;A)) = (U \times X)\,;A$. Thus $U \times (X \oplus A) = (U \times X)\,;A$. The other equality follows by a similar argument.   □

A number of studies have examined various generalizations of erosion and dilation to graphs and hypergraphs [HV93, Ste07, CNS09, BB11]. In this section, given a subgraph K and an H-relation R we take the dilation $K \oplus R$ and the erosion $R \ominus K$ to be defined exactly as for any subset of U. It can be checked that $K \oplus R$ and $R \ominus K$ are themselves subgraphs given the assumptions about K and R.

5.2   Subgraphs as Relations

In this section the properties of erosion and dilation for hypergraphs are investigated by embedding H-Sub in H-Rel via $K \mapsto U \times K$ and using the properties of the relational operations established in the previous sections. Note that

Proposition 21. *for* $X \subseteq U$, $U \times X$ *is an* H-*relation iff* X *is a subgraph.*   □

Furthermore, the mapping $U \times \_$ is an isomorphism of posets between H-Sub and the subset of H-Rel consisting of all H-relations of the form $U \times X$ for some X.

Next consider how the pseudocomplement and its dual in H-Sub correspond to the operations in H-Rel. These operations in H-Sub will be denoted by $\neg K$ and $\lrcorner K$, and $-X$ denotes the complement of $X \subset U$. This overloading of the symbols for operations on subgraphs and on H-relations should not cause any confusion.

Proposition 22. $U \times (\neg K) = \neg(U \times K)$ *and* $U \times (\lrcorner K) = \lrcorner(U \times K)$.

Proof. From the description of $\neg K$ as the largest subgraph disjoint from K we get $\neg K = -(K \oplus \smile H)$. Thus $U \times -(K \oplus \smile H) = -(U \times (K \oplus \smile H)) = -((U \times K)\,;\smile H) = -(\smile H\,;(U \times K)\,;\smile H) = H \setminus -(U \times K)\,/\,H = \neg(U \times K)$. For the other equality, start from $\lrcorner K = (-K) \oplus H$.   □

5.3   Properties of Morphological Operations on Hypergraphs

For relations on a set and dilation and erosion on subsets a basic identity is

$$X \oplus \smile A = -(A \ominus -X).$$

Since both complement and converse are involutions this has many equivalent forms such as $-(-X \oplus A) = \smile A \ominus X$. In the case of H-relations and subgraphs the generalizations of these forms are no longer straightforwardly equivalent because of the weaker properties of the left and right converses and the pseudocomplement and its dual. However, we can establish such results using the technique of showing subgraphs K_1 and K_2 to be equal by lifting them to H-relations $U \times K_1$ and $U \times K_2$ and showing these are equal in H-Rel. This is demonstrated in Theorems 23 and 24. The proofs of these use the observation that if A is any reflexive relation then $A \setminus (U \times X) = U \times X = A \,; (U \times X)$.

Theorem 23. *For any subgraph K and H-relation R, $K \oplus \smile R = \lrcorner(R \ominus \neg K)$.*

Proof. To show that $U \times (K \oplus \smile R) = U \times (\lrcorner(R \ominus \neg K))$ in H-Rel we need $(U \times K) \,; \smile R = \lrcorner(\neg(U \times K) \,/\, R)$.

We have $(U \times K) \,; \smile R = (U \times K) \,; H \,; \smile R \,; H = (U \times K) \,; \smile R \,; H$ since $U \times K$ is an H-relation. Expanding $\lrcorner(\neg(U \times K) \,/\, R)$ yields $H \,; \smile H \,; (U \times K) \,; \smile R \,; H = (U \times K) \,; \smile R \,; H$ as required.   □

Theorem 24. *For any subgraph K and H-relation R, $\smile R \ominus K = \neg((\lrcorner K) \oplus R)$.*

Proof. $U \times \neg(\lrcorner K \oplus R) = \neg(H \,; (U \times -K) \,; H \,; R) = \neg((U \times -K) \,; R)$ since R is an H-relation. Expanding this gives $H \setminus ((U \times K)\dagger -R)/H = \frown H \dagger (U \times K)\dagger -R\dagger \frown H = (U \times K)\dagger -R\dagger \frown H = (U \times K)\dagger \frown H \dagger -R\dagger \frown H$ since $U \times K$ is an H-relation. This is equal to $(U \times K)\dagger \neg R$, but $\neg R = \frown \smile R$ by Proposition 16 so the result follows from Proposition 20.   □

We saw earlier that the identity $X \oplus \smile A = -(A \ominus -X)$ connecting erosion, dilation, converse and complement for a subset X and a relation A could be deduced immediately from the more general fact that $B \,; \smile A = -(-B \,/\, A)$ for arbitrary relations A and B on a set U. To see that morphology on graphs is more subtle than this consider Theorem 23. Generalizing this to a statement about arbitrary H-relations R and S we would get

$$R \,; \smile S = \lrcorner(\neg R \,/\, S).$$

If this holds in general then $H \,; \smile H = \lrcorner(\neg H \,/\, H)$ which implies $\smile H = \lrcorner \neg H$. But $\lrcorner \neg H = \smile \smile H$ by Proposition 16 so we would have $\smile H = \smile \smile H$. This is certainly not the case as is shown by taking $U = \{u, v, w, x, y\}$ and H to be the reflexive closure of $\{(x, u), (x, v), (y, v), (y, w)\}$.

6   Conclusions and Further Work

We have investigated some of the main operations available on H-relations and demonstrated their applications to mathematical morphology for hypergraphs. Although previous work has used H-relations, this appears to be the first account that deals with all the relational operations considered here and the interactions between them.

Relation algebras abstract the essential properties of relations on a set, so one can ask whether it would be useful to generalize from relation algebras to structures supporting the operations on H-relations. One could certainly start with equations specifying a bi-Heyting algebra and it is likely that the operation $R \mapsto \frown R$ would need to be present. The need for further operations and what equations should connect them remain to be investigated.

It is clear that while hypergraphs were the initial motivation for this work, many of the results will hold in a more general setting where the relation H need only be transitive and reflexive (a preorder). More generally still, we can move from preorders to categories and the notion of a distributor or profunctor [Bor94]. As we have already noted, H-relations are indeed a special case of profunctors, but it is not clear whether the weaker notions of converse introduced here are a special case of a well-known construction for profunctors.

References

[BB11]   Bloch, I., Bretto, A.: Mathematical Morphology on Hypergraphs: Prelim-
             inary Definitions and Results. In: Debled-Rennesson, I., Domenjoud, E.,
             Kerautret, B., Even, P. (eds.) DGCI 2011. LNCS, vol. 6607, pp. 429–440.
             Springer, Heidelberg (2011)
[BD74]   Balbes, R., Dwinger, P.: Distributive Lattices. University of Missouri Press
             (1974)
[BHR07]   Bloch, I., Heijmans, H.J.A.M., Ronse, C.: Mathematical morphology. In:
             Aiello, M., Pratt-Hartmann, I., van Benthem, J. (eds.) Handbook of Spatial
             Logics, ch. 14, pp. 857–944. Springer (2007)
[BMSW06]   Brown, R., Morris, I., Shrimpton, J., Wensley, C.D.: Graphs of Morphisms
             of Graphs. Bangor Mathematics Preprint 06.04, Mathematics Department,
             University of Wales, Bangor (2006)
[Bor94]   Borceux, F.: Handbook of Categorical Algebra. Basic Category Theory,
             vol. 1. Cambridge University Press (1994)
[CNS09]   Cousty, J., Najman, L., Serra, J.: Some Morphological Operators in Graph
             Spaces. In: Wilkinson, M.H.F., Roerdink, J.B.T.M. (eds.) ISMM 2009.
             LNCS, vol. 5720, pp. 149–160. Springer, Heidelberg (2009)
[HV93]   Heijmans, H., Vincent, L.: Graph Morphology in Image Analysis. In:
             Dougherty, E.R. (ed.) Mathematical Morphology in Image Processing, ch.
             6, pp. 171–203. Marcel Dekker (1993)
[Law86]   Lawvere, F.W.: Introduction. In: Lawvere, F.W., Schanuel, S.H. (eds.) Cat-
             egories in Continuum Physics. Lecture Notes in Mathematics, vol. 1174,
             pp. 1–16. Springer (1986)
[Law91]   Lawvere, F.W.: Intrinsic co-Heyting boundaries and the Leibniz rule in
             certain toposes. In: Carboni, A., et al. (eds.) Category Theory, Proceed-
             ings, Como 1990. Lecture Notes in Mathematics, vol. 1488, pp. 279–281.
             Springer (1991)
[LM92]   Mac Lane, S., Moerdijk, I.: Sheaves in Geometry and Logic. A First Intro-
             duction to Topos Theory. Springer (1992)
[Paw82]   Pawlak, Z.: Rough sets. International Journal of Computer and Informa-
             tion Sciences 11, 341–356 (1982)
[Ros90]   Rosenthal, K.I.: Quantales and their applications. Pitman Research Notes
             in Mathematics, vol. 234. Longman (1990)
[Sch11]   Schmidt, G.: Relational Mathematics. Encyclopedia of Mathematics and
             its Applications, vol. 132. Cambridge University Press (2011)
[Ser82]   Serra, J.: Image Analysis and Mathematical Morphology. Academic Press,
             London (1982)

[Ste07]   Stell, J.G.: Relations in Mathematical Morphology with Applications to Graphs and Rough Sets. In: Winter, S., Duckham, M., Kulik, L., Kuipers, B. (eds.) COSIT 2007. LNCS, vol. 4736, pp. 438–454. Springer, Heidelberg (2007)

[Ste10]   Stell, J.G.: Relational Granularity for Hypergraphs. In: Szczuka, M., Kryszkiewicz, M., Ramanna, S., Jensen, R., Hu, Q. (eds.) RSCTC 2010. LNCS (LNAI), vol. 6086, pp. 267–276. Springer, Heidelberg (2010)

[Tay99]   Taylor, P.: Practical Foundations of Mathematics. Cambridge University Press (1999)

Extension Properties
of Boolean Contact Algebras

Ivo Düntsch[1,*] and Sanjiang Li[2,*]

[1] Department of Computer Science
Brock University
St. Catharines, Ontario, Canada, L2S 3A1
duentsch@brocku.ca
[2] Centre for Quantum Computation and Intelligent Systems
University of Technology Sydney
Broadway NSW 2007, Australia
Sanjiang.Li@uts.edu.au

Abstract. We show that the class of Boolean contact algebras has the joint embedding property and the amalgamation property, and that the class of connected Boolean contact algebras has the joint embedding property but not the amalgamation property.

1 Introduction

Boolean contact algebras (BCAs) arise in spatial–temporal reasoning, and are hybrid algebraic–relational structures. Their history goes back to the 1920's, augmenting Leśniewski's mereology – whose models may be regarded as Boolean algebras with the least element removed – with a binary predicate of "being in contact". A contact relation on a Boolean algebra B is, loosely speaking, a symmetric and reflexive relation $\mathcal{C}$ on the nonzero elements of B with additional compatibility properties. Standard models are Boolean algebras of regular closed sets of some topological space $\mathcal{X}$ where two such sets are said to be in contact if their intersection is not empty. The elements of a BCA are abstractions of regions, and thus they are in the tradition of point–free spatial reasoning.

Given a BCA $\langle B, \mathcal{C} \rangle$, from the mereological "part of" relation P – which is the Boolean order of B – and the contact relation $\mathcal{C}$, several other binary relations can be defined as indicated in Table 1. In particular, the relations $DC, EC, PO, TPP, TPP^\smile, NTPP, NTPP^\smile$ along with the identity relation $1'$, known as the *RCC-8 relations*, have achieved some prominence in constraint–based spatial reasoning (see [16]). The binary relation algebra generated by $\mathcal{C}$ and P on a BCA B is called the *contact relation algebra (CRA)* of B. The problem arises, how many relations are there in the CRA of B, in particular, if there is a non–trivial BCA with a finite CRA? Only partial answers have been obtained for this problem. Although it was known that the CRA of the standard model defined on $\mathbb{R}^n$ contains infinitely many relations for $n \geq 1$ [13], this result cannot

---

* The ordering of authors is alphabetical and equal authorship is implied.

W. Kahl and T.G. Griffin (Eds.): RAMiCS 2012, LNCS 7560, pp. 342–356, 2012.
© Springer-Verlag Berlin Heidelberg 2012

Table 1. Relations defined from $\mathcal{C}$ and $\leq$

(1.1)	$P := \leq$	part of
(1.2)	$PP := P \cap -1'$.	proper part of
(1.3)	$O := P^{\smile} \; ; \; P$	overlap
(1.4)	$PO := O \cap -(P \cup P^{\smile})$	partial overlap
(1.5)	$EC := \mathcal{C} \cap - O$	external contact
(1.6)	$TPP := PP \cap (EC \; ; \; EC)$	tangential proper part
(1.7)	$NTPP := PP \cap -TPP$	non–tangential proper part
(1.8)	$DC := -\mathcal{C}$	disconnected

be generalized to other non–trivial BCAs. It turns out that an answer to this problem is connected to the existence of a countable BCA with an ω – categorical theory. Here, a first order theory is ω – *categorical* if all its countably infinite models are isomorphic, and a structure is called ω – categorical if its first order theory is ω – categorical. Primary examples are the theory of atomless Boolean algebras and the theory of dense linear orders.

If $\langle B, \mathcal{C} \rangle$ is an ω – categorical BCA, then, by the Engeler, Ryll–Nardzewski, Svenonius Theorem (see e.g. [9, Theorem 7.3.1]), its group H of automorphisms has only finitely many orbits in its action on $B \times B$. Because the CRA of B is contained in the binary relation algebra generated by the orbits of H (see e.g. [10]), this implies that the CRA of such B is finite. So, the question is whether there exists an ω – categorical BCA. Because the class **K** of finite BCAs is uniformly finitely generated – i.e. there is a function $f : \omega \to \omega$ such that every n–generated structure in **K** has cardinality at most $f(n)$ – this is equivalent to ask if there is a countable homogeneous BCA. Here, a first order structure $\mathfrak{D}$ is called *homogeneous* if every automorphism between finite substructures of $\mathfrak{D}$ can be extended over $\mathfrak{D}$. By the Fraïssé Theorem [9, Theorem 7.3.2], to answer the last question, we need to answer if **K** has the hereditary property, the joint embedding property and the amalgamation property. For a survey of the amalgamation property in algebraic logic we invite the reader to consult [14].

Motivated by the above observation, we start such investigation in the present paper and show that the class of BCAs has the hereditary property, the joint embedding property, and the amalgamation property, and that the class of connected BCAs does not have the amalgamation property. Our tools are free constructions of Boolean algebras and two forms of graph products which have been studied since the 1960s. A more detailed study of the homogeneous BCA and its CRA is left to a companion paper [5].

Related to this investigation, Bodirsky and Wölfl [3] have recently shown that a class of finite models of RCC-8 theory has both the joint embedding property and the amalgamation property. Thus, by Fraïssé's method, there is a countable homogeneous structure into which all finite RCC-8 models can be embedded. It follows from their investigations that for any k, the network consistency problem for RCC-8 restricted to networks of treewidth at most k can be solved in polynomial time. It is worth noting that an RCC-8 model in the sense of [3] has no Boolean structure and constraints like "a is contained in the union of b and c" are not supported by the RCC-8 constraint language they use. In [21] Wolter and Zakharyaschev proposed a more expressive constraint language, called BRCC-8, which supports Boolean-terms in the formation of RCC-8 constraints. The existence of the homogeneous BCA will suggest the possibility of a similar polynomial result for the consistency problem of BRCC-8 networks with bounded treewidth.

2   Basic Definitions and Notation

If $\langle U, R \rangle$ and $\langle V, S \rangle$ are binary relational structures, and $f : U \to V$ is a mapping, then f is called a *homomorphism* if aRb implies $f(a)Sf(b)$. In relational parlance, this is just R ; $f = f$; S. We call f a *strong homomorphism*, if f is onto and for each $s, t \in V$ with sSt there are $a, b \in U$ such that aRb and $f(a) = s$, $f(b) = t$. For $X \subseteq U$ we set $f[X] := \{f(a) : a \in X\}$. If $a \in U$, we let $R(a) = \{b \in U : aRb\}$ be *the range of a under R*. The restriction of R to X is denoted by $R \upharpoonright X$, i.e. $R \upharpoonright X = R \cap (X \times X)$. R is called *connected* if for all $a, b \in U$ there are $p_0, p_1, \ldots, p_n \in U$ such that $p_0 = a$, $p_n = b$, and $p_0 R p_1 \ldots p_{n-1} R p_n$. A *graph* is a reflexive and symmetric binary relation.

Fix a countable first order language $\mathcal{L}$, and let $\mathfrak{D}$ be an $\mathcal{L}$ – structure with universe D. An $\mathcal{L}$ – structure $\mathfrak{D}'$ with universe D' is called a *substructure of* $\mathfrak{D}$, if $D' \subseteq D$, each designated function on D' is the restriction of the corresponding function on D, and each designated n–ary relation on D' is the intersection of the corresponding relation on D with D'^n. A class $\mathbf{K}$ of $\mathcal{L}$ – structures has the

1. *Hereditary property* (HP) if $\mathbf{K}$ is closed under substructures.

2. *Joint embedding property* (JEP) if for any $A, B \in \mathbf{K}$, there are some $C \in \mathbf{K}$ and embeddings $e_A : A \to C$, $e_B : B \to C$. Such C is called a *free product of A and B* (with respect to $\mathbf{K}$) if for every $M \in \mathbf{K}$ and homomorphisms $f_A : A \to M$, $f_B : B \to M$ there is a unique homomorphism $f : C \to M$ such that $f \circ e_A = f_A$ and $f \circ e_B = f_B$. Readers familiar with category theory will recognize the free product as the *co–product*, sometimes called *direct sum*, of A and B in the appropriate category.

3. *Amalgamation property* (AP) if for all $A, B, C \in \mathbf{K}$ such that C is (isomorphic to) a common substructure of A and B, say, with embeddings $h_B : C \to B$ and $h_A : C \to A$, there are some $D \in \mathbf{K}$ and embeddings $e_A : A \to D$ and $e_B : B \to D$ such that $e_A \circ h_A = e_B \circ h_B$, see below. D is called an *amalgamated product of A and B over C*.

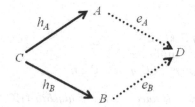

4. *Disjoint amalgamation property* (DAP) – also called *strong amalgamation property* – if it has the AP and the amalgamated product may be chosen in such a way that $e_A[A] \cap e_B[B] = (e_A \circ h_A)[C]$. If we assume that C is a common substructure of A and B, then an amalgamated product D of A and B over C containing A and B as substructures may be chosen in such a way that $A \cap B = C$.

5. *Free amalgamation property* (FAP) if it has the AP as above and for every $\mathcal{L}$-structure M and homomorphisms $f_A : A \to M$, $f_B : B \to M$ with $f_A \circ h_A = f_B \circ h_B$ there is a unique homomorphism $f : D \to M$ such that $f \circ e_A = f \circ e_B$. A free amalgamated product together with the morphisms e_A and e_B constitutes a *pushout* in the sense of category theory.

Throughout, $\langle B, +, \cdot, -, 0, 1 \rangle$ will denote a Boolean algebra with natural ordering $\leq$. With some abuse of language, we shall usually identify algebras with their base set. We denote by B^+ the set of all nonzero elements of B. If A is a (Boolean) subalgebra of B, we write $A \leq B$. A *filter of B* is a nonempty set $F \subseteq B$ such that $a \in F$ and $a \leq b$ imply $b \in F$, and $a, b \in F$ imply $a \cdot b \in F$. Each filter is contained in a maximal filter, commonly called an ultrafilter. If $M \subseteq B$ we denote by $[M]^B_{\mathrm{filt}}$ the filter of B generated by M. The *Stone space of B* is the set $\mathrm{Ult}(B)$ of ultrafilters of B together with the topology with an open base of the sets of the form $\{F \in \mathrm{Ult}(B) : a \in F\}$. The set of closed–open (clopen) sets of this topology is a subalgebra of 2^{2^B} isomorphic to B [11].

It is well known that the class of all Boolean algebras has the free amalgamation property:

Theorem 1. *[11, Theorem 11.19.] If A, B, C are Boolean algebras and $h_A : C \to A$ and $h_B : C \to B$ are embeddings, then there is (up to isomorphism) a unique amalgamated free product of A and B over C.*

Reminiscent of the "direct sum", we shall denote the (Boolean) free amalgamated product of A and B over C by the standard notation $A \oplus_C B$. If C is the two element Boolean algebra, we just write $A \oplus B$ and call it the *free product of A and B*. The existence of $A \oplus B$ shows that the class of all Boolean algebras has the JEP as well.

3   Boolean Contact Algebras

A contact relation $\mathcal{C}$ on a Boolean algebra B is defined by the following five properties:

C_0. $x \, \mathcal{C} \, y$ implies $x, y \neq 0$.
C_1. $x \neq 0$ implies $x \, \mathcal{C} \, x$.
C_2. $x \, \mathcal{C} \, y$ implies $y \, \mathcal{C} \, x$.   (symmetry)
C_3. $x \, \mathcal{C} \, y$ and $y \leq z$ implies $x \, \mathcal{C} \, z$.   (monotonicity)
C_4. $x \, \mathcal{C} \, (y + z)$ implies $(x \, \mathcal{C} \, y$ or $x \, \mathcal{C} \, z)$.   (distributivity)

A pair $\langle B, \mathcal{C}_B \rangle$ is called a *Boolean contact algebra* (BCA). This is the definition of contact algebras given in [2] in the Handbook of Spatial Logics, and by Dimov and Vakarelov [4] in their seminal papers on proximity structures.

Other properties have been considered in the literature: Contact algebras which satisfy

C_5. $\mathcal{C}(x) = \mathcal{C}(y)$ implies $x = y$.   (extensionality)
C_6. If $(\forall z)[x \, \mathcal{C} \, z$ or $y \, \mathcal{C} \, (-z)]$ then $x \, \mathcal{C} \, y$.   (interpolation)

are called *connection algebras* in [19] in the context of proximity structures. Finally, in [18] a Boolean connection algebra is a BCA $\langle B, \mathcal{C} \rangle$ which satisfies C_5, and, in addition, the following condition:

C_7. If $x \notin \{0, 1\}$, then $x \, \mathcal{C} \, (-x)$.   (connectivity)

BCAs which satisfy C_5 and C_7 are sometimes called *RCC algebras* after the Region Connection Calculus of [15], and BCAs which satisfy C_7 have been called *generalized RCC algebras* in [12].

The smallest contact relation is given by $x \, \mathcal{C}_{\min} \, y \iff x \cdot y \neq 0$. Usually, $\mathcal{C}_{\min}$ is called the *overlap relation*, denoted by $\mathcal{O}$; clearly, $\mathcal{O}$ satisfies C_5. The largest contact relation is $\mathcal{C}_{\max} = B^+ \times B^+$; it satisfies C_7. The four element Boolean algebra has exactly the contact relations $\mathcal{C}_{\min}$ and $\mathcal{C}_{\max}$. The only extensional contact relation on a finite Boolean algebra is the overlap relation $\mathcal{O}$.

Note that a BCA homomorphism is injective: If a Boolean homomorphism $f : \mathfrak{B} \to \mathfrak{B}_0$ is not injective, then there is some $x \in B^+$ such that $f(x) = 0$. It follows from C_2 that $x \, \mathcal{C} \, x$, but $f(x)(-\mathcal{C}_0)f(x)$ by C_0.

There is a well developed duality theory for BCAs. Generalizing results of [8] in the finite case and of [6] in the area of discrete proximity algebras, in [7] the following duality between contact relations and graphs on its space of ultrafilters is exhibited:

Theorem 2. *Let B be a Boolean algebra. There is a bijective order preserving correspondence between the contact relations on B and the reflexive and symmetric relations on $\mathrm{Ult}(B)$ which are closed in the product topology of $\mathrm{Ult}(B)$.*

Proof. We just give the construction; details can be found in [7]. If R is a reflexive and symmetric closed relation on $\mathrm{Ult}(B)$, set $\mathcal{C}_R = \bigcup \{F \times G : \langle F, G \rangle \in R\}$. If $\mathcal{C}$ is a contact relation on B, let $R_{\mathcal{C}} = \{\langle F, G \rangle \in \mathrm{Ult}(B)^2 : F \times G \subseteq \mathcal{C}\}$.   □

$R_{\mathcal{C}}$ is called the *dual of* $\mathcal{C}$, and $\mathcal{C}_R$ is called the *dual of* R. Our proofs will sometimes be based on this duality, as graphs on ultrafilters are often easier to handle than contact relations. As an application we present a result from [7]:

Theorem 3. *Let $\langle B, \mathcal{C} \rangle$ be a BCA.*

1. $\mathcal{C}$ satisfies C_6 if and only if $R_{\mathcal{C}}$ is an equivalence relation.
2. If B is finite, then $\mathcal{C}$ satisfies C_7 if and only if $R_{\mathcal{C}}$ is connected.

Our final aim in this section is to describe the dual of the situation that $\langle A, \mathcal{C}_A \rangle$ is a substructure of $\langle B, \mathcal{C}_B \rangle$. To this end, we first quote Proposition 3.2 of [4]:

Lemma 1. *If $\langle B, \mathcal{C} \rangle$ is a BCA, and F', G' are filters of B such that $F' \times G' \subseteq \mathcal{C}$, then there are ultrafilters F, G of B such that $F' \subseteq F$, $G' \subseteq G$, and $F \times G \subseteq \mathcal{C}$.*

Theorem 4. *Let $\langle A, \mathcal{C}_A \rangle$ and $\langle B, \mathcal{C}_B \rangle$ be BCAs with dual relations R_A and R_B, and suppose that $A \leq B$. Then, $\mathcal{C}_A = \mathcal{C}_B \restriction A$ if and only if*

$$R_A = \{ \langle F \cap A, G \cap A \rangle : F, G \in \mathrm{Ult}(B), \langle F, G \rangle \in R_B \}. \tag{3.1}$$

Proof. "$\Rightarrow$": For the "$\subseteq$" direction, suppose that $\langle F', G' \rangle \in R_A$. We need to find $F, G \in \mathrm{Ult}(B)$ such that $F' = F \cap A$, $G' = G \cap A$, and $\langle F, G \rangle \in R_B$.

By definition, $\langle F', G' \rangle \in R_A$ implies that $F' \times G' \subseteq \mathcal{C}_A$, and $\mathcal{C}_A \subseteq \mathcal{C}_B \restriction A$ implies that $F' \times G' \subseteq \mathcal{C}_B$. Let F_0 be the filter of B generated by F', and G_0 the filter of B generated by G'. Since F' is a filter of A, it is closed under $\cdot$, and therefore, $F_0 = \{ b \in B : (\exists a)[a \in F' \text{ and } a \leq b] \}$. Similarly, $G_0 = \{ b \in B : (\exists a)[a \in G' \text{ and } a \leq b] \}$. Since $F' \times G' \subseteq \mathcal{C}_B$ and by C_3, it follows that $F_0 \times G_0 \subseteq \mathcal{C}_B$. Now, by Lemma 1, there are ultrafilters F, G of B such that $F_0 \subseteq F$, $G_0 \subseteq G$, and $F \times G \subseteq \mathcal{C}_B$, i.e. $\langle F, G \rangle \in R_B$. Since $F' \subseteq F, G' \subseteq G$, and F', G' are ultrafilters of A, we have $F' = F \cap A$ and $G' = G \cap A$.

For "$\supseteq$", let $F, G \in \mathrm{Ult}(B)$ and $\langle F, G \rangle \in R_B$, i.e. $F \times G \subseteq \mathcal{C}_B$. Since $\mathcal{C}_B \restriction A \subseteq \mathcal{C}_A$ we have $(F \cap A) \times (G \cap A) \subseteq \mathcal{C}_A$ which shows $\langle F \cap A, G \cap A \rangle \in R_A$.

"$\Leftarrow$": Let $a, b \in A$ and $a \, \mathcal{C}_A \, b$; we show that $a \, \mathcal{C}_B \, b$. Since $a \, \mathcal{C}_A \, b$, by Lemma 1 there are $F', G' \in \mathrm{Ult}(A)$ such that $a \in F', b \in G'$ and $\langle F', G' \rangle \in R_A$. By the hypothesis, there are $F, G \in \mathrm{Ult}(B)$ such that $F' = F \cap A, G' = G \cap A$ and $F \times G \subseteq \mathcal{C}_B$. Hence, $a \, \mathcal{C}_B \, b$.

Conversely, let $a, b \in A$ and $a \, \mathcal{C}_B \, b$. Then, by Lemma 1 there are ultrafilters F, G of B with $a \in F, b \in G$, and $F \times G \subseteq \mathcal{C}_B$. By the hypothesis, $\langle F \cap A, G \cap A \rangle \in R_A$, and thus, $(F \cap A) \times (G \cap A) \subseteq \mathcal{C}_A$. It follows from $a, b \in A$ that $a \, \mathcal{C}_A \, b$. $\square$

Corollary 1. *Let $\langle A, \mathcal{C}_A \rangle$ be a BCA, and $A \leq B$. Suppose that R_A is the dual of $\mathcal{C}_A$, and define a relation R_B on $\mathrm{Ult}(B)$ by*

$$\langle F, G \rangle \in R_B \iff \langle F \cap A, G \cap A \rangle \in R_A. \tag{3.2}$$

Then, the dual $\mathcal{C}_B$ of R_B is the largest contact relation $\mathcal{C}$ on B such that $\langle A, \mathcal{C}_A \rangle$ is a substructure of $\langle B, \mathcal{C} \rangle$.

Proof. Since R_B satisfies (3.1), $\langle A, \mathcal{C}_A \rangle$ is a substructure of $\langle B, \mathcal{C}_B \rangle$. Next, suppose that $\langle A, \mathcal{C}_A \rangle$ is a substructure of $\langle B, \mathcal{C} \rangle$, and that R is the dual of $\mathcal{C}$. Suppose that $\langle F, G \rangle \in R$; then, by (3.1) $\langle F \cap A, G \cap A \rangle \in R_A$, and therefore, $\langle F \cap A, G \cap A \rangle \in R_B$ by the "$\Leftarrow$" direction of (3.2). Thus, $R \subseteq R_B$, and it follows from Theorem 2 that $\mathcal{C} \subseteq \mathcal{C}_B$. $\square$

The result of Theorem 4 comes as no surprise: If A is a subalgebra of B, then by duality for Boolean algebras the mapping $p : \mathrm{Ult}(B) \to \mathrm{Ult}(A)$ defined by $H \mapsto H \cap A$ is onto. If, furthermore, $\langle A, \mathcal{C}_A \rangle$ and $\langle B, \mathcal{C}_B \rangle$ are BCAs with dual relations R_A and R_B, then Theorem 4 shows that $\mathcal{C}_B$ is an extension of $\mathcal{C}_A$ if and only if $R_A = \{\langle p(F), p(G)\rangle : \langle F, G\rangle \in R_B\}$; in other words, R_A is the image of R_B under the action of p on $\mathrm{Ult}(B) \times \mathrm{Ult}(B)$.

With some abuse of language we call R_B an extension of R_A if $A \le B$, and (3.1) holds.

Since the Stone space of a finite Boolean algebra B is discrete, and each ultrafilter is determined by an atom of B, the construction can be somewhat simplified in this case: For a finite BCA $\langle B, \mathcal{C}\rangle$ we let $\Gamma(\langle B, \mathcal{C}\rangle)$ be the structure $\langle U, E\rangle$ where $U = At(B)$, and $\langle a, b\rangle$ is an edge in E if and only if $a\,\mathcal{C}\,b$. Conversely, if E is a graph on U we let $\Delta(\langle U, E\rangle)$ be the structure $\langle B, \mathcal{C}\rangle$ where B is the powerset algebra of U and $\mathcal{C}$ is the relation on B defined by $a\,\mathcal{C}\,b$ if and only if there are $x, y \in U$ with $x \in a$, $y \in b$ and xEy. Theorem 2 and Theorem 4 can now be reformulated as follows:

Theorem 5. *Let $\mathfrak{B} = \langle B, \mathcal{C}\rangle$ be a finite BCA, and $\mathfrak{U} = \langle U, E\rangle$ be a finite graph, i.e. $|U| < \omega$.*

1. *$\Gamma(\mathfrak{B})$ is a graph, and $\Delta(\mathfrak{U})$ is a BCA.*
2. *$\Delta(\Gamma(\mathfrak{B})) \cong \mathfrak{B}$, and $\Gamma(\Delta(\mathfrak{U})) \cong \mathfrak{U}$.*
3. *If $\mathfrak{U} = \langle A, \mathcal{C}_A\rangle$ and $\mathfrak{B} = \langle B, \mathcal{C}_B\rangle$ are finite BCAs and $e : \mathfrak{U} \to \mathfrak{B}$ is a BCA embedding, then $h : \Gamma(\mathfrak{B}) \to \Gamma(\mathfrak{U})$ is a strong graph homomorphism, where for $b \in At(B)$, $h(b)$ is defined as follows: If $b \in At(B)$, then the smallest element of $[\{b\}]^B_{\mathrm{filt}} \cap e[A]$ is an atom of $e[A]$, and we set $h(b) = a$. Since e is an embedding, h is well defined.*

Example 1. Let B be the Boolean algebra generated by its atoms $At(B) = \{b_0, b_1, b_2, b_3\}$, and set $R = \{\langle b_0, b_1\rangle, \langle b_1, b_2\rangle, \langle b_2, b_3\rangle\}$. If $x, y \in B \setminus \{0, 1\}$, $x = b_{i_0} + \ldots + b_{i_k}$, $y = b_{j_0} + \ldots + b_{j_n}$ let

$$x\,\mathcal{C}_B\,y \iff (\exists i_r, j_s)[i_r = j_s \text{ or } b_{i_r} R b_{j_s} \text{ or } b_{j_s} R b_{i_r}].$$

Then, $\mathcal{C}_B$ is a contact relation on B, and $\Delta(\mathfrak{B})$ is the graph $R \cup R^\smile \cup 1'$ on $\{a_0, \ldots, a_3\}$. Suppose that A is the Boolean algebra generated by its atoms $AT(A) = \{d_0, d_1, d_2\}$, $S = \{\langle d_0, d_1\rangle, \langle d_1, d_2\rangle\}$, and for $x = d_{i_0} + \ldots + d_{i_k}$, $y = d_{j_0} + \ldots + d_{j_n}$ let

$$x\,\mathcal{C}_A\,y \iff (\exists i_r, j_s)[i_r = j_s \text{ or } d_{i_r} S d_{j_s} \text{ or } d_{j_s} R d_{i_r}].$$

Then, $\mathcal{C}_A$ is a contact relation on A, and $\Delta(\mathfrak{U})$ is the graph $S \cup S^\smile \cup 1'$ on $\{d_0, d_1, d_2\}$. Now, suppose that $f : At(B) \to At(A)$ is defined by, see below.

$$f(b_0) = f(b_2) = d_1, \quad f(b_1) = d_2, \quad f(b_3) = d_0,$$

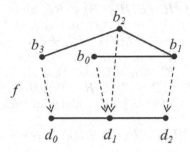

Then, f is a strong homomorphism and its dual embedding e of $\mathfrak{A}$ into $\mathfrak{B}$ is determined by $e(d_0) = b_3, e(d_1) = b_0 + b_2$, and $e(d_2) = b_1$.   □

4   Extension Properties of Boolean Contact Algebras

Let $\mathbf{K}$ be the class of all BCAs, and $\mathbf{K}_{\mathcal{C}_6}$, respectively, $\mathbf{K}_{\mathcal{C}_7}$, be the collection of all BCAs which satisfy C_6, respectively, C_7. Since the axioms $C_0 - C_4$ and C_7 are universal, both $\mathbf{K}$ and $\mathbf{K}_{\mathcal{C}_7}$ possess HP.

Theorem 6. $\mathbf{K}$, $\mathbf{K}_{\mathcal{C}_6}$, and $\mathbf{K}_{\mathcal{C}_7}$ have the JEP.

Proof. We shall exhibit two different ways to embed two BCAs into a third one, both arising from familiar constructions in graph theory. Let $\mathfrak{A} := \langle A, \mathcal{C}_A \rangle$ and $\mathfrak{B} := \langle B, \mathcal{C}_B \rangle$ be BCAs with dual relations R_A, respectively, R_B, and let $D := A \oplus B$ be the (Boolean) free product of A and B. $h : D \to 2^{\mathrm{Ult}(D)}$ is the Stone embedding, i.e. $h(a) = \{F \in \mathrm{Ult}(D) : a \in F\}$. The collection $\{h(a) : a \in D\}$ is a clopen basis for the Stone topology on $\mathrm{Ult}(D)$. Suppose w.l.o.g. that $A, B \le D$, and $H, H' \in \mathrm{Ult}(D)$ with $H \sim \langle F, G \rangle$, $H' \sim \langle F', G' \rangle$, i.e. $H \cap A = F$, $H \cap B = G$ and $H' \cap A = F'$, $H' \cap B = G'$.

Construction 1. Our first construction is based on the *direct product of graphs* [17]: Set

$$H R_D^0 H' \text{ if and only if } F R_A F' \text{ and } G = G' \text{ or } F = F' \text{ and } G R_B G'. \quad (4.1)$$

Clearly, R_D^0 is reflexive and symmetric, since R_A and R_B have these properties. Next, we show that R_D^0 is closed in the product topology of $\mathrm{Ult}(D) \times \mathrm{Ult}(D)$: Suppose that $\langle H, H' \rangle \in \mathrm{cl}(R_D^0)$, and assume that $\langle H, H' \rangle \notin R_D^0$. There are three cases:

1. $H \cap A \ne H' \cap A$ and $H \cap B \ne H' \cap B$: Then, there are $a \in A$, $b \in B$ such that $a \in H \cap A$, $-a \in H' \cap A$, and $b \in H \cap B$, $-b \in H' \cap B$. It follows that $a \cdot b \in H$ and $-a \cdot -b \in H'$. Since $\langle H, H' \rangle \in \mathrm{cl}(R_D^0)$, $(h(a \cdot b) \times h(-a \cdot -b)) \cap R_D^0 \ne \emptyset$, and therefore there are $P, P' \subset \mathrm{Ult}(D)$ such that $a \cdot b \subset P$, $-a \cdot -b \subset P'$, and $\langle P, P' \rangle \in R_D^0$. Let w.l.o.g. $P \cap A = P' \cap A$. Now, $a \cdot b \in P$ implies $a \in P \cap A$, and $-a \cdot -b \in P'$ implies $-a \in P' \cap A$. This, however, contradicts $P \cap A = P' \cap A$.

2. $H \cap A = H' \cap A$ and $\langle H \cap B, H' \cap B \rangle \notin R_B$: Since $\langle H \cap B, H' \cap B \rangle \notin R_B$ there are $b \in H \cap B$ and $b' \in H' \cap B$ such that $\langle b, b' \rangle \notin C_B$. It follows from $\langle H, H' \rangle \in \mathrm{cl}(R_D^0)$ that $(h(b) \times h(b')) \cap R_D^0 \neq \emptyset$, i.e. there are $P, P' \in \mathrm{Ult}(D)$ such that $b \in P$, $b' \in P'$ and $\langle P, P' \rangle \in R_D^0$. Since $P \cap B$ and $P' \cap B$ are ultrafilters of B and $\langle b, b' \rangle \notin C_B$, it follows that $P \cap B \neq P' \cap B$. By definition of R_D^0, we have $\langle P \cap B, P' \cap B \rangle \in R_B$. But then, by definition of R_B, it follows that $(P \cap B) \times (P' \cap B) \subseteq C_B$, which in turn implies that $b \ C_B \ b'$, a contradiction.

3. $H \cap B = H' \cap B$ and $\langle H \cap A, H' \cap A \rangle \notin R_A$: This is analogous to the previous case.

To prove that R_D^0 extends both R_A and R_B we use Theorem 4 and show that

$$R_A = \{\langle H \cap A, H' \cap A \rangle : H, H' \in \mathrm{Ult}(D) \text{ and } \langle H, H' \rangle \in R_D^0\}, \qquad (4.2)$$

$$R_B = \{\langle H \cap B, H' \cap B \rangle : H, H' \in \mathrm{Ult}(D) \text{ and } \langle H, H' \rangle \in R_D^0\}. \qquad (4.3)$$

Without loss of generality we prove only (4.2):

"$\subseteq$": Let $F, F' \in \mathrm{Ult}(A)$ with $\langle F, F' \rangle \in R_A$. Choose any $G \in \mathrm{Ult}(B)$. If $H \sim \langle F, G \rangle$, $H' \sim \langle F', G \rangle$, then $\langle H, H' \rangle \in R_D^0$ and $F = H \cap A$, $F' = H' \cap A$.

"$\supseteq$": Let $\langle H, H' \rangle \in R_D^0$, $H \sim \langle F, G \rangle$, $H' \sim \langle F', G' \rangle$, and w.l.o.g. $G = G'$. Then, $F = H \cap A$, $F' = H' \cap A$, and $\langle F, F' \rangle \in R_A$ by definition of R_D^0.

Up to now we have shown that $\mathbf{K}$ has the JEP. To prove that $\mathbf{K}_{C_7}$ has the JEP, suppose that both C_A and C_B satisfy C_7, and let $x \in D$, $x \neq 0, 1$. Then, there are $a_0, \ldots, a_n \in A$, $a_i \neq 0$, and $b_0, \ldots, b_n \in B$, $b_j \neq 0$, such that $x = a_0 \cdot b_0 + \ldots + a_n \cdot b_n$ and

$$a_i \cdot b_i \not\leq a_0 \cdot b_0 + \ldots + a_{i-1} \cdot b_{i-1} + a_{i+1} \cdot b_{i+1} + \ldots + a_n \cdot b_n,$$

in particular, $a_i \cdot b_i \neq 0$ for all $i \leq n$. By distributivity,

$$-x = \sum \left\{ \prod \{-a_i : i \in I\} \cdot \prod \{-b_i : i \in \{0, \ldots, n\} \setminus I\} : I \subseteq \{0, \ldots, n\} \right\}.$$

Since $-x \neq 0$, there is some $I \subseteq \{0, \ldots, n\}$ such that

$$\prod \{-a_i : i \in I\} \cdot \prod \{-b_i : i \in \{0, \ldots, n\} \setminus I\} \neq 0.$$

We consider three cases:

1. Suppose that $I = \{0, \ldots, n\}$. Then, $\prod \{-a_i : i \leq n\} = -(a_0 + \ldots + a_n) \neq 0$, and thus, $a_0 + \ldots + a_n \neq 1$. Therefore, $(a_0 + \ldots + a_n) \ C_A \ -(a_0 + \ldots + a_n)$, and so there are ultrafilters F, F' of A such that

$$a_0 + \ldots + a_n \in F, \ -(a_0 + \ldots + a_n) \in F', \langle F, F' \rangle \in R_A.$$

Since F is an ultrafilter, suppose w.l.o.g. that $a_0 \in F$, and choose some $G \in \mathrm{Ult}(B)$ with $b_0 \in G$; such G exists, since $b_0 \neq 0$. Setting $H \sim \langle F, G \rangle$ and $H' \sim \langle F', G \rangle$, we obtain $\langle H, H' \rangle \in R_D^0$, and $a_0 \cdot b_0 \in H$. Since $a_0 \cdot b_0 \leq x$ we obtain $x \in H$. Furthermore, $-(a_0 + \ldots + a_n) \in F'$ and $-(a_0 + \ldots + a_n) \leq -x$ imply that $-x \in H'$; hence, $x \ C_D^0 \ -x$.

2. If $I = \emptyset$, then $b_0 + \ldots + b_n \neq 1$ and we proceed analogously.

3. Let $I \neq \emptyset, \{0, \ldots, n\}$, and choose $I = \{i_0, \ldots, i_k\}, J = \{i_{k+1}, \ldots, i_n\}$ such that $I \cup J = \{0, \ldots, n\}, I \cap J = \emptyset$, and I is maximal with the property

$$-a_{i_0} \cdot \ldots \cdot -a_{i_k} \cdot -b_{i_{k+1}} \cdot \ldots \cdot -b_{i_n} \neq 0. \tag{4.4}$$

Possibly permuting the indices, we may suppose that $I = \{0, \ldots, k\}$. By our previous considerations we have $0 \leq k < n$, and the maximality of k implies that for every j with $k < j \leq n$,

$$-a_0 \cdot \ldots \cdot -a_k \cdot -a_j \cdot \prod \{-b_i : k < i \leq n, i \neq j\} = 0.$$

The independence of A and B now implies that

$$-a_0 \cdot \ldots \cdot -a_k \cdot -a_j = 0,$$

see e.g. [11], Lemma 11.7. Hence, for every $k < j \leq n$ we have in particular

$$-a_0 \cdot \ldots \cdot -a_k \cdot a_j \neq 0. \tag{4.5}$$

It follows from (4.4) that $-b_{k+1} \cdot \ldots \cdot -b_n \neq 0$, and from $b_j \neq 0$ we obtain $-b_{k+1} \cdot \ldots \cdot -b_n \neq 1$.

Since $\mathcal{C}_B$ satisfies C_7, $(-b_{k+1} \cdot \ldots \cdot -b_n) \, \mathcal{C}_B \, (b_{k+1} + \ldots + b_n)$, and so there are $G, G' \in \mathrm{Ult}(B)$ such that $-b_{k+1} \cdot \ldots \cdot -b_n \in G$, $b_{k+1} + \ldots + b_n \in G'$ and $\langle G, G' \rangle \in R_B$. Since G' is an ultrafilter, let w.l.o.g. $b_n \in G'$. Choose some ultrafilter F of A with $-a_0 \cdot \ldots \cdot -a_k \cdot a_n \in F$, and let $H \sim \langle F, G \rangle$ and $H' \sim \langle F, G' \rangle$. Then,

$$\langle H, H' \rangle \in R_D^0, \quad -a_0 \cdot \ldots \cdot -a_k \cdot a_n \cdot -b_{k+1} \cdot \ldots \cdot -b_n \in H, \quad a_n \cdot b_n \in H'.$$

It follows from $a_n \cdot b_n \leq x$ and $-a_0 \cdot \ldots \cdot -a_k \cdot a_n \cdot -b_{k+1} \cdot \ldots \cdot -b_n \leq -x$ that $x \in H$ and $-x \in H'$, hence, $x \, \mathcal{C}_D^0 \, -x$.

Concerning C_6, observe that the fact that R_A and R_B are transitive does not imply that R_D is transitive: If $\langle F_0, F_1 \rangle \in R_A, F_0 \neq F_1, \langle G_1, G_2 \rangle \in R_B, G_1 \neq G_2$, then $\langle \langle F_0, G_1 \rangle, \langle F_1, G_1 \rangle \rangle \in R_D$, $\langle \langle F_1, G_1 \rangle, \langle F_1, G_2 \rangle \rangle \in R_D$, on the other hand, $\langle \langle F_0, G_1 \rangle, \langle F_1, G_2 \rangle \rangle \notin R_D$.

Construction 2. The second construction is related to the *Kronecker product of graphs* [20]: Set

$$\langle H, H' \rangle \in R_D^1 \iff \langle H \cap A, H' \cap A \rangle \in R_A \text{ and } \langle H \cap B, H' \cap B \rangle \in R_B. \tag{4.6}$$

Since both R_A and R_B are reflexive and symmetric, so is R_D^1. To show that R_D^1 is closed, let $\langle H, H' \rangle \in \mathrm{cl}(R_D^1)$. Assume that $\langle H, H' \rangle \notin R_D^1$; then, $\langle H \cap A, H' \cap A \rangle \notin R_A$ or $\langle H \cap B, H' \cap B \rangle \notin R_B$. Let w.l.o.g. $\langle H \cap A, H' \cap A \rangle \notin R_A$; then, there are $a \in H \cap A$ and $a' \in H' \cap A$ such that $a(- \mathcal{C}_A)a'$. Since $\langle H, H' \rangle \in h(a) \times h(a')$

and $\langle H, H' \rangle \in \mathrm{cl}(R_D^1)$, it follows that $(h(a) \times h(a')) \cap R_D^1 \neq \emptyset$, and therefore there are $P, P' \in \mathrm{Ult}(D)$ such that $a \in P$, $a' \in P'$, and $\langle P, P' \rangle \in R_D^1$. But then, $\langle P \cap A, P' \cap A \rangle \in R_A$, i.e. $(P \cap A) \times (P' \cap A) \subseteq C_A$. Hence, $a\, C_A\, a'$, contradicting our assumption.

It is straightforward to show as above that C_D^1 extends both C_A and C_B. If both R_A and R_B are transitive, then, clearly, so is R_D^1 and thus, C_D^1 satisfies C_6 in this case by Theorem 3. Finally, $C_D^0 \subseteq C_D^1$ shows that C_D^1 satisfies C_7, if C_A and C_B satisfy C_7. □

We prepare the proof of the amalgamation property by two lemmas. The first is a characterization of the amalgamated free product of Boolean algebras from [11]:

Lemma 2. *Suppose that $A, B \leq D$, and that C is a subalgebra of both A and B. Then, D is the amalgamated free product of A and B over C, if $A \cup B$ generates D, and for all $a \in A, b \in B$ with $a \cdot b = 0$ there is some $c \in C$ such that $a \leq c$ and $b \leq -c$.*

The following lemma describes the ultrafilters of $A \oplus_C B$ in terms of $\mathrm{Ult}(A)$ and $\mathrm{Ult}(B)$ see [11], Ch. 4, Exercise 9. We give a proof for the convenience of the reader:

Lemma 3. *Suppose that D is the amalgamated free product of A, B over C. Then, each ultrafilter of D is of the form $[F \cup G]_{\mathrm{filt}}^D$ where $F \in \mathrm{Ult}(A)$, $G \in \mathrm{Ult}(B)$, and $F \cap C = G \cap C$. Furthermore, if $H' \in \mathrm{Ult}(D)$ such that $H' \cap A = F$ and $H' \cap B = G$, then $H = H'$.*

Proof. Since ultrafilters are maximal filters it is enough to show that $H := [F \cup G]_{\mathrm{filt}}^D$ is an ultrafilter in D when $F \in \mathrm{Ult}(A)$, $G \in \mathrm{Ult}(B)$, and $F \cap C = G \cap C$. We first show that H is proper: If there are $a \in F, b \in G$ such that $a \cdot b = 0$, with Lemma 2 choose some $c \in C$ with $a \leq c$ and $b \leq -c$. Then, $c \in F \cap C$ and $-c \in G \cap C$, and thus, by the hypothesis $F \cap C = G \cap C$, we have $c, -c \in F \cap G \cap C$, contradicting that F and G are proper filters.

Assume that there is some $a \in D$ such that $a, -a \notin H$. Since D is generated by $A \cup B$, there are $a_0, \ldots, a_n \in A$, $b_0, \ldots, b_n \in B$ such that $a = a_0 \cdot b_0 + \ldots + a_n \cdot b_n$; in particular, there is some $i \leq n$ with $a_0 \cdot b_0 \notin H$ and $-(a_0 \cdot b_0) \notin H$. It follows that $a_0 \notin F$ or $b_0 \notin G$, and $-a_0 \notin F$ and $-b_0 \notin G$. This contradicts that $F \in \mathrm{Ult}(A)$ and $G \in \mathrm{Ult}(B)$. Finally, suppose that $H' \in \mathrm{Ult}(D)$ such that $H' \cap A = F$ and $H' \cap B = G$, and assume that $H \neq H'$. Then, there are $x \in H, y \in H'$ such that $x \cdot y = 0$. Choose some $c \in C$ such that $x \leq c$, $y \leq -c$ and proceed as in the first part of the proof. □

If $H \in \mathrm{Ult}(D)$ corresponds to $\langle F, G \rangle$ we will indicate this by $H \sim \langle F, G \rangle$.

Theorem 7. $\mathbf{K}$ *and* $\mathbf{K}_{C_6}$ *have the disjoint AP.*

Proof. Suppose that $\langle A, C_A \rangle$, $\langle B, C_B \rangle$, $\langle C, C_C \rangle \in \mathbf{K}$ with $\langle C, C_C \rangle$ a substructure of both $\langle A, C_A \rangle$ and $\langle B, C_B \rangle$, and let D be the Boolean amalgamated free product of A and B over C; then, $A \cap B = C$ [11].

We first note that the first construction of Theorem 6 need not work: Suppose that $\langle F, F' \rangle \in R_A$ such that $F \cap C \neq F' \cap C$; then, there is no $G \in \text{Ult}(B)$ for which $F \cap C = G \cap C$ and $F' \cap C = G \cap C$. Thus, we use the second construction and define for $H, H' \in \text{Ult}(D)$

$$\langle H, H' \rangle \in R \iff \langle H \cap A, H' \cap A \rangle \in R_A \text{ and } \langle H \cap B, H' \cap B \rangle \in R_B. \quad (4.7)$$

As in proof of Theorem 6 one shows that R is reflexive, symmetric, and closed. To show that R extends both R_A and R_B we use Theorem 4, in other words, we show that

$$R_A = \{\langle H \cap A, H' \cap A \rangle : H, H' \in \text{Ult}(D) \text{ and } \langle H, H' \rangle \in R\}, \quad (4.8)$$

$$R_B = \{\langle H \cap B, H' \cap B \rangle : H, H' \in \text{Ult}(D) \text{ and } \langle H, H' \rangle \in R\}. \quad (4.9)$$

Without loss of generality we show only (4.8):

"$\subseteq$": Let $F, F' \in \text{Ult}(A)$ with $\langle F, F' \rangle \in R_A$. We need to find $H, H' \in \text{Ult}(D)$ such that $H \cap A = F$, $H' \cap A = F'$, and $\langle H, H' \rangle \in R$, i.e. additionally $\langle H \cap B, H' \cap B \rangle \in R_B$. Let $P := F \cap C$ and $P' := F' \cap C$. Since $\langle C, \mathcal{C}_C \rangle$ is a substructure of $\langle A, \mathcal{C}_A \rangle$, we have $\langle P, P' \rangle \in \mathcal{C}_C$ by Theorem 4. Since $\langle C, \mathcal{C}_C \rangle$ is a substructure of $\langle B, \mathcal{C}_B \rangle$ as well, there are $G, G' \in \text{Ult}(B)$ such that $P = G \cap C, P' = G' \cap C$, and $\langle G, G' \rangle \in R_B$.

Suppose that $H := [F \cup G]^D_{\text{filt}}$ and $H' := [F' \cup G']^D_{\text{filt}}$. Since $F \cap C = G \cap C$ and $F' \cap C = G' \cap C$, both H and H' are ultrafilters of D by Lemma 3, and the choice of G and G' shows that H and H' fulfill the desired condition.

"$\supseteq$": Let $\langle H, H' \rangle \in R$. By definition of R, $\langle H \cap A, H' \cap A \rangle \in R_A$.

Clearly, if both R_A and R_B are transitive, then so is R, and thus, if C_A and C_B satisfy C_6 so does C_R by Theorem 3. $\qquad \square$

Theorem 8. $\mathbf{K}_{C_7}$ *does not have AP.*

Proof. Our strategy is as follows: We will define three connected graphs, $\mathfrak{U} = \langle U, E \rangle, \mathfrak{U}' = \langle U', E' \rangle, \mathfrak{V} = \langle V, R \rangle$, and strong homomorphisms $f : \mathfrak{U} \to \mathfrak{V}$, $g : \mathfrak{U}' \to \mathfrak{V}$. Then we shall show that there is no connected graph $\mathfrak{W} = \langle W, L \rangle$ for which there exist strong homomorphisms $f' : \mathfrak{W} \to \mathfrak{U}$, $g' : \mathfrak{W} \to \mathfrak{U}'$ which satisfy $f \circ f' = g \circ g'$. The result then follows from the duality theorem 5 and Theorem 3(2).

Consider the graph $\langle U, E \rangle$ where $U = \{0, 1, 2, 3\}$ and E is the smallest graph on U containing the chain $0E1E2E3$; observe that E is connected. Suppose that $\langle U', E' \rangle$ is a copy of $\langle U, E \rangle$, where $U' = \{0', 1', 2', 3'\}$, and $xE'y$ if and only if xEy. Let $V = \{a, b, c\}$, and R be smallest graph on V containing the chain $aRbRc$. Let $f : U \to V$ and $g : U' \to V$ be defined by

$$f(0) = f(2) = b, \qquad f(1) = c, \qquad f(3) = a,$$
$$g(0') = g(2') = b, \qquad g(1') = a, \qquad g(3') = c.$$

1. f *is a graph homomorphism*:

$$f(\langle 0, 1 \rangle) = \langle b, c \rangle \in R,$$
$$f(\langle 1, 2 \rangle) = \langle c, b \rangle \in R,$$
$$f(\langle 2, 3 \rangle) = \langle b, a \rangle \in R.$$

The rest follows from symmetry of E and R.

2. *f is a strong homomorphism, i.e. each edge of R is the image of an edge of E:*

$$\langle a, b \rangle = \langle f(3), f(2) \rangle \text{ and } \langle 3, 2 \rangle \in E,$$
$$\langle b, c \rangle = \langle f(0), f(1) \rangle \text{ and } \langle 0, 1 \rangle \in E.$$

The rest follows from symmetry of E and R.

3. *g is a graph homomorphism:*

$$g(\langle 0', 1' \rangle) = \langle b, a \rangle \in R,$$
$$f(\langle 1', 2' \rangle) = \langle a, b \rangle \in R,$$
$$f(\langle 2', 3' \rangle) = \langle b, c \rangle \in R.$$

The rest follows from symmetry of E' and R.

4. *g is a strong homomorphism, i.e. each edge of R is the image of an edge of E:*

$$\langle a, b \rangle = \langle g(1'), g(2') \rangle \text{ and } \langle 1', 2' \rangle \in E',$$
$$\langle b, c \rangle = \langle g(2'), g(3') \rangle \text{ and } \langle 2', 3' \rangle \in E'.$$

The rest follows from symmetry of E and R.

Thus, the duals of the graphs and mappings satisfy the conditions of Theorem 5, i.e. $\Delta(\mathfrak{V})$ is isomorphic to a substructure of $\Delta(\mathfrak{U})$ and $\Delta(\mathfrak{U}')$.

Suppose that there are a graph $\langle W, L \rangle$ and strong homomorphisms $f' : \langle W, L \rangle \to \langle U, E \rangle$, $g' : \langle W, L \rangle \to \langle U', E' \rangle$ such that $f \circ f' = g \circ g'$. We shall prove that $\langle W, L \rangle$ is not connected. Our first step is to prove that $f'^{-1}(0) \cap g'^{-1}(0') \neq \emptyset$ or $f'^{-1}(2) \cap g'^{-1}(2') \neq \emptyset$: Assume otherwise. If $f'(x) = 0$, then $f(f'(x)) = b$, and $g^{-1}(b) = \{0', 2'\}$ together with the hypothesis imply that $g'(x) = 2'$. Similarly, if $g'(x) = 2'$, then $g(g'(x)) = b$ and $f^{-1}(b) = \{0, 2\}$ imply that $f'(x) = 0$. Therefore, we have $f'^{-1}(0) = g'^{-1}(2')$, and, similarly

$$f'^{-1}(2) = g'^{-1}(0'), \ f'^{-1}(1) = g'^{-1}(3'), \ f'^{-1}(3) = g'^{-1}(1'), \qquad (4.10)$$

the last three cases being analogous. Let $x, y \in W$ such that $f'(x) = 1$, $f'(y) = 2$, and xLy. These exist, since $1E2$, and f' is a strong homomorphism. Then, $g'(x) = 3'$, $g'(y) = 0'$ by (4.10). Since xLy and g' preserves L, we must have $3'E'0'$, contradicting the definition of E'.

Thus, we have $f'^{-1}(0) \cap g'^{-1}(0') \neq \emptyset$ or $f'^{-1}(2) \cap g'^{-1}(2') \neq \emptyset$. As the cases are analogous, we only consider the former. Our aim is to show that if $x \in f'^{-1}(0) \cap g'^{-1}(0')$ and $y \notin f'^{-1}(0) \cap g'^{-1}(0')$, then $x(-L)y$. Thus, let $f'(x) = 0$, $g'(x) = 0'$, and $f'(y) \neq 0$ or $g'(y) \neq 0'$, and assume that xLy. There are three cases:

Case 1: $f'(y) = 0$, $g'(y) \neq 0'$: By the hypothesis, $f(f'(y)) = f(0) = b = g(g'(y))$. Since $g^{-1}(b) = \{0', 2'\}$, and $g'(y) \neq 0'$, we must have $g'(y) = 2'$. Thus, if xLy, then $0'E'2'$ which is not the case. $\qquad \square$

Case 2: $f'(y) \neq 0$, $g'(y) = 0'$: By the hypothesis, $g(g'(y)) = g(0') = b = f(f'(y))$. Since $f^{-1}(b) = \{0,2\}$, and $f'(y) \neq 0$, we must have $f'(y) = 2$. Thus, if xLy, then $0E2$ which is not the case.   □

Case 3: $f'(y) \neq 0$, $g'(y) \neq 0'$: If xLy, then $\langle f'(x), f'(y) \rangle \in E$ and $E(0) = \{1\}$ imply that $f'(y) = 1$; similarly, $g'(y) = 1'$. Thus, $f(f'(y)) = f(1) = c$ and $g(g'(y)) = a$, a contradiction.   □

Hence, $f'^{-1}(0) \cap g'^{-1}(0')$ and its complement in W are disconnected. The other case is analogous, and it follows now from duality that $\Delta(\mathfrak{U})$ and $\Delta(\mathfrak{U}')$ cannot be amalgamated over $\Delta(\mathfrak{V})$.   □

5 Outlook

If $\mathbf{K}$ is the class of all finite BCAs, then since $\mathbf{K}$ has HP, JEP, and AP, there is a countable universal homogeneous BCA by the Fraïssé construction. In related work [5] we investigate this algebra and the relation algebra generated by its contact relation. In particular, we show the CRA of this algebra is finite.

Acknowledgments. Ivo Düntsch acknowledges support by the Natural Sciences and Engineering Research Council of Canada and by the Bulgarian National Fund of Science, contract number DID02/32/2009. He is also grateful to the University of Technology Sydney for the opportunity to visit Prof. Sanjiang Li. The work of Sanjiang Li was supported by an ARC Future Fellowship (FT0990811). We would like to thank the anonymous referees for careful reading and helpful suggestions.

References

1. Aiello, M., Pratt-Hartmann, I., van Benthem, J. (eds.): Handbook of Spatial Logics. Springer, Dordrecht (2007)
2. Bennett, B., Düntsch, I.: Algebras, axioms, and topology. In: Aiello, et al. [1], pp. 99–159
3. Bodirsky, M., Wölfl, S.: RCC8 is polynomial on networks of bounded treewidth. In: Proceedings of the 22nd International Joint Conference on Artificial Intelligence, pp. 756–761 (2011)
4. Dimov, G., Vakarelov, D.: Contact algebras and region–based theory of space: A proximity approach – I, II. Fundamenta Informaticae 74, 209–282 (2006)
5. Düntsch, I., Li, S.: On the homogeneous countable Boolean contact algebra (2012) (preprint)
6. Düntsch, I., Vakarelov, D.: Region–based theory of discrete spaces: A proximity approach. Annals of Mathematics and Artificial Intelligence 49, 5–14 (2007)
7. Düntsch, I., Winter, M.: The Lattice of Contact Relations on a Boolean Algebra. In: Berghammer, R., Möller, B., Struth, G. (eds.) RelMiCS/AKA 2008. LNCS, vol. 4988, pp. 99–109. Springer, Heidelberg (2008)
8. Galton, A.: The Mereotopology of Discrete Space. In: Freksa, C., Mark, D.M. (eds.) COSIT 1999. LNCS, vol. 1661, pp. 251–266. Springer, Heidelberg (1999)

9. Hodges, W.: Model theory. Cambridge University Press, Cambridge (1993)
10. Jónsson, B.: Maximal algebras of binary relations. Contemporary Mathematics 33, 299–307 (1984)
11. Koppelberg, S.: General Theory of Boolean Algebras. In: Handbook of Boolean Algebras, vol. 1. North–Holland (1989)
12. Li, S., Ying, M.: Generalized Region Connection Calculus. Artificial Intelligence 160(1-2), 1–34 (2004)
13. Li, Y., Li, S., Ying, M.: Relational reasoning in the region connection calculus (2003), http://arxiv.org/abs/cs.AI/0505041
14. Madarász, J., Sayed Ahmed, T.: Amalgamation, interpolation and epimorphisms in alegraic logic. In: Andréka, H., Ferenczi, M., Németi, I. (eds.) Cylindric-like Algebras and Algebraic Logic, Bolay Society Mathematical Studies, vol. 22, pp. 65–77. North–Holland (2012)
15. Randell, D.A., Cui, Z., Cohn, A.G.: A spatial logic based on regions and connection. In: Nebel, B., Swartout, W., Rich, C. (eds.) Proceedings of the 3rd International Conference on Knowledge Representation and Reasoning, pp. 165–176. Morgan Kaufmann (1992)
16. Renz, J., Nebel, B.: Qualitative spatial reasoning using constraint calculi. In: Aiello, et al. [1], pp. 161–215
17. Sabidussi, G.: Graph multiplication. Mathematische Zeitschrift 72(1), 446–457 (1959)
18. Stell, J.: Boolean connection algebras: A new approach to the Region Connection Calculus. Artificial Intelligence 122, 111–136 (2000)
19. Vakarelov, D., Dimov, G., Düntsch, I., Bennett, B.: A proximity approach to some region–based theories of space. J. Appl. Non-Classical Logics 12, 527–529 (2002)
20. Weichsel, P.M.: The Kronecker Product of Graphs. Proceedings of the American Mathematical Society 13, 47–52 (1962)
21. Wolter, F., Zakharyaschev, M.: Spatial reasoning in RCC-8 with Boolean region terms. In: Horn, W. (ed.) ECAI, pp. 244–250. IOS Press (2000)

Author Index